**Flow Simulation
with High-Performance
Computers II**

Edited by
Ernst Heinrich Hirschel

Notes on Numerical Fluid Mechanics (NNFM)   Volume 52

Series Editors: Ernst Heinrich Hirschel, München (General Editor)
               Kozo Fujii, Tokyo
               Bram van Leer, Ann Arbor
               Michael Leschziner, Manchester
               Maurizio Pandolfi, Torino
               Arthur Rizzi, Stockholm
               Bernard Roux, Marseille

Volume 51   Numerical Treatment of Coupled Systems. Proceedings of the Eleventh GAMM-Seminar, K January 20–22, 1995 (W. Hackbusch / G. Wittum, Eds.)
Volume 50   Computational Fluid Dynamics on Parallel Systems. Proceedings of a CNRS-DFG Symposi in Stuttgart, December 9 and 10, 1993 (S. Wagner, Ed.)
Volume 49   Fast Solvers for Flow Problems. Proceedings of the Tenth GAMM-Seminar, Kiel, January 14–16, 1994 (W. Hackbusch / G. Wittum, Eds.)
Volume 48   Numerical Simulation in Science and Engineering. Proceedings of the FORTWIHR Symposium on High Performance Scientific Computing, München, June 17–18, 1993 (M. Griebel / Ch. Zenger, Eds.)
Volume 47   Numerical Methods for the Navier-Stokes Equations (F.-K. Hebeker, R. Rannacher, G. Wittum, Eds.)
Volume 46   Adaptive Methods – Algorithms, Theory, and Applications. Proceedings of the Ninth GAM Seminar, Kiel, January 22–24, 1993 (W. Hackbusch / G. Wittum, Eds.)
Volume 45   Numerical Methods for Advection – Diffusion Problems (C. B. Vreugdenhil / B. Koren, Eds
Volume 44   Multiblock Grid Generation – Results of the EC/BRITE-EURAM Project EUROMESH, 1990–1992 (N. P. Weatherill / M. J. Marchant / D. A. King, Eds.)
Volume 43   Nonlinear Hyperbolic Problems: Theoretical, Applied, and Computational Aspects Proceedings of the Fourth International Conference on Hyperbolic Problems, Taormina, Italy, April 3 to 8, 1992 (A. Donato / F. Oliveri, Eds.)
Volume 42   EUROVAL – A European Initiative on Validation of CFD Codes (W. Haase / F. Brandsma / E. Elsholz / M. Leschziner / D. Schwamborn, Eds.)
Volume 41   Incomplete Decompositions (ILU) – Algorithms, Theory, and Applications (W. Hackbusch / G. Wittum, Eds.)
Volume 40   Physics of Separated Flow – Numerical, Experimental, and Theoretical Aspects (K. Gersten, Ed.)
Volume 39   3-D Computation of Incompressible Internal Flows (G. Sottas / I. L. Ryhming, Eds.)
Volume 38   Flow Simulation on High-Performance Computers I (E. H. Hirschel, Ed.)
Volume 37   Supercomputers and Their Performance in Computational Fluid Mechanics (K. Fujii, Ed.)
Volume 36   Numerical Simulation of 3-D Incompressible Unsteady Viscous Laminar Flows (M. Deville / T.-H. Lê / Y. Morchoisne, Eds.)
Volume 35   Proceedings of the Ninth GAMM-Conference on Numerical Methods in Fluid Mechani (J. B. Vos / A. Rizzi / I. L. Ryhming, Eds.)
Volume 34   Numerical Solutions of the Euler Equations for Steady Flow Problems (A. Eberle / A. Rizzi / E. H. Hirschel)
Volume 33   Numerical Techniques for Boundary Element Methods (W. Hackbusch, Ed.)
Volume 32   Adaptive Finite Element Solution Algorithm for the Euler Equations (R. A. Shapiro)
Volume 31   Parallel Algorithms for Partial Differential Equations (W. Hackbusch, Ed.)

Volumes 1 to 25, 45 are out of print.
The addresses of the Editors and further titles of the series are listed at the end of the book.

Flow Simulation with High-Performance Computers II

DFG Priority Research Programme
Results 1993–1995

Edited by
Ernst Heinrich Hirschel

Die Deutsche Bibliothek – CIP-Einheitsaufnahme

Flow simulation with high performance computers:
DFG priority research programme results / ed. by
Ernst Heinrich Hirschel. – Braunschweig; Wiesbaden:
Vieweg.
  Früher u. d. T.: Flow simulation on supercomputers
NE: Hirschel, Ernst Heinrich [Hrsg.]:
Deutsche Forschungsgemeinschaft
2. 1993–1995. – 1996
  (Notes on numerical fluid mechanics; Vol. 52)
  ISBN 3-528-07652-6

NE: GT

All rights reserved
© Friedr. Vieweg & Sohn Verlagsgesellschaft mbH, Braunschweig/Wiesbaden, 1996

Vieweg ist a subsidiary company of Bertelsmann Professional Information.

No part of this publication may be reproduced, stored in a retrieval
system or transmitted, mechanical, photocopying or otherwise,
without prior permission of the copyright holder.

Produced by Langelüddecke, Braunschweig
Printed on acid-free paper
Printed in Germany

ISSN 0179-9614
ISBN 3-528-07652-6

Foreword

This volume contains thirty-seven reports on work, which was conducted between 1993 and 1995 in the Priority Research Programme "Flow Simulation with High-Performance Computers" of the Deutsche Forschungsgemeinschaft (DFG, German Research Society), 1989 to 1995.

The main purpose of this publication is to give an overview over the work conducted in the second half of the programme, and to make the results obtained available to the public. The reports are grouped under the four headings "Flow Simulation with Massively Parallel Systems", "Direct and Large-Eddy Simulation of Turbulence", "Mathematical Foundations, General Solution Techniques and Applications" and "Results of Benchmark Computations". All contributions to this publication have been reviewed by a board consisting of F. Durst (Erlangen), R. Friedrich (München), D. Hänel (Duisburg), R. Rautmann (Paderborn), H. Wengle (München), and the editor. The responsibility for the contents of the reports nevertheless lies with the authors.

E.H. Hirschel
Editor

Preface

The Deutsche Forschungsgemeinschaft (DFG) sponsored the development of numerical simulation techniques in fluid mechanics since 1989 in a Priority Research Program "Flow Simulation with High-performance Computers". The major results obtained in this program until 1992 were published in summarizing articles in Volume 38 of the "Notes on Numerical Fluid Mechanics" of the Vieweg Verlag. The present volume summarizes the results of the second half of the program, which completed its investigations December 1995.

The problems studied included developments of solution techniques as well as applications. Altogether 35 projects were supported. It is worthwhile to point out, that 13 projects concentrated on the parallelization of algorithms and related work. The program thereby stimulated research in this relatively new branch of computational fluid dynamics in Germany with marked success. Development of general solution techniques and their application were pursued in 17 projects, and the remainder 5 projects were devoted to the development of simulation techniques for transitional and turbulent flows.

The work on parallel solution techniques included development of a time and space multi-grid method, of parallel finite element methods for the solution of the Navier-Stokes equations, performance enhancement of parallelized algorithms, adaptive operator techniques, and parallel interactive and integrated visualization techniques. The applications of parallelized algorithms comprised simulation of turbulent flow, of various viscous incompressible flows, of unsteady flows in turbomachinery, of chemically reacting flow, and of aerodynamic flows.

The work on other solution techniques included approximations in high order norms, low Mach number solutions based on asymptotic analysis, solutions based on the artificial compressibility concept, on higher order upwind schemes on unstructured grids, and others. The applications included simulation of aerodynamic and of hypersonic flows, of flows in turbomachinery and other complex internal flows. The investigations of transitional and turbulent flows were aimed at direct simulation of internal compressed turbulent flow and of separated flows; at large-eddy simulation of near-wall turbulence, of turbulent flows in curved pipes, and of turbulent boundary-layer flow over a hemisphere.

The cooperation with the Groupment de Recherche Mécanique des Fluides Numérique of the French Centre National de la Recherche Scientifique (CNRS) was continued. The second joint workshop was held May 3 - 5 1993 in Lambrecht (Pfalz) on the topic "Three-Dimensional Flow - Alternative Formulations and Solution of the Conservation Equations". At this occasion several representatives of the CNRS and the DFG under the chairmanship of one of the vice presidents of the DFG, Prof. S. Wittig, met and discussed possibilities of cooperation. Agreement was reached on the following points: Cooperation in 10 joint projects within the frame of existing programs; participation of a French representative in the meetings of the German reviewing board and vice versa; organization of a joint meeting in Sophia-Antipolis at the end of 1994 with the aim of preparing a joint CNRS-DFG research program on computational fluid dynamics.

A third joint CNRS-DFG workshop was organized December 9 - 10, 1993 at Stuttgart University under the topic "Computational Fluid Dynamics on Parallel Systems". The contributions were published in volume 50 of the "Notes on Numerical Fluid Dynamics" of the Vieweg Verlag under the title of the workshop. They were edited by S. Wagner of Stuttgart University. The fourth workshop was held in Sophia-Antipolis November 25 - 26, 1994. As proposed in Lambrecht one year earlier, 20 projects for a prospective joint program entitled "Numerical Flow Simulation, A French - German Research Initiative" were discussed in the presence of official representatives of the CNRS and the DFG, and in March 1995 a proposal for such a program was submitted to the CNRS and the DFG under the title "Joint French - German Research Program: Numerical Flow Simulation". The individual proposals were reviewed by a French - German reviewing board November 27, 1995 and submitted for final decision to the CNRS and the DFG.

The DFG Priority Research Program "Flow Simulation on High-Performance Computers" substantially stimulated and fostered research in this field over a long period of time. It was a safe stepping stone for initiating and supporting work on paralellization. It brought together engineers and applied mathematicians in fruitful cooperation. The program helped to maintain international competiveness in flow research and markedly fastened the ties to the corresponding French program.

It is with great pleasure, that the undersigned take this opportunity to thank the DFG for supporting work on numerical flow simulation over more than six years. The efforts of the reviewers, Profs. Dr. G. Böhme and Dr. R. Rannacher are gratefully acknowledged. Their invaluable expertise helped to shape the program in many ways. They stimulated interdisciplinary discussion between engineers and mathematicians, who participated in the program. Their efforts are reflected in the articles published in this volume. Without the continuous help and far-sighted administering of the program by Dr. W. Lachenmeier it would have been impossible to maintain continuity in the research activities over the years. We thank him for his efforts.

Finally the Vieweg Verlag is gratefully acknowledged for publishing the results in the Notes on Numerical Fluid Mechanics, and Prof. Dr. E. H. Hirschel for editing this volume.

Bonn-Bad Godesberg, January 1996

E. Krause　　　　　　　　　　　　　　　　　　E. A. Müller

CONTENTS

Page

I. FLOW SIMULATION WITH MASSIVELY PARALLEL SYSTEMS 1

M. Schäfer: Introduction to Part I 3

J. Burmeister, W. Hackbusch: On a Time and Space Parallel Multi-Grid Method Including Remarks on Filtering Techniques 5

O. Dorok, V. John, U. Risch, F. Schieweck, L. Tobiska: Parallel Finite Element Methods for the Incompressible Navier-Stokes Equations 20

M. Griebel, W. Huber, C. Zenger: Numerical Turbulence Simulation on a Parallel Computer Using the Combination Method 34

F. Lohmeyer, O. Vornberger: CFD with Adaptive FEM on Massively Parallel Systems 48

M. Lenke, A. Bode, T. Michl, S. Wagner: On the Performance Enhancements of a CFD Algorithm in High Performance Computing 64

F. Durst, M. Schäfer, K. Wechsler: Efficient Simulation of Incompressible Viscous Flows on Parallel Computers 87

J. Hofhaus, M. Meinke, E. Krause: Parallelization of Solution Schemes for the Navier-Stokes Equations 102

K. Engel, F. Eulitz, S. Pokorny, M. Faden: 3-D Navier-Stokes Solver for the Simulation of the Unsteady Turbomachinery Flow on a Massively Parallel Hardware Architecture 117

H. Rentz-Reichert, G. Wittum: A Comparison of Smoothers and Numbering Strategies for Laminar Flow Around a Cylinder 134

G. Bader, E. Gehrke: Simulation of Detailed Chemistry Stationary Diffusion Flames on Parallel Computers 150

I.S. Doltsinis, J. Urban: An Adaptive Operator Technique for Hypersonic Flow Simulation on Parallel Computers 165

J. Argyris, H.U. Schlageter: Parallel Interactive and Integrated Visualisation 184

C. Meiselbach, R. Bruckschen: Interactive Visualization: On the Way to a Virtual Wind Tunnel 203

CONTENTS (continued)

Page

II. DIRECT AND LARGE-EDDY SIMULATION OF TURBULENCE 209

C. Härtel: Introduction to Part II 211

E. Güntsch, R. Friedrich: Direct Numerical Simulation of Turbulence Compresssed in a Cylinder 213

C. Maaß, U. Schumann: Direct Numerical Simulation of Separated Turbulent Flow Over a Wavy Boundary 227

C. Härtel, L. Kleiser: Large-Eddy Simulation of Near-Wall Turbulence 242

M. Breuer, W. Rodi: Large Eddy Simulation for Complex Turbulent Flows of Practical Interest 258

M. Manhart, H. Wengle: Large-Eddy Simulation and Eigenmode Decomposition of Turbulent Boundary Layer Flow Over a Hemisphere 275

III. MATHEMATICAL FOUNDATIONS, GENERAL SOLUTION TECHNIQUES AND APPLICATIONS 291

M. Meinke: Introduction to Part III 293

R. Rautmann: Navier-Stokes Approximations in High Order Norms 295

S. Blazy, W. Borchers, U. Dralle: Parallelization Methods for a Characteristic's Pressure Correction Scheme 305

K. Steiner: Weighted Particle Method Solving Kinetic Equations for Dilute Ionized Gases 322

K.J. Geratz, R. Klein, C.D. Munz, S. Roller: Multiple Pressure Variable (MPV) Approach for Low Mach Number Flows Based on Asymptotic Analysis 340

M. Weimer, M. Meinke, E. Krause: Numerical Simulation of Incompressible Flows with the Method of Artificial Compressibility 355

M. Wierse, D. Kröner: Higher Order Upwind Schemes on Unstructured Grids for the Nonstationary Compressible Navier-Stokes Equations in Complex Timedependent Geometries in 3-D 369

CONTENTS (continued)

Page

L. Xue, T. Rung, F. Thiele: Improvement and Application of a Two Stream-Function Formulation Navier-Stokes Procedure 385

R. Hentschel, E.H. Hirschel: AMRFLEX3D-Flow Simulation Using a Three-Dimensional Self-Adaptive, Structured Multi-Block Grid System 400

Ž. Lilek, M. Perić, V. Seidl: Development and Application of a Finite Volume Method for the Prediction of Complex Flows 416

R. Vilsmeier, D. Hänel: Computational Aspects of Flow Simulation on 3-D, Unstructured, Adaptive Grids 431

C. Roehl, H. Simon: Flow Simulations in Aerodynamically Highly Loaded Turbomachines Using Unstructured Adaptive Grids 446

W. Evers, M. Heinrich, I. Teipel, A.R. Wiedermann: Flow Simulation in a High Loaded Radial Compressor 461

H. Greza, S. Bikker, W. Koschel: Efficient FEM Flow Simulation on Unstructured Adaptive Meshes 476

J. Grönner, E. von Lavante, M. Hilgenstock, M. Kallenberg: Numerical Methods for Simulating Supersonic Combustion 491

D. Nellessen, G. Britten, S. Schlechtriem, J. Ballmann: Aeroelastic Computations of a Fokker-Type Wing in Transonic Flow 501

E. Laurin, J. Wiesbaum: Three-Dimensional Numerical Simulation of the Aerothermodynamic Reentry 517

S. Brück, G. Brenner, D. Rues, D. Schwamborn: Investigations of Hypersonic Flows Past Blunt Bodies at Angle of Attack 530

IV. RESULTS OF BENCHMARK COMPUTATIONS 545

M. Schäfer, S. Turek (and F. Durst, E. Krause, R. Rannacher): Benchmark Computations of Laminar Flow Around a Cylinder 547

F. Durst, M. Fischer, J. Jovanović, H. Kikura, C. Lange: LDA Measurements in the Wake of a Circular Cylinder 556

I. FLOW SIMULATION WITH MASSIVELY PARALLEL SYSTEMS

INTRODUCTION TO PART I

by

M. Schäfer

Institute of Fluid Mechanics
University of Erlangen-Nürnberg
Cauerstr. 4, D-91058 Erlangen, Germany

In recent years intensive research has been undertaken (and is still continuing) to improve the performance of computer codes for flow computations by employing both more efficient solution algorithms and more powerful computer systems. Concerning the computer aspects, it is nowadays widely accepted that only parallel computers will satisfy the future demands of computational fluid mechanics, because solely parallel processing can provide the necessary scalability in computing power and memory for the increasing requirements in this field. Here, the combined usage of modern numerical techniques and high-performance parallel computers significantly enlarges the practical capabilities of computational techniques for complex fluid mechanical tasks.

Due to the aspects mentioned above, flow simulation with massively parallel systems has been one of the key topics inside the present Priority Research Programme of the "Deutsche Forschungsgemeinschaft". To this subject 13 out of 35 projects can be associated, and the papers in this part of the volume give an overview of the corresponding major results that could be achieved within these projects. The composition of the participating institutes, which constitutes a well balanced mixture of research groups from mathematics, computer science and engineering, reflects the high importance of interdisciplinarity for a successful work on this topic. This multidisciplinary cooperation has turned out to be of substantial benefit for the individual research and development projects, not only, but in particular, for the projects reported in this part.

When considering the parallelization of flow computations, it is very important to take into account its interaction to other aspects of the problem, because the type of flow, the discretization technique, the solution algorithms as well as the characteristics of the parallel machine usually have a strong influence on the overall efficiency of the computation. As contributions to this aspect of the work, a quite representative set of combinations of techniques has been treated within the projects. Here, also the close link to the research efforts described in Parts II and III of the present volume should be emphasized.

Concerning the approaches employed for the parallelization, the contributions comprise the most common and promising methods applicable to computational fluid mechanics: grid partitioning, domain decomposition, time parallelization and combination methods. Implementations and results are reported for a variety of parallel systems including the use of both distributed and shared memory programming models.

The flow problems considered, cover turbulent and laminar flows, steady and unsteady flows, compressible and incompressible flows as well as problems with heat and mass transfer involving chemical reactions. As some examplary applications, problems from aerodynamics, turbomachinery and combustion are considered in the publications in this section.

Spatial discretization approaches included in the considerations, comprise finite element and finite volume methods on unstructured, block-structured and hierarchically structured grids, and both explicit and implicit methods (with strong emphasis to the latter) are discussed for time discretization. In this context, adaptivity with respect to space and time is a topic to which special attention is paid. Various solution algorithms, including pressure-correction schemes, artificial compressibility methods, preconditioned conjugate gradient methods and, in particular, the powerful multigrid methods are considered with respect to their performance in computational fluid mechanics and their interaction with the parallelization.

In the papers detailed investigations concerning the performance of the different approaches are reported. These studies cover questions of static and dynamic load balancing, scalability, numerical and parallel efficiency, robustness and portability. This also includes aspects of the influence of arithmetic and communication performance of parallel platforms and operating systems on the efficiency of the simulations.

In addition to the issues related to computational efficiency, also adequate user interfaces and pre- and postprocessing facilities play an important role for the routinely use of parallel flow simulations for complex applications. In particular, the visualization of the huge amount of (time-dependent) numerical flow data within a parallel computer environment requires special attention. These aspects are also addressed within this section.

Altogether, the papers give a comprehensive overview of the outcome of the research and development efforts in the Priority Research Programme in the field of flow simulation with massively parallel systems and they can also be regarded as a summary of the current state of the art in this field. A lot of useful information can be extracted, concerning how to exploit the computing power provided by modern parallel computer architectures in an efficient way for numerical flow simulations.

From the papers, it becomes obvious that the use of massively parallel systems, if combined with advanced numerical methods and appropriate accompanying tools, has a significant impact towards an efficient, accurate and reliable solution of practical flow problems in engineering and science. The work reported constitutes a well founded basis for future developments which can be foreseen in this field and which will yield a further significant improvement of the capabilities of computational fluid mechanics in practice.

In summary, the results provided in this section clearly document the success of the Priority Research Programme with respect to the activities in the field of flow computations on massively parallel systems.

ON A TIME AND SPACE PARALLEL MULTI-GRID METHOD INCLUDING REMARKS ON FILTERING TECHNIQUES

Jens Burmeister, Wolfgang Hackbusch
Mathematisches Seminar II
Lehrstuhl Praktische Mathematik
Universität Kiel, Hermann–Rodewald–Straße 3/1
24098 Kiel, Germany

SUMMARY

In this paper we discuss the numerical treatment of parabolic problems by multi-grid methods under the aspect of parallelisation. Reflecting the concept to treat the time and space variables independently, the time–parallel multi-grid method is combined with a space–parallel multi-grid method. The space–parallel multi-grid method could be interpreted as a global multi-grid with a special domain decomposition smoother. The smoother requires approximations for the SCHUR complement. This question leads to the discussion of filtering techniques in a more general situation. Adaptivity in time is achieved by using extrapolation techniques which offers a third source of parallelism in addition to the time and space parallelism.

INTRODUCTION

We consider parabolic problems defined by an elliptic operator L^{ell}

$$u_t + L^{ell}u = q(x,t), \quad x = (x_1, x_2) \in \Omega \subset \mathbb{R}^2, \quad t_0 < t \leq T, \tag{1}$$

with initial values for $t = t_0$ and boundary conditions on $\partial\Omega \times [t_0, T]$. The discretisation in space (method of lines) leads to a system of ordinary differential equations

$$u'_h = -L_h^{ell} u_h + f_h, \quad \text{with} \quad \psi_0 := u_h(t_0) \quad \text{given.} \tag{2}$$

The unknown vector function u_h is defined on a space grid Ω_h. h represents the space discretisation parameter. Every component of u_h is a time–dependent function.
Due to the natural stiffness of system (2) the time integration will be done by implicit formula. A suitable choice is the implicit EULER scheme of consistency order 1 or the

implicit trapezoidal rule of order 2 which are both one–step–formula. The time integration method "proceeds" along the time axis. The evaluation points are denoted by $t_k, k = 0, 1, \ldots$. The corresponding sequence of time discretisation parameters is defined by $\Delta t_k := t_k - t_{k-1}$ for $k = 1, \ldots$.

The paper is organised in the following way. Reflecting the concept of treating time and space variables independently, we start with the discussion of the time–parallel multi-grid method ($\rightarrow$ TIME PARALLELISM). The treatment of adaptivity in time is based on extrapolation techniques ($\rightarrow$ TIME–ADAPTIVE SOLUTION). The space–parallel multi-grid method ($\rightarrow$ SPACE PARALLELISM) discussed here is a global multi-grid method with a domain-decomposition-type smoother ($\rightarrow$ APPROXIMATE BLOCK-LDU-DECOMPOSITION). The smoother requires an approximation of a SCHUR-complement which can be determined by using filtering techniques ($\rightarrow$ REMARKS ON FILTERING TECHNIQUES). For the space–parallel method we present numerical results.

We refer to [19] for an introduction to the Finite-Element methods for parabolic problems, to the monography [12] for an introduction to the multi-grid theory, to [13] for a state-of-the-art overview of iteration methods (chapter 10 is about multi-grid methods) and to [4], [5] for other publications from this project. Related topics are discussed in [7] and [16]. A comparison of different smoothers in a multi-grid approach to flow problems can be found in [18].

TIME PARALLELISM

The standard solution procedure for (2) is characterised by solving discrete problems timestep by timestep in a very strongly sequential manner. A different solution approach has been proposed by Hackbusch in [11]. The basic idea of the so-called *time–parallel multi-grid method* is to build up a collection of m successive time evaluation points $[t_{k+1}, \ldots, t_{k+m}]$ and to solve the corresponding blocksystem by means of multi-grid techniques. Using the implicit EULER scheme the blocksystem has the following bidiagonal structure:

$$\begin{bmatrix} L_{h,k+1} & & & 0 \\ -\frac{1}{\Delta t_{k+2}}I_h & L_{h,k+2} & & \\ & \ddots & \ddots & \\ 0 & & -\frac{1}{\Delta t_{k+m}}I_h & L_{h,k+m} \end{bmatrix} \begin{bmatrix} u_{h,k+1} \\ u_{h,k+2} \\ \vdots \\ u_{h,k+m} \end{bmatrix} = \begin{bmatrix} f_{h,k+1} + \frac{1}{\Delta t_{k+1}}u_{h,k} \\ f_{h,k+2} \\ \vdots \\ f_{h,k+m} \end{bmatrix} \quad (3)$$

with I_h identity matrix and

$$L_{h,k+j} = \frac{1}{\Delta t_{k+j}}I_h + L^{ell}_{h,k+j} \quad j = 1, \ldots, m \quad .$$

The gridfunction $u_{h,k+j}$ is an approximation of the solution u on the grid Ω_h at time t_{k+j}. Using the implicit trapezoidal scheme as time–integration method leads to a similar bidiagonal blocksystem. The time-parallel method has been discussed in many articles. We recall the basic properties:

- The time-parallel method is an iterative solver for the bidiagonal blocksystem (3).

- In [4] a convergence result for the model problem ($L^{ell} = -\Delta$) has been proved, i.e., the spectral radius of the iteration matrix is bounded strictly below 1 by a constant independently of the meshsize h, the time discretisation parameter Δt **and** the number of timesteps m, which will be solved simultaneously.

- Parallel implementations are discussed in [2],[4] and [15].

- Numerical experiments with the incompressible Navier-Stokes equations are documented in [3] and [15].

The parabolic problem (1) defines an unknown space and time dependent function $u = u(x,t)$. For practical reasons the irregular grid Ω_h (represented by the space discretisation parameter h) and the time evaluation points t_k should be chosen with respect to the regularity of the problem and by the solution procedure itself. This leads to the discussion of stepsize control or more generally of adaptivity.

A special time stepping strategy based on extrapolation techniques is discussed in the next chapter. The necessary existence of asymptotic expansions and a practicable error estimator is presented in [1] and [5]. The parallel implementation benefits mainly from the fact that the time-parallel multi-grid method is embedded in a natural way.

TIME–ADAPTIVE SOLUTION

Let a time-evaluation point $t := t_k$ and a gridfunction $u_h(t_k)$ be given (by initial values or already be calculated). A time-stepping strategy is a mechanism to find a "satisfactory" new timestep $\Delta t := t_{k+1} - t_k$ to proceed in time.

In general one starts with an initial guess Δt and decides whether Δt it is acceptable or not.

To make a decision one has to estimate the error based on observed data (in our case these data will be produced by extrapolation techniques). If the guess Δt is not acceptable, Δt will be decreased and tested again.

EXTRAPOLATION TECHNIQUES

Let $\mathcal{F} = (n_i)_{i \in \mathbb{N}}$ denote a sequence of positive numbers ($n_i \in \mathbb{N}$). The ROMBERG sequence $\mathcal{F}_R$ and the BULIRSCH sequence $\mathcal{F}_B$ are given by

$$\mathcal{F}_R := \{1, 2, 4, 8, 16, \ldots\}, \quad \mathcal{F}_B := \{1, 2, 3, 4, 6, 8, 12, \ldots\} \quad . \tag{4}$$

Only subsets of finite dimension $d+1$ with starting index i are of practical interest:

$$\mathcal{F}^{[i,i+d]} := \{n_{i+j} \in \mathcal{F} \mid 0 \leq j \leq d\}; \tag{5}$$

for example $\mathcal{F}_R^{[3,5]} = \{8, 16, 32\}$.

Let a subset $\mathcal{F}^{[i,i+d]}$ be given. The auxiliary timesteps

$$\tau_{i+j} := \frac{\Delta t}{n_{i+j}}, \quad n_{i+j} \in \mathcal{F}^{[i,i+d]}, \quad 0 \leq j \leq d \tag{6}$$

correspond to $d+1$ equidistant subdivisions of the time interval $[t, t+\Delta t]$. Every subdivision defines a blocksystem of dimension $m = n_{i+k}$ with bidiagonal structure (compare (3)). The solution of every blocksystem produces approximations $v_h(t + \Delta t; \tau_{i+j}), j = 0, \ldots, d$ of $u_h(t + \Delta t)$ which are written down in the first column of the extrapolation tableau

$$\begin{array}{c|cccccc}
\tau_{i+0} & v_h(t + \Delta t; \tau_{i+0}) & =: & T_{i+0,0} & & & \\
\tau_{i+1} & v_h(t + \Delta t; \tau_{i+1}) & =: & T_{i+1,0} & T_{i+1,1} & & \\
\tau_{i+2} & v_h(t + \Delta t; \tau_{i+2}) & =: & T_{i+2,0} & T_{i+2,1} & T_{i+2,2} & \\
\vdots & \vdots & & \vdots & \vdots & \vdots & \ddots \\
\tau_{i+d} & v_h(t + \Delta t; \tau_{i+d}) & =: & T_{i+d,0} & T_{i+d,1} & \ldots & \ldots & T_{i+d,d}
\end{array} \tag{7}$$

The extrapolated values $T_{i+j,k}$ are computed by

$$T_{i+j,k} := T_{i+j,k-1} + \frac{T_{i+j,k-1} - T_{i+j-1,k-1}}{(n_{i+j}/n_{i+j-k})^\gamma - 1}, \quad 0 < k \leq j \leq d \tag{8}$$

and are also approximations of $u_h(t+\Delta t)$. γ depends on the consistency order of the time integration method; $\gamma = 1$ in the case of implicit EULER, $\gamma = 2$ for the trapezoidal rule.

Remark: *The time-parallel multi-grid method is used to calculate the approximations* $v_h(t+\Delta t; \tau_{i+j}), \ j = 0, \ldots, d.$

ALGORITHM AND PROCESSOR TOPOLOGY

The time–adaptive solution of parabolic problems (1) needs the specification of the following components:

- a space discretisation method defining (2),
- a time integration method of consistency order γ,
- a sequence $\mathcal{F}^{[i,i+d]}$ defining an extrapolation tableau of depth d,
- an accuracy *eps* and an initial guess Δt.

The resulting algorithm reads as follows:

1. Set $t = t_0$ and $\psi = \psi_0$.

2. Define τ_{i+j}, $j = 0, \ldots, d$ (see (6)) and calculate the entries of the first column $T_{i+j,0}$, $j = 0, \ldots, d$ (see (7)).

3. Calculate extrapolated values $T_{i+j,k}$, $0 < k \leq j \leq d$ (see (8)).

4. Calculate the subdiagonal error estimator and a new timestep

$$\bar{\epsilon}_{i+d,d-1} := ||T_{i+d,d-1} - T_{i+d,d}||, \quad \Delta t' := \Delta t \left(\frac{eps}{\bar{\epsilon}_{i+d,d-1}} \right)^{\frac{1}{\gamma d}} . \quad (9)$$

5. If $\bar{\epsilon}_{i+d,d-1} \gg eps$ then define $\Delta t := \Delta t'$ and goto 2., otherwise set $t := t + \Delta t, \Delta t := \Delta t', \psi := T_{i+d,d}$.

6. If $t = T$ then return the solution stored in ψ, otherwise goto 2.

The above algorithm offers two sources of parallelism. One comes from the time-parallel method and one from the extrapolation tableau (the entries $T_{i+j,0}$, $j = 0, \ldots, d$ can be calculated independently).
The total number of processors $\#P$ used depends on the choice of $\mathcal{F}^{[i,i+d]}$ and is given by

$$\#P = \sum_{j=i}^{i+d} n_j, \quad n_j \in \mathcal{F}^{[i,i+d]} . \quad (10)$$

The processor network consists logically of $d+1$ clusters. In every cluster the processors are connected in form of a single-linked cyclic list. A head processor is marked in every cluster and it is the gateway for information flow between clusters. The head processor of cluster d is logically the master processor. Figure 1 shows the processor topology corresponding to $\mathcal{F}_R^{[2,4]}$.

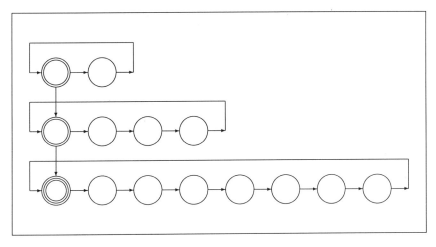

Figure 1: Processor topology for $\mathcal{F}_R^{[2,4]}$

All processors are involved in the calculation of the entries $T_{i+j,0}, j = 0, \ldots, d$ of the first column of the extrapolation tableau. For this part of the algorithm no communication between clusters is needed; therefore the clusters are working independently. In every cluster a time–parallel multi-grid method is working. The results are located in the head processors.

Remark: *For $j \in \{0, \ldots, d\}$ let w_j denote the computational time for cluster j. Due to basic properties of the time-parallel algorithm we have*

$$w_0 \approx w_1 \approx \ldots \approx w_d \quad .$$

The head processors are calculating the extrapolated values while all other processors are idle. The master processor calculates the subdiagonal error estimator. Because the arithmetic work of part 1 (first column entries) dominates part 2 (extrapolation and error estimation) load balancing is not disturbed significantly.

Up to now we assumed that each processor in Figure 1 computes the spatial problem. In the following we explain, how this part can be parallelised, too.

SPACE PARALLELISM

The space-parallel multi-grid method discussed in the following is classical in the sense that it is a combination of a smoothing step and a coarse-grid correction. The components of the coarse-grid correction are standard whereas the smoothing step is new.

The smoother is based on a nonoverlapping domain decomposition. We denote the internal boundaries by the synonym *skeleton*. The set of unknowns can be decomposed into those "living in the subdomains" and those "living on the skeleton".

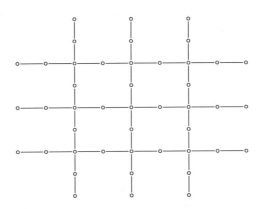

Figure 2: Skeleton with 4 elementary cycles

Thus we are able to discuss the smoother in a pure algebraic framework. Let

$$Kx = b \quad \text{with} \quad K \in \mathbb{R}^{n \times n}, \quad x, b \in \mathbb{R}^n \tag{11}$$

represent the discrete spatial problem at time-evaluation point t_k. The matrix K is assumed to be symmetric positive definite with 2×2-block structure

$$K := \begin{bmatrix} A & B \\ B^T & C \end{bmatrix} \tag{12}$$

corresponding to the above mentioned decomposition of the unknowns. In this context we say that A is acting on the subdomains and C on the skeleton. B and B^T are transfer operators between the subdomains and the skeleton.

Most of the methods treating the 2×2-blocksystem (12) can be described formally by means of an approximate block-LDL^T decomposition.

APPROXIMATE BLOCK-LDU DECOMPOSITION

The (exact) block-LDL^T decomposition of the 2×2-blocksystem (12) is given by

$$K = LDL^T \quad \text{with} \quad L = \begin{bmatrix} I & \\ B^T A^{-1} & I \end{bmatrix} \quad \text{and} \quad D = \begin{bmatrix} A & \\ & S \end{bmatrix}. \tag{13}$$

The matrix $S = C - B^T A^{-1} B$ is called the SCHUR complement (of A in K). Under the given assumptions S is symmetric positive definite and 'lives' on the skeleton.

First we introduce an additive splitting

$$A = W_L - R_L \tag{14}$$

of the submatrix A. The approximation W_L is assumed to be symmetric positive definite. W_L is used to define an approximation of L (this justifies the subscript L) in the following way

$$\tilde{L} = \begin{bmatrix} I & \\ B^T W_L^{-1} & I \end{bmatrix}. \tag{15}$$

Instead of analysing the original discrete operator K, we study the properties of

$$\tilde{K} = \tilde{L}^{-1} K \tilde{L}^{-T}, \tag{16}$$

representing a W_L-dependent left and right transformation of K.

Remark:

$$\tilde{K} = \begin{bmatrix} A & A(A^{-1} - W_L^{-1})B \\ B^T(A^{-1} - W_L^{-1})A & S + T \end{bmatrix} \tag{17}$$

with
$$T := B^T(A^{-1} - W_L^{-1})A(A^{-1} - W_L^{-1})B \quad . \tag{18}$$

By applying the transformation
$$W = \tilde{L}\tilde{W}\tilde{L}^T \tag{19}$$
every approximation $\tilde{W}$ of $\tilde{K}$ leads directly to an approximation W of K.

Of special interest is the block-JACOBI approximation
$$\tilde{W}_{JAC} = \begin{bmatrix} A & \\ & S+T \end{bmatrix} \tag{20}$$
of $\tilde{K}$ and, more generally, the modified block-JACOBI approximation
$$\tilde{W}_{modJAC} = \begin{bmatrix} W_D & \\ & W_{S+T} \end{bmatrix} \tag{21}$$
with
$$W_D \quad \text{symm., pos. def.} \quad , \quad A = W_D - R_D \tag{22}$$
and
$$W_{S+T} \quad \text{symm., pos. def.} \quad , \quad S+T = W_{S+T} - R_{S+T} \quad . \tag{23}$$

Definition: The approximate block-LDL^T decomposition of K given in terms of W_L, W_D and W_{S+T} is defined as
$$W = \begin{bmatrix} I & \\ B^T W_L^{-1} & I \end{bmatrix} \begin{bmatrix} W_D & \\ & W_{S+T} \end{bmatrix} \begin{bmatrix} I & W_L^{-1^T}B \\ & I \end{bmatrix} \quad . \tag{24}$$

Remark: W_L, W_D and W_{S+T} symmetric positive definite implies W to be symmetric positive definite.

Remark: Let $x^{(0)}$ be given. The approximate block-LDL^T iteration is defined as
$$x^{(i+1)} = x^{(i)} - W^{-1}(Kx^{(i)} - b), \quad i = 1, 2, \ldots \tag{25}$$
with the approximate block-LDL^T decomposition W from (24).

A lot of articles are published concerning the question how to approximate the SCHUR complement $S = C - B^T A^{-1} B$ (see [6] for an overview). The choice $W_{S+T} = B$ is one of simplest approximation. We define a slightly modified approximation $W_{S+T} = B^{\text{local}}$ given by the coefficients
$$B_{\alpha\beta}^{\text{local}} := \begin{cases} 0, & \alpha \text{ crosspoint} \wedge (\alpha \neq \beta) \\ B_{\alpha\beta}, & \text{otherwise} \end{cases} \quad . \tag{26}$$

B^{local} is used in our numerical experiments. The results are documented below.

A more sophisticated approach is to find approximations W_{S+T} by means of few values Sz_i or $(S+T)z_i$. Such a method is the filtering technique by Wittum [22] which is explained in a more general setting in the following.

REMARKS ON FILTERING TECHNIQUES

The following discussion generalises the tridiagonal matrices used in [22]. Tridiagonal matrices yield a chain-like matrix graph. However, the skeleton leads to graphs like in Figure 2 or in Figure 3.

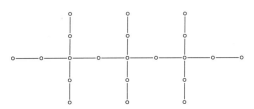

Figure 3: Skeleton with 3 cross points, no elementary cycle

Let I be an index set. We recall the definition of the *graph of a matrix*

$$\text{Graph}(A) := \{(\iota,\kappa) \in I \times I \mid a_{\iota\kappa} \neq 0\} \tag{27}$$

for a given matrix $A \in \mathbb{R}^{I \times I}$. The elements of I are called *knots* whereas $(\iota,\kappa) \in \text{Graph}(A)$ is called a *(directed) edge* from knot ι to knot κ. If the matrix A has symmetric structure ($a_{\iota\kappa} \neq 0 \iff a_{\kappa\iota} \neq 0$), Graph(A) is called *undirected* due to the fact that both edges between knot ι and knot κ are elements of Graph(A).

Let $G \subset I \times I$ be an undirected graph. We define the set of matrices

$$\mathcal{M}_G := \{T \in \mathbb{R}^{I \times I} \mid T \text{ symmetric, } \text{Graph}(T) \subset G\} \tag{28}$$

depending on G. Let a quadrupel of vectors

$$e, g, e', g' \quad \in \mathbb{R}^I \tag{29}$$

be given which fulfils the *symmetry condition*

$$<e, g'> = <e', g> \quad . \tag{30}$$

The vectors e and e' will be referred from now on as *testvectors*.

In the following we discuss the solution of the problem:

> Find a symmetric matrix T which fulfils the *graph condition*
> $$T \in \mathcal{M}_G \qquad (31)$$
> and the *filter condition*
> $$Te = g, \quad Te' = g' \ . \qquad (32)$$

Remark: *Let M be a symmetric matrix. Multiplying the pair of testvectors $\{e, e'\}$ by M produces*
$$g := Me, \quad g' := Me' \ . \qquad (33)$$
Then the symmetry of M implies the symmetry condition (30) and the filter condition (32) reads
$$Me = Te, \quad Me' = Te' \ .$$
The latter equations show that (the approximation) T has the same action on $\mathrm{span}\{e, e'\}$ as (the exact matrix) M.

We define the local determinant
$$\Delta_{\iota\kappa} := e_\iota e'_\kappa - e_\kappa e'_\iota = \det \begin{bmatrix} e_\iota & e'_\iota \\ e_\kappa & e'_\kappa \end{bmatrix}$$
for every edge $(\iota, \kappa) \in G$. In addition to the graph condition and the filter condition the pair of testvectors $\{e, e'\}$ has to fulfil the *local determinant condition*
$$\Delta_{\iota\kappa} \neq 0 \quad \text{for all edges} \quad (\iota, \kappa) \in G \ . \qquad (34)$$

Remark: *For every edge $(\iota, \kappa) \in G$ holds*
$$\Delta_{\iota\kappa} = -\Delta_{\kappa\iota} \ . \qquad (35)$$

In terms of the matrix entries
$$\delta_\iota := T_{\iota\iota} \quad \text{(diagonal entries)}, \qquad (36)$$
$$t_{\iota\kappa} := T_{\iota\kappa} \quad \text{(offdiagonal entries)} \qquad (37)$$
the filter condition reads componentwise as
$$\delta_\iota e_\iota + \sum_{\kappa \neq \iota} t_{\iota\kappa} e_\kappa = g_\iota \ , \qquad (38)$$
$$\delta_\iota e'_\iota + \sum_{\kappa \neq \iota} t_{\iota\kappa} e'_\kappa = g'_\iota \ . \qquad (39)$$
Elimination of δ_ι leads to
$$\sum_\kappa t_{\iota\kappa} \Delta_{\iota\kappa} = \gamma_\iota \quad \text{with} \quad \gamma_\iota := g'_\iota e_\iota - g_\iota e'_\iota \ . \qquad (40)$$
Instead of the unknowns $t_{\iota\kappa}$ we define the unknowns $\tau_{\iota\kappa}$:
$$\tau_{\iota\kappa} := t_{\iota\kappa} \Delta_{\iota\kappa} \ . \qquad (41)$$

The equation (40) changes to

$$\sum_{\kappa} \tau_{\iota\kappa} = \gamma_\iota \quad . \tag{42}$$

Remark: *For every edge* $(\iota, \kappa) \in G$ *holds*

$$\tau_{\iota\kappa} = -\tau_{\kappa\iota} \quad . \tag{43}$$

Remark:

(a) The offdiagonal entries $t_{\iota\kappa}$ of T can be calculated by solving (42) and using the variable transformation (41).

(b) If the offdiagonal entries are known the diagonal entries δ_ι can be calculated from (38) or (39).

Remark: *The symmetry condition (30) is equivalent to*

$$\sum_{\iota} \gamma_\iota = 0 \quad . \tag{44}$$

To discuss the solution of problem (31), (32) in more details we need some more notations from the graph theory.
The tuple $(\iota_0, \iota_1, \ldots, \iota_k)$ is called a *path* in graph G if $\iota_0, \iota_1, \ldots, \iota_k \in I$ and $(\iota_{i-1}, \iota_i) \in G, i = 1, \ldots, k$. A path is a *cycle* if $\iota_0 = \iota_k$. A cycle is called *elementary* if the cycle contains no other cycle(s) (see Figure 2 for an example).

Remark (Homogeneous problem): *We consider the homogeneous problem*

$$\sum_{\kappa} \tau_{\iota\kappa} = 0 \tag{45}$$

(compare (42)) and assume that an elementary cycle $G_\nu \subset G$ exists. The undirected cycle G_ν could be decomposed into the disjoint union of the set of counterclockwise (ccw) directed edges G_ν^{ccw} and the set of clockwise (cw) directed edges G_ν^{cw}:

$$G_\nu = G_\nu^{ccw} \uplus G_\nu^{cw} \quad . \tag{46}$$

Set

$$\tau_{\iota\kappa} := \begin{cases} 1 & , \text{if } (\iota, \kappa) \in G_\nu^{ccw}, \\ 0 & , \text{otherwise}, \end{cases} \tag{47}$$

and extend the definition of $\tau_{\iota\kappa}$ to all edges of G_ν by applying (43) and set $\tau_{\iota\kappa} = 0$ whenever $\iota \notin G_\nu$ or $\kappa \notin G_\nu$. These coefficients define a matrix T_ν. We conclude that T_ν is uniquely determined and solves the homogeneous problem (31), (32) ($g = g' = 0$).

Theorem: *Let $k \geq 0$ be the number of elementary cycles in G. Assume that the symmetry condition (30) and the local determinant condition (34) are valid. Then all solutions of problem (31), (32) belong to the following affine subspace:*

$$T = T_0 + \sum_{\nu=1}^{k} \alpha_\nu T_\nu, \quad \alpha_\nu \in \mathbb{R} \quad . \tag{48}$$

Here, T_0 represents a solution of the inhomogeneous problem (31), (32) and T_ν is the solution of the homogeneous problem (45) defined above for the elementary cycle G_ν.

In the presence of $k \geq 1$ elementary cycles the solution is not uniquely determined. In the following we discuss two different approaches to reduce the k degrees of freedom. Both are based on orthogonal relations between the unknowns.

Remark (τ-orthogonality): Due to (48) the $\tau_{\iota\kappa}$ unknowns can be written as

$$\tau_{\iota\kappa} = \tau_{\iota\kappa}^{(0)} + \sum_{\nu=1}^{k} \alpha_\nu \tau_{\iota\kappa}^{(\nu)} \quad . \tag{49}$$

We propose to reduce the degrees of freedom requiring the orthogonal relations

$$<\tau_{\iota\kappa}, \tau_{\iota\kappa}^{(\nu)}> = 0, \quad \nu = 1, \ldots, k \quad . \tag{50}$$

Remark (t-orthogonality): Due to (48) the $t_{\iota\kappa}$ unknowns can be written as

$$t_{\iota\kappa} = t_{\iota\kappa}^{(0)} + \sum_{\nu=1}^{k} \alpha_\nu t_{\iota\kappa}^{(\nu)} \quad . \tag{51}$$

We propose to reduce the degrees of freedom requiring the orthogonal relations

$$<t_{\iota\kappa}, t_{\iota\kappa}^{(\nu)}> = 0, \quad \nu = 1, \ldots, k \quad . \tag{52}$$

Numerical experiments indicate that there is no significant difference between both approaches.

Remark:

- T tridiagonal leads to a chain-like matrix graph without elementary cycle(s) $\Rightarrow k = 0$ in (48); i.e., the solution is uniquely determined.

- A more general situation is considered in Figure 3. The graph contains no elementary cycle(s) but has a finite number of crosspoints $\Rightarrow k = 0$ in (48); i.e., the solution is uniquely determined.

Extensions of the filtering technique to the unsymmetric case, still restricted to tridiagonal structures, are discussed in [20].

NUMERICAL RESULTS

We consider the two–dimensional model equation

$$-\mathbf{div}\ \phi(x,y)\ \mathbf{grad}\ u(x,y) = rhs$$

on the unit square $\Omega = (0,1)^2$. The solution is fixed to $u(x,y) = x^2 + y^2$ defining the right-hand side rhs and DIRICHLET boundary conditions. Our 4 testproblems differ in the choice of the coefficient function $\phi(x,y)$ (see [22]).

Test problem	Coefficient function $\phi(x,y)$
2	$\tan(\frac{\pi}{2}x)$
3	$1 - e^{-xy}$
4a	$100[\sin(\frac{\pi}{2}\mu x)\sin(\frac{\pi}{2}\mu y) + 1]$ with $\mu = 1$
4b	$100[\sin(\frac{\pi}{2}\mu x)\sin(\frac{\pi}{2}\mu y) + 1]$ with $\mu = 8$

The domain decomposition is defined with respect to a decomposition of the unit square in $p = m \times m$ subdomains of equal size (array-topology). Numerical results are performed for $m \in \{1,2,3,4,5\}$, i.e., up to $p = 25$ processors are used.
We discretise the model problem by linear finite elements on a regular triangular mesh with northeast-southwest diagonals. In every subdomain the coarsest grid (level 0) contains 25 knots and 24 triangles. Regular refinement (four times) gives the finest grid. The unknowns in the subdomains are ordered in a lexicographical manner.

Remark: *The number of unknowns in every subdomain on level l is proportional to 2^{4+2l}. The total number of unknowns is proportional to $2^{4+2l}p$. For a 5×5 problem the total number of unknowns on the finest grid is 160000.*

The smoother is an approximate block-LDL^T iteration defined by the components (compare (24))
$$W_L = 2W^{\text{Jac}}, \quad W_D = W^{\text{ILU}} \quad \text{and} \quad W_S = B^{\text{local}}$$
i.e., W_L represents a damped JACOBI approximate of A, W_D represents a 7-point-ILU approximate (see [13]) and W_S from (26).
We have performed one pre- and one post-smoothing step. Standard restriction and prolongation operators are the remaining components of the parallel multi-grid with V-cycle. The start vector is chosen equal to zero. We measure the defect in the Euclidean norm
$$||d||^2 = \frac{1}{\#I}\sum_{\alpha \in I} d_\alpha^2$$
and iterate until $||d|| \leq eps = 10^{-6}$. The convergence rate ρ is defined as the geometric average of the error reduction factors.
The runs are performed on a PARSETYC SuperCluster with 32 T800-components. Dividing the overall runtime by the numbers of iterations we get the runtime for one iteration step. In order to allow a hardware and software independent statement define the time unit (100 percent) by the time of the longest run (5×5 problem).

Test problem 2: Convergence rates ρ and relative runtimes (in percent)

Level \ topology	1x1	2x2	3x3	4x4	5x5
Level 1	0.041	0.081	0.085	0.088	0.102
(h=1/8)	2%	36%	53%	79%	100%
Level 2	0.059	0.104	0.108	0.110	0.111
(h=1/16)	8%	37%	61%	83%	100%
Level 3	0.075	0.109	0.113	0.121	0.122
(h=1/32)	21%	49%	67%	87%	100%
Level 4	0.083	0.110	0.121	0.122	0.123
(h=1/64)	37%	67%	88%	94%	100%

Test problem 3: Convergence rates ρ and relative runtimes (in percent)

Level \ topology	1x1	2x2	3x3	4x4	5x5
Level 1 (h=1/8)	0.053 / 3%	0.117 / 28%	0.118 / 68%	0.116 / 76%	0.115 / 100%
Level 2 (h=1/16)	0.068 / 9%	0.086 / 34%	0.090 / 57%	0.090 / 81%	0.091 / 100%
Level 3 (h=1/32)	0.077 / 24%	0.087 / 51%	0.090 / 70%	0.091 / 85%	0.092 / 100%
Level 4 (h=1/64)	0.080 / 46%	0.085 / 73%	0.088 / 86%	0.088 / 93%	0.088 / 100%

Test problem 4a: Convergence rates ρ and relative runtimes (in percent)

Level \ topology	1x1	2x2	3x3	4x4	5x5
Level 1 (h=1/8)	0.052 / 3%	0.184 / 35%	0.184 / 59%	0.182 / 80%	0.168 / 100%
Level 2 (h=1/16)	0.060 / 9%	0.150 / 44%	0.150 / 66%	0.140 / 81%	0.139 / 100%
Level 3 (h=1/32)	0.060 / 26%	0.121 / 66%	0.122 / 84%	0.115 / 89%	0.115 / 100%
Level 4 (h=1/64)	0.058 / 45%	0.109 / 79%	0.111 / 90%	0.112 / 96%	0.113 / 100%

Test problem 4b: Convergence rates ρ and relative runtimes (in percent)

Level \ topology	1x1	2x2	3x3	4x4	5x5
Level 1 (h=1/8)	0.071 / 3%	0.154 / 31%	0.200 / 57%	0.175 / 77%	0.175 / 100%
Level 2 (h=1/16)	0.126 / 11%	0.160 / 43%	0.141 / 64%	0.144 / 80%	0.144 / 100%
Level 3 (h=1/32)	0.133 / 31%	0.145 / 67%	0.122 / 77%	0.118 / 89%	0.121 / 100%
Level 4 (h=1/64)	0.118 / 49%	0.123 / 83%	0.116 / 90%	0.112 / 95%	0.116 / 100%

The numerical results presented above indicate a h-independent convergence rate which is not influenced by the underlying array topology (parameter p). The practical implementation uses ideas published in [17]. The slow increase of the percent values is explained by the increase of communication between processors. The communication at all is dominated by the communication which is necessary to compute the solution on the coarsest grid (a broadcast and concentrate procedure).

REFERENCES

[1] AUZINGER, W., FRANK, R., MACSEK, F.: "Asymptotic Error Expansion for Stiff Equations: The implicit Euler Scheme", SIAM J. of Num. Anal. <u>27</u> No. 1 (1990) pp. 66-104.

[2] BASTIAN, P., BURMEISTER, J., HORTON, G.: "Implementation of a parallel multigrid method for parabolic partial differential equations" (1991) pp. 18–27 in [9].

[3] BURMEISTER, J., HORTON, G.: "Time-parallel multi-grid solution of the Navier-Stokes equations", (1991) pp. 155-166 in [10].

[4] BURMEISTER, J.: "Time-Parallel Multi-Grid Methods", (1993) pp. 56-66 in [14].

[5] BURMEISTER, J, PAUL, R.: "Time-adaptive solution of discrete parabolic problems with time-parallel multi-grid methods", (1995) pp. 49-58 in [21].

[6] CHAN, T.F., MATHEW, T.P.: "Domain decomposition algorithms", Acta Numerica (1994) pp. 61-143, Cambridge University Press.

[7] DOROK, O., JOHN, V., RISCH, U., SCHIEWECK, F., TOBISKA, L.: "Parallel Finite Element Methods for the Incompressible Navier-Stokes Equations", in this publication.

[8] GLOWINSKI, R., LIONS, J.-R. (Editors): "Computing methods in applied sciences and engineering", VI. Proc. of the 6th International Symposium on Comp. Methods in Applied Sciences and Engineering. Versaille, France, Dec. 12-16, 1983, North Holland, 1984.

[9] HACKBUSCH, W. (Editor): "Parallel Algorithms for Partial Differential Equations", Proceedings of the Sixth GAMM-Seminar, Kiel, January 19-21, 1990, Notes on Numerical Fluid Mechanics, Volume 31, Vieweg-Verlag, Braunschweig, 1991.

[10] HACKBUSCH, W., TROTTENBERG, U. (Editors): "Multi-grid Methods III", Proceedings of the 3rd European Conference on Multi-grid Methods, Bonn, October 1-4, 1990, International Series of Numerical Mathematics, Vol. 98, Birkhäuser Verlag, Basel, 1991.

[11] HACKBUSCH, W.: "Parabolic multi-grid methods", (1984) in [8].

[12] HACKBUSCH, W.: "Multi-Grid Methods and Applications", Springer Series in Computational Mathematics $\underline{4}$, Springer-Verlag, Berlin, Heidelberg, 1985.

[13] HACKBUSCH, W.: "Iterative Solution of Large Sparse Systems of Equations", Applied Mathematical Sciences $\underline{95}$, Springer-Verlag, New York, 1993.

[14] HIRSCHEL, E.H. (Editor): "Flow Simulation with High-Performance Computers I", Notes on Numerical Fluid Mechanics, Volume 38, Vieweg-Verlag, Braunschweig, 1993.

[15] HORTON, G.: "Ein zeitparalleles Lösungsverfahren für die Navier-Stokes-Gleichungen", doctoral thesis, University of Erlangen-Nürnberg, 1991.

[16] LILEK, Ž., PERIĆ, M., SCHRECK, E.: "Parallelization of Implicit Methods for Flow Simulation", (1993) pp. 135-146 in [21].

[17] MEYER, A.: "A parallel preconditioned conjugate gradient method using domain decomposition and inexact solvers on each subdomain", Computing $\underline{45}$ (1990) pp. 217-234.

[18] RENTZ-REICHERT, H., WITTUM, G.: "A Comparison of Smoothers and Numbering Strategies for Laminar Flow around a Cylinder", in this publication.

[19] THOMÉE, V.: "Galerkin Finite Element Methods for Parabolic Problems", Lecture Notes in Mathematics $\underline{1054}$, Springer-Verlag, Berlin, Heidelberg, 1984.

[20] WAGNER, CH.: "Frequenzfilternde Zerlegungen für unsymmetrische Matrizen und Matrizen mit stark variierenden Koeffizienten", doctoral thesis, University of Stuttgart, 1995.

[21] WAGNER, S. (Editor): "Computational Fluid Dynamics on Parallel Systems", Notes on Numerical Fluid Mechanics, Volume 50, Vieweg-Verlag, Braunschweig, 1995.

[22] WITTUM, G.: "Filternde Zerlegungen. Schnelle Löser für große Gleichungssysteme", Teubner Skripten zur Numerik, B.G. Teubner Verlag, Stuttgart, 1992.

PARALLEL FINITE ELEMENT METHODS FOR THE INCOMPRESSIBLE NAVIER–STOKES EQUATIONS

O. Dorok, V. John, U. Risch, F. Schieweck, L. Tobiska
Institut für Analysis und Numerik
Otto-von-Guericke Universität Magdeburg
Postfach 4120, D-39016 Magdeburg

SUMMARY

We consider parallel and adaptive algorithms for the incompressible Navier-Stokes equations discretized by an upwind type finite element method. Two parallelization concepts are used, a first one based on a static domain decomposition into macroelements and a second one based on a dynamic load balancing strategy. We investigate questions of the scalability up to the massive parallel case and the use of a posteriori error estimators. The arising discrete systems are solved by parallelized multigrid methods which are applied either directly to the coupled system or within a projection method.

1. INTRODUCTION

The reliable numerical prediction of complex flows requires the use of the most efficient numerical methods together with the most powerful computer architectures. Concerning the hardware, only parallel computers can offer the necessary computational performance and memory capacity for future needs to provide solutions with acceptable accuracy. During the last years multigrid methods have become a powerful technique for solving the corresponding algebraic equations and adaptive finite elements have been used to get a better resolution of boundary and interiour layer regions.

In this paper we consider efficient finite element methods for solving the incompressible Navier–Stokes equations in a two-dimensional domain Ω

$$-\frac{1}{\mathrm{Re}}\Delta u + (u\cdot\nabla)u + \nabla p = f, \quad \nabla\cdot u = 0 \quad \text{in } \Omega, \tag{1}$$

with given velocity at the Lipschitz continuous boundary $\partial\Omega$ in the case of higher Reynolds numbers and their parallel implementation on MIMD-machines. For other publications from this project see [3],[13].

We use two different approaches of parallelization. A first one, presented in Section 3, is based on a fixed decomposition of the domain into macroelements which are uniformly refined and assigned to different processors. The second one, desribed in Section 4, uses a posteriori error estimates for a local mesh refinement and requires dynamic load balancing strategies. The implementation of this full adaptive algorithm on a parallel machine has been realized using modules of the package *ugp1.0* [1] which is also used in other projects of the DFG Priority Research Program "Flow Simulation with High-Performance Computers" [7],[8].

2. DISCRETIZATION METHOD

For the discretization of the weak formulation of the Navier–Stokes equations (1) we use two non–conforming elements, namely the P_1/P_0 triangular element of Crouzeix/Raviart [2] and the rotated bilinear quadrilateral Q_1/Q_0 element of Rannacher/Turek [9]. Both element pairs guarantee the discrete Babuška–Brezzi stability condition. Furthermore, in both cases the unknowns are the velocity values at the midpoints of the element edges and the pressure values at the element centres. This fact is advantageous for parallelization since a velocity node belongs to only two elements and a pressure node to only one element which leads to a very cheap local communication. In order to handle the effect of dominated convection we use an upwind stabilization of the convective term which has been developed in [12] for a triangular mesh. This technique contains both the simple first order and the Samarskij upwinding.

3. A PARALLELIZATION STRATEGY BASED ON MACRO ELEMENTS

Here, we use our finite element approximation of upwind type [12] on a quadrilateral mesh with nonconforming rotated bilinear elements for the velocity and piecewise constant elements for the pressure [9]. The algorithmical basis for our parallel Navier–Stokes solver is a robust and efficient sequential multigrid method [10], [11]. We use the domain decomposition into macroelements $\{M_i\}$ and distribute the whole work and data of the sequential algorithm to different tasks $\{T_i\}$ on a processor network where the task T_i is associated with the macroelement M_i. To avoid recursive data dependencies a modification has to be made in the smoothing procedure. However, this has just a slight influence on the convergence rate of the parallel algorithm compared to the sequential one (see [11]). Thus, we have a parallel solver with a good numerical efficiency up to the massive parallel case. We discuss the influence of the communication overhead and the coarsest grid solver on the parallel efficiency for an increasing number of macroelements and processors.

Sequential Multigrid Solver

For the definition of a multilevel grid sequence we start with a decomposition of the domain Ω into quadrilateral elements M_i which we call *macroelements* and which define the finite element mesh on the coarsest grid level $\ell=1$. Then, all finer grid levels are defined by uniform refinement with the usual modification at curved boundary parts.

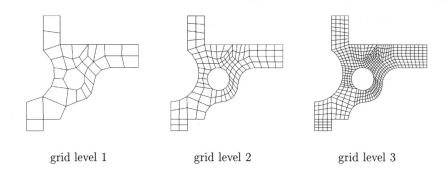

grid level 1　　　　　　　grid level 2　　　　　　　grid level 3

Figure 1: Example of a multilevel grid sequence

This can be done in parallel on each macroelement. Figure 1 shows an example of such a grid generation. For this kind of blockstructured grids we have on the one hand a relatively high flexibility for complex geometries and on the other hand the local regularity of the grid on the macroelements. This opens the possibility for vectorization on each macroelement and it guarantees the regularity assumption for the global mesh which is necessary for stability and convergence of our element pair [9]. Furthermore, we get a natural and nearly optimal load balancing for the parallelization if we assign one processor to each macroelement.

Now we want to solve the discrete Navier–Stokes problem on some finest grid level $\ell=k$, i.e. we want to find on this grid a solution vector $y=(u,p)^\mathsf{T}$, consisting of the velocity vector u and the pressure vector p, which satisfies the following nonlinear system of equations

$$L(y)\, y = F \tag{2}$$

with

$$L(y) := \begin{pmatrix} A(u) & B \\ B^\mathsf{T} & 0 \end{pmatrix}, \quad y := \begin{pmatrix} u \\ p \end{pmatrix}, \quad F := \begin{pmatrix} f \\ g \end{pmatrix} \tag{3}$$

where the matrix $A(u)$ contains the upwind-discretization of the nonlinear convection term. To solve this nonlinear problem we use a fixed-point iteration where in the n-th step we calculate $\bar{y}^n$ as an approximate solution of the linear system

$$L(y^{n-1})\, \bar{y}^n = F \tag{4}$$

and take the new iterate

$$y^n = y^{n-1} + \omega(\bar{y}^n - y^{n-1})$$

with some damping parameter ω (in numerical calculations we found the value $\omega=0.9$ to be suitable).

For solving the linear problem (4) we apply a certain number of multigrid cycles. This multigrid iteration is stopped if the initial defect is damped by a factor of 0.1 or after a maximum of 5 iterations. We will only briefly mention the components of the multigrid algorithm (for details see [10]). We take an F–cycle where the coarse grid correction is damped by a factor of 0.8. For the grid transfer we use standard finite element prolongation and restriction with a natural modification at points where the nonconforming finite element functions are discontinuous. The coarsest grid problems are solved by an augmented Lagrange algorithm. We use a blockwise Gauss–Seidel smoother similar to that proposed by Vanka [15].

In the following we want to explain the smoother in a little more detail. In order to perform one smoothing iteration on some actual grid level we pass through all elements K of that grid and update for each K the corresponding local block vector $y_K = (u_K, p_K)^\mathsf{T}$ which consists of all unknowns belonging to that element K (these are the 8 velocity values at the 4 midpoints of the edges of K and the one constant pressure value p_K on element K). We update y_K by solving a modified local system (see [10])

$$\tilde{L}_K y_K = \tilde{F}_K \qquad (5)$$

which is derived from the global system (4) on the current grid level by considering the equations that are associated with element K. Let us note that for this element loop we use a Gauss-Seidel-like update mechanism, i.e. within one smoothing step we can have a spreading of new updated information over the whole domain. At the end of each smoothing iteration we apply an underrelaxation step.

Parallelization

At first we construct a task network $\{T_i\}$ which is topologically equivalent to the macroelement decomposition $\{M_i\}$ of our domain (see Figure 2). That means we have a mapping

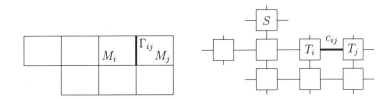

Figure 2: Macroelement decomposition $\{M_i\}$ and the related task network $\{T_i\}$ with an extra solver task S

$M_i \rightarrow T_i$ where any two neighbouring macroelements M_i and M_j having a common edge Γ_{ij} are assigned to neighbouring tasks T_i and T_j connected by a communication channel c_{ij}. Then, we map the set of all tasks $\{T_i\}$ to a given set of processors $\{P_j\}$. If we have enough processors we will assign one task to one processor otherwise we have a load balancing problem to get a suitable mapping. Now our parallelization is characterized as follows. For the algorithm we use the sequential one with a slight modification in the smoother (see below). Each task T_i is responsible for all operations of the algorithm with data belonging to macroelement M_i. For operations with ∂M_i–data on the boundary of M_i the task T_i exchanges its Γ_{ij}–data with all neighbouring tasks T_j via the corresponding channels c_{ij}. We call this *local communication*. For operations with global Ω–data (as for instance the calculation of a global defect norm or the solution of the coarsest grid problem) the task T_i exchanges data with the whole task network $\{T_i\}$. This kind of communication we will call *global communication*.

Parallel Smoother

In order to avoid recursive data dependencies between the macroelement tasks we split the one loop of the sequential smoothing procedure over all elements in the whole domain Ω into several loops over the elements in the macroelements M_i which can be executed independently and in parallel. That means all tasks T_i update at the same time their own unknowns on M_i. As a consequence, neighbouring tasks T_i and T_j produce different velocity values at the nodes x on the common macroelement boundary $\Gamma_{ij}=\partial M_i \cap \partial M_j$. Therefore, the different values are exchanged by local communication and then each two tasks T_i and T_j compute for all nodes $x \in \Gamma_{ij}$ the following equalized value

$$u(x) := \frac{1}{2}\left(u(x)_{|M_i} + u(x)_{|M_j}\right). \tag{6}$$

So, our parallel smoother differs from the sequential one by the fact that within one smoothing iteration the spreading of new updated information is restricted only to the macroelements. Therefore, we would expect a slower convergence of the parallel multigrid algorithm compared to the sequential one. However, numerical tests show (see [11]) that there is only a small difference in the convergence rates of both methods such that we achieve a good numerical efficiency for our parallel solver even in the case of a large number of macroelements.

COARSEST GRID SOLVER

In order to solve the global coarsest grid problems we add an extra solver task S on an extra processor P_S to the processor network such that all data dependencies are concentrated on S. At the beginning of the n-th nonlinear step the task S waits for the restricted velocity vector $u^{\text{old}}=u^{n-1}$ on grid level $\ell=1$ which will be transmitted by a global communication from the task network $\{T_i\}$. Once u^{old} has been received, the matrix $A=A(u^{\text{old}})$ and the augmented Lagrange matrix $C=A+rBB^\mathsf{T}$ are generated and the LU–factorization of C is computed. There is an additional parallelism between the solver task S and the macroelement tasks T_i in the following sense. At the same time, when S deals with the generation of A and C and the factorization of C, the tasks T_i can continue their work within the first multigrid cycle until the first coarsest grid problem occurs. Then, within the multigrid cycle, task S waits to get from the task network $\{T_i\}$ the global right hand side vector of the coarsest grid problem. After this vector has been received, the solution vector is computed by the augmented Lagrange iteration and will be sent back to all other tasks T_i via global communication.

RESULTS

In order to study the scalability properties of our parallelization we consider as a test example the driven cavity problem with Reynolds number $Re=2000$ where the macroelements are defined by an $(N \times N)$-decomposition of the unit square Ω. Then, the total number of elements on the finest grid level ℓ is $N^2 4^{\ell-1}$ and the number of used processors is N^2+1, i.e. with N we simultaneously increase the problem size and the number of processors.

In Table 1 we show the behaviour of our parallel code for two different hardware situations, the Parsytec machines GCel with processor T805 and the GCPP with processor MPC601. t_{mg} denotes the CPU-time in seconds for one multigrid cycle on the finest grid level $\ell=6$. t_l, t_g and t_s are the times for local communication, global communication and for solving the coarsest grid problems within one multigrid cycle. t_g only measures the time for computing global norms whereas the time for the global communication of coarsest grid

Table 1: CPU times on two different hardware systems for level $\ell=6$

NxN	GCel					GCPP				
	t^{all}	t_{mg}	t_l	t_g	t_s	t^{all}	t_{mg}	t_l	t_g	t_s
2x2	893	10.7	.50	.003	.09	41	.50	.17	.001	.01
3x3	953	11.2	.88	.008	.25	54	.64	.30	.003	.03
4x4	1144	10.5	.96	.012	.47	73	.68	.37	.004	.04
5x5	1274	11.6	1.08	.017	1.04	88	.81	.47	.007	.07
6x6	1397	13.3	1.18	.019	2.12	95	.91	.50	.008	.12
8x8	1936	18.6	1.61	.031	6.97	109	1.03	.53	.009	.21

vectors between the macroelement-tasks and the solver-task is contained in t_s. By t^{all} we denote the total CPU-time for solving the driven cavity problem. Switching from the GCel to the GCPP the communication becomes between 2 and 3 times faster and the speed of arithmetical operations is increased by a factor of about 34. Therefore, on the GCPP the time t_l for local communication is the dominating part of t_{mg} whereas on the GCel for an increasing number of macroelements the time t_s becomes the largest portion since the one solver-task needs more and more time to solve the coarsest grid problems of increasing size. Nevertheless on the GCPP the time t_s would become dominant too if we would further increase the number of macroelements.

Table 2: Loss of efficiency on the GCel

NxN	Level 4			Level 5			Level 6		
	β_c	β_s	$\tilde{e}$	β_c	β_s	$\tilde{e}$	β_c	β_s	$\tilde{e}$
2x2	0.16	0.06	0.77	0.09	0.02	0.89	0.05	0.01	0.95
3x3	0.23	0.14	0.63	0.14	0.06	0.80	0.08	0.02	0.90
4x4	0.24	0.24	0.52	0.16	0.11	0.73	0.09	0.04	0.87
5x5	0.18	0.45	0.37	0.15	0.20	0.65	0.09	0.09	0.82
6x6	0.12	0.65	0.23	0.12	0.37	0.51	0.09	0.15	0.76
8x8	0.06	0.85	0.09	0.09	0.66	0.25	0.09	0.37	0.55

Table 3: Loss of efficiency on the GCPP

NxN	Level 6			Level 7			Level 8		
	β_c	β_s	$\tilde{e}$	β_c	β_s	$\tilde{e}$	β_c	β_s	$\tilde{e}$
2x2	0.33	0.025	0.65	0.15	0.009	0.84	0.07	0.003	0.93
3x3	0.46	0.045	0.49	0.25	0.018	0.73	0.11	0.006	0.88
4x4	0.54	0.057	0.41	0.30	0.024	0.68	0.15	0.008	0.84
5x5	0.57	0.084	0.34	0.34	0.039	0.62	0.16	0.014	0.82
6x6	0.54	0.127	0.33	0.34	0.061	0.60	0.17	0.022	0.80
8x8	0.51	0.198	0.29	0.34	0.105	0.56	0.18	0.039	0.78

In Table 2 and Table 3 we show for different finest grid levels the loss of efficiency on the GCel and the GCPP, respectively. Here, β_c is defined as the portion of t^{all} for local and global communication (i.e. $\beta_c t^{all}$ is the time for communication). β_s denotes the portion for solving the coarsest grid problems. The quantities β_c and β_s can be regarded approximately as the losses of efficiency (see [10]) such that the value $\tilde{e} := 1 - \beta_c - \beta_s$ gives a rough measure for the parallel efficiency. We see that the losses of efficiency β_c and β_s are reduced if we have a larger finest grid level ℓ. That means, in order to achieve a good parallel effieciency we have to assign a sufficiently large amount of arithmetical work to each processor. If we have a large number of macroelements and a small finest grid level ℓ then it would be more efficient to map the macroelement-tasks to a smaller number of processors. Furthermore, we see that for fixed finest grid level ℓ and increasing number of macroelements the solver portion β_s is the most rapidly increasing part among the losses of efficiency. On the GCel β_s becomes dominant already for a relatively small number of macroelements whereas on the GCPP this would happen only for a much larger number.

THE MASSIVE PARALLEL CASE

Now we want to consider the case when we have a large number of macroelements. We have seen that the most important thing is to reduce the time for the coarsest grid solver.

One way could be to choose a better numerical method for this solver. Table 4 shows the total CPU-time t^{all} for the finest grid level $\ell=6$ if we choose the following coarsest grid solvers: (a) the above defined smoother "Vanka", (b) the SIMPLE-iteration "Simple", (c) the augmented Lagrange method "augm.L." and (d) the void solver "no-slv" which does nothing, i.e. where the coarsest-grid-solver-step is omitted.

We see that for a larger number of macroelements the SIMPLE-solver beats the augmented Lagrange method. However, in the massive parallel case all coarsest grid solvers on a single solver-task are too expensive such that the do-nothing-version despite of its poor convergence rate gives better results.

Table 4: Influence of the solver-method for one solver-task on the CPU-time for $\ell=6$

NxN	Vanka	Simple	augm.L.	no-slv
2x2	897	888	893	960
3x3	882	832	953	1238
4x4	1165	1160	1144	1684
5x5	1429	1311	1274	2102
6x6	1648	1478	1397	2533
8x8	2606	1698	1936	3357
10x10	4507	2214	3998	4138
20x20	42990	11910	*	7360

Another way is to use more than one processor to solve the coarsest grid problems. By recursive bisection of the set of all macroelements $\{M_i\}$ we subdivide the domain Ω into 2^s solver-task-regions $\Omega_n := \bigcup_{j \in I_n} M_j$. Then, we split the work and data of the one solver-task S into 2^s solver-tasks $\{S_n\}$ where S_n is responsible for all operations of the solver with data associated with Ω_n. The task S_n is mapped to the same processor as the root-task $T_{r(n)}$ of the subtree of all macroelement-tasks T_i with $i \in I_n$ such that the total number of processors used for the tasks $\{T_i\}$ is not increased.

Table 5: Total CPU-times for single and parallelized coarsest grid solvers for $\ell=6$

NxN	Vanka-1 t^{all}	Vanka-par #S	Vanka-par t^{all}	Simple-1 t^{all}	Simple-par #S	Simple-par t^{all}	Vanka-all t^{all}
2x2	897	4	885	888	2	897	969
3x3	882	4	848	832	2	841	980
4x4	1165	8	1095	1160	4	1134	1250
5x5	1429	8	1246	1311	2	1228	1380
6x6	1648	16	1302	1478	4	1369	1324
8x8	2606	32	1452	1698	4	1433	1488
10x10	4507	32	1933	2214	4	1774	1766
20x20	42990	128	2873	11910	16	3578	2347
24x24	*	256	2274	19150	64	3747	1979
32x32	*	256	2661	*	512	3694	2374

Table 5 shows the total CPU-times t^{all} for solving the driven cavity problem for the Vanka-solver with one solver-task ("Vanka-1"), the parallelized Vanka-solver with $\#S$ solver-tasks ("Vanka-par"), the SIMPLE-solver with one solver-task ("Simple-1"), the parallelized SIMPLE-solver ("Simple-par") and the version ("Vanka-all") where the work of the Vanka-solver is distributed to all macroelement-tasks $\{T_i\}$ without using a solver task. With $\#S$ we present the number of used solver-tasks that gave the best results, i.e. the minimum total time t^{all}.

We see that for a large number of macroelements the parallelization of the coarsest grid solver leads to an essential improvement of the parallel efficiency. The parallelized SIMPLE-solver is not so efficient as the parallelized Vanka-solver since it needs much more communication. It is somewhat surprising that the version "Vanka-all" gives the best results in the case of a large number of macroelements.

4. ADAPTIVE PARALLEL METHOD

Adaptive methods are used in order to reduce the size of the arising system of equations. The philosophy is to put the computational work, i.e. the unknowns, into regions where a large local error of the computed solution is assumed. Accordingly, for these methods one needs additional components like a posteriori error estimators, local mesh refinement and in the parallel context dynamic load balancing.

A POSTERIORI ERROR ESTIMATORS

Here we use the non–conforming $P1/P0$ triangular element pair. Elements which should be refined are divided into four smaller triangles by connecting the midpoints of the edges ("red refinement"). Thus, we obtain hanging nodes which are handled by several closure rules ("green refinement").

In order to decide which triangles have to be refined an a posteriori error estimator is needed. Residual based error estimators for the Navier–Stokes equations and non–conforming elements have been proposed and analyzed in [16], [17].

Let n_E be the unit outer normal vector on an edge E of an element T. Then $[u]_E$ denotes the jump of a function u across E in the direction of n_E

$$[u]_E = \lim_{t \to +0} u(x + t n_E) - \lim_{t \to +0} u(x - t n_E) \quad \forall x \in E.$$

Moreover, f_h is an approximation to f obtained by numerical integration, h_T the size of T and h_E the length of E.

If (u, p) is the solution of the Navier–Stokes equations (1) and (u_h, p_h) the computed discrete solution then the error estimate for the graph norm becomes

$$\|u - u_h\|_1 + \|p - p_h\|_0 \leq C \left(\sum_{T \in \mathcal{T}} \eta_T^2 \right)^{1/2} \tag{7}$$

with

$$\eta_T^2 = h_T^2 \|f_h - (u_h \cdot \nabla) u_h\|_{0,T}^2 + \|\nabla \cdot u_h\|_{0,T}^2 + \sum_{E \in \partial T} h_E \left\| \left[\frac{1}{Re} n_E \cdot \nabla u_h - p_h n_E \right]_E \right\|_{0,E}^2.$$

The error in the L^2–norm for the velocity can be estimated in the form

$$\|u - u_h\|_0 \leq C \left(\sum_{T \in \mathcal{T}} \tilde{\eta}_T^2 \right)^{1/2} \tag{8}$$

with

$$\tilde{\eta}_T^2 = h_T^4 \|f_h - (u_h \cdot \nabla) u_h\|_{0,T}^2 + h_T^2 \|\nabla \cdot u_h\|_{0,T}^2 + \sum_{E \in \partial T} h_E^3 \left\| \left[\frac{1}{Re} n_E \cdot \nabla u_h - p_h n_E \right]_E \right\|_{0,E}^2.$$

Both estimators η_T and $\tilde{\eta}_T$ contain three terms. The first and the second one are the residuals of the momentum and continuity equation of the strong formulation of the Navier–Stokes equations. The third term, which contains the jumps across edges, is a connection between the weak and strong formulation for non–smooth functions (like the discrete solution).

The above a posteriori error estimators have been used only to generate adaptively refined meshes. If we want to use them really for estimating the error we need quantitative estimates for the constants C in (7) and (8). The great difficulties connected with this problem have been focussed [4] and [6], the research in this direction is just at the beginning. Both error estimators, η_T and $\tilde{\eta}_T$, which are different only in the powers of h_T and h_E, in general produce different meshes with different solutions. This can be seen already for convection diffusion problems. As an example we show in Figure 3 the mesh and the resolution of an interior boundary layer for the model problem

$$\begin{aligned} -10^{-6} \Delta u + (3,1) \cdot \nabla u + u &= 100 r(r - 0.5)(r - \sqrt{2}/2) & \text{in} & \quad \Omega \\ u &= 0 & \text{on} & \quad \partial \Omega \end{aligned}$$

where Ω is an L-shaped domain and $r = \sqrt{(x - 0.5)^2 + (y - 0.5)^2}$.

A more complete study for different error estimators and error indicators for convection diffusion problems is contained in [5].

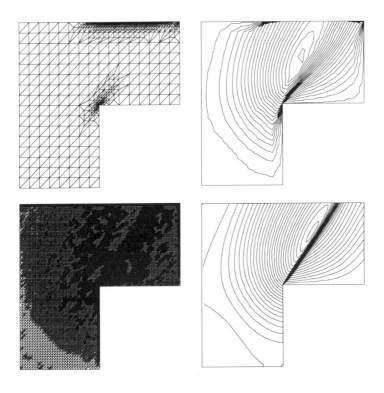

Figure 3: Comparison between grids and solution produced with H^1–norm error estimator (above) and a L^2–norm error estimator (below)

THE SOLVER

We use the fixed–point iteration (4) as the outer nonlinear iteration. Alternatively to the above approach where the multigrid technique is applied directly to the coupled system we use a different concept which seems to be more suitable for non–stationary problems, [14].

We consider the Schur complement system with respect to the pressure and solve the arising equations by a preconditioned Richardson iteration

$$p^{l+1} = p^l + \alpha(B^T C^{-1} B)^{-1}(-B^T A^{-1} B p^l + B^T A^{-1} f - g), \quad l = 0, 1, \ldots, L-1, \qquad (9)$$

where C is an invertible matrix and $\alpha > 0$ a damping factor. In each iteration step one has to solve a linear system of the abstract form

$$B^T C^{-1} B x = s. \qquad (10)$$

We choose $C = \hat{A}$ where $\hat{A}^{-1}$ corresponds to a number of multigrid cycles applied to a system of the form $Ay = b$ and we solve (10) by GMRES. All components of the solver, multigrid as well as GMRES, are well suited for parallel computing. After having stopped the iteration (9) with p^L we compute u from the equation

$$Au = f - Bp^L. \tag{11}$$

An important ingredient of the solver is the multigrid method for approximating A^{-1} in the right hand side of (9) and for solving (11). As grid transfer operation we use the L^2–projection between meshes. The smoother is a damped block Jacobi with inner ILU(β) iterations. The number and form of the blocks depend on the distribution of the elements on the processors. Unknowns on the interfaces between processors can be handled by one processor only or by both and afterwards the results are averaged. The coarse grid system is solved either directly or iteratively, depending on its size.

The implementation of the algorithm described above has been done using the program package *ugp1.0* developed by P. Bastian [1]. In particular data structures, load balancing routines and the parallel overhead of this program have been used. However, the existing data structures had to be extended in order to handle multidimensional data and non–conforming elements.

5. EXTENSIONS

Extensions of the proposed method to the Boussinesq approximation of the Navier–Stokes equations and to the non-stationary case have been studied too. In the first case, additionally the energy equation

$$-\frac{1}{\text{RePr}}\Delta T + u \cdot \nabla T = 0 \quad \text{in } \Omega \tag{12}$$

has to be included and the right hand side f in (1) is given by the buoyancy force

$$f = \frac{\text{Ra}}{\text{Re}^2\text{Pr}} T\vec{g}. \tag{13}$$

The resulting coupled problem for the unknowns u, p, T is solved by a multigrid method which is similar to the one described in Section 3. As an example the flow through a channel over a hot obstacle has been calculated for two different Rayleigh numbers $\text{Ra} = 10^6$ and $\text{Ra} = 10^8$ where $\text{Pr} = 0.71$ and $\text{Ra} = \text{Re}\,\text{Pr}^2$. In Figure 4 and 5 the streamlines of the velocity field and the isolines for the temperature, respectively, are shown.

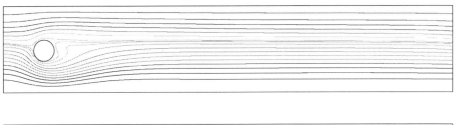

Figure 4: Streamlines, $Ra = 10^6$ and $Ra = 10^8$

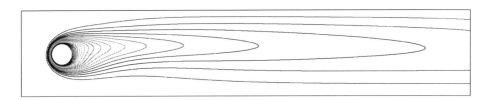
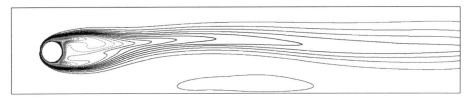

Figure 5: Iso-temperature lines, $Ra = 10^6$ and $Ra = 10^8$

For solving the non-stationary Navier–Stokes equations, an implicit second order in time discretization based on the backward difference formula

$$\left.\frac{\partial u}{\partial t}\right|_{t=t_n} \approx \frac{3u(t_n) - 4u(t_{n-1}) + u(t_{n-2})}{2\Delta t}. \tag{14}$$

is used. The arising problem for $(u(t_n), p(t_n))$ is solved by the stationary solver described in Section 4.

As an example we present the flow through the channel $[0,4] \times [0,1]$ with $Re = 100$, changing inflow parts and fixed outflow part. At the outflow part $\{(4,y) : 0.25 \leq y \leq 0.75\}$ we assume a fixed developed parabolic velocity profile u_o. The inflow parts are $\{(0,y) : 0.25 \leq y \leq 0.75\}$ with the inflow profile $(1-t)\,u_o$ and $\{(x,0) : 1 \leq x \leq 1.5\}$ with $t\,u_o$ for $t \in [0,1]$. For $t \geq 1$ the inflow does not change any longer. We have started with the solution of the stationary equations. The mesh has been adapted by an L^2-error estimator similar to that presented in Section 4 with an additional term that estimates the error in time (for more details see [17]). The resulting meshes for some time steps are shown in Figure 6 together with the corresponding vector plots of the velocity.

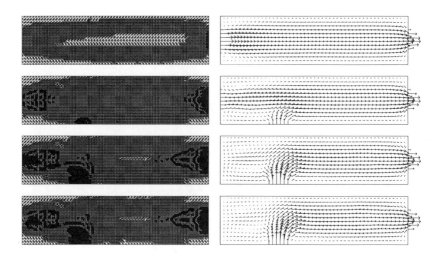

Figure 6: Mesh and vector plot of the velocity for $t = 0.0, t = 0.5, t = 1.0$ and $t = 1.25$

ACKNOWLEDGEMENT

For our numerical tests we used the GCel-1024 of the Paderborn Center for Parallel Computing. Some of the pictures have been produced by the package GRAPE.

REFERENCES

[1] P. Bastian, *Parallele adaptive Mehrgitterverfahren.* PhD thesis, Universität Heidelberg, 1994.

[2] M. Crouzeix, P.-A. Raviart, Conforming and nonconforming Finite Element Methods for solving the stationary Stokes equations. *RAIRO Anal. Numer.* **7**, 33-76 (1973).

[3] O. Dorok, G. Lube, U. Risch, F. Schieweck, L. Tobiska, Finite Element discretization of the Navier–Stokes equations. In: E.H. Hirschel (Ed.): Flow Simulation with High–Performance Computers I, *Notes on Numerical Fluid Mechanics*, Vol. 38, Vieweg-Verlag 1993, pp. 67-78.

[4] K. Eriksson, D. Estep, P. Hansbo, C. Johnson, Introduction to adaptive methods for differential equations. *Acta Numerica*, 1-54 (1995).

[5] V. John, A comparison of some error estimators for convection diffusion problems on a parallel computer. Preprint 12/94, Otto-von-Guericke Universität Magdeburg, Fakultät für Mathematik, 1994.

[6] C. Johnson, On computability and error control in CFD. *Int. Jour. Num. Meth. Fluids* **20**, 777-788 (1995).

[7] H. Reichert, G. Wittum, Solving the Navier-Stokes equations on unstructured grids. In: E.H. Hirschel (Ed.): Flow Simulation with High–Performance Computers I, *Notes on Numerical Fluid Mechanics*, Vol. 38, Vieweg-Verlag 1993, pp. 321-333.

[8] H. Rentz-Reichert, G. Wittum, A comparison of smoothers and numbering strategies for laminar flow around a cylinder. In this publication.

[9] R. Rannacher, S. Turek, Simple Nonconforming Quadrilateral Stokes Element. *Numer. Methods Partial Differential Equations* **8**, 97-111 (1992).

[10] F. Schieweck, A parallel multigrid algorithm for solving the Navier–Stokes equations. *IMPACT Comput. Sci. Engrg.* **5**, 345-378 (1993).

[11] F. Schieweck, Multigrid Convergence Rates of a Sequential and a Parallel Navier–Stokes Solver. In: Hackbusch, W., Wittum, G. (Eds.): Fast Solvers for Flow Problems, *Notes on Numerical Fluid Mechanics*, Vol.49, Vieweg-Verlag 1995, pp. 251-262.

[12] F. Schieweck, L. Tobiska, A nonconforming finite element method of upstream type applied to the stationary Navier-Stokes equation. *RAIRO Modél. Math. Anal. Numér.* **23**, 627–647 (1989).

[13] L. Tobiska, A note on the artificial viscosity of numerical schemes. *Comp. Fluid Dynamics* **5**, 281–290 (1995).

[14] St. Turek, A comparative study of time stepping techniques for the incompressible Navier-Stokes equations: From fully implicit nonlinear schemes to semi-implicit projection methods. to appear, 1995.

[15] S. Vanka, Block-implicit multigrid calculation of two-dimensional recirculating flows. *Comput. Methods Appl. Mech. Engrg.* **59**, 29-48 (1986).

[16] R. Verfürth, A posteriori error estimators for the Stokes Stokes equations. II Nonconforming discretizations. *Numer. Math.* **60**, 235-249 (1991).

[17] R. Verfürth, A posteriori error estimates for nonlinear problems. Finite element discretizations of parabolic equations. Bericht 180, Ruhr-Universität Bochum, Fakultät für Mathematik, 1995.

NUMERICAL TURBULENCE SIMULATION ON A PARALLEL COMPUTER USING THE COMBINATION METHOD

M. Griebel, W. Huber and C. Zenger

Institut für Informatik, Technische Universität München,
Arcisstrasse 21, D-80290 München,
Germany

SUMMARY

The parallel numerical solution of the Navier-Stokes equations with the sparse grid combination method was studied. This algorithmic concept is based on the independent solution of many problems with reduced size and their linear combination. The algorithm for three-dimensional problems is described and its application to turbulence simulation is reported. Statistical results on a pipe flow for Reynolds number $Re_{cl} = 6950$ are presented and compared with results obtained from other numerical simulations and physical experiments. Its parallel implementation on an IBM SP2 computer is also discussed.

INTRODUCTION

So-called sparse grid techniques are promising for the solution of linear PDEs. There, for three-dimensional problems, only $O(h_m^{-1}(\log_2(h_m^{-1}))^2)$ grid points are needed instead of $O(h_m^{-3})$ grid points as in the conventional full grid case. Here, $h_m = 2^{-m}$ denotes the mesh size in the unit cube. However, the accuracy of the sparse grid solution is of the order $O(h_m^2(\log_2(h_m^{-1}))^2)$ (with respect to the L_2- and $L_\infty-$norm), provided that the solution is sufficiently smooth (i.e. $|\frac{\delta^6 u(x,y,z)}{\delta x^2 \delta y^2 \delta z^2}| \leq \infty$). This is only slightly worse than the order $O(h_m^2)$ obtained for the usual full grid solution. For further details on sparse grids, see [2, 7, 6, 16].

For the solution of problems arising from the sparse grid discretization approach, two different methods have been developed in recent years: first, multilevel-type solvers for the system that results from a tensor-product hierarchical-basis-like finite element/finite volume or finite difference discretization (FE, FV, FD) and second, the so-called *combination method*. There, the solution is obtained on a sparse grid by a certain linear combination of discrete solutions on different meshes. For details on the combination method in Cartesian coordinates, see [10].

For related techniques on new numerical methods for PDEs, see [4, 12].

SPARSE GRIDS AND THE COMBINATION METHOD IN A PIPE GEOMETRY

Our aim is turbulence simulation in a pipe geometry. Therefore, we use cylindrical coordinates instead. The Navier-Stokes equations are considered on the unit cylinder $\Omega =]0,1[\times]0,2\pi[\times]0,1[\subset \mathbb{R}^3$ in the coordinate system (r,φ,z), where r denotes the radius, φ the angle and z the length. Compare also Figure 1. For reasons of simplicity, we first discuss the main ideas of our method for a partial differential equation $Lu = f$ in a unit cylinder with a linear, elliptic operator L of second order and appropriate boundary conditions. The usual approach is to discretize the problem by an FE, FV or FD method on the grid Ω_{h_m,h_m,h_m} with mesh sizes 2^{-m} in the r- and z-directions, respectively, and mesh size $2\pi \cdot 2^{-m}$ in the φ-direction. Note that the discretization of a partial differential equation using cylindrical coordinates is somewhat involved because a singularity appears on the middle axis of the cylinder. However, there exist methods that are able to deal with this problem, see [1, 14]. They result in a stable discretization of second-order accuracy.

Then, the system of linear equations $L_{h_m,h_m,h_m} u_{h_m,h_m,h_m} = f_{h_m,h_m,h_m}$ obtained must be solved. We obtain a solution u_{h_m,h_m,h_m} with error $e_{h_m,h_m,h_m} = u - u_{h_m,h_m,h_m}$ of the order $O(h_m^2)$, if u is sufficiently smooth. Here, we assume that u_{h_m,h_m,h_m} represents an appropriate interpolant defined by the values of the discrete solution on grid Ω_{h_m,h_m,h_m}.

Extending this standard approach, we now study linear combinations of discrete solutions of the problem on different cylindrical grids. To this end, let Ω_{h_i,h_j,h_k} denote a grid on the unit cylinder with mesh sizes $h_i = 2^{-i}$ in the r-direction, $h_j = 2\pi \cdot 2^{-j}$ in the φ-direction and $h_k = 2^{-k}$ in the z-direction. In [10], the *combination method* was introduced. We recall the definition of the combined solution $u^c_{h_m,h_m,h_m}$ (but now for cylindrical coordinates):

$$u^c_{h_m,h_m,h_m} := \sum_{i+j+k=m+2} u_{h_i,h_j,h_k} - 2 \sum_{i+j+k=m+1} u_{h_i,h_j,h_k} + \sum_{i+j+k=m} u_{h_i,h_j,h_k}. \quad (1)$$

Here, i,j,k range from 1 to m, where m is defined by the mesh size $h_m = 2^{-m}$ of the associated full grid. Thus, we have to solve $(m+1) \cdot m/2$ problems $L_{h_i,h_j,h_k} u_{h_i,h_j,h_k} = f_{h_i,h_j,h_k}$ with $i+j+k = m+2$, each with about 2^m unknowns, $m \cdot (m-1)/2$ problems $L_{h_i,h_j,h_k} u_{h_i,h_j,h_k} = f_{h_i,h_j,h_k}$ with $i+j+k = m+1$, each with about 2^{m-1} unknowns, and $(m-1) \cdot (m-2)/2$ problems $L_{h_i,h_j,h_k} u_{h_i,h_j,h_k} = f_{h_i,h_j,h_k}$ with $i+j+k = m$, each with about 2^{m-2} unknowns. Then, we combine their tri-linearly interpolated solutions. Note that, in the combination process, the tri-linear interpolation of the solutions u_{h_i,h_j,h_k} to some finer grids must be properly adjusted to cylindrical coordinates. This leads to a solution defined on the sparse grid $\Omega^s_{h_m,h_m,h_m}$ in cylindrical coordinates (see Figure 2). The sparse grid $\Omega^s_{h_m,h_m,h_m}$ is a subset of the associated full grid Ω_{h_m,h_m,h_m}. Its grid points are defined to be the union of the grid points of the grids Ω_{h_i,h_j,h_k}, where $i+j+k = m+2$ and i,j and k range from 1 to m. Altogether, the combination method involves $O(h_m^{-1}(\log_2(h_m^{-1}))^2)$ unknowns, in contrast to $O(h_m^{-3})$ unknowns for the conventional full grid approach. Additionally, the combination solution $u^c_{h_m,h_m,h_m}$ is nearly as accurate as the solution u_{h_m,h_m,h_m} on the associated full grid provided that the solution u is sufficiently smooth. This is only slightly worse than for the associated full grid, where the error is of the order $O(h_m^2)$. For details on sparse grids and the combination method, see [2, 8, 10, 16].

To apply the combination method also in the case where *different* amounts of grid points for the different coordinate directions must be used, we have to generalize Eqn.

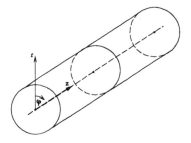

Fig. 1 Cylindrical coordinates in a unit cylinder.

Fig. 2 Front views ($r - \varphi$–direction) of the sparse grid $\Omega^s_{h_m,h_m,h_m}$ and the associated full grid Ω_{h_m,h_m,h_m} for the case $m = 4$.

(1). To this end, we have to change the notation. Let L, M, N denote the number of volume cells in the r-, φ- and z-directions, respectively, and let $\Omega_{(L,M,N)}$ denote the corresponding grid. Without loss of generality, let L, M, N be even and positive. Now, we seek the *factorization* of L, M and N, where

$$L =: 2^l \cdot \kappa_l, \quad M =: 2^m \cdot \kappa_m, \quad N =: 2^n \cdot \kappa_n, \qquad (2)$$

with $\kappa_l, \kappa_m, \kappa_n, l, m, n \in \mathbb{N}, \kappa_l, \kappa_m, \kappa_n \geq 2$ and l, m, n maximal. The value p, which will be needed later in Eqn. (5), is determined by

$$p := min(l,m,n) + 1. \qquad (3)$$

Now, we define the mesh sizes h_i, h_j, h_k of a grid Ω_{h_i,h_j,h_k} to be

$$h_i = k_L^{-1} \cdot 2^{-i}, \quad h_j = k_M^{-1} \cdot 2^{-j} \cdot 2\pi, \quad h_k = k_N^{-1} \cdot 2^{-k}, \qquad (4)$$

with $k_L, k_M, k_N \in \mathbb{R}$. The mesh sizes h_i, h_j and h_k then take the values k_L, k_M and k_N into account, which are defined via $L =: 2^p \cdot k_L, M =: 2^p \cdot k_M, N =: 2^p \cdot k_N$. The corresponding solutions are described by u_{h_i,h_j,h_k}. In this way, we are able to give a generalized combination formula for the case of different amounts of grid points/cells for the different coordinate directions by

$$u^c_{(L,M,N)} := \sum_{i+j+k=p+2} u_{h_i,h_j,h_k} - 2 \sum_{i+j+k=p+1} u_{h_i,h_j,h_k} + \sum_{i+j+k=p} u_{h_i,h_j,h_k}. \qquad (5)$$

Now, i, j, k range from 1 to p, where p is given by Eqn. (3). $u^c_{(L,M,N)}$ provides a solution on the *generalized* sparse grid $\Omega^s_{(L,M,N)}$ which is associated with $\Omega_{(L,M,N)}$ (by Eqns. (4) and (5)). Its grid points are now defined to be the union of the points of the grids Ω_{h_i,h_j,h_k} according to Eqns. (4) and (2), where $i + j + k = p + 2$ and each index i, j and k ranges from 1 to p.

The use of extremely distorted grids with a very high aspect ratio of the mesh sizes might cause certain problems (sensitive discretization scheme, slowing down of the solver, no resolution of physical features). Therefore, we must be able to limit the range of solutions involved and to use only a subset of all possible grids. We achieve this by introducing a new parameter $q \in \mathbb{N}$, $1 \leq q \leq p$, into the combination method. We define

$$u_{(L,M,N)}^{c,q} := \sum_{i+j+k=q+2} u_{h_i,h_j,h_k} - 2 \sum_{i+j+k=q+1} u_{h_i,h_j,h_k} + \sum_{i+j+k=q} u_{h_i,h_j,h_k} \qquad (6)$$

where each index i, j, k now ranges only from 1 to q. The mesh sizes h_i, h_j and h_k for grid Ω_{h_i,h_j,h_k} are now defined by $h_i = k_L^{-1} \cdot 2^{-i}, h_j = k_M^{-1} \cdot 2^{-j} \cdot 2\pi$ and $h_k = k_N^{-1} \cdot 2^{-k}$, where k_L, k_M and k_N are defined by

$$k_L = L/2^q, k_M = M/2^q, k_N = N/2^q. \qquad (7)$$

For the case $q = p$, we obtain the earlier Eqn. (5), but for values $q < p$ we limit the range of the combination method and thus exclude solutions u_{h_i,h_j,h_k} on extremely distorted grids. In the case $q = 1$, we just obtain the full grid solution $u_{(L,M,N)}$ and no combination takes place. The corresponding generalized sparse grid $\Omega_{(L,M,N)}^{s,q}$ consists of the union of the points of the grids Ω_{h_i,h_j,h_k} according to Eqns. (4) and (7), where $i + j + k = q + 2$ and i, j, k range from 1 to q.

Now, in Eqn. (6), we have to solve $q \cdot (q+1)/2$ problems $L_{h_i,h_j,h_k} u_{h_i,h_j,h_k} = f_{h_i,h_j,h_k}$, $i + j + k = q + 2$, each with about $D := L \cdot M \cdot N \cdot 2^{2-2q}$ unknowns, $q \cdot (q-1)/2$ problems $L_{h_i,h_j,h_k} u_{h_i,h_j,h_k} = f_{h_i,h_j,h_k}$, $i + j + k = q + 1$, each with about $D/2$ unknowns, and $(q-1) \cdot (q-2)/2$ problems $L_{h_i,h_j,h_k} u_{h_i,h_j,h_k} = f_{h_i,h_j,h_k}$, $i + j + k = q$, each with about $D/4$ unknowns.

APPLICATION TO TURBULENCE SIMULATION

From the Department of Fluid Dynamics of the TU München, we obtained the DNS-code FLOWSI (for details, see [15]). The code FLOWSI is written in Fortran77. It uses a finite volume discretization method on staggered grids. For the discretization of the Navier-Stokes equations in space, a central-difference method is employed, and for the discretization in time, a Leap Frog method is used. The discretization error is of second order in space and time (for further details, see [13]). For a direct numerical simulation of a pipe flow with a maximal Reynolds number $Re_{cl} = 6950$, a full grid $\Omega_{(L,M,N)}$ with $L = 96, M = 128$ and $N = 256$ grid points in the r-,φ- and z-directions would be necessary.

Now, we use the FLOWSI-code in the combination method to obtain a new, overall Navier-Stokes solver, that works on a sparse grid only. It possess all the advantages of the sparse grid approach, especially with respect to storage requirements and, as we will see in the following section, there is evidence that the computed turbulent flow is statistically as accurate as a direct numerical simulation on the corresponding full grid.

To this end, we proceed as follows: At a time step t_i we use the Navier-Stokes code FLOWSI to treat the different discretized and unsteady problems that arise in the combination method (Eqn. (6)) on the grids Ω_{h_i,h_j,h_k}, where $i + j + k = q + 2$, $i + j + k = q + 1$ and $i + j + k = q$. We stay on these grids for T successive time steps Δt and perform the necessary computations with FLOWSI. Then, after T time steps, we assemble the combination solution $u_{(L,M,N)}^{c,q}$ for time step t_{i+T} from the different computed iterates $u_{h_i,h_j,h_k}^{t_i+T}$. After that, we project new iterates $u_{h_i,h_j,h_k}^{t_i+T,new}$ for all the grids Ω_{h_i,h_j,h_k} involved in Eqn. (6) from it and repeat this process with $t_i := t_{i+T}$. This is illustrated in Figure 3. The projection process is done by weighted restriction. Note

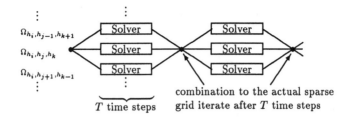

Fig. 3 Integration of the combination method in FLOWSI.

that the weights must be properly adjusted to cylindrical coordinates, see [11]. In this way, we obtain an outer time cycle with step size $T \cdot \Delta t$, where the actual iterates on the sparse grid $\Omega^{s,q}_{(L,M,N)}$ are assembled, and an inner time cycle where on every grid Ω_{h_i,h_j,h_k} an instance of the code FLOWSI operates concurrently for T time steps Δt. Altogether, we obtain a new algorithm for the treatment of the Navier-Stokes equations based on the sparse grid approach.

Our new method has two parameters, q and T, that directly and indirectly influence the performance of the solver with respect to the number of operations involved, the memory requirement and the accuracy of the results obtained for the turbulence simulation.

Recall that for $q = 1$ only the full grid $\Omega_{(L,M,N)}$ is involved. In this case, we directly obtain the DNS simulation of FLOWSI on grid $\Omega_{(L,M,N)}$. However, for $q > 1$, we obtain, first, a sparse grid to work with, i.e. $\Omega^{s,q}_{(L,M,N)}$ in the outer iteration, and second, the smaller problems on the different grids Ω_{h_i,h_j,h_k} in the inner iteration. For $q = p$, we would exploit the full power of the combination method, but then we must also deal with distorted grids.

TURBULENCE STATISTICS

We now report on the results of turbulence simulations obtained with the newly developed technique. We consider the flow through a pipe with a geometry as given in Figure 4. The normalized Reynolds number is $Re_{cl} = u_{cl} \cdot D/\nu = 6950$ (compare Figure 4). Thus, for a direct numerical simulation, as was performed in [5], a grid with $96 \times 128 \times 256$ cells is necessary. There, 41 samples were computed for the evaluation of different statistical moments. Now, our aim is to evaluate the new algorithm described

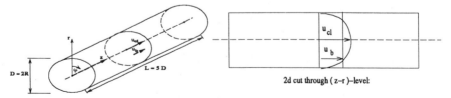

Fig. 4 Pipe geometry.

in the previous section for different values of its parameters q and T. The value of Δt is chosen to be 1/50 of the Kolmogorov time-scale. Regarding T, we have performed experiments with $T = 1$, $T = 5$, $T = 10$, $T = 25$ and $T = 50$ so far, corresponding to 1/50, 1/10, 1/5, 1/2 and 1 of the Kolmogorov time-scale, respectively. Regarding q, we applied our method for the values $q = 2, 3, 4$ and 5. Then, in the combination method, $4, 10, 19$ and 31 grids are involved. We started the simulations with a fully developed turbulent flow field as initial data. All computations were performed in single precision. In our experiments, for time reasons, we have only computed 20 samples so far.

Table 1 shows the mean flow property u_b/u_τ resulting from the DNS experiments in [5] and from the combination method algorithm (CM) as described in the previous section. Here, u_{cl} denotes the centerline velocity, u_b the mean velocity and u_τ the shear stress velocity. Recall that the results of the DNS are equivalent to CM with $q = 1$. Note, further, that the combination method with $q = 3$ and the large $T = 50$ gave bad results. The flux in the main direction dropped after a few samples. However, for more moderate values of T, we obtained values for u_b/u_τ that are in accordance with the results of the DNS. The same holds for $q = 4, T = 10$, $q = 5, T = 1$ and $q = 5, T = 5$. For comparison purposes, we give the results from physical measurements of real turbulent pipe flows in Table 2. The figures were obtained from the literature, see [5].

Table 1 Numerical experiments: mean flow U_b/U_τ.

	Re_{cl}	Re_τ	U_b/U_τ
DNS Unger	6950	360	14.74
DNS Eggels	6950	360	14.72
CM q=2, T=25	6950	360	14.72
CM q=2, T=50	6950	360	14.79
CM q=3, T=25	6950	360	14.76
CM q=3, T=50	6950	360	–
CM q=4, T=5	6950	360	14.83
CM q=4, T=10	6950	360	14.84
CM q=5, T=1	6950	360	14.86
CM q=5, T=5	6950	360	14.85

Table 2 Physical experiments: mean flow U_b/U_τ.

	Re_{cl}	Re_τ	U_b/U_τ
HWA Weiss	7350	379	14.76
PIV Westerweel	7100	366	14.88
LDA Westerweel	7200	371	14.68
LDA Durst	9600	500	14.70

In Figure 5, we plot the mean velocity profile divided by the wall friction velocity u_τ in the z-direction. It can be seen clearly that the profiles obtained with the combination algorithm for $q = 2$, $T = 25$ and $T = 50$ coincide with the profile obtained from the

DNS on the full grid. Also, the graphs for $q = 3$, $T = 25$, $q = 4$, $T = 5$, $q = 4$, $T = 10$ $q = 5$, $T = 1$ and $q = 5$, $T = 5$ are in good agreement with the DNS results.

Figure 6 shows the mean velocity profiles divided by the wall friction velocity that had been obtained from physical measurements of real turbulent pipe flows [5]. We see that the numerically computed data and the measured data do not differ significantly. For studying the fluctuations, we focus on the root-mean-square (rms) values of the

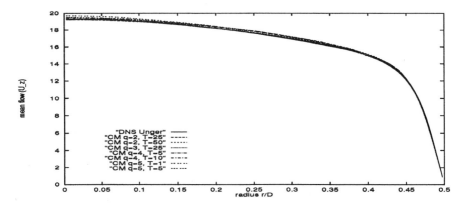

Fig. 5 Numerical simulations: axial mean velocity U_z divided by the wall friction velocity.

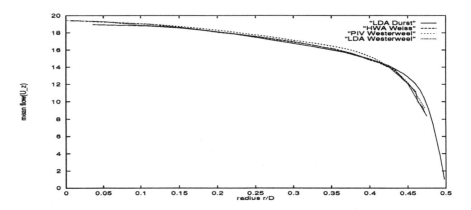

Fig. 6 Physical experiments: axial mean velocity U_z divided by the wall friction velocity.

velocity in the z-direction. The rms values are more critical for discretization errors than the first statistical moments. In Figure 7, we plot the rms profiles in the axial direction divided by the wall friction velocity u_τ on grid $\Omega_{(96/q,128/q,256/q)}$. Figure 8 also shows the rms profiles from measurements of physical experiments [5]. The rms profiles that have been obtained with the combination method for $q = 2$, 3, 4 and $T = 50$, 25, 10, 5 are in good agreement with that of the DNS, whereas for the case $q = 5$ and $T = 5$, 1, we obtain slightly worse results. The small differences near the center of the cylinder $r = 0$ might stem from the influence of the numerical treatment of the middle axis singularity in the Navier-Stokes equations in cylindrical coordinates on the combination method. Whether our reduced number of gathered samples is the reason for the slight differences away from the middle axis is an open question. We also tried to compute

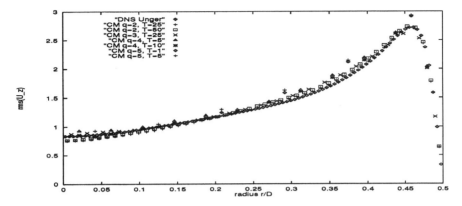

Fig. 7 Numerical simulations: axial rms values divided by the wall friction velocity, represented on grid $\Omega_{(96/q, 128/q, 256/q)}$.

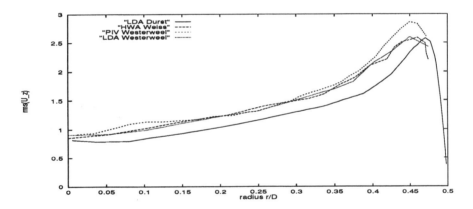

Fig. 8 Physical experiments: Axial rms-values divided by the wall friction velocity.

third statistical moments such as the skewness and the flatness. However, so far we have only gathered 20 samples, which is not sufficient for the sound evaluation of a third or fourth statistical moment. In a qualitative way, however, the present skewness results (see Figures 9 and 10) and flatness results (see Figures 11 and 12) obtained from the combination method with $q = 2$ ($T = 25$, $T = 50$), $q = 3$ ($T = 25$), $q = 4$ ($T = 5$, $T = 10$) and $q = 5$ ($T = 5$, $T = 1$) are in good agreement with the results given in [5]. For more details on numerical results, see [11].

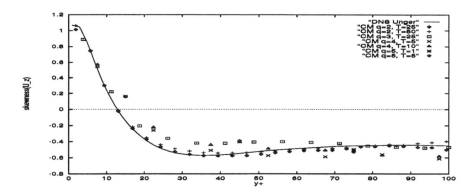

Fig. 9 Numerical simulations: axial skewness values on grid $\Omega_{(96/q,128/q,256/q)}$.

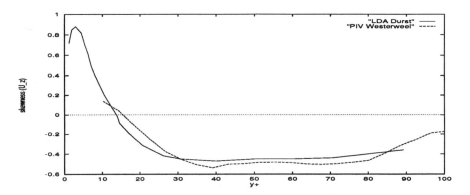

Fig. 10 Physical experiments: axial skewness values.

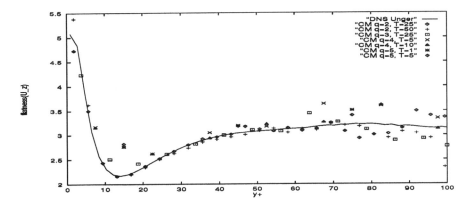

Fig. 11 Numerical simulations: axial flatness values on grid $\Omega_{(96/q, 128/q, 256/q)}$.

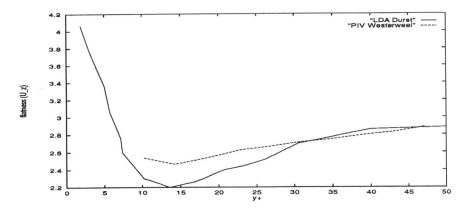

Fig. 12 Physical experiments: axial flatness values.

PARALLELIZATION ASPECTS

The parallelization of our combination algorithm can be achieved straightforwardly on a relative coarse grain level. The computations for the different subproblems that are associated with the grids Ω_{h_i,h_j,h_k} with $i+j+k = q+2$, $i+j+k = q+1$ and $i+j+k = q$ (compare Eqn. (6)) can be performed *fully in parallel*. Since the size of these grids and thus the amount of operations are known in advance, load balancing is simple and can be done statically.

However, in our combination algorithm, an iterate on the associated sparse grid must be assembled after T time steps, compare Figure 3. Thus, it seems that data structures have to be set up and stored for the sparse grid iterate (on one processor), and the data belonging to the grids Ω_{h_i,h_j,h_k} that are associated with the combination formula can be assigned to different processors. This results in a master-slave-type topology [8]. Then, the combination of the sparse grid iterate (from the iterates on all the grids Ω_{h_i,h_j,h_k} involved in Eqn. (6)) must be done in a synchronized $P:1$ communication, which might cause a bottleneck. Here, P denotes the number of processors. The same holds for the projection process to the new iterates $u_{h_i,h_j,h_k}^{t_i+T,new}$, compare Figure 3.

However, the *explicit* storage and assemblage of the sparse grid iterate are not really necessary. In [3, 9] it is described how the explicit combination to the sparse grid and the reprojection to the different grids Ω_{h_i,h_j,h_k} with $i+j+k = q+2$, $i+j+k = q+1$ and $i+j+k = q$, can be avoided. This is achieved by *directly* exchanging relevant data between the processors that store the data associated with the grids Ω_{h_i,h_j,h_k} ($i+j+k = q+2$, $i+j+k = q+1$ and $i+j+k = q$) and by updating the data on these grids. For details, see [9, 11]. In this way, the master-slave-type topology with synchronization bottleneck can be overcome to some extent. The storage necessary for the explicit assemblage of the sparse grid iterate is avoided and we obtain a parallel implementation with a distributed parallel storage concept. We implemented both parallel versions on an IBM SP2. On the IBM machine we use PVM 3.3.7 and PVMe 3.2. Further details are given in [11].

For different values of q, Table 3 shows the fraction of the amount of memory necessary to store the data belonging to *one*, i.e. the largest grid arising in Eqn. (6), versus the amount of memory necessary to store the data belonging to the full grid $\Omega_{(L,M,N)}$. This resembles the case that a sufficient amount of processors is available.

Table 3 Requirements for distributed storing.

q	Number of grids involved in the combination method	Storage requirements in comparison with the full grid
1	1	1
2	4	1/4
3	10	1/16
4	19	1/64
5	31	1/256
6	46	1/1024
7	64	1/4096
8	85	1/16384

First, we measured the run time of the sequential version of our code (which means the full grid version for $q = 1$) on an SP2 and on a Cray Y-MP at the Leibniz Computing Center in Munich. In Table 4 it can be seen that a node of the vector computer is about three times faster than a node of the SP2.

Table 4 Overall run times for 500 time steps in sec. of the full grid version.

	IBM SP2	Cray Y-MP
Run time	15500	4350

We now give the results of the parallelized version obtained on the IBM SP2. Running our code with PVMe we obtained the values given in Table 5. We also used the public domain version of PVM. Then, of course, the run times are slightly longer than when using PVMe, seen in Table 6. Using PVMe it is not possible to run more than one process on one node. Hence, for $q = 4$ we have to use 19 nodes and for $q = 5$ we have to use 31 nodes. It should be mentioned that both parallel implementations are not optimized. Now, the overall computing time drops very slowly on increasing the number of processors (see also Table 6). The non optimized version of our code is the reason for the poor speed up.

Table 5 Overall run times for 500 time steps in sec. using PVMe 3.2 on P processors.

P=	19	31
$q = 4, T = 10$	582	-
$q = 5, T = 5$	-	550

Table 6 Overall run times for 500 time steps in sec. using PVM 3.3.7 on P processors.

P=	1	2	4	8	10	19	31
$q = 4, T = 10$	4050	2520	1500	1200	1020	956	-
$q = 5, T = 5$	2400	2040	1620	1260	1200	1185	1065

Comparing Tables 4 and 5 or Tables 4 and 6, the parallel implementation of the combination method on an SP2 is more than seven times faster (using PVMe) and more than four times faster (using PVM) than the sequential version on a single node of a Cray Y-MP.

CONCLUSION

We have presented a variant of the combination method that results, together with a given conventional Navier-Stokes solver such as FLOWSI, in a solver for the Navier-Stokes equations on sparse grids. We applied this new method to the simulation of turbulent pipe flow with $Re_{cl} = 6950$. It turned out that we obtained for first and second statistical moments practically the same results as for a DNS, but with far less computational costs and lower storage requirements.

Furthermore, our new algorithm can easily be parallelized on a relative coarse grain level. The computations for the different subproblems arising in our method in a natural way can be performed *fully in parallel*. Since the size of these grids and thus the amount of operations are known in advance, load balancing is simple and can be done statically. We implemented different parallel versions of our method on an IBM SP2 using PVM 3.3.7 and PVMe 3.2. For a detailed presentation of further results, see [11].

This work shows the potential and perspectives of our method for future computations with higher Reynolds numbers on finer grids which currently cannot be handled using the conventional full grid approach owing to its huge memory requirements.

Acknowledgements: We thank Prof. R. Friedrich and his group, especially F. Unger, K. Wagner and E. Guentsch, for providing the code FLOWSI and the results of their DNS experiments.

REFERENCES

[1] S. BARROS, *Optimierte Mehrgitterverfahren für zwei- und dreidimensionale elliptische Randwertaufgaben in Kugelkoordinaten*, GMD Bericht Nr.178, R. Oldenbourg Verlag, München, Wien Nr.178, 1989.

[2] H. BUNGARTZ, *Dünne Gitter und deren Anwendung bei der adaptiven Lösung der dreidimensionalen Poisson-Gleichung*, Dissertation, Institut für Informatik, TU München, 1992.

[3] H. BUNGARTZ AND W. HUBER, *First experiments with turbulence simulation on workstation networks using sparse grid methods*, in Notes on Numerical Fluid Mechanics: Computational Fluid Dynamics on Parallel Systems, S. Wagner, ed., Vieweg, 1995.

[4] J. BURMEISTER AND W. HACKBUSCH, *On a time and space parallel multi-grid method including remarks on filtering techniques*, in this publication, Vieweg, 1996.

[5] J. EGGELS, F. UNGER, M. WEISS, J. WESTERWEEL, R. ADRIAN, R. FRIEDRICH, AND F. NIEUWSTADT, *Fully developed turbulent pipe flow: a comparison between direct numerical simulation and experiments*, Journal of Fluid Mechanics, 268, pp. 175-209, (1994).

[6] M. GRIEBEL, *Parallel Multigrid Methods on Sparse Grids*, International Series of Numerical Mathematics, 98 (1991), pp. 211-221.

[7] ———, *A parallelizable and vectorizable multi-level algorithm on sparse grids*, in Proceedings of the Sixth GAMM-Seminar, W. Hackbusch, ed., vol. 31, Vieweg, Braunschweig, Notes on Numerical Fluid Mechanics, 1991, pp. 94-100.

[8] M. GRIEBEL, W. HUBER, U. RÜDE, AND T. STÖRTKUHL, *The combination technique for parallel sparse-grid-preconditioning and -solution of PDEs on multiprocessor machines and workstation networks*, in Proceedings of the Second Joint International Conference on Vector and Parallel Processing CONPAR/VAPP V 92, 634, pp. 217-228, L. Bougé, M. Cosnard, Y. Robert, D. Trystram, ed., Springer Verlag, Berlin, Heidelberg, New York, 1992.

[9] M. GRIEBEL, W. HUBER, T. STÖRTKUHL, AND C. ZENGER, *On the parallel solution of 3D PDEs on a network of workstations and on vector computers*, in Lecture Notes in Computer Science,732, pp. 276-291, Computer Architecture: Theory, Hardware, Software, Applications, A. Bode and M. Dal Cin, ed., Springer Verlag, Berlin, Heidelberg, New York, 1993.

[10] M. GRIEBEL, M. SCHNEIDER, AND C. ZENGER, *A combination technique for the solution of sparse grid problems*, in Proceedings of the IMACS International Symposium on Iterative Methods in Linear Algebra, pp. 263-281, P. de Groen and R. Beauwens, ed., Elsevier, Amsterdam, 1992.

[11] W. HUBER, *Turbulenzsimulation mit der Kombinationsmethode auf Workstation-Netzen und Parallelrechnern*, Dissertation, Institut für Informatik, TU München, 1996.

[12] M. MANHART AND H. WENGLE, *Large-eddy simulation and eigenmode decomposition of turbulent boundary layer flow over a hemisphere*, in this publication, Vieweg, 1996.

[13] L. SCHMITT, *Numerische Simulation Turbulenter Grenzschichten* , Tech. Report 82/2, Lehrstuhl für Fluidmechanik, TU München, 1982.

[14] G. SMITH, *Numerische Lösung von partiellen Differentialgleichungen*, Vieweg Verlag, Braunschweig, 1970.

[15] F. UNGER, *Numerische Simulation turbulenter Rohrströmungen*, Dissertation, Lehrstuhl für Fluidmechanik, TU München, 1994.

[16] C. ZENGER, *Sparse grids*, in Parallel Algorithms for Partial Differential Equations, Proceedings of the Sixth GAMM-Seminar, Kiel, 1990, Notes on Numerical Fluid Mechanics, 31, W. Hackbusch, ed., Vieweg, Braunschweig, 1991.

CFD with adaptive FEM on massively parallel systems

Frank Lohmeyer, Oliver Vornberger

Department of Mathematics and Computer Science
University of Osnabrück, D-49069 Osnabrück, Germany

Summary

An explicit finite element scheme based on a two-step Taylor-Galerkin algorithm allows the solution of the Euler and Navier-Stokes equations for a wide variety of flow problems. To obtain useful results for realistic problems, one has to use grids with an extremely high density to obtain a good resolution of the interesting parts of a given flow. Since these details are often limited to small regions of the calculation domain, it is efficient to use unstructured grids to reduce the number of elements and grid points. As such calculations are very time consuming and inherently parallel, the use of multiprocessor systems for this task seems to be a very natural idea. A common approach for parallelization is the division of a given grid, where the problem is the increasing complexity of this task for growing processor numbers. Some general ideas for this kind of parallelization and details of a Parix implementation for Transputer networks are presented. To improve the quality of the calculated solutions, an adaptive grid refinement procedure was included. This extension leads to the need for a dynamic load balancing for the parallel version. An effective strategy for this task is presented and results for up to 1024 processors show the general suitability of this approach for massively parallel systems.

Introduction

The introduction of the computer into engineering techniques has resulted in the growth of a completely new field, termed *computational fluid dynamics* (CFD). This field has led to the development of new mathematical methods for solving the equations of fluid mechanics. These improved methods have permitted advanced simulations of flow phenomena on the computer for a wide variety of applications. This leads to a demand for computers which can manage these extremely time-consuming calculations within acceptable run times. Many of the numerical methods used in computational fluid dynamics are inherently parallel, so that the appearance of parallel computers makes them a promising candidate for this task.

One problem arising when implementing parallel algorithms is the lack of standards on both the hardware and software sides. As aspects such as processor topology, parallel operating system and programming languages have a much greater influence on parallel than on sequential algorithms, one has to choose an environment where it is possible to obtain results which can be generalized to a larger set of other environments. We think that future supercomputers will be massively parallel systems of the MIMD class with distributed memory and strong communication capabilities. On the software side, it seems that some standards can be established in the near future. As our algorithm is designed for message-passing environments, the appropriate standard is MPI (Message Passing Interface).

In the CFD field, there is another important point: the numerical methods for the solution of the given equations. As we are mainly computer scientists, we decided not to invent new mathematical concepts but to develop an efficient parallel version of an algorithm which was developed by experienced engineers for sequential computers and which is suitable for the solution of problems in the field of turbomachinery [1]. The hardware platforms which are availiable to us are Transputer systems of different sizes, which fulfil the demands mentioned above. At the time the algorithm was developed, there was no MPI environment availiable for our transputer systems, so here we will present a version using Parix (Parallel extensions to Unix, a parallel runtime system for Parsytec machines) that needs only a small number of parallel routines, which are common in most message-passing environments. Most of these routines are hidden inside a few communication procedures, so that they can be replaced easily, when changing the parallel environment. The following two sections give a very short overview of the physical and mathematical foundations of the used numerical methods (for a detailed description, see [1,2]) and a brief outline of the general parallelization strategy (see also [3,4,5]). A similar approach, especially for the numerical methods, can be found in [6]. The next section describes in detail some grid division algorithms which are a very important part for this kind of parallel algorithms, because they determine the load balancing between processors. The special subjects of adaptive refinements and dynamic load balancing are discussed in a separate section. Then some results are presented, and the last section closes with a conclusion and suggestions for further research.

Foundations

For our flow calculations on unstructured grids with the finite element method, we use Navier-Stokes equations for viscous flow and Euler equations for inviscid flow. The Navier-Stokes (Euler) equations can be written in the following form:

$$\frac{\partial U}{\partial t} + \frac{\partial F}{\partial x} + \frac{\partial G}{\partial y} = 0, \qquad (1)$$

where U, F and G are four-dimensional vectors. U describes mass, impulses and energy, F and G are flow vectors. The vectors for the Navier-Stokes equations are

$$U = \begin{pmatrix} \rho \\ \rho u \\ \rho v \\ \rho e \end{pmatrix}, \qquad (2)$$

$$F = \begin{pmatrix} \rho u \\ \rho u^2 + \sigma_x \\ \rho u v + \tau_{xy} \\ (\rho e + \sigma_x) u + \tau_{yx} v - k \frac{\partial T}{\partial x} \end{pmatrix}, \qquad (3)$$

$$G = \begin{pmatrix} \rho v \\ \rho u v + \tau_{xy} \\ \rho v^2 + \sigma_y \\ (\rho e + \sigma_y)v + \tau_{xy}u - k\frac{\partial T}{\partial y} \end{pmatrix}, \qquad (4)$$

for the Euler equations the σ, τ and k terms can be neglected; in both cases we have to add two equations to close the system.

The solution of these differential equations is calculated with an explicit Taylor-Galerkin two-step algorithm. In the first step (the so-called predictor step), an intermediate result

$$U^{n+1/2} = U^n - \frac{\Delta t}{2}\left(\frac{\partial F^n}{\partial x} + \frac{\partial G^n}{\partial y}\right) \qquad (5)$$

is calculated and in the second step (the corrector step) this result is used to calculate

$$\Delta U = -\Delta t \left(\frac{\partial F^{n+1/2}}{\partial x} + \frac{\partial G^{n+1/2}}{\partial y}\right). \qquad (6)$$

The differential equations can be expressed in a weighted residual formulation using triangular finite elements with linear shape functions. Therefore, in the first step the balance areas of the convective flows for one element have to be calculated on the nodes of each element. In the second step the balance area for one node is calculated with the help of all elements which are defined with this node. A pictorial description of these balance areas of the two steps is given in Figure 1.

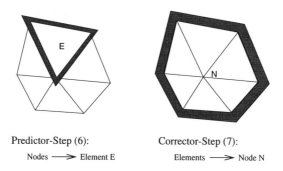

Predictor-Step (6):　　　　　Corrector-Step (7):
Nodes ⟶ Element E　　　Elements ⟶ Node N

Figure 1: Balance areas.

The calculation with the finite element method, which divides the calculation area into triangles, leads to the characteristic summation of the element matrices into the global mass matrix M and to the following equation system:

$$M\,\Delta U = \Delta t\, R_S(U^n), \qquad (7)$$

where R_S is the abbreviation for the summation of the right-hand sides of Eqn. (6) for all elements. The inversion of the matrix M is very time consuming and therefore we use, with the help of the so-called lumped mass matrix M_L, the following iteration steps:

$$\Delta U^0 = \frac{\Delta t\, R_S}{M_L}, \tag{8}$$

$$\Delta U^{\nu+1} = \Delta U^\nu + \frac{\Delta t\, R_S - M \Delta U^\nu}{M_L}. \tag{9}$$

For the determination of ΔU three iteration steps are sufficient. If we consider stationary flow problems, only the initial iteration has to be calculated.

The time step Δt must be adjusted in a way such that the flow of information does not exceed the boundaries of the neighbouring elements of a node. This leads to small time steps if non-stationary problems are to be solved (in the case of stationary problems we use a local time step for each element). In both cases the solution of a problem requires the calculation of many time steps, so that the steps (5), (6), (8) and (9) are carried out many times for a given problem. The resulting structure for the algorithm is a loop over the number of time steps, where the body of this loop consists of one or more major loops over all elements and some minor loops over nodes and boundaries (major and minor in this context reflects the different run times spent in the different calculations).

Another important characteristic of this method is the use of unstructured grids. Such grids are characterized by various densities of the finite elements for different parts of the calculation area. The elements of an unstructured grid differ in both size and number of adjacent elements, which can result in a very complex grid topology. This fact is one main reason for the difficulties arising in constructing an efficient parallel algorithm.

The main advantage of unstructured grids is their ability to adapt a given flow. To obtain a high resolution of the details of a flow, the density of the grids must only be increased in the interesting parts of the domain. This leads to a very efficient use of a given number of elements. One problem arising in this context is the fact that in most cases the details of a flow are the subject of investigations, so that it is impossible to predict the exact regions, where the density of the grid has to be increased. A solution of this problem is a so-called adaptive grid refinement, where the calculations start with a grid with no or few refinements. As the calculations proceed, it is now possible to detect regions where the density of the grid is not sufficient. These parts of the grid will then be refined and the calculations proceed with the refined grid. This refinement step is repeated until the quality of the solution is sufficient.

Parallelization

If we are looking for parallelism in this algorithm, we observe that the results of one time step are the input for the next time step, so the outer loop has to be calculated in sequential order. This is not the case for the inner loops over elements, nodes and boundaries which can be carried out simultaneously. Hence the basic idea for a parallel version of this algorithm is a distributed calculation of the inner loops. This can be achieved by a so-called grid division, where the finite element grid is partitioned into sub-grids. Every processor in the parallel system is then responsible for the calculations on one of these sub-grids. Figure 2 shows the implication of this strategy for the balance areas of the calculation steps.

The distribution of the elements is non-overlapping, whereas the nodes on the border between the two partitions are doubled. This means that the parallel algorithm carries out the same number of element-based calculations as in the sequential case, but some node-based calculations are carried out twice (or even more times, if we think of more complex divisions for a larger number of processors). Since the major loops are element based, this strategy should lead to parallel calculations with a nearly linear speed-up. One remaining problem is the construction of a global solution out of the local sub-grid solutions. In the predictor step the flow of control is from the nodes to the elements, which can be carried out independently. But in the corrector step we have to deal with balancing areas which are based on nodes which have perhaps multiple incarnations. Each of these incarnations of a node sums up the results from the elements of its sub-grid, whereas the correct value is the sum over all adjacent elements. As Figure 3 shows, the solution is an additional communication step where all processors exchange the values of common nodes and correct their local results with these values.

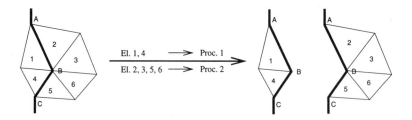

Figure 2: Grid division.

This approach, where the sequential algorithm is used on distributed parts of the data sets and where the parallel and the sequential version are arithmetically equivalent, is usually described with the keyword *data decomposition*. Other *domain decomposition* approaches have to deal with numerically different calculations in different parallel cases, and have to pay special attention to numerical stability. In the case of implicit algorithms, it is common to make a division of the grid nodes, due to the structure of the resulting system of linear equations, which have to be solved in parallel. The main advantage of the explicit algorithm used here is the totally local communication structure, which results in a higher parallel efficiency, especially for large numbers of processors.

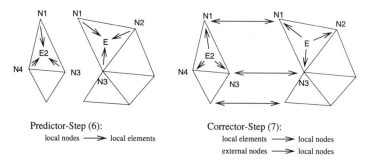

Figure 3: Parallel calculations.

This structure implies an MIMD architecture and the locality of data is exploited best with a distributed memory system together with a message-passing environment. This special algorithm has a very high communication demand, because in every time step for every element loop an additional communication step occurs. An alternative approach in this context is an overlapping distribution, where the subgrids have common elements around the borders. This decreases the number of necessary communications but leads to redundant numerical calculations. We decided to use non-overlapping divisions for two reasons: first they are more efficient for large numbers of subgrids (and are therefore better suited for massively parallel systems), and second we want to use adaptive grids. The required dynamic load balancing would be a much more difficult task for overlapping subgrids. The only drawback of our approach is that to obtain high efficiencies a parallel system with high communication performance is required, so it will not work, e.g., on workstation clusters. Our current implementation is for Transputer systems and uses the Parix programming environment, which supplies a very flexible and comfortable interprocessor communication library. This is necessary if we think of unstructured grids which have to be distributed over large processor networks leading to very complex communication structures.

Grid division

If we now consider the implementation of the parallel algorithm, two modules have to be constructed. One is the algorithm running on every processor of the parallel system. This algorithm consists of the sequential algorithm operating on a local data set and additional routines for the interprocessor communication. These routines depend on the general logical processor topology, so that the appropriate choice of this parameter is important for the whole parallel algorithm. In Parix this logical topology has to be mapped onto the physical topology which is realized as a two-dimensional grid. For two-dimensional problems there are two possible logical topologies: one-dimensional pipeline and two-dimensional grid. They can be mapped in a canonical way onto the physical topology, so that we have implemented versions of our algorithm for both alternatives.

The second module we had to implement is a decomposition algorithm which reads in the global data structures and calculates a division of the grid and distributes the corresponding local data sets to the appropriate processor of the parallel system. Such a division can require interprocessor connections, which are not supplied by the basic logical topology. These connections are built dynamically with the so-called virtual links of Parix and collected in a virtual topology.

The essential part of the whole program is the division algorithm which determinates the quality of the resulting distribution. This algorithm has to take different facts into consideration to achieve efficient parallel calculations. First it must ensure that all processors have nearly equal calculation times, because idle processors reduce the speed-up. To achieve this, it is necessary first to distribute the elements as evenly as possible and then minimize the overhead caused by double calculated nodes and the resulting communications. A second point is the time consumed by the division algorithm itself. This time must be considerably less than that of the actual flow calculation. Therefore, we cannot use an optimal division algorithm, because the problem is NP-complete and such an algorithm would take more time than the whole flow calculation. For this reason, we had to develop a good heuristic for the determination of a grid division. This task is mostly sequential and, as the program has to deal with the whole global data sets, we decided to map this process onto a workstation outside the Transputer system. Since nowadays such a workstation is much faster than a single Transputer, this is no patched-up solution, and the performance of the whole calculation even increases.

According to the two versions for the parallel module, we also implemented two versions for the division algorithm. Since the version for a one-dimensional pipeline is a building block for the two-dimensional case, we present this algorithm first (see Figure 4). The division process is done in several phases here: an initialization phase (0) calculates additional information for each element. The weight of an element represents the calculation time for this special element type (these times are varying because of special requirements of, e.g., border elements). The virtual coordinates reflect the position where in the processor topology this element should roughly be placed (therefore the dimension of this *virtual space* equals the dimension of the logical topology). These virtual coordinates (here it is actually only one coordinate) can be derived from the real coordinates of the geometry or from special relations between groups of elements. An example of the latter case is elements belonging together because of the use of periodic borders. In this case nodes on opposite sides of the calculation domain are strongly coupled, and this fact should be reflected in the given virtual coordinates.

```
Phase 0:  calculate element weights
          calculate virtual coordinates

Phase 1:  find element ordering with small bandwidth
          a)   use virtual coordinates for initial ordering
          b)   optimize bandwidth of ordering

Phase 2:  find good element division using ordering and weights

Phase 3:  optimize element division using communication analysis
```

Figure 4: One-dimensional division algorithm.

Before the actual division, an ordering of the elements with a small bandwidth is calculated (phase 1). This bandwidth is defined as the maximum distance (that is, the difference of indices in the ordering) of two adjacent elements. A small bandwidth is a requirement for the following division step. Finding such an ordering is again an NP-complete problem, so we cannot obtain an optimal solution. We use a heuristic, which calculates the ordering in two steps. First we need a simple method to obtain an initial ordering (a). In our case we use a sorting of elements according to their virtual coordinates. In the second step (b), this ordering is optimized, e.g., by exchanging pairs of elements if this improves the bandwidth until is no more exchange is possible.

With the obtained ordering and the element weights, the actual division is now calculated. First the elements are divided into ordered parts with equal weights (phase 2). Then this division is analysed in terms of resulting borders and communications and is optimized by reducing the border length and number of communication steps by exchanging elements with equal weights between two partitions (phase 3).

If we now want to construct a division algorithm for the two-dimensional grid topology, we can use the algorithm described above as a building block. The resulting algorithm has the structure shown in Figure 5. The only difference between the one- and the two-dimensional versions of the initialization phase is the number of virtual coordinates which here, of course, is two. Phase 3 has the same task, which is much more complex in the two-dimensional case. The middle phase here is a two-stage use of the one-dimensional strategy, where the grid is first cut in the x-dimension and then all pieces are cut in the y-dimension. This strategy can be substituted by a sort of recursive bisectioning, where in every step the grid is cut into two pieces in the larger dimension and both pieces are cut further using the same strategy.

```
Phase 0 (initialization) similar to 1D-algorithm

for #processors in x-dimension do

        calculate meta-division M
        using phases 1 and 2 of 1D-algorithm

        divide meta-division M in y-dimension
        using phases 1 and 2 of 1D-algorithm

Phase 3 (optimization) similar to 1D-algorithm
```

Figure 5: Two-dimensional division algorithm.

Adaptive refinement and dynamic optimization

The parallelization approach described in the last two sections is well suited for fixed grids, which remain constant through all calculation steps. We will now introduce a simple but effective method for an adaptive grid refinement and an improvement of the parallel algorithm which takes into account that the work load for each processor has changed after every refinement step. Before we can describe the algorithms, some questions have to be answered: what does refinement mean exactly, which parts should be refined and how can we construct the new, refined grid.

- Refinement in our case means the splitting of elements into smaller elements, which replace the original elements. An additional type of refinement is the replacement of a group of small elements by a smaller group of larger elements. This procedure is useful if the density of a part of the grid is higher than necessary. At the moment only the first type of refinement is implemented, so in the following we will only describe the splitting of elements.

- The question of which parts should be refined now turns into the selection of elements that should be refined. Therefore, we choose for a given flow problem a characteristic function, e.g. the pressure field. We then look for elements where the gradient of this function exceeds a given bound.

- How should these elements be split into smaller ones? In Figure 6 two possibilities are shown: using the left alternative leads to numerical problems, caused by the shape of the resulting triangles. Especially if an element is refined several times, the new grid nodes will be placed near to the remaining sides, leading to very flat triangles, which should be avoided. Hence we must use the right alternative, which leads to the problem that the new nodes are placed on the edges of a triangle. This would result in an inconsistent grid, because all nodes must be corners of elements. The solution of this problem is an additional splitting of elements with such edges. Elements with two or three refined edges must be split into four elements (as if they were originally selected for refinement). Elements with only one refined edge must be split into two elements in a canonical way. This additional splitting of elements can lead to new elements with nodes on their edges, so the process has to be repeated until no more splitting is necessary.

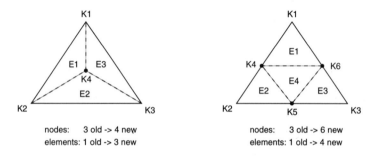

Figure 6: Splitting of an element.

- How can we construct the new grid? Every split element will be replaced by one of the new elements and the remaining new elements will be added to the element list. All new nodes are added to the node list and all new border nodes and elements are added to the appropriate lists. A new local time step for every element must be calculated and in the case of an non-stationary solution a new global time step must be calculated from the local time steps. After this, some derived values such as element sizes have to be reinitialized and then the calculations can continue.

The resulting refinement algorithm is shown in Figure 7. All steps can be implemented in a straightforward manner with one exception: the colouring of the elements that must be refined to obtain a consistent grid. This can be done using the recursive algorithm shown in Figure 8.

```
calculate reference function F
Delta = (Max(F) - Min(F)) * refine_rate

for (El in element-list)
   dF = local gradient of F in El

   if (dF > Delta)
      mark El with red
   end-if
end-for

mark all elements with nodes on edges:
   with yellow for full refinement
   with green for half refinement

refine all elements full or half according to their colour
```

Figure 7: Refinement algorithm.

```
for (El in element-list)
   if (El is marked red)
      mark_neighbours(El)
   end-if
end-for


mark_neighbours(El):

   for (E in neighbour-elements(El))

      if (E is marked white)
         mark E with green
      else if (E is marked green)
         mark E with yellow
         mark_neighbours(E)
      end-if

   end-for

end-mark_neighbours
```

Figure 8: Colouring algorithm.

To use this adaptive refinement in the parallel algorithm, we have to analyse the different parts for parallelism:

```
calculate reference function F
     local data, fully parallel, no communication
Delta = (Max(F) - Min(F)) * refine_rate
     local data, global min/max, mostly parallel, global communication
mark elements with red or white
     local data, fully parallel, no communication
mark additional elements with yellow or green
     global data, sequential, global communication (collection)
refinement of elements
     global data, sequential, global communication (broadcast)
construct new node- and element-lists
     local data, fully parallel, no communication
reinitialize all dependent variables
     local data, fully parallel, local communications
```

We can see that most of the parts can be performed in parallel with no or little communication. The only exception is the additional colouring of elements and the construction of new elements and nodes. These parts operate on global data structures, so that a parallel version of them must lead to a very high degree of global communication. As these parts need only very little of the time spent on the complete refinement step, we decided to keep this part sequential.

For a typical flow calculation up to five refinement steps are sufficient in most cases, so that the lack of parallelism in the refinement step decribed above is not very problematic. Much more important is the fact that after a refinement step the work load of every processor has changed. As the refinement takes place in only small regions, there are a few processors with a load that is much higher than the load of most of the other processors. This would not only slow down the calculations, but also can lead to memory problems if further refinements in the same region take place. The solution of this problem is a dynamic load balancing, where parts of the sub-grids are exchanged between processors until equal work load and nearly equal memory consumption is reached.

The load is obtained by simply measuring the CPU time needed for one time step including communication times but excluding idle times. The items that could be exchanged between processors are single elements including their nodes and all the related data. To achieve this efficiently, we had to use dynamic data structures for all element and node data. There is one data block for every node and every element. These blocks are linked together in many different dynamic lists. The exchange of one element between two processors is therefore a complex operation: the element has to be removed from all lists on one processor and must be included in all lists on the other processor. Since nodes can be used on more than one processor, the nodes belonging to that element must not be exchanged in every case. It must be checked whether they are already on the target processor and if other elements on the sending processor will need them, too. Nodes not availiable on the target processor must be sent to it and nodes which are no longer needed on the sending processor must be deleted there.

One remaining problem is to find an efficient strategy for the exchange of elements. A good strategy should deliver an even load balance after only a few steps and every step should be finished in a short time. As elements can only be exchanged between direct neighbours, our first approach was a local exchange between pairs of processors. This resulted in fast exchange steps, but showed a bad convergence behaviour. We will present here a global strategy, where the balancing is carried out along the rows and columns of the processor grid. The areas for load balancing are alternating all rows and all columns, where every row (column) is treated independently of all other rows (columns). If we interpret a single row (column) as a tree with the middle processor as the root, we can use a modification of a tree-balancing algorithm developed for combinatorial optimization problems [7].

```
if (root)
    receive load_sub_l (load of left subtree)
    receive load_sub_r (load of right subtree)

    global_load = local_load + loads of subtrees
    load_opt = global_load / num_procs

    send load_opt to subtrees

    load_move_l = load_opt * num_sub_l - load_sub_l
    load_move_r = load_opt * num_sub_r - load_sub_r

else

    receive load_sub from subtree (if not leaf)
    load_sub_n_me += load_sub
    send load_sub_n_me to parent node

    receive load_opt from parent node
    send load_opt to subtree (if not leaf)

    load_move_sub = load_opt * num_sub - load_sub
    load_move_top = load_sub_n_me - (num_sub + 1) * load_opt
end-if
```

Figure 9: Dynamic optimization (step 1).

This algorithm uses two steps: in a first step information about the local loads of the tree is moving up to the root and the computed optimal load value is propagated down the tree. In a second step the actual exchange is done according to the optimal loads found in the first step. The structure of the first step is shown in Figure 9, where num_procs is the number of processors in the tree (= row or column) and load_move_*direction* is the load that has to be moved in that direction in the second step.

The second step is very simple: all processors translate the loads they have to move into the appropriate number of elements and exchange these elements. For the decision, which element to send in a specific direction, the virtual coordinates of this element are looked up and the element with the greatest value is chosen. With this strategy the resulting sub-grids of the new grid division will have nearly optimal shapes.

Results

The algorithms described in the previous sections were tested with many different grids for various flow problems. As a kind of benchmark problem we use the non-stationary calculation of inviscid flow behind a cylinder, resulting in a vortex street. One grid for this problem was used for all our implementations of the parallel calculations. This grid has a size of about 12 000 grid points which are forming nearly 20 000 elements (P1). Other problems used with the adaptive refinement procedure are stationary turbine flows with grids of different sizes, all of them using periodic borders.

All measurements were made with a 1024 processor system located at the $(PC)^2$ of the University of Paderborn. It consists of T805 Transputers, each of them equipped with 4 Mbyte local memory and coupled together as a two-dimensional grid. Our algorithms are all coded in Ansi-C using the Parix communication library.

First we will present some results for the dynamic load balancing. Figure 10 shows the different convergence behaviour of two different strategies. To investigate this, we used a start division of our reference problem with an extremely bad load balancing (because of the memory needs of the start division the following results are obtained with a smaller grid with 5000 elements and an 8x8 processor array — larger grids on larger arrays show the same behaviour, but cannot be started with this extreme imbalance). The left-hand picture shows the results of a local strategy, where the balance improves in the first steps, but stays away from the optimum for a large number of optimization steps. The right-hand picture shows the effects of the described global balancing, where after two steps the balance is nearly optimal.

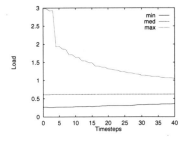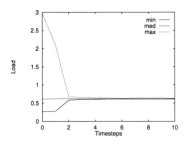

Figure 10: Different load balancing strategies.

An example for the adaptive refinement procedure is shown in Figure 11, where the grid and the pressure field around a turbine are shown after five refinement steps. One can see the high resolution of the two shocks made visible by the adaptive refinements. Without refinement one of these effects can only be guessed and the other is missing completely.

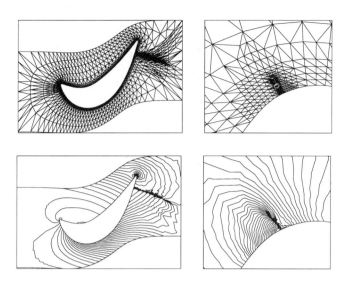

Figure 11: Grid and isobars after five refinement steps (complete and zoomed).

The development of the corresponding loads can be seen in Figure 12. After each refinement step, four balancing steps were carried out, using the global strategy. The picture shows the efficient and fast balancing of this method.

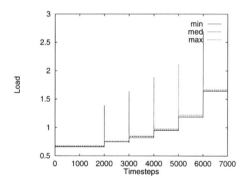

Figure 12: Load balancing for adaptive refinements.

Finally, we will present results for large processor numbers. In Figure 13 the speed-ups for some parameter settings of our reference problem are shown. In the speed-up curves the difference between the logical topologies 1D-pipeline and 2D-grid is shown. In the left part of the picture we can see that for up to 256 processors we achieve nearly linear speed-up with the grid topology, whereas the pipe topology is linear for a maximum of only 128 processors. If we increase the size of the problem (P2), the speed-ups are closer to the theoretical values, which proves the scalability of the parallelization approach.

61

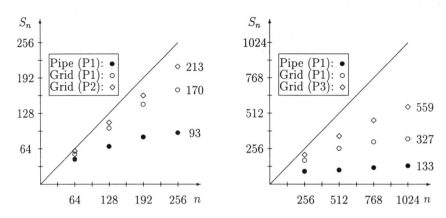

Figure 13: Speed-ups for different topologies.

If we increase the number of processors (right-hand picture), we observe that the grid topology again is superior to the pipe topology, but the increase of speed-up is no longer linear. It is a common problem for most parallel algorithms that for a fixed problem size there is always a number of processors where the speed-up is no longer increasing proportionally to the number of processors. If we want to obtain the same efficiencies as for 256 processors, we have to use grids with approximately 50 000 elements. This was impossible on the T8 system used, because such problems are too large for it. The largest problem that fits into the 4 Mbyte nodes has about 32 000 elements (P3), so that the speed-ups for 1024 processors are limited on this machine. Nevertheless, the increase of speed-up to 559 again shows the scalability of our algorithm.

Conclusion

In this paper we have introduced a parallelization for the calculation of fluid flow problems on unstructured grids. An existing sequential algorithm has been adjusted for Transputer systems under Parix and investigations on the parallelization of this problem have been made. For two logical processor topologies we have developed different grid division algorithms and compared them for some benchmark problems. The grid topology has shown its superiority over the pipe topology. This was expected since a two-dimensional topology must be better suited for two-dimensional grids than a one-dimensional topology which is not scalable for large processor numbers. The speed-up measurements on a 1024 Transputer cluster showed the general usefulness of the choosen approach for massively parallel systems.

Further, we presented an adaptive refinement procedure which is used for the solution of flow problems with a priori unknown local effects. For the parallel version of this procedure we showed the need for a dynamic load balancing. A global strategy for this balancing was described in detail. We presented results for the performance of this strategy and compared it with a local strategy. We showed the excellent convergence behaviour of our strategy and the usefulness of the dynamic load balancing together with the adaptive refinement.

Dynamic load balancing is fully parallel and hardware independent, so that changes of the basic hardware nodes can be done without changing the developed algorithm. To exploit this advantage of our algorithms, they must be implemented in as portable a manner as possible. To achieve this, our further research will concentrate on porting the current implementation to MPI and studying the resulting performance on different hardware platforms.

References

[1] A. Vornberger. *Strömungsberechnung auf unstrukturierten Netzen mit der Methode der finiten Elemente*. Ph.D. Thesis, RWTH Aachen, 1989.

[2] W. Koschel and A. Vornberger. *Turbomachinery Flow Calculation on Unstructured Grids Using the Finite Element Method*. Finite Approximations in Fluid Mechanics II, Notes on Numerical Fluid Mechanics, Vol. 25, pp. 236-248, Aachen, 1989.

[3] F. Lohmeyer, O. Vornberger, K. Zeppenfeld and A. Vornberger. *Parallel Flow Calculations on Transputers*. International Journal of Numerical Methods for Heat and Fluid Flow, Vol. 1, pp. 159-169, 1991.

[4] F. Lohmeyer and O. Vornberger. *CFD with parallel FEM on Transputer Networks*. Flow Simulation with High-Performance Computers I, Notes on Numerical Fluid Mechanics, Vol. 38, pp. 124-137, Vieweg, Braunschweig, 1993

[5] F. Lohmeyer and O. Vornberger. *Flow Simulation with FEM on Massively Parallel Systems*. Computational Fluid Dynamics on Parallel Systems, Notes on Numerical Fluid Mechanics, Vol. 50, pp. 147-156, Vieweg, Braunschweig, 1995.

[6] H. Greza, S. Bikker and W. Koschel. *Efficient FEM Flow Simulation on Unstructured Adaptive Meshes*. In this publication.

[7] M. Böhm and E. Speckenmeier. *Effiziente Lastausgleichsalgorithmen*. Proceedings of TAT-94, Aachen, 1994.

On the Performance Enhancements of a CFD Algorithm in High Performance Computing

Michael Lenke	Thomas Michl
Arndt Bode	Siegfried Wagner
Lehrstuhl für Rechnertechnik und	Institut für Aerodynamik und
Rechnerorganisation	Gasdynamik
TU München	Universität Stuttgart
W–80290 München	W–70550 Stuttgart

SUMMARY

This report deals with aspects concerning HPC with respect to the industrial CFD solver NSFLEX for aerodynamic problems. Since 1992, this solver has been parallelized and tested on different parallel computer architectures and different communication paradigms. Good performance results could be achieved. Based on that parallel solver, some more aspect are considered in this report to improve its performance behavior. Multigrid techniques and cg-like methods are implemented as well as the use of new programming paradigms like FORTRAN 90 and HPF (High Performance Fortran). Again, performance improvement can be achieved. Furthermore, one very important aspect concerning the application cycle of high parallel solvers is introduced and an interactive tool concept is proposed which enables online monitoring, online analyzing and online steering of parallel simulations to the user.

INTRODUCTION

High-Performance Computing (HPC) is of great importance to science and engineering and therefore of interest in many research projects. Appropriately, from the point at which the potential of computational simulation was recognized, the demand for CPU-time and storage has increased accordingly. The restrictions of the capacity of single-processor systems can only be overcome with parallel computer systems. But that is still not sufficient. Principles, techniques, and tools must support the application designers in their efforts to develop efficient parallel simulation algorithm and the application users in their test and production work with parallel simulation algorithms. Both must be set on in a similar manner to master the great challenge of parallelism.

Within the Priority Research Program "Flow Simulation On High Performance Computers" of the DFG (Deutsche Forschungsgemeinschaft, DFG-PRP) engineers and computer scientists have cooperated in the projects Bo 818/2 and Wa 424/9 to solve interdisciplinary problems in High Performance Computing regarding the industrial Euler and Navier-Stokes solver NSFLEX. In the last decade, NSFLEX was developed by MBB (Messerschmitt Bölkow Blohm, today DASA) and has been disposable for the projects Bo 818/2 and Wa 424/9.

One advantage of this interdisciplinary cooperation was that a wide range of aspects concerning High Performance Computing could be considered. The aspects were: (i) parallelizations of the solver on parallel computer systems with distributed and shared memory architectures, and the use of different communication paradigms to accelerate the performance of NSFLEX, (ii) the integration of modern numerical methods (multigrid and conjugate gradient (cg) techniques) to accelerate the numerical part of NSFLEX, (iii) the code conversion from FORTRAN 77 to Fortran 90 and HPF (High Performance Fortran) to take into account software engineering aspects and to improve the basis of the solver for further investigations in High Performance Computing, and (iv) the implementation of a tool concept which improve the productivity of the work with parallel simulation algorithms in High Performance Computing. Results of this research work have been documented in several publications: [2], [24], [19] and [29] in 1993, [18] in 1994, [22], [20], [26], [27], [36], [25] and [16] in 1995.

This paper shortly describes the governing equations of NSFLEX, the parallel structure and the message based communication model of the parallelized NSFLEX which is now called PARNSFLEX. The implementation of a FAS multigrid method in NSFLEX90 and the results are given. NSFLEX90 is the FORTRAN 90 conversion of NSFLEX. Results which are achieved with NSFLEX90 and its parallel version PARNSFLEX90 are given, too. After that, the the work and results with CGNS, which is a parallel cg-like method, are discussed. Research works and results concerning the Fortran 90 and HPF transformation of NSFLEX are shown. Finally, the concept of the tool environment VIPER which improve the productivity of the work with parallel simulation algorithms in High Performance Computing is presented.

GOVERNING EQUATIONS

The basic equations are the time-dependent Reynolds–averaged Navier–Stokes equations. Conservations laws are used with body–fitted arbitrary coordinates ξ, η, ζ building a structured Finite–Volume grid. In Cartesian coordinates, using vector notations, the basic equations read:

$$
\begin{aligned}
\frac{\partial \rho}{\partial t} + \nabla (\rho \mathbf{v}) &= 0 \\
\frac{\partial (\rho u)_i}{\partial t} + \nabla [(\rho u)_i \mathbf{v}] - \frac{\partial}{\partial x_j}(\tau_{ij} - p\delta_{ij}) &= 0 \\
\frac{\partial (\rho e)}{\partial t} + \nabla [(\rho e) \mathbf{v}] - \frac{\partial}{\partial x_j}\left(k\frac{\partial T}{\partial x_j} - pu_j + u_i\tau_{ij}\right) &= 0
\end{aligned}
\quad (1)
$$

where e is the internal energy, τ is the stress–tensor, δ is the Kronecker–symbol. The symbols ρ, p, T, k denote density, static pressure, temperature and heat conductivity, respectively. Turbulent viscosity is computed from the algebraic turbulence–model of Baldwin & Lomax [1].

Rewriting the Navier–Stokes equations with $U^T = (\rho, \rho u, \rho v, \rho w, \rho e)$ and E, F, G representing the fluxes and source terms in ξ, η, ζ and first order discretization in time leads to the backward Euler implicit scheme:

$$
D\frac{U^{n+1} - U^n}{\Delta t} + E_\xi^{n+1} + F_\eta^{n+1} + G_\zeta^{n+1} = 0. \quad (2)
$$

D is the cell volume. Newton linearization of the fluxes at the time level $n+1$,

$$E_\xi^{n+1} = E_\xi^n + (A^n \Delta U)_\xi, \text{ with the Jacobian matrix: } A^n = \frac{\partial E^n}{\partial U^n} \quad (3)$$

leads to the linearized flow-equation:

$$\frac{D}{\Delta t}\Delta U + (A^n \Delta U)_\xi + (B^n \Delta U)_\eta + (C^n \Delta U)_\zeta = -(E_\xi + F_\eta + G_\zeta)^n. \quad (4)$$

The inviscid part of the Jacobian on the left–hand–side is discretized with an upwind scheme, which is based on the split van Leer fluxes [37]. The viscous part of the Jacobian is discretized by central differences. The inviscid fluxes on the right-hand-side are discretized up to third order in space. Shock capturing capabilities are improved by a modified van Leer flux–vector–splitting method [9].

First order space–discretization on the left–hand–side leads to a linear system of equations,

$$A\mathbf{c} = \mathbf{b} \quad (5)$$

where A is a sparse, non–symmetric, block–banded and positive–definite matrix, $\mathbf{c}$ represents the time–difference of the solution vector $\mathbf{U}$, $\mathbf{b}$ the right–hand–side of equation (5), which can be relaxed by a factor ω.

Parallel Implementation

Specification

One standard of the parallel solver was simple transferability, which is guaranteed by Fortran 77 / Fortran90 standard language with some really common extensions and general message passing calls.

The code should have good performance on almost any computer system. It should be fully vectorizeable. And, in addition, it should perform very good on almost any parallel computer, even on those with slow networks, like workstation clusters.

Strategy

A multiblock strategy, where the grid is partitioned in blocks, satisfies all specifications. The blocks overlap each other with two cellrows (figure 2). These two cellrows store the internal boundary conditions for overlapping blocks, such as the conservative solution vector $\mathbf{U}$ for the flux calculation and the time–difference of the conservative solution vector $\Delta\mathbf{U}$, which is needed in the implicit scheme. These time–differences can be relaxed by some factor ω to make the internal boundary condition more weak. This information exchange must only be done one time per time–step.

As in the cg–like methods some global scalar values have to be determined, there is some extra communication needed, concerning α, ω, β and the convergence criterion. These communication of scalar values is needed some times per cg–step (figure 6).

Grid partitioning is done by a small sequential program (figure 1).

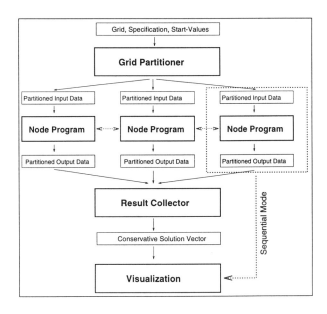

Figure 1: Parallel strategy for a SPMD programming model.

A SPMD programming model was chosen, where each node runs the same program. A host program is only necessary, if a special distribution of programs to nodes is wanted, or if instance numbers are not definitively set with the load of the program. Otherwise each node program identifies its block number with the instance number. Interconnections to other blocks are given with the grid.

Running a parallel node program, which also should run sequentially, results in some double computations concerning some general variables and the fluxes at the overlapping boundaries. These double computations are neglectable, because the implicit part of the solver consumes under all conditions the dominant part of the CPU–time. With this strategy parallelism is in most cases coarse–grain, which enables high speed-ups.

If possible file–I/O is done in parallel, i. e. if the operation system of the multiprocessor can handle that.

Sequential runs (on a vector supercomputer) are possible (dotted box in figure 1). Then the blocks are computed one after the other. Communication between blocks is now simply done by a copy. As in the parallel run the implicit solver works blockwise.

The parallelly written result files are collected and *sequentialized* by a collector. Afterwards data can be visualized.

MESSAGE BASED COMMUNICATION CODE

Two routines handle communication. The first one scans the overlapping boundaries of the block's domain and creates a 1–D receive–table. This table contains all necessary information for communication, such as needed information from neighboring blocks and their correlation with the local block. Needed information is packed in demands that are

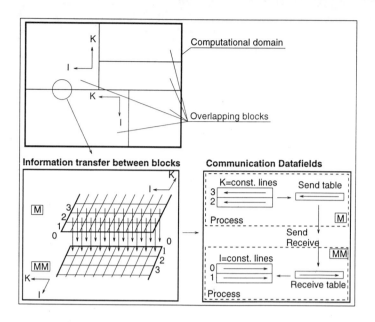

Figure 2: Blocks and communication.

sent to the according processes. With these demands each process creates a 1–D send-table. These tables allow a simple and secure control of all messages that are to be sent or to be received. The second routine is sending and receiving all messages during the iterative process (figure 2).
Both routines use general send and receive calls. Differences are introduced in the interface which translates send and receive calls to machine–specific message passing statements.
The developed message based model allows the computation of almost any block structured grid, which is derived from split O–, C–(C–)type, H–(H–, O–)type, or lego–type grid. The I, J, K orientation of the blocks is arbitrary. Several independent connections between two blocks are possible. Hence, the solver works on structured grids but on unstructured blocks.
Another parallelization approach had been done within the German POPINDA project in 1995. Again, the aim was to develop a portable parallel code of NSFLEX. Here, subroutines of the communication library CLIC map the interprocess communication to parallel computer systems (for more details see [10]). The achieved performance results of the POPINDA approach are similar successfully as with PARNSFLEX of our project since 1991 (see [2], [24], [19], [29], [18], [22], [20], [26] and [27]).

MULTIGRID SOLVER BASED ON NSFLEX

The principles of multigrid methods have been established for linear systems in the late 1970s and early 1980s, principally by BRANDT [3], HACKBUSCH [8] and STÜBEN & TROTTENBERG [33]. The central property of this technique is its ability to achieve optimal convergence rates on solving elliptic problems like Poisson equation. Much work has been done in the 1980s to extend this technique for more complicated pde's like the Euler or Navier-Stokes equations.

It could be demonstrated in cases of laminar flows in simple rectilinear geometries that in linear cases the favorable convergence properties carry over to the flow describing equations. Speed-up factors of the order 10 up to 100 were reported in publications, especially by solving the time–independent Euler or Navier-Stokes equations [4, 5, 11, 14], which were, however, still elliptical.

In case of nonlinear hyperbolic time–dependent equations it has to be distinguished between *direct* and *indirect* multigrid methods. The direct method follows the Full Approximation Scheme (FAS) concept *in space and time*. Consequently, one multigrid cycle extends over more than one time step by using the direct multigrid method. On the other side the indirect multigrid method works only at each single time step. The indirect method can be used to accelerate the matrix inversion of the implicit method. In this case the iterative procedure corresponds to a solution of a discrete quasi-elliptic system *only in space*. So far, there are results published, concerning (i) implementations of *direct* multigrid technique in *explicit* solvers [9, 38] or, respectively, (ii) implementations of *indirect* multigrid technique in *implicit* solvers [9, 13] of the time-dependent Euler or Navier-Stokes equations. In both cases no impressive performance speedups could be reached with factors of 3 up to 5.

The challenge now is to implement a *direct* multigrid technique in an *implicit* solver.

The integration of multigrid techniques is based on the NSFLEX90 code which is the Fortran 90 implementation of NSFLEX. It is stressed that this new Fortran standard allows the recursive code formulation. This progressive programming technique is used in MG-Nsflex90, which enormously simplifies the design of multigrid algorithms. The integration work was started with the implementation of an indirect FMG-CS (Full Multigrid Correction Scheme) algorithm into the complex *implicit* solver NSFLEX90 which is based on the *time-dependent* Navier-Stokes equations. Then a direct multigrid method is applied with a simplified FAS concept using a new proposed approach. For this work *standard-coarsening* (i.e. doubling the mesh spacing in all three directions from one grid to the next coarsest grid) is used to employ the hierarchy of grid levels to accelerate the convergence of the iterative computation method. Both V-cycles and W-cycles are implemented in MG-Nsflex90. The same transfer operators can be used in FAS-mode of MG-Nsflex90 as in the CS-mode. The boundary conditions are, however, treated in a different way in each multigrid mode. A new bilinear volume-weighted prolongation is used. The steady flow state solution is reached with sufficient accuracy, essentially less computational efforts and good performance speed-ups.

In this section, only the much more interesting FAS multigrid implementation and the achieved results are discussed. Further information will be found in [21]. It is mentionable, that after some few code modification both direct (applied in time) and indirect (applied in space) multigrid method could be combined with each other in one code. This interesting method combination is not tested yet because of lack of suitable grids.

THEORY OF FULL APPROXIMATION SCHEME (FAS)

For accelerating schemes Brandt [3] has proposed the FAS concept. This concept is suited to nonlinear equations and, therefore, adapted in the current solution scheme of the time-dependent Navier-Stokes equations in time and space. An arbitrary discretized nonlinear equation

$$L_k U_k = f_k \qquad (6)$$

on the k-th grid level with the iteration variable U, L a nonlinear operator, can be approximated on the next coarser grid by a modified difference approximation

$$L_{k+1} U_{k+1} = f_{k+1} + \tau_k^{k+1} \qquad (7)$$

where τ is the relative discretization error caused by the problem transferring from one grid level to another. It maintains the truncation error of the finest grid on all other grids in the used grid hierarchy. After some solution steps on the $k+1$-th grid the correction between the transfered fine grid solution and the new coarse grid solution U_{k+1} is transfered to the fine grid and the former fine grid solution is corrected according to

$$U_{k,new} = U_{k,old} + I_{k+1}^k (U_{k+1} - I_k^{k+1} U_{k,old}). \qquad (8)$$

Integration of a direct multigrid method

Solving Euler or Navier-Stokes equations is a nonlinear problem, so a Full Approximation Scheme (FAS) must be used to implement a multigrid method working *in space and time*. According to equations (7) the following equation (9) has to be solved (DIAG, ODIAG and RHS are NSFLEX specific terms, which describe the system vectors)

$$DIAG_k \Delta U_k = -\omega RHS_k - ODIAG_k + \tau_k^{k+1}. \qquad (9)$$

It is very difficult to find the suitable τ. In fact, we did not find it. But in case of hyperbolical problems it seems to be not so necessary to know the exact value of τ as in case of elliptical problems. The computation of the discretization error τ_k^{k+1} is even essential for solving elliptical problems or for time accurate solving of non-steady hyperbolical equations.

Brandt recognized in [3] that the absolute high-frequency changes in each time step are negligible because they are small compared to the high frequencies themselves. Following this idea it is not necessary to determine an exact result in each time step. With the introduction of time stepping into the NSFLEX-solver the results during the computation are even more inexact but still exact enough for a stable computation of steady Navier-Stokes equations with the time-dependent method. The robustness of the time-dependent method for solving steady hyperbolic equations allows an experimental implementation of a simplified FAS-algorithm where τ is neglected (τ is set to 0).

New bilinear prolongation operator

The new *bilinear volume-weighted prolongation* operator (proposed by Lenke and Oliva in [21]) provides the value transfer of a variable to a next finer grid level. This operator uses cell volumes to get a more exact value for a variable transferred to a finer grid (see figure 3).

According to figure 3, $V_{2i,2j}$ is the value of the any discrete variable V in cell $(2i, 2j)$. Then with the formula

$$V_{2i,2j} = \sum_{\alpha=i-1}^{i} \sum_{\beta=j-1}^{j} C_{\alpha,\beta} V_{\alpha,\beta} \qquad (10)$$

the coefficients C for the usual bilinear prolongation are

$$C_{i,j} = \frac{9}{16} \qquad C_{i-1,j} = \frac{3}{16}$$

$$C_{i,j-1} = \frac{3}{16} \qquad C_{i-1,j-1} = \frac{1}{16}$$

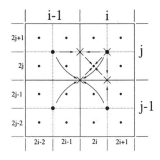

Figure 3: Scheme of the bilinear prolongation

and for the volume-weighted bilinear prolongation with $J_{i,j}$ as the volume of the cell (i,j)

$$C_{i-1,j} = \frac{J_{i,j} + 2J_{i,j-1}}{4\sum_{\alpha=i-1}^{i}\sum_{\beta=j-1}^{j} J_{\alpha,\beta}} \qquad C_{i,j} = \frac{J_{i,j} + 2J_{i,j-1} + 2J_{i-1,j} + 4J_{i-1,j-1}}{4\sum_{\alpha=i-1}^{i}\sum_{\beta=j-1}^{j} J_{\alpha,\beta}}$$

$$C_{i-1,j-1} = \frac{J_{i,j}}{4\sum_{\alpha=i-1}^{i}\sum_{\beta=j-1}^{j} J_{\alpha,\beta}} \qquad C_{i,j-1} = \frac{J_{i,j} + 2J_{i-1,j}}{4\sum_{\alpha=i-1}^{i}\sum_{\beta=j-1}^{j} J_{\alpha,\beta}} .$$

PERFORMANCE RESULTS WITH MG-NSFLEX90 AND MG-PARNSFLEX90

The performance results of the method is tested for flow problems around the airfoil NACA 0012. The flows are computed for subsonic (Ma=0.07) and transonic (Ma=0.7) inflows using the Euler method. In order to compare the efficiency of the solvers, a Work Unit (WU) is defined as the CPU-time needed for the single grid solver (without multigrid) to perform a time step. An Euclidean norm of the residual vector for the finest mesh is used to evaluate the results which is defined as

$$\frac{\sum_{i,j,k}(|\Delta\rho| + |\Delta\rho u| + |\Delta\rho v| + |\Delta\rho w| + |\Delta e|)_{i,j,k}}{N} .$$

Still another criterion for the convergence quality of MG-Nsflex90 is suggested and used in this paper that is called *error*. It is based on the knowledge of one reference result for the considered flow case. For this paper the flow distribution after 20 000 time step iterations on a single grid are accepted as the reference result. Denoting the reference result with U_{ref} and the considered multigrid result with U, the error e is defined as

$$e = \sum_{i}(U_{ref} - U) \qquad (11)$$

with index i running over all volumes. For presentation of results in this paper the Euclidean norm of error was used

$$\|e\|_2 := \left[\sum_{k=1}^{n} u_k^2\right]^{1/2}. \tag{12}$$

The computations for the NACA 0012 airfoil are carried out on a 256x64 O-mesh with Euler equations. With this profile grid it was possible to use up to three coarser grid levels. The essentially improved history of convergence for the multigrid method compared with the single grid method is easy to recognize.

SERIAL PERFORMANCE RESULTS WITH MG-NSFLEX90

Table 1 shows the performance results which are achieved with serial simulation runs of MG-NSFLEX90 on an IBM workstation (Ma=0.7, Euler method). There are two speedup factor (su-t and su-i) which build the total serial speedup factor su-s. The run time speedup factor is called su-t. As it can be seen the run time performance is improved up to the factor 8.14, in case of four grid levels. The speedup factor regarding the error is called su-i. As it can be seen the performance regarding the error is improved up to 2.05. Summarizing both speedup factors to su-s, total performance speedup factors up to 16.69 could be achieved, in case of four grid levels. Regarding the complexity of this multigrid problem these results are very impressive.

Table 1: NACA 0012 — serial computation after 2 000 time step iterations

# grids	time (sec)	error	su-t	su-i	su-s
1	9790	15.42	1.00	1.00	1.00
2	4761	7.67	2.06	2.01	4.14
3	2374	6.28	4.12	2.46	10.14
4	1203	7.51	8.14	2.05	16.69

PARALLEL PERFORMANCE RESULTS WITH MG-PARNSFLEX90

MG-NSFLEX90 has been parallelized to MG-PARNSFLEX90. The same parallelization strategies are used as in PARNSFLEX. Table 2 shows the performance results of MG-PARNSFLEX with the use of four processors on an IBM workstation cluster (Ma=0.7, Euler method). The factor su-p gives the speedup due to the use of that four processors. The factor su tells the total speedup with respect to the multigrid method performance, the error performance and the parallel computing performance.

The result of the overall effort is that the NACA 0012 profile grid can now be simulated in 8:08 min (formerly 2:43:10 h). That means a speedup of a factor 41.22.

Figure 4 gives the error distribution of the density after 2 000 time step iterations of the single grid method and the 4-level-mg-method. It could be noticed that in the multigrid solution there are still errors around the shock shape $I \approx 50$ (at the top of the NACA 0012 profile) and at the end of the NACA 0012 profile $I \approx 170$. It can be concluded that with a 4-level-mg-method these errors could essentially not be eliminated any more.

Table 2: NACA 0012 — parallel computation on four processors after 2000 time step iterations

# grids	time (sec)	su-p	su-s	su
1	2874	3.41	1.00	3.41
2	1432	3.32	4.14	13.74
3	810	2.93	10.14	29.71
4	488	2.47	16.69	41.22

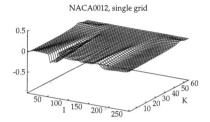
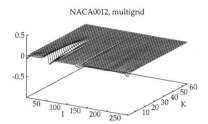

Figure 4: NACA 0012 — error distribution of the density after 2 000 time step iterations

Figure 5 depicts the results of the above tables in work units and progressions of the residuum. Regarding the progression in figure 5 a new mg-strategy can be recommended: To reach the best residuum (log(residuum) ≈ −4.5) with the native 2-level-mg-method at 2 100 work units) the first 100 work units can be done by the 4-level-mg-method up to log(residuum) ≈ −3.8, followed by the 3-level-mg-method for the next 400 work units up to log(residuum) ≈ −4.2 and, finally, replaced by the 2-level-mg-method for 800 work units to reach log(residuum) ≈ −4.5. Totally, only 1300 work units have to be used against 2 100 in the native 2-level-mg-method. To sum up, if it is possible to determine the optimal switch points in advance a further speedup factor will be achieved.

CGNS, A PARALLEL NAVIER–STOKES SOLVER, BASED ON BI–CGSTAB

PARALLEL CG–LIKE METHOD

Linear, non–symmetric and positive–definite systems of equations (like the flow–equation 5) can efficiently be solved by cg–like methods.
Out of a wide variety of cg–like methods, we had chosen a right–preconditioned variant of BI–CGSTAB [35]. BI–CGSTAB is based on BCG [7] and respects some experiences made with round–of–errors [12] from another variant of BCG, the CGS–method [32]. BI–CGSTAB has comparable convergence–rates to GMRES [31, 35], but has the advantage

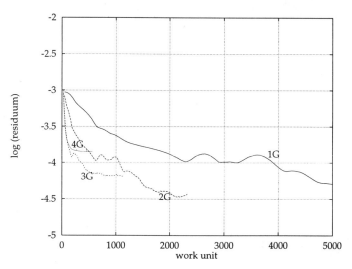

Figure 5: NACA 0012 — progression of residuum

over GMRES, that computing time and storage requirements do not increase during iteration.

Numerical experiments with left- and right-preconditioned variants of BI-CGSTAB showed that the modified right-preconditioned variant (figure 6) has a better convergence characteristic and, additionally, is by a factor of two to three faster than the left-preconditioned one [23]. One reason for this is that the right-preconditioned variant needs only two preconditioning steps instead of three for the left-preconditioned version.

Smoothing the incoming residual vector r_0 further improved the convergence rate and stabilized the algorithm.

The numerical efficiency of BI-CGSTAB is directly linked to efficient preconditioning. The task of preconditioning is to reduce the condition–number of the coefficient Matrix A (equation 5). Preconditioning is usually costly, and one has to find a good balance between computational effort and the clustering of Eigen-values by preconditioning. Suitable preconditioners are ILU-type methods – especially the stabilized–ILU method [34] showed good performance – and a blockdiagonal preconditioner, which is preferable for some stiff cases. Both kinds of preconditioners are implemented blockwise, which means that they run independently on each block. Due to the loss of coupling between the blocks one theoretically looses numerical efficiency. On the other hand one saves an additional communication effort in an inner loop of the iterative process. The numerical efficiency of recursive preconditioners, like ILU-type methods, is linked to the numbering of the elements. Presently the implemented ILU-preconditioners do not respect a special numbering for different flow-cases. Running these types of preconditioners uncoupled on each block leads, from a global point of view, to different numberings of the elements for different partitionings of the grid.

Comparing the implemented BI-CGSTAB method for the Navier-Stokes equations with the previously implemented Gauß–Seidel-method, showed a speedup of about 2.5 to 9 [23] – depending on the point of view and the investigated flow-case.

$$r_0 = A - c_0$$

Smooth r_0

Set: $p_{-1} = q_{-1} = 0$, and $\alpha_{-1} = \rho_{-1} = \omega_{-1} = 1$

do k = 1,...,max. number of iterations

$\rightarrow \rho_k = (r_0, r_k)$

$\rightarrow \beta_k = \frac{\rho_k}{\rho_{k-1}} \frac{\alpha_{k-1}}{\omega_{k-1}}$

$p_k = r_k + \beta_k (p_{k-1} - \omega_{k-1} t_{k-1})$

- Solve: $M y_k = p_k$ for y_k

$q_k = A y_k$

$\rightarrow \alpha_k = \frac{\rho_k}{(r_0, q_k)}$

$s_k = r_k - \alpha_k q_k$

- Solve: $M z_k = s_k$ for z_k

$t_k = A z_k$

$\rightarrow \omega_k = \frac{(t_k, s_k)}{(t_k, t_k)}$

$c_{k+1} = c_k + \alpha_k y_k + \omega_k z_k$

$\rightarrow$ If c_{k+1} accurate enough, exit loop.

$r_{k+1} = s_k - \omega_k t_k$

end do

- indicates preconditioning.
- $\rightarrow$ indicates global operation on a multiprocessor.

Figure 6: Modified right–preconditioned BI–CGSTAB algorithm.

NACA0012 FLOWFIELD RESULT

The flow around a NACA0012–profile is computed at a Mach–number of 0.5 and a Reynolds–number of 5 000 (angle of attack $\alpha = 0°$). The resulting flow–field leads to small separation bubbles at the trailing edge of the profile. The test–case makes numerical diffusion and stability of Navier–Stokes solvers evident. Large diffusion leads to late, or even no separation. A small amount of numerical diffusion can lead to pure convergence. The computation was done with a C–type grid provided on the ECARP–database. The grid has 280 x 50 cells. The boundary–layer is resolved within about 10 cells at the trailing edge and about five cells at the leading edge. Due the low number of cells in the boundary–layer, the linear region is not resolved, which requires a higher–order evaluation of the friction–coefficient c_f.

Separation is computed at 82.7 % of the chord–length (figure 7), which is in good agreement with literature (81.4 % to 83.5 %) [28]. The flow is absolutely symmetric (in figure 7

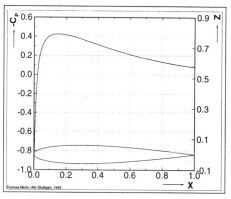 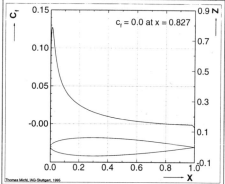

Figure 7: Pressure and friction for NACA0012 (Ma = 0.5, Re = 5 000).

the results of the upper and lower side are plotted).

PERFORMANCE ON A MULTIPROCESSOR SYSTEM

Presently CGNS runs on Intel Paragon XP/S-5 with the NX communication library and on Cray T3D with pvm 3.2 for the communication of boundary–values and virtual shared memory for global operations.

Presented are the runs on the Cray T3D system. The whole implicit part of the code is written in Fortran90 and is compiled with the option -*O3* (compiler version: cf90 1.0.2.0), while the driving parts and flux–calculation routines are still written in Fortran77 (compiler option: -*Wf" -o aggress" -Wl"-Dreadahead=on"*). Presently the Fortran90 code leads to significantly longer run–times than the Fortran77 code. Hence, comparisons of Fortran90–codes to Fortran77–codes should be treated with care.

The runs on the Cray T3D for all partitionings of the grid were done with a CFL–number of 10 and a CFL–number of four for the boundary–layer. The used preconditioner is a Stabilized–ILU method. Convergence is gained within 1 600 iterations (figure 8), which takes 720 [s] on 32 processors.

The partitioning of the grid is done along lines I=const., which means along grid–lines orthogonal to the profile. As the grid has 280 lines in that directions, the grid is partitioned in blocks which are only approximately balanced. The partitioning strategy leads to blocks, which have at most one line (which means 50 cells for that grid) more than others. Hence, the computation is slightly unbalanced.

The speedup of table 3 is referenced to the 4–block–run. The speedup is up to 16 processors in the linear range and superlinear for the 32–block case. This is an effect from numbering in preconditioning. For the partitioning in 32 blocks the preconditioner is more efficient, which leads to a reduced number of cg-like iterations per time-step.

The time spent for the communication of boundary–values is for all cases below 1.5 % of the computing time, and therefore negligible. The time spent for global operations (mainly global sums from global scalar products within BI-CGSTAB) increases from 0.1 % for 4 processors to 4.7 % for 32 processors. Nevertheless the increase is moderate and will only limit the parallel efficiency if the problem gets too small.

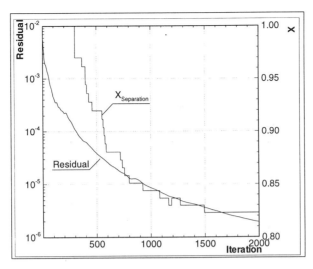

Figure 8: Convergence history for CGNS on NACA0012.

Table 3: Run–times until convergence of CGNS on Cray T3D.

Number of processors	[-]	4	8	16	32
Iterations	[-]	1 600	1 600	1 600	1 600
Elapsed time	[s]	6 087	3 148	1 577	720
Communication of boundary values	[s]	12.2	15.7	15.8	10.1
Global communication	[s]	6.1	9.4	20.5	33.8
Total communication	[s]	18.3	25.1	36.3	43.9
Speed–up	[-]	4	7.7	15.4	33.8

Evaluation of new programming paradigms in High Performance Computing with respect to Fortran 90 and HPF (High Performance Fortran)

FORTRAN (FORmula TRANslation) is the principle language used in fields of scientific, numerical, and engineering programming. On the one hand it requires much less computer science knowledge by the programmer than other languages, and it is much easier to use. On the other hand fast object codes can be generated, which is very important for scientific, numerical and engineering applications. Therefore, on almost every computer system compilers with preprocessors are available for this language.

But FORTRAN 77 does not fulfill requirements of the newer computer architectures, modern programming paradigms, and software-engineering aspects. Because of lacks in standard FORTRAN 77 the technical committee X3J3, responsible for the development of FORTRAN standards, brought out a new, urgently-needed modern version of that language — FORTRAN 90:

Architecture support The meaning of the use of vector computers, parallel computers, and parallel vector computers to solve the Grand Challenges, e.g. in CFD (computational fluid dynamic), will increase steadily. By means of array statements and array operations FORTRAN 90 supports pipeline and vector processing in an easy and natural way.

Among others the attribute facility supports compiler processing to generate much faster codes for the target system.

But there are no features which support explicit parallelism. Implicit parallelism exists by means of data independent array statement sequences.

Programming paradigm support Each language promotes programming styles and techniques. Procedural programming, the use of information hiding by encapsulation, data abstraction, and object–oriented programming are some keywords.

A technique is supported by a language if there are elements and mechanisms which are applicable in an easy, safe, and efficient way. Also compiler and run–time system are able to supervise the compliance with the paradigm. A technique is not supported if strains are necessary to realize the paradigm.

FORTRAN 77, as well as Algol, Algol 66, Pascal, and C, belongs to the procedural programming languages where as FORTRAN 90 supports, among others, a safe method of encapsulating derived data types using modules.

Therefore the new standard does conform to the trend changing main emphasis from procedure design to data design.

Software engineering support Software must be developed with respect to some concepts [6]. During their whole life time the following issues are meaningful for software packages: Software maintenance, reverse engineering, further development, software design and redesign, verification and validation, and portability. Those aspects should be supported by the implementation language.

FORTRAN 90 supports the effort of problem–matched data types and problem–oriented programming by the use of modules. Design, redesign, and testing are made easier. Portability is guaranteed.

For the code transformation from NSFLEX (which was written in FORTRAN 77) to NSFLEX90 (which was written in FORTRAN 90) some aspects of newer computer architectures, of modern programming paradigms, and of software engineering are considered. The transformation is done without any changes of the numerical features but with a complete redesign of the program structure: (i) The low-level storage addressing function which is a characteristic of the NSFLEX code (to speed up the performance) is removed to eliminate (without any loss of performance) the overall indirect addressing and to enable direct addressing. (ii) User-defined and problem-matched data structures are introduced as well as 1145 array statements. (iii) Dynamic memory allocation is integrated as well as pointer references to the most important data structures. Furthermore, data encapsulating by using the module concept is implemented for a total of 63 FORTRAN 90 modules. (iv) Each COMMON block becomes a single module in which each data structure gets its own storage generator and de-generator and is declared as PUBLIC. Consequently, each module can be used from external procedure units or other module units via the USE mechanism. (v) The used IMPLICIT NONE statement is forcing explicit declarations of data objects for safer programming. (vi) Attributes are used wherever it was possible in order to help the compiler and the runtime system to detect errors. (vii) The WHERE statement is used to perform array operations only for certain elements. (viii) Transformation functions are implemented to reduce arrays as well as overloading and generic functions (forces the CONTAIN mechanism). (ix) Auxiliary arrays are eliminated to save processor resources, scalars are expanded to arrays, statement sequences are extracted out of loops and nested loops are exchanged if necessary. Loop labels are replaced by loop names, GOTO statements are eliminated and replaced by structured programming constructs. (x) Redundant code fragments are removed and invariant statements are extracted out of loops. (xi) IF sequences are changed to SELECT CASE structures and some other more. Some features of FORTRAN 90 are not considered in that code version of NSFLEX90, although they were possible and useful (NAMELIST I/O, alias statements, direct-access files).

Some other statements are intentionally avoided (GOTO, BLOCK, DATA and COMMON blocks). To guarantee a safe and correct transition without any loss of semantics there are mainly three complete transformation steps needed, each with different main emphasis. As a result of the successful transition to NSFLEX90 *more code quality* and *much better performance options* are achieved with (as already mentioned) incorporated aspects of newer computer architectures, modern programming paradigms and software engineering. Each program designer in high performance computing is recommended to make himself familiar with and to have a critical look at the full potential of FORTRAN 90. Consequently old FORTRAN 77 programming styles must be overcome and a new point of view has to be acquired.

Since FORTRAN 90 is a full subset of HPF, the effort on a code adaption is low. High Performance Fortran additionally introduces not only parallel program constructs but also data placement directives. The main intention behind HPF was to allow the programmer to specify parallelism on a very high level of abstraction. The error-prone task of transforming this parallelism into explicit message passing code was meant to be performed by a compiler. Therefore the programming should become rather architecture independent. On the other side, with the possibility of explicit data placement, HPF programs should become reasonably efficient for a wide range of parallel computers.

HPF programs are *data-parallel* programs following the SPMD paradigm. The same program runs on all processors but only on a subset of the data structures. The update of the data is performed according to the owner computes rule where communication is introduced whenever data is required that resides on another processor. Therefore, the mapping of data onto the processors is a global optimization problem. The distribution of the data heavily determines the overall computation time of the parallel program.

HPF offers *data distribution* directives for the programmer to control data locality. With the PROCESSORS directive one can specify the size and shape of an array of abstract processors. The ALIGN directive allows elements of different arrays to be aligned with each other indicating that they ought to be distributed onto the processors the same way. DISTRIBUTE is used to simply map the data on a number of processors. This can be done in a *block*, *cyclic* or *block cyclic* manner, where the block distribution always gives a processor a contiguous sub-array or single elements of the specified array, respectively. The TEMPLATE directive finally may be necessary in cases when none of the program's arrays really has the right shape for a distribution.

Concurrency in HPF can be expressed via the following language constructs: The FORALL statement (and also the FORALL construct, a more complicated version) which is a generalization of the array assignment and WHERE statements of FORTRAN 90, looks at first a bit like a DO loop. It is quite different though in that it assigns to a block of array elements and does not iterate over the array in any specific order. All right hand sides are computed before any assignments are made. Another possibility for the programmer in HPF to express parallelism is to use the INDEPENDENT directive. With this, the programmer can declare a loop to be parallel which the compiler itself would not be able to detect.

HPF contains a lot more than the above described features which are all in the officially defined *subset HPF*. The intention of the High Performance Fortran Forum, which is the organization responsible for the definition of the HPF standard, to define a subset standard, was to encourage early releases of HPF compilers. The whole language specification itself though contains a lot more than what has been mentioned here. There are, however, still topics not covered by HPF which strongly limit its parallelization possibilities. An ongoing effort of the High Performance Fortran Forum is therefore to present a new specification (HPF 2) where many of these subjects are addressed.

NSFLEXHPF evolved from NSFLEX90 and was developed in 1995 by Rathmayer, Oliva, Krammer and Lenke. As described above, some explicit parallel language constructs and compiler directives are used to support a compiler-driven parallel execution (known as implicit parallelization technique). Today, a full HPF compiler system is still lacking. Therefore, NSFLEXHPF has to be adapted and evaluated in the first stage to existing HPF subset supported like ihpf, xhpf, ADAPTOR and PREPARE compiler systems (see [30]). Full HPF will be used if available later. It must be remarked that a lot of fine tuning work has to be done to adapt NSFLEXHPF to the existing subset compilers to get satisfied run time performance results (see [30]).

A TOOL CONCEPT FOR AN INTERACTIVE TEST AND PRODUCTION CYCLE

Typical applications of the so-called *Grand Challenges* need massively parallel computer system architectures. Tools like parallel debuggers, performance analyzers and visualizers help the code designer to develop efficient parallel algorithms. Such tools merely support the development cycle. But the technical and scientific engineers who make use of parallel high-performance computing applications, e. g., numerical simulation algorithms in Computational Fluid Dynamics (CFD), must be supported in their engineering work (test and production activities) by another kind of tool. A tool for the *application cycle* is required because old, conventional suggestions regarding the arrangement for the *application cycle* rely on strictly sequential procedures. They are due to the heritage of traditional work on former vector computers. That formative influence is still felt in todays arrangements for the *application cycle*, prevents a more efficient engineering work and, therefore, it must be overcome.

New tool conceptions have to be introduced to enable *on-line interaction* between the technical and scientific engineer and his running parallel simulation. It is stressed that the be-all and end-all for the technical and scientific engineer is the possibility to interact on-line with the running simulation. The technical and scientific engineer is only interested in physical parameters of the mathematical model (e. g., stress, density, velocity, energy, temperature and the like) and parameters of the numerical method (e. g., error distribution, residual, convergence acceleration factors, convergence rates, numerical methods and the like). Information about those parameters allow to judge the progress and the quality of the simulation run. He is not interested in the serial or parallel code of the simulation algorithm. To interact with these parameters will be the only way to accelerate the application cycle and to increase in general the productive power of simulation methods. In this case, the technical and scientific engineer has full control over the parallel simulation and he can suit the running simulation to his examination aspect. Doing so, the response time for the technical and scientific questions (answered by passing the knowledge gain cycle) are cut down, and system resources (e. g., CPU time, main and disk memory, system buses, network capacities, operating system components and the like) are saved. There can be no discussion about the advantages of interactive tools in HPC. The questions rather are, (i) how will it be possible to form the interesting simulation parameters as observeable and modifiable objects of a graphical user interface GUI and (ii) how efficiently can such a tool implementation be in its practical use on parallel HPC applications.

VIPER stands for **VI**sualization of **P**arallel numerical simulation algorithms for **E**xtended **R**esearch and offers physical parameters of the mathematical model and parameters of the numerical method as objects of a graphical user tool interface for online observation and online modification. The VIPER prototype is applied on PARNSFLEX up to a 64 block Cast 7 profile grid. As a test parallel computer system an Intel Paragon XP/S with 64 processors was selected.

Figure 9 shows the designed process architecture of VIPER. It is assumed that the graphical user tool interface and the parallel simulation are running on different computer systems which are interconnected via LAN and WAN. The separation of visualization code and number crunching code is essential. That is due to the availability of X11-server processes for graphical display representations. As a rule, such X11-server processes are not available on processors of parallel or vector computer systems. Therefore a special *client-server-client* process architecture implementation is concepted.

The so-called *client-server-client* process architecture with the DUAL SERVER combination *VIPER-D* (see figure 9) is the heart of the *VIPER* implementation. One client is the simulation with its parallel processes which are running on a remote parallel computer system. The other client is the graphical user tool interface process *VIPER-GUI* which is running on a workstation. Both clients are the inherent active partners of the DUAL SERVER which offers different services to its clients. *VIPER-D* exists as a distributed process combination on the parallel computer system as well as in the LAN of the graphic workstation.

VIPER can be characterized by the following features (most of them are already implemented):

- The power of the tool becomes available for almost all existing applications by an easy source code instrumentation via library procedures of *VIPER-PM*.

- The technical and scientific engineer is able to observe and modify on-line physical parameters of the mathematical model and parameters of the numerical method as objects of a graphical user interface. Thereby, he can control the running parallel simulation algorithm at any time.

- Interaction points (IPs) and the associated objects can be switched in an active or deactive state at any time. Thereby, the working with the running simulation can be adjusted to the examination aspect.

- Only active IPs and objects are processed. Thereby, a large amount of data are not

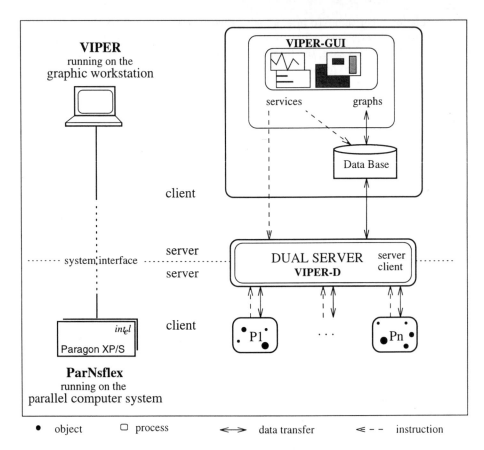

Figure 9: Process architecture of VIPER

to be exported and transfered which can extremely save system resources and effect the tool performance positively.

- It becomes possible to switch between various mathematical methods or computational models.

- Objects of the data base *VIPER-DB* can be displayed in various ways, analyzed and processed via operators applied on these objects.

- The annoying effect of object observation (read transaction) and object modification (write transaction) is reduced to the mere moments of data export and import. This ensures a minimum of runtime delays for optimized parallel simulation algorithms.

- The technical and scientific engineer becomes able to attach/detach the running parallel simulation at any desired time. If no *VIPER-GUI* process is attached or no IPs or rather no objects are in an active state the DUAL SERVER *VIPER-D* sleeps and exerts no worth mentioning runtime delays to the running simulation.

- It is suggestable to attach multiple *VIPER-GUIs*, running on different workstations, to one and the same *VIPER-D* to observe the same or different objects of one and the same parallel simulation run.

- A strict separation is intended between the code designer who knows the internals of the parallel simulation algorithm and the engineer who knows the engineering aspects of the theoretical model.

- The parallel simulation algorithm can be based on message-passing or/and shared-variables models.

First evaluations indicate the superiority of VIPER conception against conventional batch-oriented procedures. For more details see [16], [17] and [15]. It is stressed again, that the efficiency of the work with high parallel simulation algorithm is essentially depending on the formation of the application cycle. Old, batch oriented postprocessing must be overcome and replaced with interactive online monitoring, online analyzing and online steering.

Conclusion

Undoubtedly, CFD will take a dominate part in HPC with parallel computer architectures. Alike, principles, techniques and tools must support application designers and application users to master the great challenge of parallelism. To solve the problems for this, requires interdisciplinary methods. One step was done by the projects Bo 818/2 and Wa 424/9 in the DFG-PRP.

The development of ParNsflex, CGNS and MG-ParNsflex90 etc. showed, that the cooperation between information–science and engineering–science leads to highly efficient methods for the solution of the Navier–Stokes equations on a wide variety of multiprocessor systems.

The implementation of direct multigrid methods in the implicit Euler solver lead to speedups up to a factor of 41.22 by the use of the 4-level-mg- method and four processors. (The results of the Navier-Stokes solver did not reach that high speedup factor of the Euler solver.) The drawback that coarse grid levels could not resolve the physics in critical areas (e. g. the shock shape) can be overcome by switching back the coarsest grid level to a finer one to reduce again the numerical computation error and to arise the accuracy.

Implementing cg–like methods with suitable preconditioning lead to speedups (compared to Gauß–Seidel) by a factor of 2.5 to 9. The presented method CGNS is a research code. Its design aims on the flexible and easy implementation of new algorithmic parts. It is not tuned to special hardware and takes some double computations into count to support investigations on algorithmic parts. Presently the code gets only some 5 % out of the theoretical peak–performance of the Cray T3D. A special adaption of the code to individual system will lead to further significant speed–ups.

With the implementation of the VIPER concept an important step could be done to support the application users in their test and production cycle with parallel CFD applications.

All implementations achieve a favorable basis for further research work.

References

[1] BALDWIN, B. S., UND LOMAX, H. Thin Layer Approximation and Algebraic Model for Separated Turbulent Flow. *Report AIAA 78-0257* (1978).

[2] BODE, A., LENKE, M., WAGNER, S., UND MICHL, T. Implicit Euler Solver on Alliant FX/2800 and Intel iPSC/860 Multiprocessors. In *Flow Simulation with High-Performance Computers I*, E. H. Hirschel, Ed., vol. 38 of *Notes on Numerical Fluid Mechanics (NNFM)*. Vieweg, Braunschweig, 1993, Seiten 41–55.

[3] BRANDT, A. *Multigrid Techniques: 1984 Guide with Applications to Fluid Dynamics.* GMD-Studien 85, 1984.

[4] DICK, E. A multigrid method for steady incompressible navier-stokes equations based on partial flux splitting. In *International journal for numerical methods in fluids* (1989), vol. 9, Seiten 113–120.

[5] DICK, E. Multigrid methods for steady Euler- and Navier-Stokes equations vased on polynomial flux-difference splitting. In *International Series of Numerical Mathematics* (1991), vol. 98.

[6] FAIRLEY, R. E. *Software Engineering Concepts.* Mc Graw-Hill series in software engineering and technology. McGraw-Hill, 1985.

[7] FLECHTER, R. Conjugate Gradient Methods for Indefinite Systems. In *Lecture Notes in Math. 506*, H. Watson, Ed. Springer, Berlin, 1976.

[8] HACKBUSCH, W. *Multi-Grid Methods and Applications.* Springer-Verlag, 1985.

[9] HÄNEL, D., UND SCHWANE, R. An Implicit Flux Vector Splitting Scheme for the Computation of Viscous Hypersonic Flows. *Report AIAA 89-0274* (1989).

[10] HÖLD, R. K., UND RITZDORF, H. Portable Parallelization of the Navier-Stokes Code NSFLEX. POPINDA project report, 1995.

[11] HORTMANN, M., PERIC, M., UND SCHEUERER, G. Finite volume multigrid prediction of laminar natural convection: bench-mark solutions. In *International journal for numerical methods in fluids* (1990), vol. 11, Seiten 189–207.

[12] HOWARD, D., CONNOLLEY, W. M., UND ROLLETT, J. S. Solution of Finite Element Fluid Matrix Systems by Unsymmetric Conjugate Gradient and Direct Methods on the Cray–XMP and Cray 2. In *Conference Proceedings on Application of Supercomputers in Engineering: Fluid Flow and Stress Analysis Applications* (Amsterdam, 1989), C. A. Brebbia, Ed., Elsevier.

[13] KANARACHOS, A., UND VOURNAS, I. Multigrid solutions for the compressible euler equations by an implicit characteristic-flux-averaging. In *Computational Fluid Dynamics '92* (1992), vol. 2, Elsevier Science Publishers.

[14] KOREN, B. Multigrid and Defect Correction for the Steady Navier-Stokes Equations. In *Journal of Computational Physics* (1990), vol. 87, Seiten 25–46.

[15] LENKE, M. An interactive engineering tool for parallel HPC applications. *Advances in Engenineering Software (AES).* (accepted 1995).

[16] LENKE, M. VIPER — A tool concept to control parallel CFD applications. In *Applications of High-Performance Computing in Engineering IV* (June 1995), H. Power, Ed., Wessex Institute of Technology, UK, Computational Mechanics Publications, Southampton Boston, Seiten 179–186. ASE'95.

[17] LENKE, M. VIPER — Konzept eines interaktionsvitalen Werkzeuges zur Leistungssteigerung im Anwendungszyklus paralleler Simulationsalgorithmen. In *Luft- und Raumfahrt — Schlüsseltechnologien für das 21. Jahrhundert* (1995). Annual Meeting 1995, Bonn - Bad Godesberg, 26.-29. September 1995.

[18] LENKE, M., UND BODE, A. Methoden zur Leistungssteigerung von "real–world" Simulationsanwendungen in der Luft- und Raumfahrt für das technisch-wissenschaftliche Hochleistungsrechnen. In *Basistechnologien für neue Herausforderungen in der Luft- und Raumfahrt* (1994), Seiten 465–474. Annual Meeting 1994, Erlangen, Oktober 4-7.

[19] LENKE, M., BODE, A., MICHL, T., MAIER, S., UND WAGNER, S. Dataparallel Navier-Stokes Solutions on Different Multiprocessors. In *Applications of Supercomputers in Engineering III*, C. Brebbia und H.Power, Eds. Computational Mechanics Publications, Sept. 1993, Seiten 263–277. ASE'93 in Bath (UK).

[20] LENKE, M., BODE, A., MICHL, T., UND WAGNER, S. Nsflex90 - A 3d Euler and Navier-Sokes solver in Fortran 90. In *Computational Fluid Dynamics on Parallel System* (1995), S. Wagner, Ed., vol. 50 of *Notes on Numerical Fluid Mechanics (NNFM)*, Vieweg Braunschweig, Seiten 125–134.

[21] LENKE, M., UND OLIVA, L. Implementation of a direct FAS-similar methode in an implicit solver for the Navier-Stokes equations. (to be submitted 1996).

[22] LENKE, M., RATHMAYER, S., BODE, A., MICHL, M., UND WAGNER, S. Parallelizations with a real-world CFD application on different parallel architectures. In *High-performance Computing in Engineering*, C. B. H. Power, Ed., vol. 2. Computational Mechanics Publications, Southhampton, UK, 1995, ch. 4, Seiten 119 - 166.

[23] MICHL, T. *Effiziente Euler– und Navier–Stokes–Löser für den Einsatz auf Vektor–Hochleistungsrechnern und massiv–parallelen Systemen*. Dissertation, Institut für Aerodynamik und Gasdynamik der Universität Stuttgart, 1995.

[24] MICHL, T., MAIER, S., WAGNER, S., LENKE, M., UND BODE, A. The Parallel Implementation of a 3d Navier-Stokes Solver on Intel Multiprocessor Systems. In *Applications on Massively Parallel Systems* (Sept. 1993), vol. 7 of *SPEEDUP*, Proceedings of 14th Workshop; Zurich, Seiten 19–23.

[25] MICHL, T., UND WAGNER, S. A Parallel Navier-Stokes Solver Based on BI-CGSTAB on Different Multiprocessors. In *International Symposium on Computational Fluid Dynamics* (Sept. 1995), vol. II, Seiten 827—832.

[26] MICHL, T., WAGNER, S., LENKE, M., UND BODE, A. Big Computation with a 3-D Parallel Navier-Stokes Solver on Different Multiprocessors. In *Computational Fluid Dynamics on Parallel Systems* (1995), S. Wagner, Ed., vol. 50 of *Notes on Numerical Fluid Mechanics (NNFM)*, Vieweg Braunschweig, Seiten 157–166.

[27] MICHL, T., WAGNER, S., LENKE, M., UND BODE, A. Dataparallel Implicit 3-D Navier-Stokes Solver on Different Multiprocessors. In *Parallel Computational Fluid Dynamics - New Trends and Advances* (May 1995), P. L. A. Ecer, J. Hauser und J. Periaux, Eds., Elsevier, North-Holland, Seiten 133–140.

[28] PAILLERE, H., DECONINCK, H., UND BONFIGLIOLI, A. A Linearity Preserving Wave-Model for the Solution of the Euler Equations on Unstructured Meshes. In *Computational Fluid Dynamics '94* (Chisester, 1994), S. Wagner, E. Hirschel, J. Periaux, und R. Piva, Eds., John Wiley & Sons.

[29] PETERS, A., BABOVSKY, H., DANIELS, H., MICHL, T., MAIER, S., WAGNER, S., BODE, A., UND LENKE, M. Parallel Implementation of Nsflex on an IBM RS/6000 Cluster. In *Supercomputer Applications in the Automotive Industries* (Sept. 1993), Seiten 289–296.

[30] RATHMAYER, S., OLIVA, L., KRAMMER, J., UND LENKE, M. Experiences with HPF compiler systems with respect to the real world simulation algorithm NsflexHPF. (submitted).

[31] SAAD, Y., UND SCHULTZ, M. H. GMRES: A Generalized Minimal Residual Algorithm for Solving Nonsymmetric Linear Systems. *SIAM J. Sci. Stat. Comput. 7* (1986), 856–869.

[32] SONNEFELD, P. CGS, a fast Lanczos–type Solver for Nonsymmetric Linear Systems. *SIAM J. Sci. Stat. Comput. 10* (1989), 36–52.

[33] STÜBEN, K., UND TROTTENBERG, U. *Multigrid methods.* GMD-Studien 96, 1985.

[34] VAN DER VORST, H. A. Iterative Solution Methods for Certain Sparse Linear Systems with a Non–Symmetric Matrix Arising from PDE Problems. *Journal of Computational Physics 44* (1981), 1–19.

[35] VAN DER VORST, H. A. BI–CGSTAB: A Fast and Smoothly Converging Variant of BI–CG for the Solution of Nonsymmetric Linear Systems. *SIAM J. Sci. Stat. Comput. 13*, 2 (1992), 631–644.

[36] WAGNER, S. Computational Fluid Dynamics on Parallel Systems. vol. 50. Vieweg und Sohn, Braunschweig/Wiesbaden, 1995.

[37] WIRTZ, R. *Eine Methodik für die aerodynamische Ventwurfsrechnung im Hyperschall.* Dissertation, Institut für Aerodynamik und Gasdynamik der Universität Stuttgart, 1993.

[38] ZIEGLER, H.-J. Einsatz von Mehrgittermethoden zur Konvergenzbeschleunigung bei Euler-Lösern. Diplomarbeit, Universität der Bundeswehr, Munich, 1992.

EFFICIENT SIMULATION OF INCOMPRESSIBLE VISCOUS FLOWS ON PARALLEL COMPUTERS

F. Durst, M. Schäfer and K. Wechsler

Institute of Fluid Mechanics
University of Erlangen-Nürnberg
Cauerstr. 4, D-91058 Erlangen, Germany

SUMMARY

A parallel multigrid method for the prediction of laminar and turbulent flows in complex geometries is described. Geometrical complexity is handled by a block structuring technique, which also constitutes the base for the parallelization of the method by grid partitioning. Automatic load balancing is implemented through a special mapping procedure. High numerical efficiency is obtained by a global nonlinear multigrid method with a pressure-correction smoother also ensuring only slight deteriotation of the convergence rate with increasing processor numbers. By various numerical experiments the method is investigated with respect to its numerical and parallel efficiency. The results illustrate that the high performance of the underlying sequential multigrid algorithm can be largely retained in the parallel implementation and that the proposed method is well suited for solving complex flow problems on parallel computers with high efficiency.

INTRODUCTION

The numerical simulation of practically relevant flows often involves the handling of complex geometries and complex physical and chemical phenomena requiring the use of very fine grids and small time steps in order to achieve the necessary numerical accuracy. In recent years intensive research has been undertaken to improve the performance of flow computations in order to extend their applicability to a cost-effective solution of practically relevant flow problems. These improvements concerned both acceleration by the use of more efficient solution algorithms such as multigrid methods (e.g. [1, 10, 15]) and the acceleration by the use of more efficient computer hardware such as high-performance parallel computers (e.g. [18, 23]). In this paper a numerical solution method for the incompressible Navier-Stokes equations is presented, which combines efficient numerical techniques and parallel computing. A similar procedure for steady flows in simple orthogonal geometries is described in [22].

The underlying numerical scheme is based on a procedure described by Perić [19], consisting of a fully conservative second-order finite volume space discretization with a colocated arrangement of variables on non-orthogonal grids a pressure-correction method of SIMPLE type for the iterative coupling of velocity and pressure, and an iterative ILU decomposition method for the solution of the sparse linear systems for the different variables. For time discretization, implicit second-order schemes (fully implicit or Crank-Nicolson) are employed by means of a generalized three-level scheme. A nonlinear multigrid scheme, in which the pressure-correction method acts as a smoother on the different grid levels, is used for convergence acceleration. The general solution scheme is closely related to the work reported by Lilek et al. [16].

For the treatment of complex geometries, and as a base for the parallelization of the method, the concept of block-structured grids is used. With this approach complex geometries can be modelled easily, numerically efficient "structured" algorithms can be used within each block, regions with different material properties (e.g. solid/liquid problems) or problems requiring various grid systems moving against each other can be handled in a straightforward way and a natural basis for the parallelization of the solution method by grid partitioning is provided. Depending on the number of processors, the block structure suggested by the geometry is restructured by an automatic load-balancing procedure such that the resulting subdomains can be assigned suitably with the individual processors. The major objective of the parallelization strategy was to preserve the high numerical efficiency of the sequential method in the parallel implementation. Therefore, in particular, the parallel multigrid method is implemented globally, i.e. without being affected by the grid partitioning. This ensures a close coupling of the subdomains and only a slight deterioration of the numerical efficiency compared to the corresponding sequential algorithm can be observed.

The performance of the parallel algorithm was studied by a variety of numerical experiments for steady and unsteady flows These include investigations of the interdependence of the grid size, the time step size, the number of processors, the multigrid algorithm and grid partitioning. In this paper we concentrate more on the methodological aspects of the method to illustrate the intrinsic properties of the solution techniques employed. It should be noted that in addition to these studies, the code was also intensively validated by comparison with experimental results, and it was applied for the investigation of various practical fluid mechanical applications. Corresponding results can be found elsewhere, e.g. [3, 4, 6, 11, 12, 28], and will therefore not reported here. In general, the results indicate that parallel computers, combined with advanced numerical methods, can yield the computational performance required for an efficient, accurate and reliable solution of practical flow problems in engineering and science.

GOVERNING EQUATIONS AND BASIC NUMERICAL METHOD

We consider the time-dependent flow of an incompressible Newtonian fluid in an arbitrary three-dimensional domain. The basic conservation equations governing transport of mass and momentum are given by

$$\frac{\partial (\rho v_j)}{\partial x_j} = 0, \quad (1)$$

$$\frac{\partial (\rho v_i)}{\partial t} + \frac{\partial}{\partial x_j}\left(\rho v_j v_i - \mu \frac{\partial v_i}{\partial x_j}\right) + \frac{\partial p}{\partial x_i} - \frac{\partial}{\partial x_j}\left[\mu \left(\frac{\partial v_j}{\partial x_i}\right)\right] = s_{v_i}. \quad (2)$$

where $v = (v_1, v_2, v_3)$ is the velocity vector with respect to the Cartesian coordinates (x_1, x_2, x_3), t is the time, ρ is the density, μ is the dynamic viscosity, s_{v_i} are external forces and p is the pressure. In most applications, the above equations have to be complemented by one or more scalar equations in order to describe fully the flow problem. These are of the general form

$$\frac{\partial (\rho \phi)}{\partial t} + \frac{\partial}{\partial x_j}\left(\rho v_j \phi - D_\phi \frac{\partial \phi}{\partial x_j}\right) = s_\phi, \quad (3)$$

where ϕ is the transported variable (e.g. temperature or concentrations of species), s_ϕ are sources or sinks of ϕ and D_ϕ is a diffusion coefficient. When using a turbulence model ϕ

may represent turbulence quantities such as the turbulence kinetic energy, its dissipation rate or Reynolds stresses (see, e.g., [29]). Boundary and initial conditions for v and ϕ are required for the completion of the definition of the flow problem.

For the spatial discretization of Eqns. (1)-(3) a finite volume method with a colocated arrangement of variables is employed. The basic procedure was described in detail by Demirdžić and Perić [5], so only a brief summary is given here. The solution domain is discretized into hexahedral (in general non-orthogonal) finite volume cells. The convection and diffusion contribution to the fluxes are evaluated using a central differencing scheme, which for the convective part is implemented using the deferred correction approach proposed by Khosla and Rubin [13].

For the time discretization a generalized three-level scheme, which as special cases also includes common two-level schemes, is employed (e.g. [9]). Applying the scheme to the system of ordinary differential equations resulting from the spatial discretization, approximations v_h^n, p_h^n, and ϕ_h^n to the solution of Eqns. (1)-(3) at the time level $t_n = n\Delta t$ ($n = 1, 2, \ldots$) are defined as solutions of nonlinear algebraic systems of the form

$$\mathbf{L}_h v_h^n = 0, \tag{4}$$

$$(1+\xi)v_h^n + \theta \Delta t \left[\mathbf{A}_h(v_h^n)v_h^n + \mathbf{G}_h p_h^n\right] = (1+2\xi)v_h^{n-1} - \xi v_h^{n-2} + \Delta t \mathbf{F}_{vh} +$$
$$(\theta - 1)\Delta t \left[\mathbf{A}_h(v_h^{n-1})v_h^{n-1} + \mathbf{G}_h p_h^{n-1}\right], \tag{5}$$

$$(1+\xi)\phi_h^n + \theta \Delta t \mathbf{B}_h(v_h^n)\phi_h^n = (1+2\xi)\phi_h^{n-1} - \xi \phi_h^{n-2} + \Delta t \mathbf{F}_{\phi h}$$
$$(\theta - 1)\Delta t \mathbf{B}_h(v_h^{n-1})\phi_h^{n-1}. \tag{6}$$

The discrete operators $\mathbf{A}_h$, $\mathbf{B}_h$, $\mathbf{G}_h$, $\mathbf{F}_{vh}$, $\mathbf{F}_{\phi h}$ and $\mathbf{L}_h$ are defined according to the spatial discretization including the corresponding discretization of the boundary conditions. The parameter h is a measure of the spatial resolution (e.g. the maximum CV diameter) and $\Delta t > 0$ is the time step. The parameters θ and ξ are blending factors giving the possibility of controlling the time discretization. We note that the cases $(\theta, \xi) = (1, 0)$, $(\theta, \xi) = (1/2, 0)$ and $(\theta, \xi) = (1, 1/2)$ correspond to the first-order fully implicit Euler scheme, the second-order implicit Crank-Nicolson scheme and the second-order fully implicit scheme, respectively. All three methods are unconditionally stable, but it is well known that a different behaviour for spatially non-smooth solutions may occur (e.g. [9]).

The time stepping process defined by Eqns. (4)-(6) is started from initial values v_h^0 and ϕ_h^0 which are suitable spatial approximations of the initial values given for v and ϕ. Assuming that v_h^{n-1}, p_h^{n-1}, and ϕ_h^{n-1} have already been computed, we are faced with the problem of solving the nonlinear system of Eqns. (4)-(6) for v_h^n, p_h^n, and ϕ_h^n. For this a nonlinear full approximation multigrid scheme with a pressure-correction smoother is employed.

The smoothing procedure is based on the well-known SIMPLE algorithm proposed by Patankar and Spalding [17], but various modifications to the original method are introduced (a similar procedure for steady problems is described in [21]). The determination of $v_h^{n,k}$, $p_h^{n,k}$, and $\phi_h^{n,k}$ assuming that the corresponding values of the previous level $k-1$ have already been computed, is done in several steps leading to a decoupling with respect to the different variables.

In the first step, an intermediate approximation $v_h^{n,k-1/2}$ to $v_h^{n,k}$ is obtained by solving the momentum equation (5) with the pressure term, the source term and the matrix coefficients formed with values of the preceding iteration $k-1$, treating the non-orthogonal contribution in $\mathbf{A}_h^{n,k-1}$ explicitly, and introducing an under-relaxation. In the second step, corrections $\tilde{p}_h^{n,k}$ and $\tilde{v}_h^{n,k}$ are sought to obtain the new pressure $p_h^{n,k} = p_h^{n,k-1} + \tilde{p}_h^{n,k}$ and the new velocity $v_h^{n,k} = v_h^{n,k-1/2} + \tilde{v}_h^{n,k}$ exactly fulfilling the discrete continuity equation (4). By considering a modified momentum equation together with Eqn. (4), an equation for the pressure correction $\tilde{p}_h^{n,k}$ is derived, where a selective interpolation technique is used for making the cell face velocities dependent on the nodal pressures, which is necessary to avoid oscillatory solutions that may occur owing to the non-staggered grid arrangement (see [24]). To keep the structure of the pressure-correction equation the same as for the discrete momentum equation and to improve its diagonal dominance the contributions due to grid non-orthogonality are neglected (see, e.g., [20]). For the computation of the new pressure an under-relaxation is also employed.

Finally, the iteration step is completed by solving Eqn. (6) for $\phi_h^{n,k}$, where, as in the momentum equations, the non-orthogonal contribution in $\mathbf{B}_h^{n,k-1}$ is treated explicitly, and an under-relaxation is introduced.

The structure and the solution of the linear systems of equations for the different unknowns are discussed in the next section within the framework of the parallelization technique employed. Theoretical results concerning the convergence and smoothing properties of the considered pressure-correction approach were discussed by Wittum [30] in the more general setting of transforming smoothers.

The above pressure-correction procedure removes efficiently only those Fourier components of the error whose wavelengths are comparable to the grid spacing, which usually results in a quadratic increase in computing time with the grid spacing. To keep this increase close to linear, a full approximation multigrid scheme is applied directly to the nonlinear system of Eqns. (4)-(6), with the above pressure-correction scheme acting as a smoother for the different grid levels. For the movement through the grid levels the well known V-cycle strategy is employed, which in the case of steady problems is combined with the nested iteration technique (full multigrid) to improve the initial guesses on the finer grids (e.g. [8]). The problem on the coarsest grid is not solved exactly, but approximated by carrying out some pressure-correction iterations. Since the pressure-correction algorithm employed is not only a good smoother, but also a good solver, this does not cause a significant deterioration of the rate of convergence of the multigrid procedure. For prolongation and restriction, bilinear interpolation is used. A detailed decription of the multigrid technique employed can be found in [7] and is thus omitted here.

BLOCK-STRUCTURED GRID PARTITIONING AND LOAD BALANCING

Having discussed the general global solution procedure, we shall now look in more detail at the employed block-structured grid partitioning approach also used for the parallelization of the method. The principal idea is to transform the block structure, which results from the requirements to model the geometry (geometrical block structure), by some suitable mapping process to a new block structure (parallel block structure), which, in addition to the geometrical ones, also meets the requirements for efficient implementation on a parallel computer.

In general, depending on the number of geometrical blocks M and the number of available processors P, for the mapping process two situations have to be distinguished:

- if $M > P$, the geometrical blocks are suitably grouped together to have as many groups as processors and the resulting groups are assigned to the individual processors;

- if $M < P$, the geometrical blocks are partitioned in order to obtain finally a block structure with as many blocks as processors and the resulting blocks are assigned to the individual processors.

In the case $M = P$ no mapping has to be performed and the parallel block structure is taken to be identical with the geometrical structure.

For the parallel block structure, several requirements with respect to obtain an efficient parallel implementation can be formulated:

- a similar number of control volumes per processor to ensure good load balancing on the parallel machine;

- a small number of neighbouring blocks located on other processors to have few communication processes;

- short block interfaces along blocks located on different processors to have a small amount of data to transfer;

- avoidance of steep flow gradients across block interfaces in order to retain good coupling of the subdomains for a high numerical efficiency;

- fully automatic mapping process working for arbitrary M and P, which is not too time consuming compared with the flow computation.

It is obvious that, for general flow problems as considered in this paper, not all of these requirements can optimally be fulfilled simultaneously and, therefore, some compromise has to be found. Our approach is mainly based on the first and last of the above criteria. Topological and flow-specific aspects are not taken into account.

In the case $M > P$, the geometrical blocks are grouped into P subsets, such that the number of CVs is similar in all subsets. This is done by taking the lexicographically first block and adding as many other blocks provided that the number of CVs in this first subset is smaller than N/P, where N is the total number of CVs. Then the next block is assigned to the next subset and the procedure is repeated successively until P subsets are formed (for $P = 1$, this procedure also includes the serial case).

For the partitioning in the case $M < P$, two different strategies are implemented:

- Direct decomposition:
  The processors are assigned to the geometrical blocks according to the number of CVs in the blocks and the blocks are subdivided one-dimensionally in the coordinate direction with the largest number of CVs.

- Recursive decomposition:
  The block with the largest number of CVs, is halved in the direction with the largest number of CVs resulting in a new block structure. The process is repeated until the number of blocks equals the number of processors.

Which strategy is preferable depends on the problem geometry, the block structure used to model it and the number of available processors. An advantage of the recursive decomposition is that, at least theoretically, as many processors as there are control volumes on the coarsest grid can be used for the parallel computation. If direct decomposition is used, the number of processors assigned to a block may not be larger than the largest number of control volumes in one direction in this block. (However, from the numerical point of view, it is not reasonable to use too many processors for a grid of given size.) Results concerning a comparison of direct and recursive mapping can be found in [25]. Although the mapping approaches considered are relatively simple, they are very fast and have turned out to ensure good load balancing for a wide range of applications.

In addition to the determination of the parallel block structure, with respect to efficient parallelization the coupling of the blocks along the block interfaces, i.e. the transfer of information among neighbouring blocks, requires special attention. For this, along the block interfaces auxiliary control volumes containing the corresponding boundary values of the neighbouring block are introduced (see Fig. 1) which have to be updated in a suitable way. The crucial point in this respect is the parallelization of the sparse linear system solver. Since there are no recurrencies and owing to the auxiliary CVs, the parallelization of the other components of the method (assembly of the systems, prolongation, restriction, etc.) is straightforward. These operations can be done (in principle) simultaneously for all CVs and locally on the individual processors.

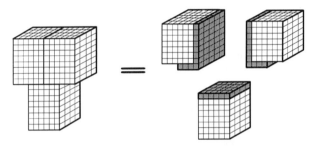

Figure 1: Block-structured grid partitioning and computational interface handling with auxiliary control volumes.

Numbering the unknowns in a natural way according to the block structure, i.e. lexicographically within each block, and then employing block by block subsequential numbering, the coefficient matrices of all linear systems (let us denote it commonly by $\mathbf{M}y = b$) have a block structure $\{\mathbf{B}_{ij}\}_{i,j=1,...,M}$ corresponding to the block-structured grid. As a linear system solver, a parallel variant of the ILU method of Stone [27] is employed. An iteration step of this method can be written in the form

$$y_i^{n+1} = y_i^n - (\mathbf{L}_i\mathbf{U}_i)^{-1}(\sum_{j=1}^{M} \mathbf{B}_{ij}y_j^n - b_i) \qquad (7)$$

where $\mathbf{L}_i\mathbf{U}_i$ are (local) ILU decompositions of the block diagonal matrices $\mathbf{B}_{ii}$. The computations in Eqn. (7) can be carried out concurrently for all $i = 1,\ldots,M$, if y^n is available in the auxiliary CVs for computing $\mathbf{B}_{ij}y_j^n$ for $i \neq j$. For this, the boundary data in the auxiliary control volumes have to be updated in each iteration.

An aspect of major concern related to these local communications is the prevention of dead locks, i.e. to ensure that the parallel computer does not enter in a state where the processes cannot continue computations because of misarrangement of the communcation processes. This problem is solved by setting up a schedule of the communication processes for every processor instead of following a specific algorithm for the communications, which would not cope with all complex types of block-structured grids that may occur. In addition to that, using this schedule approach, one is able to improve significantly the performance of the communication processes by doing communications in parallel as often as possible, which can be achieved by rearranging the entries of the schedule.

In addition to the local exchanges, some global information transfer of residuals is required for convergence checks and for the initial distribution and final collection of the data. As an example, in Fig. 2 the flow diagram of the parallel pressure-correction procedure is given when using it for computations with a k-ϵ model.

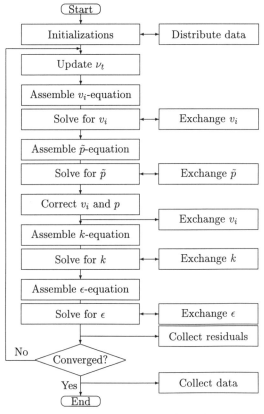

Figure 2: Flow chart of the parallel pressure-correction scheme for flow equations with k-ϵ turbulence modelling.

NUMERICAL RESULTS

a) Steady Laminar Flow

First, some results concerning the properties of our solution method for a steady problem with respect to the interaction of the multigrid technique, the parallelization and the number of CVs are presented. For this, the three-dimensional flow around a circular cylinder in a square channel at $Re = 20$ is considered, which represents one of the test

Table 1: Computing times and numbers of fine grid iterations (in parentheses) for single-grid (SG) and full multigrid (MG+NI) methods with corresponding acceleration factors (with respect to computing time) for different numbers of processors and control volumes for the flow around a circular cylinder

Method	6144 CV P=1	49152 CV P=8	393216 CV P=32	3145728 CV P=128
SG	1679(58)	6041(207)	40604(675)	–
MG+NI	624(25)	783(27)	1738(26)	2630(25)
Acc.	2.7	7.7	23.4	> 70

cases for benchmark computations reported by Schäfer and Turek [26]. In Fig. 3, part of the geometry is shown together with the computed pressure distribution on the cylinder surface, the iso-surface of zero velocity and stream ribbons giving an indication of the flow pattern.

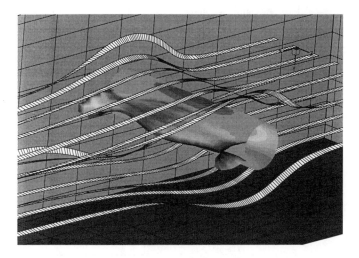

Figure 3: Pressure distribution on the cylinder surface, iso-surface of zero velocity and stream ribbons for the flow around a circular cylinder in a square channel (main flow direction from right to left).

The problem is computed with the multigrid method with nested iteration for different grid sizes, where the coarsest grid with 768 CVs and up to 5 grid levels are used. With increasing number of fine-grid CVs, the number of processors is also increased (not linearly). V-cycles with three pre-smoothing and three post-smoothing iterations on the finest grid (successively increasing by one on the coarser levels) are employed. The under-relaxation factors are 0.8 for velocities and 0.2 for pressure. In Table 1 the computing times and numbers of fine-grid iterations are given for the computations, and for comparison the corresponding values for the single-grid computation are also indicated. In Fig. 4 the computed drag and lift coefficients obtained for the different grid sizes are plotted.

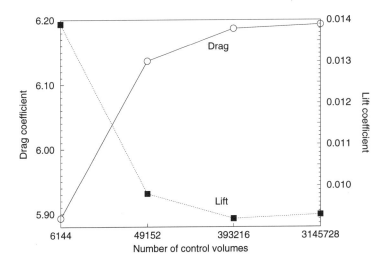

Figure 4: Computed drag and lift coefficients with different numbers of control volumes for the flow around a circular cylinder.

Several conclusions can be drawn from the results. The multigrid method is significantly superior to the corresponding single-grid computation. While the single-grid method shows a typical linear increase of iteration numbers with grid refinement, the iteration numbers for the multigrid method change only slightly. When the number of processors is increased for an increasing number of control volumes, the multigrid method gives a very good scale-up in the computing time, which, because of the increase in iteration numbers caused by the decoupling due to the grid partitioning, is not the case for the single-grid computations. If one compares, for instance, the results for $P = 8$ and $P = 32$, which correspond to grid sizes per processor differing by a factor of two, one can see that a good parallel efficiency is obtained. We have scale-ups of 97% for SG and 90% for MG+NI. The lower value when using MG is due to the increase in work on coarser grids, where the portion of communication relative to the arithmetic operations is larger. The given drag and lift coefficients, which are quantities of practical relevance in many applications, give an indication of the accuracy that can be achieved with the considered grids for such global values. Also, comparing the results in Table 1 and Fig. 4, the improved accuracy that can be achieved at the same computing time with the multigrid method compared with the single-grid method becomes apparent.

b) Unsteady Laminar Flow

For the investigation of the behaviour of our method for unsteady flows, we consider as a test problem the time-dependent two-dimensional natural convection in a square cavity with a circular obstacle. The cavity walls are at a fixed temperature and on the circle wall a time-dependent temperature $T_c(t) = 295 + sin(\pi t/50)$ is prescribed. The Prandtl number is $Pr = 6.7$ and the initial conditions at $t = 0$ are $v_1 = v_2 = T = 0$. The configuration and the corresponding complex flow structure are illustrated in Fig. 5, showing the predicted streamlines for two points of time corresponding to the temporal maximum and minimum value of the maximum vertical velocity.

Figure 5: Predicted streamlines for two points of time corresponding to the temporal maximum and minimum value of the maximum vertical velocity for natural convection flow in a square cavity with a circular obstacle.

For the multigrid method the coarsest grid with 80 CVs is employed and (3,3,3)-V-cycles are employed. The under-relaxation factors are 0.7 for velocities, 0.3 for pressure and 0.9 for temperature. In Table 2 the number of fine-grid iterations and the computing times (on a Parsytec MC-3 transputer system) are indicated for computing the flow from the initial state to $t = 200\,s$ for various grid sizes, time step sizes and processor numbers. As in the steady case, the iteration numbers increase only slightly with the processor number. The fine-grid iteration numbers decrease when the grid is refined, which seems to be due to the fact that more coarser grid levels are involved in the computation and less work has to be done on the fine grid. The smaller the time step size, the smaller is the iteration number per time step, but the total number of iterations, owing to the larger number of time steps, increases and, therefore, also the total computing time is longer. Of course, with a smaller time step also higher accuracy is obtained. Comparing, for instance, the computing times for 5120 CV and $P = 6$ with those for 20480 CV and $P = 24$, one can further see that a four times larger problem can be solved in approximately the same time with four times more processors.

For the above test problem, we also studied the performance of the parallel multigrid method in comparison with the corresponding parallel single-grid method. The flow was computed with both methods for three different processor numbers with a fixed ratio of CVs per processor. The computing times, the number of fine grid iterations and the acceleration factors obtained with the multigrid method are given in Table 3 for different time-step sizes. One can see the increase in the acceleration factor with the grid size. Again, with the multigrid method one can solve a larger problem in nearly the same amount of computing time when the processor number is increased by the same factor as the number of CVs. For larger time steps the multigrid acceleration factor does not depend significantly on the time-step size. If the time-step size becomes smaller than a certain limit, i.e. if the asymptotic range of convergence with respect to time is reached,

Table 2: Numbers of fine-grid iterations (top) and computing times (bottom) for different time-step sizes, number of CVs and processor numbers for natural convection flow in a square cavity with a circular obstacle

| | $\Delta t = 4.0$ | | | $\Delta t = 2.0$ | | | $\Delta t = 1.0$ | | |
Grid	P=6	P=12	P=24	P=6	P=12	P=24	P=6	P=12	P=24
1280 CV	1523	1523	1528	2656	2661	2674	4753	4754	4755
5120 CV	1410	1411	1423	2529	2529	2539	4617	4632	4628
20480 CV	1200	1204	1212	2001	2018	2034	3717	3721	3733
81920 CV	–	965	986	–	1801	1811	–	2847	2895

| | $\Delta t = 4.0$ | | | $\Delta t = 2.0$ | | | $\Delta t = 1.0$ | | |
Grid	P=6	P=12	P=24	P=6	P=12	P=24	P=6	P=12	P=24
1280 CV	0.6	0.4	0.4	1.0	0.7	0.7	1.9	1.3	1.2
5120 CV	2.0	1.2	0.8	3.6	2.1	1.5	6.5	3.9	2.8
20480 CV	6.4	3.5	2.1	10.5	5.7	3.4	19.8	10.8	6.4
81920 CV	–	10.4	5.7	–	19.4	10.5	–	29.9	16.3

the acceleration factor decreases with the time-step size. This limit value decreases with increasing grid size.

Table 3: Computing times (hours) and fine grid iteration numbers (in parentheses) for the multigrid and single-grid methods for different time-step sizes, number of CVs and processor numbers (the number of CVs per processor is fixed) and corresponding acceleration factors (with respect to computing time) for natural convection flow in a square cavity with a circular obstacle

Grid	Δt	SG	MG	SG/MG
20480 CV	4.0	25.5(3887)	11.2(1200)	2.3
($P = 3$)	2.0	43.1(6558)	18.6(2001)	2.3
	1.0	55.3(8417)	34.6(3717)	1.6
81920 CV	4.0	44.5(5937)	10.4(965)	4.3
($P = 12$)	2.0	87.1(11614)	19.4(1801)	4.5
	1.0	130.5(17396)	29.9(2847)	4.4
327680 CV	4.0	74.9(9727)	10.9(953)	6.9
($P = 48$)	2.0	–	19.6(1721)	–
	1.0	–	30.2(2658)	–

Results concerning the performance and accuracy of the method for three-dimensional time-dependent flows are reported in [26], where the unsteady flow around a circular cylinder in a channel is considered.

c) Turbulent Flow

To study the performance of the considered method for problems involving turbulence modelling, the turbulent flow in a axisymmetric bend is considered, which was set up as a benchmark case for the 2nd WUA-CFD Conference in May 1994. The bend consists of a circular cross-section entrance followed by an annulus in the opposite direction connected by a curved section of $180°$. Figure 6 shows the configuration together with the predicted turbulent kinetic energy and turbulent length scale when using the standard k-ϵ model with wall functions (according to [14]). The inflow conditions are $U_m = 1.0\,\text{m/s}$, $k = 0.01 U_m^2$ and $\epsilon = 10 k^{3/2}$, and the Reynolds number based on an entrance radius of $0.09\,\text{m}$ is $Re = 286000$.

The multigrid procedure is used with (20,20,20)-V-cycles with the coarsest grid of 256 CVs. This example was computed on a single processor (SUN10/20 workstation) in order to distinguish the effect of the multigrid method and the nested iteration from the effects of the parallelization (results for parallel turbulent flow computations, including also computations with second-order closure Reynolds stress modelling, can be found in [2]). The relaxation factors employed are 0.8 for velocity components, 0.2 for pressure and 0.9 for the turbulence quantities.

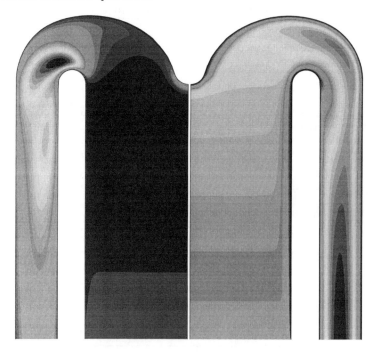

Figure 6: Predicted turbulent kinetic energy (left) and turbulent length scale (right) for the flow in the axisymmetric bend (symmetry axis in the middle).

In Fig. 7, the computing times and the numbers of fine grid iterations are given for the single-grid and multigrid methods, each with and without nested iteration, for different grid sizes. Although still significant, at least for finer grids, the multigrid acceleration is lower as in comparable laminar cases. This seems to be related to the very steep gradients occurring along the walls and to use the of the wall function approach. The acceleration factors increase nearly linearly with the grid level. The decrease in iteration numbers and computing times due to the nested iteration is larger for the single-grid than the multigrid case, and for all grids the full multigrid method is faster than the multigrid method without nested iteration.

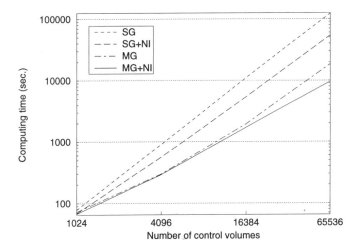

Figure 7: CPU times in minutes and numbers of fine-grid iterations (in parentheses) for single-grid (SG) and multigrid (MG) methods with and without nested iteration (NI) for different numbers of control volumes for the bend flow.

A general observation of the authors which was made during these and other computations of turbulent flows and which is worth noting is that, in general, the multigrid method stabilizes the computations (this is also the case for laminar computations). This means that the method is less sensitive with respect to the choice of under-relaxation factors, grid distortions and block decoupling than is the corresponding single-grid method.

CONCLUSIONS

We have presented a parallel implicit finite volume multigrid algorithm for the numerical prediction of laminar and turbulent flows in complex geometries. The results have shown that the applied block structuring technique is suitable for handling both complex geometries and parallel processing. In particular, it has turned out that the block-structured grid partitioning approach employed results in an efficient parallel implementation mostly retaining the high numerical efficiency of the powerful sequential multigrid solution procedure. The multigrid method, in spite of its slightly lower efficiency in terms of parallel computing, for both steady and unsteady flows is clearly superior to the corresponding single-grid method. It is also more robust with respect to the decoupling of the subdomains due to the grid partitioning and to grid distortions and under-relaxation factors, especially for turbulent flows. The problem of determining an optimal partitioning for general geometries is a very complicated task, but for a wide range of applications relatively simple strategies as considered in this work give satisfactory results.

In general, the results indicate that the increase in computer power due to MIMD parallel computers, combined with the acceleration of the multigrid technique, yields an improved computational performance, extending significantly the possibilities for reliable simulation of complex practical flow problems in engineering and science.

ACKNOWLEDGEMENTS

Financial support by the Deutsche Forschungsgemeinschaft in the Priority Research Programme Flow Simulation with High-Performance Computers and by the Bayerische Forschungsstiftung in the Bavarian Consortium of High-Performance Scientific Computing (FORTWIHR) is gratefully acknowledged.

REFERENCES

[1] L. Bai, K. Mitra, M. Fiebig, and A. Kost. A multigrid method for predicting periodically fully developed flow. *Int. J. Num. Meth. in Fluids*, 18:843–852, 1994.

[2] B. Basara, F. Durst, and M. Schäfer. A Parallel Multigrid Method for the Prediction of Turbulent Flows with Reynolds Stress Closure. In *Parallel Computational Fluid Dynamics 95*, Elsevier, Amsterdam, 1996, in press.

[3] U. Bückle, F. Durst, B. Howe, and A. Melling. Investigation of a floating element flowmeter. *Flow Meas. Instrum.*, 4:215–225, 1992.

[4] U. Bückle, Y. Katoh, M. Schäfer, K. Suzuki, and K.Takashiba. The Application of Modern Numerical Tools to the Investigation of Transport Phenomena Related to Czochralski Crystal Growth Processes. In *Proc. Int. Symp. on Heat and Mass Transfer, Kyoto*, 1994.

[5] I. Demirdžić and M. Perić. Finite volume method for prediction of fluid flow in arbitrary shaped domains with moving boundaries. *Int. J. Num. Meth. in Fluids*, 10:771–790, 1990.

[6] F. Durst, L. Kadinski, and M. Schäfer. A Multigrid Solver for Fluid Flow and Mass Transfer Coupled with Grey-Body Surface Radiation for the Numerical Simulation of Chemical Vapor Deposition Processes. *J. Cryst. Growth*, 146:202–208, 1995.

[7] F. Durst and M. Schäfer. A Parallel Blockstructured Multigrid Method for the Prediction of Incompressible Flows. *Int. J. for Num. Meth. in Fluids*, 22:1–17, 1996.

[8] W. Hackbusch. *Multi-Grid Methods and Applications*. Springer, Berlin, 1985.

[9] C. Hirsch. *Numerical Computation of Internal and External Flows*. Wiley, Chichester, 1988.

[10] M. Hortmann, M. Perić, and G. Scheuerer. Finite volume multigrid prediction of laminar natural convection: Benchmark solutions. *Int. J. Num. Meth. in Fluids*, 11:189–207, 1990.

[11] M. Hortmann, M. Pophal, M. Schäfer, and K. Wechsler. Computation of Heat Transfer with Methods of High Performance Scientific Computing. In B. Hertzberger and G. Serazzi, editors, *High-Performance Computing and Networking*, V. 919 of *Lecture Notes in Computer Science*, pp. 293–299, Springer, Berlin, 1995.

[12] M. Hortmann and M. Schäfer. Numerical prediction of laminar flow in plane, bifurcating channels. *Comput. Fluid Mech.*, 2:65–82, 1994.

[13] P. Khosla and S. Rubin. A Diagonally Dominant Second-Order Accurate Implicit Scheme. *Computers and Fluids*, 2:207–209, 1974.

[14] B. Launder and B. Spalding. The numerical computation of turbulent flows. *Comp. Meth. Appl. Mech. Eng.*, 3:269–289, 1974.

[15] F. Lien and M. Leschziner. Multigrid Acceleration for Recirculating Laminar and Turbulent Flows Computed with a Non-Orthogonal, Collocated Finite-Volume Scheme. *Comp. Meth. Appl. Mech. Eng.*, 118:351–371, 1994.

[16] Ž. Lilek, M. Perić, and V. Seidl. Development and Application of a Finite Volume Method for the Prediction of Complex Flows. In this publication.

[17] S. Patankar and B. Spalding. A calculation procedure for heat, mass and momentum transfer in three dimensional parabolic flows. *Int. J. Heat Mass Transf.*, 15:1787–1806, 1972.

[18] R. Pelz, A. Ecer, and J. Häuser (editors). *Parallel Computational Fluid Dynamics '92*. North-Holland, Amsterdam, 1993.

[19] M. Perić. *A Finite Volume Method for the Prediction of Three-Dimensional Fluid Flow in Complex Ducts*. PhD Thesis, University of London, 1985.

[20] M. Perić. Analysis of pressure-velocity coupling on nonorthogonal grids. *Num. Heat Transf., Part B*, 17:63–82, 1990.

[21] M. Perić, R. Kessler, and G. Scheuerer. Comparison of finie-volume numerical methods with staggered and colocated grids. *Computers and Fluids*, 16:389–403, 1988.

[22] M. Perić and E. Schreck. Computation of fluid flow with a parallel multigrid solver. *Int. J. Num. Meth. in Fluids*, 16:303–327, 1993.

[23] K. Reinsch, W. Schmidt, A. Ecer, J. Häuser, and J. Periaux (editors). *Parallel Computational Fluid Dynamics '91*. North-Holland, Amsterdam, 1992.

[24] C. Rhie and W. Chow. Numerical study of the turbulent flow past an airfoil with trailing edge separation. *AIAA Journal*, 21:1525–1532, 1983.

[25] M. Schäfer, E. Schreck, and K. Wechsler. An Efficient Parallel Solution Technique for the Incompressible Navier-Stokes Equations. In F.-K. Hebeker, R. Rannacher, and G. Wittum, editors, *Numerical Methods for the Navier-Stokes Equations*, V. 47 of *Notes on Numerical Fluid Mechanics*, pp. 228–238, Vieweg, Braunschweig, 1994.

[26] M. Schäfer and S. Turek. Benchmark Computations of Laminar Flow Around a Cylinder. In this publication.

[27] H. Stone. Iterative solution of implicit approximations of multi-dimensional partial differential equations. *SIAM J. Num. Anal.*, 5:530–558, 1968.

[28] K. Wechsler, I. Rangelow, Z. Borkowicz, F. Durst, L. Kadinski, and M. Schäfer. Experimental and Numerical Study of the Effects of Neutral Transport and Chemical Kinetics on Plasma Etching System CF_4/Si. In *Proc. Int. Symp. Plasma Etching*, pp. 909–914, 1993.

[29] D. Wilcox. *Turbulence Modeling for CFD*. DCW Industries, La Canada, 1993.

[30] G. Wittum. On the Convergence of Multi-Grid Methods with Transforming Smoothers. *Num. Math.*, 57:15–38, 1990.

Parallelization of Solution Schemes for the Navier-Stokes Equations

J. Hofhaus, M. Meinke, E. Krause

Aerodynamisches Institut, RWTH Aachen,
Wüllnerstraße zw. 5 u. 7,
52062 Aachen, Germany

Abstract

An explicit and implicit scheme for the solution of the Navier-Stokes equations for unsteady and three-dimensional flow problems were implemented on several parallel computer systems. The explicit scheme is based on a multi-stage Runge-Kutta scheme with multigrid acceleration, in the implicit scheme a dual time stepping scheme and a conjugate gradient with incomplete lower-upper decomposition preconditioning is applied. Two examples of complex flows were simulated and compared with experimental flow visualizations in order to demonstrate the applicability of the developed solution methods. Presented are the essential details of the solution schemes, their implementation on parallel computer architectures, and their performance for different hardware configurations.

Introduction

The numerical simulation of three-dimensional, steady, and especially unsteady flow problems requires extensive computer resources in terms of storage capacities and computational speed, so that efficient numerical methods are necessary. In the last decades the progress on both computer hardware and numerical methods was considerable. Parallel computers have the largest potential to increase the available computer power substantially in the next future. The combination of the today's hardware of supercomputers, like vector-parallel processors, with fast numerical methods, however, is in general not straightforward. The aim of this work was the parallelization of existing and the development of new algorithms for the solution of the conservation equations. Scientists of all related projects within this research program held a symposium on "Computational Fluid Dynamics on Parallel Systems" which is published in volume 50 of the present series, see [7].

Two finite difference solution methods for the Navier-Stokes equations, discretized on structured, curvilinear grids were implemented on different parallel computers with distributed memory: an explicit Runge-Kutta method with a possible multigrid acceleration to simulate three-dimensional, unsteady, and compressible flow problems and an implicit dual-time stepping scheme to solve the governing equations for incompressible fluids, based on the concept of artificial compressibility. First results of the explicit scheme for two-dimensional flow problems are given in [13], further experiences with three-dimensional flow problems and the implicit scheme are reported in [9] and [7]. In [8], several approaches for the parallel solution of tridiagonal equation systems that arise from the implicit discretization with an alternating line-relaxation method were investigated.

In this paper a parallelized version of the multigrid method for the explicit scheme and an iterative solution for the linear equation systems in the implicit scheme with a conjugate gradient method will be discussed in more detail. After a brief description of the solution methods, the spatial discretization, and the applied multigrid scheme, results of two typical three-dimensional flow phenomena are presented, which are qualitatively compared with experimental flow visualizations. Subsequently, the parallel approach through domain decomposition and a distributed algorithm for the solution of an equation system with a sparse coefficient matrix are introduced. Comparisons of the execution times on several serial and parallel computers, as well as the performance with a growing number of processors are given. The influence of the number of processors on the effect of the applied preconditioning technique is discussed.

Method of Solution

Governing Equations

In a dimensionless form the Navier-Stokes equations transformed in general, curvilinear coordinates ξ, η and ζ read:

$$\bar{A} \cdot \frac{\partial \vec{Q}}{\partial t} + \frac{\partial \vec{E}}{\partial \xi} + \frac{\partial \vec{F}}{\partial \eta} + \frac{\partial \vec{G}}{\partial \zeta} = 0 \quad . \tag{1}$$

They describe the conservation of mass, momentum, and energy in unsteady, three-dimensional and viscous flow. For a gaseous compressible fluid, $\bar{A}$ is the identity matrix, and the vector of the conservative variables multiplied by the Jacobian of the coordinate transformation J is given by:

$$\vec{Q} = J \left(\rho, \rho \vec{u}, \rho e\right)^T \quad .$$

Here, ρ denotes the fluids density, $\vec{u} = (u, v, w)^T$ the velocity vector, and e is the internal energy. The flux vectors $\vec{E}$, $\vec{F}$, and $\vec{G}$ are splitted in a convective and a viscous part, e.g.: $\vec{E} = \vec{E}_C - \vec{E}_V$, with

$$\vec{E}_C = J \begin{pmatrix} \rho U \\ \rho U u + \xi_x p \\ \rho U v + \xi_y p \\ \rho U w + \xi_z p \\ U(\rho e + p) \end{pmatrix} \text{, and } \vec{E}_V = \frac{J}{Re} \begin{pmatrix} 0 \\ \xi_x \sigma_{xx} + \xi_y \sigma_{xy} + \xi_z \sigma_{xz} \\ \xi_x \sigma_{xy} + \xi_y \sigma_{yy} + \xi_z \sigma_{yz} \\ \xi_x \sigma_{xz} + \xi_y \sigma_{yz} + \xi_z \sigma_{zz} \\ \xi_x E_{V_5} + \xi_y E_{V_5} + \xi_z E_{V_5} \end{pmatrix} \quad .$$

Herein, ξ_x, ξ_y, ξ_z are metric terms of the coordinate transformation, U the contravariant velocity, Re the Reynolds number and $\bar{\bar{\sigma}}$ the stress tensor. E_{V_5} is the dissipative part of the energy flux containing contributions of the stress tensor and the heat flux.

For incompressible flows with constant viscosity, the Navier-Stokes equations simplify significantly. The equation for energy conservation is decoupled from the equation for mass and momentum, and can be omitted, if the distribution of the fluids temperature is not of interest. The vector of the conservative variables in Eq. (1) is then reduced to:

$$\vec{Q} = J \left(p, \vec{u}\right)^T \quad .$$

The lack of a time-derivative for the pressure p in the continuity equation yields a singular matrix $\bar{A}$ and renders the integration of the governing equations more difficult. For fluids with constant density, the vectors of the convective and diffusive fluxes reduce to, e.g.:

$$\vec{E}_C = J \begin{pmatrix} U \\ Uu + \xi_x p \\ Uv + \xi_y p \\ Uw + \xi_z p \end{pmatrix} \text{, and } \vec{E}_V = \frac{J}{Re} \begin{pmatrix} 0 \\ g_1 u_\xi + g_2 u_\eta + g_3 u_\zeta \\ g_1 v_\xi + g_2 v_\eta + g_3 v_\zeta \\ g_1 w_\xi + g_2 w_\eta + g_3 w_\zeta \end{pmatrix},$$

with: $g_1 = \xi_x^2 + \xi_y^2 + \xi_z^2$, $g_2 = \xi_x \eta_x + \xi_y \eta_y + \xi_z \eta_z$, $g_3 = \xi_x \zeta_x + \xi_y \zeta_y + \xi_z \zeta_z$.

Temporal Discretization

A common technique to advance the solution in time for the simulation of compressible flows is the Runge-Kutta method. Here, a 5-step scheme was adapted for a maximum Courant number of 3.5 for the upwind scheme described below, for details see [13].

Chorin [5] proposed to introduce an artificial equation of state which couples the pressure- to an arbitrary density distribution in order to eliminate the singularity of the matrix $\bar{A}$. Hence, the continuity equation contains a time derivative for the pressure which vanishes for steady-state solutions and $\bar{A}$ in (1) is regular. In [3] this method was extended to unsteady flows by introducing an artificial time τ and adding a pseudo time derivative $\tilde{A} \cdot \partial \vec{Q}/\partial \tau$ to (1) such that $diag\{\tilde{A}\} = (1/\beta^2, 1, 1, 1)$, where β^2 controls the artificial compressibility. Thus, the pressure field is coupled to the velocity distribution and the governing equations can be integrated in a similar way as for compressible flows. Because a steady solution is computed within the pseudo time τ, the additional terms vanish, and the unsteady solution of (1) at the physical time t is obtained.

Subsequently, the Navier-Stokes equations with the artificial compressibility terms are discretized implicitly with respect to the artificial time τ. The convective and viscous fluxes are expanded in Taylor series in order to obtain a linear system of equations of the form:

$$LHS \cdot \Delta \vec{Q}^{(\nu)} = RHS \quad , \qquad (2)$$

which has to be solved in each artificial time-step. Here, $\Delta \vec{Q}^{(\nu)}$ is the change of the primitive variables within one time-step, LHS contains the discrete spatial derivatives of the Jacobian matrices which result from the linearization, and RHS contains the discrete derivatives in space of equation (1) and a second order approximation of the physical time-derivative. Details of the linearization and discretization of LHS are given in [3] and [4].

Spatial Discretization

To preserve the conservative properties in the discretized space, Eq. (1) are formulated for a finite control volume. A corresponding difference operator, e.g. δ_ξ for the determination of the flux derivatives for a control volume in a node-centered scheme at a point (i,j,k) reads:

$$(\delta_\xi \vec{E})_{i,j,k} = \frac{\vec{E}_{i+\frac{1}{2},j,k} - \vec{E}_{i-\frac{1}{2},j,k}}{\Delta \xi} \quad .$$

In both algorithms, a second-order upwind discretization for the convective terms is applied.

For compressible flows, Roe's approximate Riemann solver, [16], is used, extended by Harten's adapted flux approach [6] to achieve second order accuracy. Alternatively central difference schemes with artificial damping terms can be applied. The fluxes at the point $(i \pm \frac{1}{2}, j, k)$ for the Roe type flux difference splitting are computed by

$$\vec{E}_{i\pm\frac{1}{2},j,k} = \frac{1}{2}(\vec{E}_{i,j,k} + \vec{E}_{i\pm1,j,k}) - \frac{1}{2}\bar{R}|\bar{\Lambda}|\bar{R}^{-1}_{i\pm\frac{1}{2},j,k}(\delta_\xi \vec{Q})_{i\pm\frac{1}{2},j,k} \quad .$$

A non-linear upwind term, controlled by the product of Eigenvector matrix $\bar{R}$ times the matrix of absolute Eigenvalues $|\bar{\Lambda}|$ times the inverse of the Eigenvector matrix, is added to the averaged flux values on the surface on the control volume. Roe's average is used to determine the states at the cell interfaces.

For incompressible flows, the projection of the variables to the cell interfaces is carried out with the QUICK-scheme, proposed by Leonard [10]. In this method the upwinding is controlled by the contravariant velocities. For a simple one-dimensional model equation

$$f_t + u_0 f_x = 0$$

the projection of f to the cell face $i+1/2$ for a positive characteristic $u_0 > 0$ is given to:

$$f_{i+\frac{1}{2}} = \frac{1}{2}(f_{i+1} + f_i) - \frac{1}{8}(f_{i+1} + 2f_i - f_{i-1})$$

which can be interpreted as a linear interpolation, corrected by a term that is proportional to the curvature of f and dependent on the local direction of the characteristic u_0. The analogous extension of the QUICK-scheme to three-dimensional functions on curvilinear grids requires additional terms to describe the curvature of f in all three space-dimensions, see [15] and [17]. To avoid high-frequency oscillations in case of flows at high Reynolds numbers, a fourth-order damping term is added to the continuity equation, which is discretized with central differences, [4].

For both, compressible and incompressible flows, the Stokes stresses are discretized with central differences of second-order accuracy.

Direct Multigrid Method

The convergence rate of the explicit scheme is accelerated by a direct multigrid method described briefly as follows. For nonlinear equations, Brandt suggests in [2] the Full Approximation Storage (FAS) multigrid concept, which was adapted to the numerical solution of the Navier-Stokes equations. Considering a grid sequence $G_k, k = 1 \ldots, m$ with a step size $h_k = 2h_{k+1}$, a finite difference approximation on the finest grid G_m may be written as

$$L_m(\vec{Q}_m) = 0 \quad , \tag{3}$$

where L_m corresponds to the discretized governing equations and $\vec{Q}_m$ corresponds to the conservative variables on the grid m. If the solution on the finest grid is sufficiently smooth, equation (3) may be approximated on a coarser grid G_{k-1} by a modified difference approximation:

$$L_{k-1}(\vec{Q}_{k-1}) = \tau^m_{k-1} \quad , \tag{4}$$

where τ is the "fine to coarse defect correction", often reffered to as the "discretization error". It maintains the truncation error of the finest grid G_m on coarser grid levels and is defined as:

$$\tau^m_{k-1} = \tau^m_k + L_{k-1}(I^{k-1}_k \vec{Q}_k) - II^{k-1}_k L_k(\vec{Q}_k) \quad . \tag{5}$$

Herein, I_k^{k-1} and $II_k^{k-1} L_k$ are restriction operators from grid G_k to G_{k-1} which are applied on the conservative variables $\vec{Q}_k$ and the difference approximation $L_k(\vec{Q}_k)$, respectively. As in node-centered schemes grid points on coarser grids coincide with fine grid nodes, point-to-point injection was chosen for the variables $\vec{Q}$. For the difference approximation a weighted restriction over all 27 neighbouring nodes was applied, hence the operators read:

$$I_k^{k-1} \vec{Q}_k = \vec{Q}_k \quad \text{and} \quad II_k^{k-1} L_k(\vec{Q}_k) = \sum_{27 \text{ cells}} \beta \, L_k(\vec{Q}_k) \quad, \tag{6}$$

where the weighting factor β is proportional to the cells volume, i.e. $\beta = J_k / \sum_{27 \text{ cells}} J_k$.

When changing from coarser to finer grid levels, the variables are interpolated to the additional nodes. According to the FAS-scheme, only the correction between the "old" fine grid solution $\vec{Q}_{k+1}^{old}$ and the "new" coarse grid solution $\vec{Q}_k$ is interpolated to the finer grid to update $\vec{Q}_{k+1}^{old}$:

$$\vec{Q}_{k+1}^{new} = \vec{Q}_{k+1}^{old} + I_k^{k+1}(\vec{Q}_k - I_{k+1}^k \vec{Q}_{k+1}^{old}) . \tag{7}$$

Herein, I_k^{k+1} is an interpolation operator, for which trilinear interpolation is used. For more details on the multigrid scheme see [12] and [14].

Results of Flow Simulations

The above described methods for the solution of the Navier-Stokes equations were successfully applied to different steady, unsteady, compressible or incompressible flow phenomena in two and three space-dimensions, see e. g. [13], [9]. A steady flow in a bended tube for compressible flows and the unsteady collision of two vortex rings in incompressible flow are presented as examples for the different solution approaches.

Steady flow in a 90° bended pipe

The flow development in curved pipes with a circular cross-section has been a subject of various investigations because of its strong secondary motions in a cross-sectional plane. The secondary flow is of special interest, because on one hand it considerably enlarges the pressure losses and, therefore the costs, e.g. for large pipelines or chemical process plants. On the other hand it improves the heat transfer in refrigerating equipment or heat exchangers. In Fig. 1 streaklines visualized in an experiment are compared with the numerical simulation at a Reynolds number of 1610 (based on the mean axial velocity and the tubes diameter). The flow enters the bend on the upper left and leaves it on the right. The computations were carried out with the explicit solution method, a Mach number of 0.3, and on a grid with 65 grid points in main flow direction, and 33×33 grid points normal to it.

Due to the curvature of the stream lines, the pressure in the bend increases from the inner wall outwards, as shown in the right plot of Fig. 2, causing two different effects: the radial pressure gradient leads to an asymmetric profile of the axial velocity component, which can maintain up to 50-70 times the tubes diameter downstream of the outflow section. Furthermore, the flow close to the wall is decelerated such that the pressure gradient initiates a secondary flow towards the inner radius at the wall of the bend and back on its plane of symmetry. This results in two large counter-rotating vortices, indicated by the twisting of the streak lines downstream in Fig. 1. In Fig. 2 streaklines from the

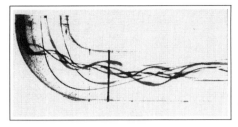 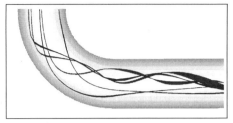

Figure 1: Streaklines in a flow through 90° curved pipe. Experiment left (Picture taken from [1]), simulation right.

experiment are compared with a vector-plot of the simulation in the outflow cross-section of the bend. Although the grid resolution was fairly coarse, the two smaller secondary vortices are also captured in the numerical solution.

 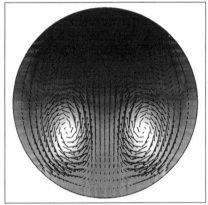

Figure 2: Flow pattern in the outflow cross section of a 90° curved pipe. The flow has entered the bend from below. Left: flow visualization with fluorescent dye in an experiment (picture taken from [1]). Right: numerical simulation: velocity vectors and pressure distribution from low (light grey) to high values (dark grey).

Oblique Collision of two vortex rings

The interest in the interaction of two vortex rings lies mainly in the examination of physical relations in flows with concentrated vorticity. Vortex rings are easily generated and forced to interact in a predetermined way. On the left-hand side of Fig. 3, pictures from an experiment described in [11] show the development of two vortex rings colliding under an angle. The rings were produced in a glass-sided water-filled tank by simultaneously pushing a certain amount of fluid through the two nozzles which can be seen in the background. In order to distinguish the two different rings, they were marked with different colours of dye (see colour plates in [11]).

As the rings approach each other, mutual cancellation of vorticity of opposite sign leads to a connection of the rings in the plane of contact, so that one larger ring is formed, Fig. 3 b. In the further development of the flow, the vortex ring assumes the

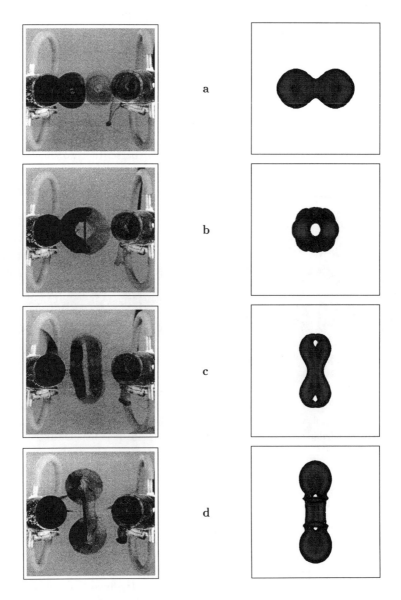

Figure 3: Collision of two vortex rings at an angle. Left: Flow visualization in an experiment (pictures taken from [11]). Right: numerical simulation: surfaces of constant pressure.

shape of an eight (3 c) and finally, two rings separate again in a plane perpendicular to the plane the original rings approached each other (3 d). This problem has a vertical plane of symmetry, that is why the final pair of rings are coloured half in their original colour and half in the colour of the other ring.

On the right-hand side of Fig. 3, three-dimensional surfaces of constant pressure of a numerical simulation of the oblique collision are plotted. The computations were obtained with the solution method for incompressible flows on a grid with $81 \times 81 \times 81$ grid points and a Reynolds number of 500, based on the radius of one initial vortex ring and its velocity of propagation. The essential mechanism of the rings connection and their subsequent separation in vertical direction are reproduced by the simulation.

Parallel Implementation

An optimal load balancing on a parallel computing environment requires a distribution of equal sized tasks on each processor. The easiest approach is the division of the domain of integration into several partitions, which are then mapped to the different processors of the multicomputer. Data dependencies at the partition boundaries require information from the neighbouring processes to compute the derivatives of the fluxes. To maintain the second order accuracy of the spatial discretization, overlapping grid lines -a boundary line on one partition corresponds to an interior line on the neighbouring partition- are established at the partition boundaries. Therefore, the structure of the parallelized program is simple, all flux integrals or derivatives can be computed independently, and the amount of data to be exchanged between the processors is reduced. However, flux integrals at the inner boundaries are computed twice by the neighbouring processors and the amount of arithmetic work increases with the number of processors. This domain splitting technique is applied to both algorithms using message passing routines.

In the explicit scheme the intermediate variables for a certain grid point for each Runge-Kutta step can be computed independently from the data of the neighbouring grid points. A parallelization of this algorithm is straightforward. For the implementation of the multigrid scheme, two different approaches are possible. For a transfer on a coarser grid, either all data from the finer grid is collected and redistributed to a possibly smaller amount of processors, or each processor generates its coarse grid locally. In the first method a multigrid cycle can be chosen exactly as if the algorithm is running on one processor, but global communication will be necessary and a perfect load balance will not always be achieved. In the second approach the maximum number of coarse grid levels that can be used is reduced by the number of partitions in each direction. This is simply due to the fact that each processor has to keep a minimum number of points on the coarsest grid. As this is only a problem when a grid with a small number of points is distributed on a large number of processors, the second method was chosen here.

In the implicit scheme, the linear equation system (2) has to be solved in each artificial time-step. The discretization on structured grids leads to a coefficient matrix which contains seven non-zero diagonals. In [3] an alternating Gauss-Seidel line relaxation method (LGS) was applied to solve (2). This requires the solution of a large number of block-tridiagonal systems, which can easily be solved in serial algorithms with the LU-decomposition. The memory requirements are comparatively small, because it is not necessary to store the complete coefficient matrix. The recursive structure of this method, however, prevents a direct parallelization. In [8], six different parallel solution methods to solve tridiagonal equation systems were implemented to the Navier-Stokes solver and compared. The results show that it is difficult to obtain an efficient solution method, especially for a large number of processors, because the number of unknowns in each tridiagonal system is fairly small.

A more efficient approach to solve sparse linear equation systems in parallel are schemes, which are based on the method of conjugate gradients. Beside an improved convergence rate, these algorithms consist of a sequence of matrix-vector and inner products of which parallelization is straightforward. As the coefficient matrix in (2) is non-symmetric, the Bi-CGStab method proposed by van der Vorst [18] was chosen. Preliminary tests confirmed that a sufficient convergence rate could only be achieved with a proper preconditioning of the equation system. For linear problems the efficiency of this method can be several orders of magnitude higher than the classical line-relaxation method. In this application, however, the solution of the linear equation system is part of an iterative procedure to solve a nonlinear problem. Herein, the linear equation system should not be solved considerable more accurate than the current residual of the nonlinear problem, namely the RHS in equation (2). Therefore, the gain in execution time for the solution of the Navier-Stokes equations compared to a line-relaxation scheme is reduced to a factor of 2-4.

The preconditioner used here is based on an Incomplete Lower-Upper decomposition (ILU) in which the entries of the triangular matrices show the same pattern as the coefficient matrix. The ILU method is recursive, but all operations in a plane with the index $i + j + k = const$ (where i, j, and k are indices for the ξ, η, and ζ direction, respectively) can be performed independently. A vectorization in these planes is therefore possible, see e.g. [19], a parallelization, however, requires a distribution of the grid normal to these index planes, which would significantly complicate the computation of the flux integrals and the communication procedures. To circumvent this problem and to maintain the initial partitioning concept, a local ILU-decomposition was implemented, in which those matrix entries, which couple the equations across the process boundaries, are neglected. Due to the local application of the preconditioning the convergence rate of the solution scheme cannot be independent of the grid partitioning and becomes worse with a growing number of processors.

The memory requirements for both the Bi-CGStab method and the ILU-preconditioning are significantly larger than for the LGS-method, because it is necessary to store the complete block coefficient matrix and its ILU-decomposition. In the artificial compressibility approach, the pressure and the velocity components are fully coupled, therefore, each entry in the matrix is a 4×4 block. The necessary amount of storage words per grid point increases in this implementation from ≈ 50 for the LGS-scheme up to ≈ 260 for the preconditioned Bi-CGStab scheme.

Results

Both solution schemes were implemented on several serial and parallel machines with distributed memory. The parallel computers used here can roughly be subdivided into two classes: computers with a small number of fast vector processors, e. g. SNI VPP500-4, which are characterized by a low ratio of communication versus arithmetic speed, and massively parallel systems with a large number of scalar processors, e.g. the Power-PC based GC-series from Parsytec, which naturally entail a finer granularity of the compu-

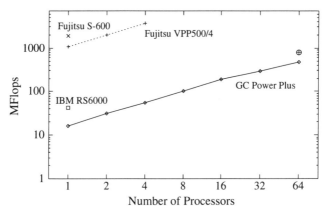

Figure 4: Computational speed for the explicit scheme

tations. In the following, the efficiency of a parallel program running on P processors is defined to be the ratio of the sequential execution time to P times the parallel execution time.

Explicit Solution Scheme

The flow in a 90° bended pipe described above served as a test problem for the explicit solution scheme. The grid size was kept constant at $34 \times 34 \times 65$ grid points. In Fig. 4 the computational speed obtained on two serial computers -the SNI S600/20 equipped with a Fujitsu vector processor and a scalar IBM RS6000 workstation- and two parallel computers -the SNI VPP500 and the Parsytec GC running PARIX 1.3- are compared.

The values for parallel tests were determined with the help of the MFlop counter and the CPU-time of the VPP500 on a single processor. Therefore the MFlop rates given in Fig. 4 do not take into account the parallelization overhead due to the double calculations. The true MFlop rate on a parallel machine can roughly be obtained by division with the efficiency value (see, e.g., point "⊕" for 64 processors of the Power PC based GC machine), because the communication overhead in explicit schemes is small compared to double calculations.

For a constant workload per processor, the efficiency of this program is nearly independent of the number of nodes, as shown in [9]. Therefore, the decisive quantity is the local problem size and the results in Fig. 4 may be used to estimate the performance for any number of grid points distributed to any number of processors.

In Fig. 5, the residuals for different grid levels and multigrid cycles are plotted versus the number of workunits needed to compute the flow in the 90° bended pipe. Obviously, significant improvements of the numerical efficiency can be achieved with the multigrid method. For the chosen grid size and the considered flow problem the best convergence rate could be achieved with 3 grid levels.

One goal of the parallelization of the multigrid scheme was to achieve convergence rates independent from the number of processors. This requires additional communication of data on the coarse grid levels. Both the location and the values of the conservative variables change for the overlapping grid lines, if the grid level is changed. Switching

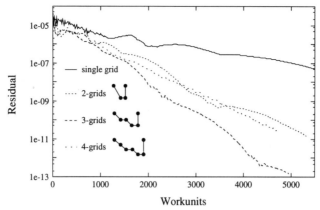

Figure 5: Convergence for different multigrid cycles

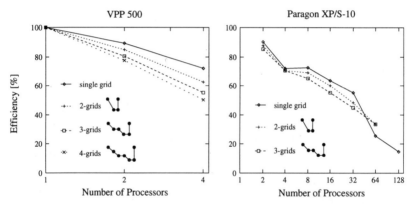

Figure 6: Efficiency of the explicit method for different multigrid cycles

from a finer to a coarser grid, an additional exchange of boundary values between the processors is necessary to compute the defect correction in (5), i.e. the discretization error of the finer grid τ_{k-1}^m, the restricted values of the variables $I_k^{k-1}\vec{Q}_k$, and the difference approximation $L_k(\vec{Q}_k)$ on the finer grid. During the switch from a coarse to a finer grid level, communication of the new variables $\vec{Q}_k$ obtained on the coarse grid is necessary to compute the interpolation in equation (7) and the new variables on the fine grid.

The efficiency of the explicit method for different multigrid cycles obtained on the SNI VPP500 and the Intel Paragon XP/S-10 are plotted in Fig. 6. The largest impact on the efficiency losses are caused by double calculations, see [9] and [13]. For more than 32 processors the parallel efficiency is drastically reduced, because the local problem size becomes very small. The difference in the results obtained on the different computers is explained by the different processor-to-communication speed. The performance losses due to the additional communication for the multigrid method is nearly proportional to the number of grid levels used. Furthermore, the reduced vector length on coarser grid levels and on smaller local grids decrease the arithmetic speed on the vector-parallel

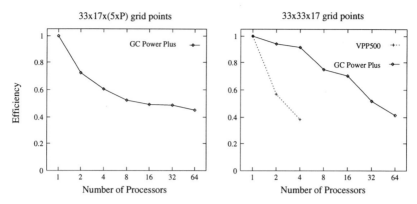

Figure 7: Efficiency of the Bi-CGStab method. Left: for a constant workload per processor. Right: for a problem of fixed size.

processors of the VPP500. The improved numerical efficiency of the multigrid method nevertheless compensates these losses many times, which is also illustrated in Table 1. Multigrid methods are, therefore, well suited to accelerate the convergence rate also on vector-parallel architectures.

Table 1: CPU-seconds on the SNI VPP500 needed to drop the residual by four orders of magnitude

grid level	1 Processor	4 Processors
single grid	2504 sec. (=1)	868 sec. (=0.35)
3 grid levels	922 sec. (=0.37)	424 sec. (=0.17)

Implicit Solution Scheme

The solution of the sparse linear equation system (2) requires the largest amount of the arithmetical work in the implicit solution method. The performance losses caused by double computations of the flux integrals in the parallel program correspond to those in the explicit scheme, but are negligible compared to the efficiency losses due to the communication in the solution of the equation systems.

Hence, the discussion in this section concentrates on the solution of the sparse equation system that has to be solved in each artificial time step. A detailed discussion of the results obtained for the different concurrent tridiagonal solvers applied to the line relaxation method are given in [8] and are not repeated here. The Bi-CGStab method with a local ILU-preconditioning was implemented on the Fujitsu VPP500 and the Parsytec GC.

As mentioned above, the method of conjugate gradients consists of matrix-vector and inner products. All floating point operations for the matrix-vector product can be computed in parallel, provided that first of all the boundary values of the vector have been exchanged between neighbouring processors. The floating point operations of the inner product of two vectors is also a local problem, the global sum, however, requires a fan-in and fan-out communication, here performed with a binary-tree topology. Since the ILU-decomposition is applied locally, no double calculations exist and performance losses are exclusively caused by communication routines.

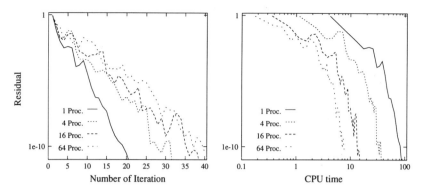

Figure 8: Residual of the iterative solution of one equation system with the Bi-CGStab method for different numbers of processors on the GC Power Plus. Left: plot over the number of iteration steps. Right: over the CPU times needed.

Test runs with a constant workload per processor were carried out for a boundary layer flow over a flat plate. The problem size was chosen to be $33 \times 17 \times (P \times 5)$ grid points, where P is the number of processors. On the left-hand side of Fig. 7 the efficiency values are plotted over the number of processors. The performance decreases until a number of eight processors is reached. This is explained by additional communication steps, because the grid is successively splitted into all three space dimensions. For partitions with more than sixteen nodes the efficiency decreases slightly, likely because the costs for the global sum of each inner product increases. The results confirm that the algorithm is scalable and suited for a large number of processors. Again, the local problem size has a decisive influence on the performance of the scheme.

On the right-hand side of Fig. 7 the efficiency values for a fixed problem size are plotted versus the number of processors. The plot shows, that the ratio of arithmetical to communicational speed has a significant impact on the performance. On the VPP500 the vector length is reduced with the growing number of processors, which further decreases the parallel efficiency. Taking into account the fairly small number of unknowns due to the limited memory equipment of one processor, the results obtained on the GC Power Plus are satisfactory.

The influence of the local ILU-decomposition on the development of the residual for the solution of one linear equation system is given in Fig. 8. The linear equation system that was solved for these plots originated from the first time step of the simulation of a boundary flow over a flat plate, where free-stream conditions were prescribed in the whole flow field except for the no-slip condition on the surface of the plate. Consequently, the coefficient matrix is strongly inhomogeneous and the convergence rate suffers from a poor preconditioning. As expected, the neglection of the coupling coefficients distinctly reduces the positive effect of the preconditioning steps with an increasing number of processors (Fig. 8 left), especially if one is interested to solve the system accurate. Using 64 processors, approximately two times the number of iterations on 1 processor were necessary to drop the residual by more than three orders of magnitude. However, as described in the previous section, it is not necessary to solve the system more accurate than the current residual RHS of the non-linear problem, which is $\approx 10^{-1}$ in this case.

The solution of the linear equation system should therefore already be stopped after a few iteration steps, so that the losses in general are considerable smaller. In fact, the overall execution time of the above stated flow problem was best, when using only one or two iteration steps to solve each linear equation system. In that case, the difference in the convergence rate for a large number of processors is small compared to the sequential program. As the plot on the right of Fig. 8 confirms, significant improvements of the execution time can nevertheless be achieved through the use of parallel computers.

Conclusion

Two different methods for the solution of the Navier-Stokes equations implemented on parallel computers were presented. A considerable reduction of the execution times through parallelization was obtained. Both solution schemes proofed to scale with the number of processors used, such that the decisive quantity for the parallel efficiency is the local problem size.

The convergence rate of the explicit solution scheme was improved by a multigrid method. To maintain the convergence acceleration in the parallel program, additional communication is necessary during the change of grid levels. Since the efficiency losses are mainly caused by double computations on overlapping grid lines, the additional communication only slightly decreases the performance of the multigrid method. Special care must be taken, however, for a sufficient performance of this method on vector-parallel machines.

The computationally most expensive part in the implicit scheme is the solution of a large sparse equation system in each physical time step. A conjugate gradient method together with a local ILU preconditioning shows a better performance as an alternating line-relaxation method at the expense of a higher memory requirement. The matrix-vector and inner products have a low arithmetic-to-communication ratio, which reduces the efficiency especially for a small local problem size. Furthermore it was shown how the local preconditioning decreases the convergence rate for an increasing number of processors. How to formulate an efficient preconditioner with a convergence rate independent from the number of processors is still an open question.

References

[1] B. Bartmann and R. Neikes. Experimentelle Untersuchung verzweigender Innenströmung II. FKM/AIF-Vorhaben Nr.113, Abschlußbericht, 1991.

[2] A. Brandt. Guide to multigrid development. In *Lecture Notes in Mathematics*, pages 220–312. Springer Verlag Berlin, 1981.

[3] M. Breuer and D. Hänel. A dual time-stepping method for 3-d, viscous, incompressible vortex flows. *Computers Fluids*, 22(4/5):467–484, 1993.

[4] Michael Breuer. *Numerische Lösung der Navier-Stokes Gleichungen für dreidimensionale inkompressible instationäre Strömungen zur Simulation des Wirbelplatzens*. Dissertation, Aerodynamisches Institut der RWTH-Aachen, Juni 1991.

[5] A.J. Chorin. A Numerical Method for Solving Incompressible Viscous Flow. *J. Comput. Phys.*, 2:12–26, 1967.

[6] A. Harten. On a Class of High Resolution Total-Variation-Stable Finite Difference Schemes for Hyperbolic Conservation Laws. *SIAM J. Numer. Anal.*, 21, 1984.

[7] J. Hofhaus and M. Meinke. Parallel solution schemes for the navier-stokes equations. In S. Wagner, editor, *Computational Fluid Dynamics on Parallel Systems*, volume 50 of *Notes on Numerical Fluid Mechanics*, pages 88–96. Vieweg Verlag, 1995.

[8] J. Hofhaus and E.F. Van De Velde. Alternating-Direction Line-Relaxation Methods on Multicomputers. *SIAM J. Sci. Comput.*, 1996. in press.

[9] E. Krause, M. Meinke, and J. Hofhaus. Experience with parallel computing in fluid mechanics. In *ECCOMAS 94*. John Wiley & Sons, Ltd., 1994.

[10] B. P. Leonard. A stable and accurate convective modelling procedure based on quadratic upstream interpolation. *Computer Methods in Applied Mechanics and Engineering*, 19:59, 1979.

[11] T. T. Lim. An Experimental Study of a Vortex Ring Interacting with an Inclined Wall. *Experiments in Fluids*, 7:453–463, 1989.

[12] M. Meinke and D. Hänel. Time accurate multigrid solutions of the Navier-Stokes equations. In W. Hackbusch and U. Trottenberg, editors, *Multigrid Methods III*, pages 289–300. Birkhäuser Verlag, 1991.

[13] M. Meinke and E. Ortner. Implementation of Explicit Navier-Stokes Solvers on Massively Parallel Systems. In E. H. Hirschel, editor, *DFG Priority Research Programme, Results 1989-1992*, volume 38 of *Notes on Numerical Fluid Mechanics*, pages 138–151. Vieweg Verlag, 1993.

[14] Matthias Meinke. *Numerische Lösung der Navier-Stokes-Gleichungen für instationäre Strömungen mit Hilfe der Mehrgittermethode*. Dissertation, Aerodynamisches Institut der RWTH-Aachen, März 1993.

[15] Y. Nakamura and Y. Takemoto. Solutions of Incompressible Flows Using a Generalized QUICK Method. In *Numerical Methods in Fluid Mechanics II, Proc. of the Int. Symp. on Comput. Fluid Dynamics, Tokyo*, Sept. 9-12 1985.

[16] P. L. Roe. Approximate riemann solvers, parameter vectors, and difference schemes. *Journal of Computational Physics*, 43, 1981.

[17] Y. Takemoto and Y. Nakamura. A three-dimensional incompressible flow solver. *Lecture Notes in Physics*, 264:594–599, 1986. Springer Verlag.

[18] H.A. Van Der Vorst. BI–CGSTAB: A Fast and Smoothly Converging Variant of BI–CG for the Solution of Nonsymmetric Linear Systems. *SIAM J. Sci. Stat. Comput.*, 13(2):631–644, 1992.

[19] S. Yoon and D. Kwak. Three-Dimensional Incompressible Navier-Stokes Solver Using Lower-Upper Symmetric-Gauss-Seidel Algorithm. *AIAA Journal*, 29(6):874–875, 1991.

3-D NAVIER-STOKES SOLVER FOR THE SIMULATION OF THE UNSTEADY TURBOMACHINERY FLOW ON A MASSIVELY PARALLEL HARDWARE ARCHITECTURE

Karl Engel, Frank Eulitz

Institute for Propulsion Technology

Stefan Pokorny, Michael Faden

Department for High Performance Computing

DLR, German Aerospace Research Establishment
51147 Cologne, Germany

SUMMARY

An interactive flow simulation system has been developed by the authors to study and analyze unsteady flow phenomena in turbomachinery components. The flow solver along with a data processing unit has been parallelized based on domain decomposition using the communication libraries PVM or MPI. The flow solver can be run on heterogeneous workstation clusters and on various parallel hardware platforms.

For full scalability, time integration is based on explicit, fully local algorithms. Stability is enhanced through a time accurate two-grid approach or a dual time stepping procedure. Convection is handled by second order TVD upwind schemes. The in- and outflow boundaries are treated with a non-reflecting boundary condition technique. Turbulence is accounted for by using the Spalart-Allmaras one-equation model.

Various examples of application are presented to demonstrate the usefulness of the developed simulation system.

INTRODUCTION

Modern compressor and turbine design trends are towards size and weight reduction. As a result of fewer stages with fewer blades per row and a narrowing of the axial gap between the blade rows, the aerodynamic loading of the blades increases. In particular, unsteady effects will gain in magnitude and in importance for the design.

Presently, many research activities are concerned with unsteady flow phenomena in turbomachinery components. However, the impact on the future design is not always clear (Adamczyk [1]) and a detailed investigation of the effects seems to be helpful for a better understanding.

Classically, experiment, theory and CFD form the basis of turbomachinery flow investigations.

In some cases the computational approach is found to be the only valid tool left for the analysis of highly non-linear and unsteady flows in complex geometries. This is particularly true when unsteady flow data of rotating configurations are needed as these can hardly be accessed with instrumentation.

The perfect numerical simulation, however is hampered by the complex flow physics of unsteady, transonic, three dimensional, viscous flows itself and the limited computational resources regarding CPU power and storage capacity.

A natural architecture for the solution of large scale problems is a parallel computer since, from the theoretical point of view, it can be scaled arbitrarily with the problem size presuming that the code is fully scalable. This is even true for turbomachinery applications: For more passages or more stages to compute, more processors with independent memory have to be added.

Based on this, a parallel and interactive flow simulation system **T.R.A.C.E.** (Turbomachinery Research Aerodynamics Computational Environment) has been developed by the authors (Engel [10], Eulitz [11], Faden [13]). This work is a part of the subproject "flow simulation with massiveley parallel systems" of the DFG research project "Strömungs-simulation mit Hochleistungsrechnern".

The flow solver along with a data processing unit has been implemented in a hardware independent manner using PVM [17]. The used explicit discretization method based on a purely local method guarantees a high degree of concurrency. Thus, parallelization is based on domain decomposition and is implemented using the message-passing model. The resulting code may be easily ported to almost any parallel machine including workstation clusters and it is especially suited for distributed memory architectures. The code runs either on IBM RS6000, SUN Sparc 20 and SGI SC 900 workstations or workstation-clusters simply by changing one program executable parameter which indicates the amount of usable hardware nodes. It was shown by the authors [6] that the parallel program implemented this way is efficient and scales over a wide range of processors.

FLOW SOLVER

The three-dimensional time dependent, Reynolds-averaged Navier-Stokes equations are solved for the compressible ideal gas in conjunction with a turbulence model. The equation system to be solved reads in general curvilinear strong conservation form in a rotating frame of reference

$$\frac{\partial \hat{Q}}{\partial t} + \frac{\partial (\hat{F} - \hat{F}_v)}{\partial \xi} + \frac{\partial (\hat{G} - \hat{G}_v)}{\partial \eta} + \frac{\partial (\hat{H} - \hat{H}_v)}{\partial \zeta} = \hat{S} \qquad (1)$$

$$\hat{Q} = \frac{1}{J}\begin{bmatrix} \rho \\ \rho u \\ \rho v \\ \rho w \\ e \end{bmatrix}, \quad F^\kappa = \begin{bmatrix} \rho U^\kappa \\ \rho u U^\kappa + \kappa_x p \\ \rho v U^\kappa + \kappa_y p \\ \rho w U^\kappa + \kappa_z p \\ (e+p) U^\kappa - \kappa_t p \end{bmatrix}, \quad F_v^\kappa = \begin{bmatrix} 0 \\ \mu m_1 u_\kappa + (\mu/3) m_2 \kappa_x \\ \mu m_1 v_\kappa + (\mu/3) m_2 \kappa_y \\ \mu m_1 w_\kappa + (\mu/3) m_2 \kappa_z \\ \mu m_1 m_3 + (\mu/3) m_2 (\kappa_x u + \kappa_y v + \kappa_z w) \end{bmatrix} \qquad (2)$$

with the contravariant velocity

$$U^\kappa = \kappa_x u + \kappa_y v + \kappa_z w + \kappa_t \qquad (3)$$

and the abbreviations

$$m_1 = \kappa_x^2 + \kappa_y^2 + \kappa_z^2$$
$$m_2 = \kappa_x u_\kappa + \kappa_y v_\kappa + \kappa_z w_\kappa \qquad (4)$$
$$m_3 = (u^2 + v^2 + w^2)_\kappa/2 + (Pr(\gamma-1))^{-1}(a^2)_\kappa .$$

Here, κ denotes any of the three curvilinear coordinates ξ, η, ζ.
The source term

$$\hat{S} = [0, 0, \rho(\Omega^2 y + 2\Omega w), \rho(\Omega^2 z - 2\Omega v), 0] . \qquad (5)$$

in the y and z direction represents the centrifugal and Coriolis forces.

The variables $t, \rho, u, v, w, e, p, T$ denote the non-dimensional time, density, Cartesian velocity components, specific total energy, pressure and temperature, respectively. The pressure

$$p = (\gamma-1)\left[e - \frac{1}{2}\rho(u^2+v^2+w^2) + \frac{1}{2}\rho\Omega^2\right] \qquad (6)$$

and the rothalpy

$$I = \frac{\gamma}{\gamma-1}\frac{p}{\rho} + \frac{1}{2}(u^2+v^2+w^2) + \frac{1}{2}\Omega^2 \qquad (7)$$

are functions of the rotation rate.

The non-dimensional molecular viscosity is calculated from Sutherland's formula,

$$\mu = T^{3/2}\frac{1 + 110/\tilde{T}_0}{T + 110/\tilde{T}_0}, \qquad (8)$$

where $\tilde{T}_0$ is the dimensional total temperature.

Numerical Model

The convective fluxes are discretized using Roe's upwind scheme [18] which is combined with van Leer's [23] MUSCL extrapolation (**M**onotonic **U**pstream **S**cheme for **C**onservation **L**aws) to obtain second order accuracy in space. The numerical flux at the cell interface is expressed as a function of the conservative states left and right of the interface:

$$\begin{aligned}\hat{F}_{j+\frac{1}{2}} = \frac{1}{2}\Bigg(&\left(\frac{\xi_x}{J}\right)_{j+\frac{1}{2}}\left[F^L_{j+\frac{1}{2}} + F^R_{j+\frac{1}{2}}\right] \\
+ &\left(\frac{\xi_y}{J}\right)_{j+\frac{1}{2}}\left[G^L_{j+\frac{1}{2}} + G^R_{j+\frac{1}{2}}\right] \\
+ &\left(\frac{\xi_z}{J}\right)_{j+\frac{1}{2}}\left[H^L_{j+\frac{1}{2}} + H^R_{j+\frac{1}{2}}\right] \\
+ &\left(\frac{\xi_t}{J}\right)_{j+\frac{1}{2}}\left[Q^L_{j+\frac{1}{2}} + Q^R_{j+\frac{1}{2}}\right] - \frac{1}{J}\hat{D}_{j+\frac{1}{2}}\Bigg). \end{aligned} \qquad (9)$$

The $\hat{G}$- and $\hat{H}$ fluxes are assembled similarly. The last of the numerical flux constituents is the upwind dissipation term which is defined at the cell interface $j+1/2$ as

$$\hat{D} = \hat{R}\Lambda\hat{R}^{-1}(Q^R - Q^L),\quad (10)$$

where the right eigenvector matrix is evaluated via Roe's averaging of the states U_j and $U_{j+1/2}$. Λ is a diagonal matrix containing the entropy-corrected eigenvalues of the inviscid part of equation system (1). According to Yee [25], one is free to choose the type of variable set to extrapolate. Here, the left and right *primitive* states $U = [\rho, u, v, w, p]^T$ are MUSCL-extrapolated, i.e.

$$U^R_{j+\frac{1}{2}} = U_{j+1} - \frac{1}{4}\left[(1-\kappa)\Delta'_{j+\frac{3}{2}} + (1+\kappa)\Delta''_{j+\frac{1}{2}}\right],$$

$$U^L_{j+\frac{1}{2}} = U_j + \frac{1}{4}\left[(1-\kappa)\Delta''_{j-\frac{1}{2}} + (1+\kappa)\Delta'_{j+\frac{1}{2}}\right]. \quad (11)$$

By variation of the parameter κ, the space discretisation can be made fully upwind ($\kappa=-1$), or upwind biased ($\kappa=1/3$ or 0). The mark indicates limited differences $\Delta_{j+1/2} = U_{j+1} - U_j$ of the flow quantities,

$$\Delta'_{j+\frac{1}{2}} = L\left(\Delta_{j+\frac{1}{2}}, \omega\Delta_{j-\frac{1}{2}}\right),$$

$$\Delta''_{j+\frac{1}{2}} = L\left(\Delta_{j+\frac{1}{2}}, \omega\Delta_{j+\frac{3}{2}}\right), \quad (12)$$

where L denotes a nonlinear limiter function. The viscous fluxes are discretized using central differences.

Time-Integration

The flow solver makes use of different time integration methods during different phases of the simulation process. To efficiently compute an initial condition for the unsteady simulation, the flow solver is run in the steady-state multistage mode which is accelerated by local time stepping and implicit residual smoothing.

For the unsteady calculation, a two-grid time accurate acceleration method following He [16], is used. This two-grid time marching method is based upon a linear multigrid method. It allows for an extension of the stability region using the following integration model to advance the solution in time (from time step n to $n+1$)

$$(U^{n+1} - U^n) = \frac{R_f}{\Delta A_f}\frac{CFL_0}{CFL}\Delta t + \frac{R_c}{\Delta A_c}\left(1 - \frac{CFL_0}{CFL}\right)\Delta t, \quad (13)$$

where R_f is the residual on the fine and R_c on the coarse mesh, respectively. This time-integration is embedded into a second-order time accurate explicit Runge-Kutta four-stage scheme. Due to the local character of this method full scalability can be maintained.

The best results for a turbulent calculation are obtained using the upwind-tuned Runge-Kutta coefficients (v. Lavante et al. [22]), for $\kappa=0$ in equation (11), $\alpha_1=0.11$, $\alpha_2=0.255$, $\alpha_3=0.46$, and $\alpha_4=1$ and by setting the coefficient $\varepsilon = 10^{-5}$ in equation (13).

Alternatively, the dual-time stepping method presented by Arnone et al. [3] can be used for the time accurate calculation. In this method, the basic equation system (1) is discretized in an implicit manner

$$\left.\frac{\partial \hat{Q}}{\partial t}\right|^{n+1} = -R_t(\tilde{Q}^{n+1}) \tag{14}$$

and then reformulated within the so called "pseudo time" τ

$$\frac{\partial \hat{Q}}{\partial \tau} = \left.\frac{\partial \hat{Q}}{\partial t}\right|^{n+1} + R_t(\tilde{Q}^{n+1}) = R_\tau(\tilde{Q}) . \tag{15}$$

In the pseudo time, the Residual $R_\tau(\tilde{Q})$ is driven to zero to satisfy the implicit three point rule

$$\frac{\partial \hat{Q}}{\partial \tau} = \left.\frac{\partial \hat{Q}}{\partial t}\right|^{n+1} + R_t(\tilde{Q}^{n+1}) = \frac{3\tilde{Q}^{n+1} - 4\tilde{Q}^n + 3\tilde{Q}^{n-1}}{2\Delta t} + R_t(\tilde{Q}^{n+1}) = R_\tau(\tilde{Q}) = 0 \tag{16}$$

for the physical time step. For this, the above mentioned steady state acceleration techniques are used.

Boundary Conditions

The treatment of the computational domain boundaries deserves special attention for time-accurate flow calculations since inappropriate boundary conditions may distort the solution considerably if not give non-physical solutions.

Solid Boundary

At the solid boundary, mirror cells are introduced to provide the physical values for the interior scheme. These are defined by the no slip condition. The wall pressure is found via the momentum equation normal to the wall.

Inflow/Outflow Boundary

Prone to unwanted artificial reflections, the inflow and outflow boundaries were found to be the most critical for the quality of the unsteady flow field.

For the steady-state multistage calculation quasi-three dimensional steady non-reflecting boundary conditions according to Saxer and Giles [19],[14] are used at the inlet and outlet boundary of the computational domain. In this technique, the solution vector at the boundary is circumferentially decomposed into Fourier modes. Whereas the zeroth mode is essentially given by the prescribed boundary conditions, the higher harmonics are modelled by the small disturbance ansatz in order to avoid unphysical reflections. In the implemented version, at the inflow the average characteristics are calculated from the requirement that the average entropy, radial and circumferential flow angle as well as the average total enthalpy have certain values.

$$\begin{aligned} \bar{S}^{n+1} &= S_{inlet} \\ \bar{\alpha}_\theta^{n+1} &= (\alpha_\theta)_{inlet} \\ \bar{\alpha}_r^{n+1} &= (\alpha_r)_{inlet} \\ \bar{H}^{n+1} &= H_{inlet} . \end{aligned} \tag{17}$$

At the exit boundary the average change of the upstream running acoustic wave is defined by the exit pressure specified at a certain radius

$$\bar{p}(r = radius) = p_{outlet} \tag{18}$$

where the radial distribution of the pressure is given by the radial equilibrium.

$$\frac{\partial \bar{p}(r)}{\partial r} = \bar{\rho}\frac{\bar{v}_\theta^2}{r}. \tag{19}$$

For the unsteady simulation the three-dimensional non-reflecting boundary conditions according to Acton, Cargill and Giles [2] are used. Here, the amplitudes of the upstream and downstream running wave patterns are calculated via a partial differential equation system in order to prevent unphysical reflections at the boundaries.

$$\frac{\partial}{\partial t}\begin{bmatrix}\tilde{c}_1\\ \tilde{c}_2\\ \tilde{c}_3\\ \tilde{c}_4\\ \tilde{c}_5\end{bmatrix} = \begin{bmatrix}\bar{v} & 0 & 0 & 0 & 0\\ 0 & \bar{v} & \frac{\bar{c}+\bar{u}}{2} & 0 & \frac{\bar{c}-\bar{u}}{2}\\ 0 & 0 & \bar{v} & 0 & 0\\ 0 & \frac{\bar{c}-\bar{u}}{2} & 0 & \bar{v} & 0\\ 0 & \bar{u} & 0 & 0 & \bar{v}\end{bmatrix}\frac{\partial}{\partial y}\begin{bmatrix}\tilde{c}_1\\ \tilde{c}_2\\ \tilde{c}_3\\ \tilde{c}_4\\ \tilde{c}_5\end{bmatrix} + \begin{bmatrix}\bar{w} & 0 & 0 & 0 & 0\\ 0 & \bar{w} & 0 & 0 & 0\\ 0 & 0 & \bar{w} & \frac{\bar{c}+\bar{u}}{2} & \frac{\bar{c}-\bar{u}}{2}\\ 0 & 0 & \frac{\bar{c}-\bar{u}}{2} & \bar{w} & 0\\ 0 & 0 & \bar{u} & 0 & \bar{w}\end{bmatrix}\frac{\partial}{\partial z}\begin{bmatrix}\tilde{c}_1\\ \tilde{c}_2\\ \tilde{c}_3\\ \tilde{c}_4\\ \tilde{c}_5\end{bmatrix}. \tag{20}$$

Mixing plane/Intergrid Boundary

During the steady-state calculation the zeroth mode is prescribed by a circumferential flux-averaging technique [15] in order to preserve mass, momentum and energy.

$$\begin{bmatrix}\partial\tilde{c}_1\\ \partial\tilde{c}_2\\ \partial\tilde{c}_3\\ \partial\tilde{c}_4\\ \partial\tilde{c}_5\end{bmatrix} = \begin{bmatrix}-\bar{c}^2 & 0 & 0 & 0 & 1\\ 0 & 0 & \bar{\rho}\bar{c} & 0 & 1\\ 0 & 0 & 0 & \bar{\rho}\bar{c} & 1\\ 0 & \bar{\rho}\bar{c} & 0 & 0 & 1\\ 0 & -\bar{\rho}\bar{c} & 0 & 0 & 1\end{bmatrix}\begin{bmatrix}\bar{p}_{up}-\bar{p}_{down}\\ \bar{v}_{x,up}-\bar{v}_{x,down}\\ (\bar{v}_\theta+\Omega r)_{up}-(\bar{v}_\theta+\Omega r)_{down}\\ \bar{v}_{r,up}-\bar{v}_{r,down}\\ \bar{p}_{up}-\bar{p}_{down}\end{bmatrix}, \tag{21}$$

For the unsteady computation, the sheared-cell technique is used. Here, the cell sides at the interface are shifted with the moving grids (Fig.1).

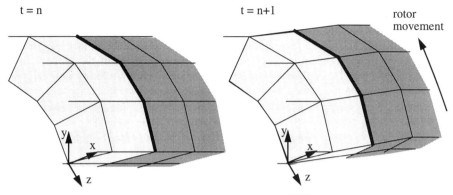

Fig. 1: Sheared cells technique between the rotor ▓ and the stator ☐ fitted grid.

This procedure allows for a time accurate coupling, preserving mass, momentum and energy.

Treatment of Turbulence

With the specific turbomachinery problem in mind a *two-layer formulation* of the Spalart and Allmaras (SA) turbulence model [21] for the eddy viscosity is proposed. It avoids effects of viscous damping of the original model where not desired. In addition the two-layer formulation will allow for a stable explicit time integration.

The version of the SA model used here is valid for free shear flows and the near-wall region outside the viscous sublayer and buffer layer,

$$\frac{Dv_t}{Dt} = c_{b1}|\Omega|v_t + \frac{1}{\sigma}[\nabla(v_t \cdot \nabla v_t) + c_{b2}(\nabla v_t)^2] - c_w f_w \left[\frac{v_t}{d}\right]^2, \qquad (22)$$

where $v_t, |\Omega|, d$ denote the eddy viscosity, magnitude of vorticity and wall distance, respectively. The negative term in equation (22) is an inviscid wall destruction term which in combination with the viscous region ensures a log layer behaviour in a classical boundary layer situation. Its function f_w reads

$$f_w = g\left[\frac{1+c_{w3}^6}{g+c_{w3}^6}\right]^{1/6}, \quad g = r + c_{w2}(r^6 - r), \quad r = \frac{v_t}{|\Omega|\kappa^2 d^2}. \qquad (23)$$

For the treatment of the near-wall region of finite Reynolds Number, i.e viscous sublayer and buffer layer, Spalart and Allmaras write the base equation (22) in terms of an auxiliary variable which is related to the eddy viscosity by a damping function. The damping function is constructed such that the new variable equals v_t except in the viscous region. This approach is not followed here as it also affects for example the outer edge of a wake migrating through several rotor and stator passages. Instead, the well established Prandtl-van Driest formulation is adopted for the viscous region

$$v_t = l^2|\Omega|, \quad l = \kappa d[1 - \exp(-y^+/A^+)]. \quad (24)$$

An improved version of (24) can also be used which takes into account effects of the pressure gradient and compressibility [5]. The inner formulation is active for $y+$ less than 8, while the outer formulation dictates the evolution of the eddy viscosity for $y+$ greater than 30. In between, the minimum of the two-formulations is taken. In effect, the matching of the inner (24) and outer formulation (22) is thus placed near the outer edge of the buffer layer. In practice, only the first 3 to 5 nodes away from the wall are given by the inner formulation. As this matching criterion may prove insufficient for complicated flows it should be accepted as provisional.

The following boundary conditions for the eddy viscosity are prescribed,

$$v_t = v_{t,freestream}, \quad \text{at the inflow boundary} \quad (25)$$

$$\frac{\partial^2 v_t}{\partial \xi^2} = 0, \quad \text{at outflow boundary } \eta = const. \quad (26)$$

All the calculations presented here were conducted with the same set of constants as given by Spalart and Allmaras,

$$c_{b1} = 0.1355, \sigma = 0.667, c_{w1} = 3.239, c_{b2} = 0.622, \kappa = 0.41, c_{w2} = 0.3, c_{w3} = 2, A^+ = 26.$$

Additional terms which provide control over laminar regions and smooth transition to turbulence are omitted here for brevity.

Equation (22) is cast into discrete form using first order upwind finite differences for the convective term and second order central differences for the diffusive term. The solution is advanced in time with the same Runge-Kutta scheme as employed by the Navier-Stokes solver. Good stability characteristics up to a CFL number of 4 to 6 are obtained when this is done in a strictly coupled manner, that is, when the flow solver and the turbulence equation are updated after each Runge-Kutta stage. So far, numerical studies indicate an acceptable insensitivity of the model to the initial condition. Regardless of the specific application, the integration is started by setting the eddy viscosity to an uniform freestream value of 10^{-4}. A precalculation with an algebraic model is thus not necessary. For an application to problems with both moving and stationary solid boundaries, equations (22) and (24) are formulated in a relative frame of reference.

A comparison of several one-equation models and the standard k-ε model for some free shear flows leads Birch [4] to the conclusion that one-equation models which solve for the eddy viscosity itself "give overall performances close to those of a standard k-ε model". The better numerical robustness of a model equation for only a single variable is obvious. The Galilean invariance property makes the SA model suitable for moving grid problems. As the model shares the treatment of the inner layer with an algebraic model it does not require any finer mesh resolution.

So far, numerical studies carried out by the authors (Eulitz [12]) indicate that the turbulence model is reasonably accurate, grid independent, and insensitive to a specified initial condition. It works well with the used acceleration techniques. Due to its local character, the implementation on a parallel local memory hardware architecture poses no additional problem.

PARALLELIZATION AND EXECUTION CONTROL STRUCTURE

Basic Design

Besides the numerical procedures to integrate the flow field and to implement the physical boundary conditions a subsystem was designed to allow for the execution within a distributed environment and to control the setup and execution of the entire simulation.

Parallelization of the flow solver is based on domain decomposition. The resulting multi-block structure of the code is extended by communication procedures to be executed on a parallel computer. Each processor of this parallel computer integrates one step on every local block and subsequently triggers exchange of the boundary values or application of the physical boundary conditions respectively.

The overall execution of the simulation process is described by a simple control language. It is used to describe and set all parameters as well as to control the execution and I/O for the entire program. The number of processors is specified on program start-up. No recompilation is necessary to run the program on a different number of processors.

The next paragraph will describe the design goals and the implementation of this subsystem.

Parallelization

The major goals of the data structures and procedures for the parallel execution of the program are:

1. Allow complete flexibility on the number of processors used. Any number from one to the number of grid blocks must be allowed.
2. Easy and efficient implementation on a parallel computer based on a message passing system.
3. Easy porting to another platform.
4. Minimize communication needs by replicating all necessary control information and system parameters over all processors.

To achieve the required flexibility all parameters describing one grid block are held in a list on every processor of the parallel computer. During startup memory for the physical values of the flow field is allocated on every processor for the blocks which are implemented on the current processor.

The connectivity information between the blocks is used to setup the interblock communication. Data exchange will then be automatically handled by the appropriate message passing system when the processor on which the destination block is implemented on differs from the location of the source block. Since all this functionality is hidden inside the parallel subsystem the numerical application remains unchanged and thus portable no matter what message passing library is used or how many processors are employed.

All the buffers needed for the asynchronous communications are allocated during the startup phase and remain available during the runtime of the system. This assures the necessary efficency and the minimal overhead caused by the additional software layers. Since all information is stored on the source and destination processors sends and receives can be paired directly without the need to communicate buffer sizes or similar parameters.

The load per block is primarily determined by the number of grid points inside the block and by the type and extent of the physical boundary conditions for that block. Optimal load balancing for a simulation is roughly achieved by evenly distributing grid points over the processors. This may be obtained in general by distributing more than one grid block on each processor so that the total number of grid points per processor is approximately constant on all processors of the parallel computer. The uniformity which is achived corresponds directly to the load balance achieved.

Solver Control

The initialization and execution of the code is implemented using a simple scripting language. All addressing of physical domains is based on the single block which will receive the appropriate command no matter where the block is actually implemented. This gives the required flexibility and independence from the number of processors used for the computation. Further flexibility is achieved by the possibility of grouping blocks. Commands send to these groups are automatically distributed to all the members of the group wherever they are implemented.

Execution of the commands is divided in three groups, immediate, delayed, and periodic commands. Immediate commands are executed as soon as they are specified. Delayed commands require an aditional parameter stating the time step at which they are to be executed. When the solver reaches the specified time step the command will automatically executed. Periodic commands are specified with two additional parameters. The first one states the first time step number at which the command will become executed the first time. The second parameter states the increment from the current active step to the next one.

This mechanism allows the flexible and easy to use specification of setup and execution of entire simulations. All addressing of commands remain independent from the actual structure of the parallel computer and the implementation of the single blocks.

VALIDATION

Reliable experimental data for unsteady turbomachinery flow as for example through a compressor or turbine are hardly available. The problem of validation of the flow solver is therefore even more critical than it is for steady-state applications where well established experiments exist. Therefore, related experiments which bear one or the other similarity to turbomachinery flow with respect to Mach number, Reynolds number, or reduced frequency are considered for validation. Typical turbomachinery flow is mainly turbulent. Consequently, also the validity of the chosen turbulence model and its combination with the flow solver needs to be assessed.

In the following, a selected validation case is presented to give an example. For more validation, the interested reader is referred to Engel [7] and Eulitz [11], [12] where various inviscid and viscous test cases with relevance for the turbomachinery application are discussed.

Self-excited shock oscillation about a DCA profile

The transonic flow about a double-circular-arc profile (DCA profile) in a two-dimensional channel is considered as sketched in figure 2. The case has been studied in great detail both experimentally and theoretically by Yamamoto and Tanida [24]. Self-excited flow oscillation, due to the interaction of the shock system with the wake and the boundary layer, can be observed for a well defined range of the channel exit pressure. The shock (mean) position which is controlled by the exit pressure determines the frequency and amplitude of the oscillation. As the experimental data are obtained in a channel of square section our two-dimensional results will be compared for the same shock mean position and not for the same exit pressure.

For the simulation, a mesh of two H-blocks of 60x160 nodes each is constructed around a DCA profile of 10% thickness and split into 8 subdomains. The Reynolds number based on entry velocity and chord length is 1,000,000. As in the experiment, laminar-turbulent transition is enforced at 10% chord length. The channel flow is choked with an entry Mach number of 0.68. A steady-state solution, generated by prescribing a stable exit pressure, serves as an initial condition for the time accurate integration.

Without any forcing or triggering, a periodic solution develops if an unstable exit pressure (0.705) is prescribed. The physical instability is solely fed by numerical round-off errors. The solution has become perfectly periodical at a normalized time of about 70 as can be seen from the residual plot in figure 4. The residual history for the lower half which is not shown here looks exactly the same except for a phase shift of 180 degrees. It is found that every two humps of the residual graph represent a cycle of the oscillation where the minima mark the reverse points of the shock motion. The instantaneous iso-density plots shown in figure 5 compare well against the Schlieren photographs by Yamamoto and Tanida. The predicted frequency of 595 Hz, as found from the residual history, is close to the corresponding experimental value of 600 Hz. In figure 7, phase plots of the pressure halfway between the channel centerline and upper wall. The present results can be further compared with calculations by other authors using the Baldwin-Lomax model [16], [24].

The case of self-excited shock oscillation was computed on 8 processors of an IBM-SP2 and an effective speed-up of 7.5 was obtained.

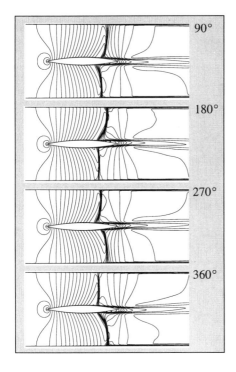

Fig. 3: Instantaneous density contours at different phase angles within a period of the self-excited shock oscillation, Re = 1,000,000.

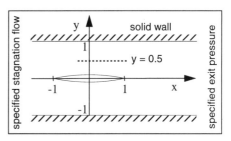

Fig. 2: Arrangement of test case.

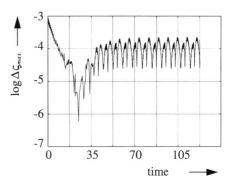

Fig. 4: Residual history for flow field on upper side of DCA profile.

APPLICATION

The flow solver system has been successfully applied to a wide range of unsteady flow problems in turbomachinery components, as for instance, to the aerodynamic interaction in a transonic compressor stage [7][9], to the wake interaction in a subsonic turbine stage [10], and to the study of rotating stall in a compressor stage and is now becoming the DLR standard code for unsteady turbomachinery flow problems.

In the following, a current application of technical relevance is presented and discussed qualitatively to demonstrate the usefulness of the developed flow simulation system. The shown results are part of a common research project with the MTU Motoren und Turbinen Union.

Investigation of Turbine Clocking

Experiments carried out at NASA MSFC during 1991 [20] indicate that the performance of a turbine can be optimized by appropriate positioning of the turbine stators, i.e. by appropriate clocking. More precisely, "clocking" terms the relative position of a stator blade to another downstream or upstream stator blade. So far, the underlying physical mechanism responsible for the efficiency variation is not fully understood. Although it may be mainly a stage interaction phenomenon, the role of secondary flow features and flow unsteadiness still has to be clarified. It is hoped that a better understanding of this phenomenon may lead to an improved turbine design.

For a detailed numerical study, a model turbine with a first stator, a rotor and a second (downstream) stator is considered as shown in figure 5.

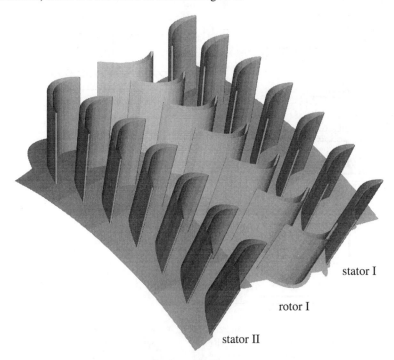

Fig. 5: Model turbine stage for the investigation of the clocking effect.

To obtain an acceptable initial solution for the unsteady simulation, a steady-state multistage calculation using the physical boundary conditions of the operating point is performed. Figure 6 shows the pressure contours in the relative frame of reference of the first stator and the first rotor for this steady state multistage calculation.

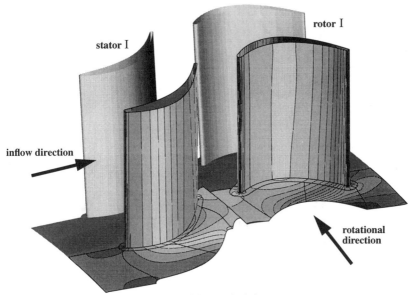

Fig. 6: Pressure contours for a steady-state multistage calculation.

With this solution as an initial guess, a time periodic does not evolve before the wake of the first stator is convected through the remaining blade rows. The evolution of the axial mass flow over roughly 5 blade passing periods in the rotor entry and exit plane, figure 7 and figure 8 respectively, indicates a fair degree of periodicity.

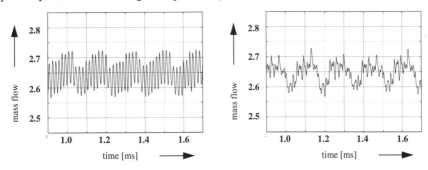

Fig. 7: Evolution of nondimensional massflow at rotor entry.

Fig. 8: Evolution of nondimensional massflow at rotor exit.

To analyse the overall effects of clocking on the multistage wake interaction and the aerodynamic losses, the specific entropy $s = \log(\kappa p) - \kappa \log(\rho)$ is considered. Iso-contours of this quantity are provided in figure 9 for the clocking position with 0% and 50% shifting at the time periodic operating point. Large scale vortex shedding at the trailing edges is observed for the first stator and the rotor. Its frequency shows up in figure 7 and figure 8 as large amplitude oscillations and is found to be independent of the rotor motion.

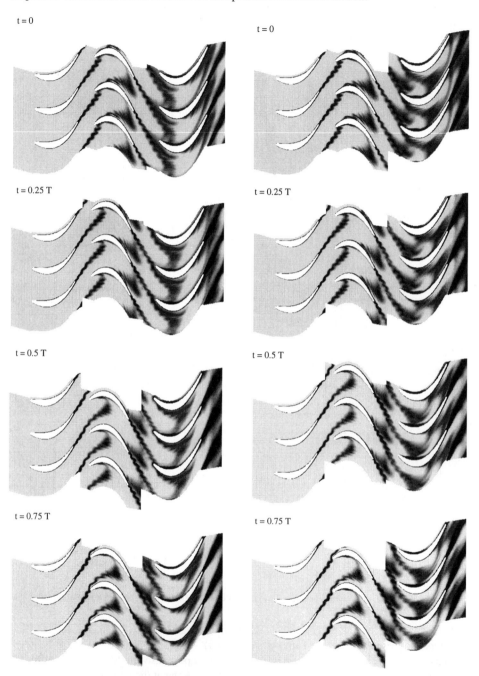

Fig. 9: Instantaneous specific entropy contours during one blade passing period fot the clocking position of 0% and 50% shift of the second stator row.

In the rotor passage, the wake is subject to a pressure gradient in axial and tangential direction and deforms accordingly. The interaction of the wake with the boundary layer on the pressure side of the rotor blade gives rise to an unsteady flow separation and reattachment. Downstream of the trailing edge, the first stator wake merges with the rotor wake. Moving further downstream, the wakes induce an oscillating separation bubble on the pressure side of the stator blade. The flow is further accelerated in the contracting stator passage. Note, that the second stator wake does not exhibit any oscillation. Possibly, the vortex shedding is suppressed by the favourable pressure gradient and interaction with the wakes from the upstream blade rows. Looking at the first stator and the rotor (in figure 9), the solution for the second clocking position is found to be in agreement with the first clocking position. In the second stator, however, the entropy distribution differs substantially from the previous case.

CONCLUSIONS

An interactive flow simulation system has been developed to study and analyze unsteady flow phenomena in turbomachinery components. The flow solver and a data processing unit have been parallelized based on domain decomposition using public domain communication libraries. This allows for use of heterogeneous workstation clusters and various parallel hardware platforms.

To guarantee a high degree of concurrency the time integration is based on explicit, fully local algorithms. To advance the efficiency a time accurate two-grid approach or a dual time stepping procedure can be used. Convection is handled by second order TVD upwind schemes. The in- and outflow boundaries are treated with a non-reflecting boundary condition technique. Turbulence is accounted for by using the Spalart-Allmaras one-equation model.

The system has proven its usefulness as a research tool for the investigation of unsteady turbomachinery flow. This is documented by recent applications of industrial relevance as for example the aerodynamic interaction in a transonic compressor stage, the wake interaction in a subsonic turbine or first studies on rotating stall.

The future development will concentrate on alternative numerical solution methods and refinement of the turbulence modeling with particular emphasis on laminar-turbulent transition.

REFERENCES

[1] ADAMCZYK, J.J., GREITZER, E.M., WISLER, D.C., "Unsteady Flow in Turbomachines: Where's the Beef ?", Proceedings of the ASME Symposium: Unsteady Flows in Aeropropulsion, 6-11 November, Chicago U.S.A., 1994.

[2] ACTON, E., and CARGILL, M., "Non-Reflecting Boundary Conditions for Computations of Unsteady Turbomachinery Flow", Proceedings of the Fourth International Symposium Unsteady Aerodynamics and Aeroalesticity of Turbomachines and Propellers, pp. 211-228, 1988.

[3] ARNONE, A., LIOU, M.-S., POVINELLI, L.A.,"Multigrid Time-Accurate Integration of Navier-Stokes Equations", AIAA-93-3361, 1993.

[4] BIRCH, S.F., "One Equation Models Revisited", AIAA-93-2903, 1993.

[5] CEBECI, T., "Calculation of Compressible Turbulent Boundary Layers with Heat and Mass Transfer", AIAA-70-741, 1970.

[6] ENGEL, K., EULITZ, F., FADEN, M., POKORNY, S., "Numerical Simulation of the Unsteady Flow on a MIMD Computer", Proceedings of the CNRS-DFG Symposium, Stuttgart, 9-10 December, 1993.

[7] ENGEL, K., EULITZ, F., FADEN, M., POKORNY, S., "Numerical Investigation of the Shock Induced Interaction in a Transonic Compressor Stage", Proceedings of the ASME Symposium: Unsteady Flows in Aeropropulsion, 6-11 November, Chicago, U.S.A., 1994.

[8] ENGEL, K., EULITZ,. F., FADEN, M., and POKORNY, S., "Validation of Different TVD-Schemes for the Calculation of the Unsteady Turbomachinery Flow", 14.th International Conference on Numerical Methods in Fluid Dynamics, ICN-MFD 94, 11-15 July, Bangalore, India, 1994.

[9] ENGEL, K., EULITZ,. F., FADEN, M., and POKORNY, S., "Numerical Investigation of the Rotor-Stator Interaction in a Transonic Compressor Stage", 30.th AIAA/ASME/SAE/ASEE Joint Propulsion Conference, June 27-29, Indianapolis, USA, 1994.

[10] ENGEL, K., EULITZ, F., GEBING, H., "Numerical Investigation of the Clocking Effects in a Multistage turbine", Proceedings of the ASME Symposium: Computational Fluid Dynamics in Aeropropulsion, 12-17 November, San Francisco, U.S.A., 1995.

[11] EULITZ, F., ENGEL, K., GEBING, H., "Numerical Investigation of Inviscid and Viscous Interaction in a Transonic Compressor", 85th AGARD-PEP-CP, 1996.

[12] EULITZ, F., ENGEL, K., GEBING, H., "Application of a one-equation eddy-viscosity model to unsteady turbomachinery flow", to be published in Proc. of the 3rd ISETMM, Editor H.W. Rodi, Elsevier, Amsterdam 1996.

[13] FADEN, M., POKORNY, S., ENGEL, K., "Unsteady Flow Simulation on a Parallel Computer", Conference Proc., 11th AIAA Computational Fluid Dynamics Conference, Orlando, FL, U.S.A., 1993.

[14] GILES, M.B., "Non-Reflecting Boundary Conditions for the Euler Equations", CFDL-TR-88-1, 1988.

[15] GILES, M.B., "UNSFLO: A Numerical Method for the Calculation of the Unsteady Flow in Turbomachinery", GTL Report #205, Gas Turbine Laboratory, MIT, 1990.

[16] HE, L., "New Two-Grid Accleration Method for Unsteady Navier-Stokes Calculations", J. Prop. and Power, 9 272., 1993.

[17] PVM 3 user's guide and reference manual. 1994, Oak Ridge National Laboratory.

[18] ROE, P.L., "Approximative Riemann Solvers, Parameter Vector and Difference Schemes", J. Comp. Phys., 43, 357, 1981.

[19] SAXER, A. P., GILES, M.B., "Quasi-3-D Non-Reflecting Boundary Conditions for the Euler Equations Calculations", AIAA 91-1603-CP.

[20] SHARMA, O.P., NI, R.H., TANRIKUT, S., "Unsteady Flows in Turbines-Impact on Design Procedure", AGARD lecture series 195, 1994.

[21] SPALART, P., R., ALLMARAS, S., R., "A One-Equation Turbulence Model for Aerodynamic Flows", AIAA-92-0439, 1992.

[22] VON LAVANTE, E., El-MILIGUI, A., CANNIZARIO, F., E., "Simple Explicit Upwind Schemes For Solving Compressible Flows", Proceedings of the 8. GAMM-Conference on Numerical Methods in Fluid Mechanics, Vieweg, Braunschweig, 29, 1990.

[23] VAN LEER, B., "Towards the Ultimate Conservation Difference Scheme V, A Second-Order Sequel to Godunov's Method", J. Comp. Phys., 32, 101, 1979.

[24] YAMAMOTO, K., TANIDA, Y., "Self-Excited Oscillation of Transonic Flow around an Airfoil in a Two-Dimensional Channel", 34th ASME, 1989.

[25] YEE, H.C., "A Class of High-Resolution Explicit and Implicit Shock Capturing Methods", VKI-Lecture Notes in Computational Fluid Dynamics, 1989.

A COMPARISON OF SMOOTHERS AND NUMBERING STRATEGIES FOR LAMINAR FLOW AROUND A CYLINDER

Henrik Rentz-Reichert, Gabriel Wittum
Institut für Computeranwendungen III, Universität Stuttgart
Pfaffenwaldring 27, D-70 569 Stuttgart

Summary

We introduce several multigrid smoothers for the incompressible Navier-Stokes equations in 2 space dimensions on unstructured grids and compare their performance for the DFG-benchmark problem of laminar flow around a cylinder depending on the kinematic viscosity. We further employ a streamwise numbering strategy of the unknowns and compare it to the hierarchical ordering the refinement module creates as well as to standard lexicographic numbering.

Introduction

The aim of the work in the present paper was to compare several smoothers for the incompressible Navier-Stokes equations. We chose a discretization which is easily extendable to three space dimensions and which yields satisfactory results in practice (cf. [RW1]). Our final goal was to find a multigrid based solution strategy which is robust with respect to geometry as well as to the (locally) dominating character — convective or diffusive — of the equations.

If this goal could be reached for the steady state case it will readily work in a fully implicit time integration scheme for the non-stationary case.

The strategy to achieve robustness in the above sense was to combine ILU_β with a streamwise numbering of the unknowns. This way we hope to marry the good, more or less geometry independent, results of multigrid together with a modified ILU smoother for elliptic problems on one hand with the effect of streamwise numbering for upwinded convection – namely to make the smoother an approximate solver in the limit – on the other hand, as indicated by experiments with a scalar convection-diffusion equation in two and three space dimensions (cf. [BW]). In difference to other authors using unstructured grids and/or multigrid components ([LPS], [VH], [DSW], [DJRST] etc.) we propose the combination of unstructured multigrid together with ILU-smoothing and numbering strategy.

In the following sections we want to discuss our discretization, solver, smoother and streamwise numbering, all of which has been implemented on the base of the programming toolbox ug3.1 developed at ICA to support unstructered locally refined multigrids on serial (parallel in preparation) computer architectures (c.f. http://www.uni-stuttgart.de/~ug). The afore mentioned limit cases of the Navier-Stokes equations have been considered in [RW2]. Here we investigate on the behaviour of the linear multigrid solver for a flow problem involving both, the dominance of diffusive transport in some parts of the domain and of convective transport in other parts of the domain.

Discretization

We discretize the incompressible Navier-Stokes equations basically following ideas of Michael Raw in [Ra] and [SR]. The base of the discretization is the quasi Newton linearization of the equations

$$-\nu \Delta \vec{u} + (\vec{u}_{old} \vec{\nabla}) \vec{u} + \vec{\nabla} p = \vec{f}$$
$$\vec{\nabla} \vec{u} = 0$$
(1)

with the unknowns $\vec{u}$, p and with a start solution or current iterant $\vec{u}_{old}$.

We use a Finite Volume Method with full upwinding for a nodal basis for the unknowns velocity and pressure. The discretization of viscous terms is of second order while the upwinding is of first order (but is shifted to central differences depending on the local Peclet number). The upwinding produces an M-matrix stencil. A special method of physical advection correction is used to achieve both a good modelling of physics and the stabilization of the mass conservation. The basic idea has much in common with the mini element method [ABF] and the stabilizing term in both cases is of the form $-ch^2 \Delta p$ with different geometry dependent constants of course.

We implemented three types of boundary conditions:

- *no slip* at solid walls (both velocities are Dirichlet, the mass conservation implies a homogenuos Neumann boundary condition for the pressure)

- *inflow* (both velocities are Dirichlet, a special treatment of the mass conservation yields a correct pressure for fully developed inflow

- *outflow* (the zero tangential velocity and the pressure are Dirichlet)

The boundary conditions are exact for a Hagen-Poiseuille pipe flow.

We also want to mention that the boundary equations are of the same form as the inner equations, i.e. we always have three unknowns and the smoothers do not have to distinguish between boundary and inner degrees of freedom. Care has only to be taken in the restriction and interpolation for Dirichlet values on the boundary because we do not want to restrict a defect or interpolate corrections to those values.

Multigrid Solver

Our steady state Navier-Stokes solver consists of a nested outer iteration, solving each level with a fixed point (or quasi Newton) iteration. The inner linear solver is a standard multigrid method but using locally refined grids in the sense that only parts of the domain are covered with the refined grid.

Smoothers

With the benchmark problem which will be defined below we tested several smoothers with many parameter configurations as there are damping factors (denoted by ω in the sequel), β-parameters for modified ILU decompositions and the number of pre- and post-smoothing steps.

Before we decribe the smoothers we have to introduce the notions of *equationwise* and *pointwise* blocking of the unknowns:

Once the ordering of the N nodes in the grid is given there are two obvious ways to remove the remaining ambiguity in the order of the $3N$ unknowns:

- $u_1, ..., u_N, v_1, ..., v_N, p_1, ..., p_N$

- $u_1, v_1, p_1, ..., u_N, v_N, p_N$

with their associated block structures, the first of which we will refer to as *equationwise* ($3\ N \times N$ blocks), the ladder as *pointwise* ($N\ 3 \times 3$ blocks). The resulting blocks we call *point blocks* and *equation blocks* and more general the size of the point blocks will not be restricted to 3×3.

We now want to decribe in detail the smoothers. The linear system we want to be solved is

$$Kx = b \qquad (2)$$

where the stiffness matrix K, the solution x and the right hand side b are blocked either pointwise or nodewise (which will be clear from the context).

With the regular decomposition

$$K = M - N \qquad (3)$$

into the *approximate inverse* M and the *rest matrix* N one iteration step reads

$$x^{(i+1)} = x^{(i)} + M^{-1} d^{(i)} \qquad (4)$$

where

$$d^{(i)} = b - Kx^{(i)} \qquad (5)$$

is the current defect.

The total number of point blocks (or nodes in the grid) will be denoted by N_{pb} and the size of the point blocks by n_{pb} (for the Navier-Stokes system this is 3). Correspondingly N_{eq} is the total number of equation blocks (or unknowns per node, here 3, namely u, v, p) and n_{eq} is the size of the equation block (or the number of nodes in the grid).

1. point block based methods (point blocks are always solved exactly):

   - **pbgs**: *damped point block Gauß-Seidel smoother*

   Let

   $$K = D + E + F \qquad (6)$$

   be a splitting of K into a point block diagonal matrix D and a strictly lower and a strictly upper point block diagonal matrix E, F resp.

   Then the approximate inverse of the pbgs is

   $$M_{pbgs} = (D + E) W^{-1} \qquad (7)$$

   where W is a diagonal damping matrix of the form

   $$W = \text{diag}(W_{pb}, ..., W_{pb}),\ W_{pb} = \text{diag}(\omega_i, i = 1, ..., n_{pb}). \qquad (8)$$

- **pbsor:** *point block SOR (successive overrelaxation) smoother*

  With the splitting of K as for the pbgs and with the same damping matrix W we have the approximate inverse of the pbsor:
  $$M_{pbsor} = (D + WE)\, W^{-1}. \qquad (9)$$

- **pbilu:** *damped β-modified point block ILU (incomplete lower-upper point-block decomposition of order 0) smoother*

  The pbilu is defined by the following algorithm (capital indices I, J and K refer to point blocks):

  $L :=$ strictly lower point block triangle of $K + I$
  $U :=$ upper point block triangle of K (including the block diagonal)
  for I=1,...,N_{pb}
     for J=I+1,...,N_{pb}
        $L_{JI} := L_{JI} U_{II}^{-1}$
        $r := 0$
        for K=I+1,...,N_{pb}
           if (L_{JK} in point block pattern of K)
              if (K<J)
                 $L_{JK} := L_{JK} - L_{JI} U_{IK}$
              else
                 $U_{JK} := U_{JK} - L_{JI} U_{IK}$
           else
              $r := r+\ Normalization(L_{JI} U_{IK})$
  $W := \mathrm{diag}(1+r_i \beta_i,\ i=1,...,n_{pb})$
  $U_{JJ} := W U_{JJ}$

  where W is a point block matrix and r as well as the result of the *Normalization* procedure are point block vectors.

  The crucial point is now the correct normalization of the neglected entries $L_{JI} U_{IK}$ which are used to increase the diagonal point blocks.

  We define
  $$(Normalization\ (L_{JI} U_{IK}))_l = \sum_{m=1}^{n_{pb}} \frac{\left|(L_{JI} U_{IK})_{lm}\right|}{\sqrt{\left|(U_{JJ})_{ll}\right|\left|(U_{JJ})_{mm}\right|}},\ l = 0, ..., n_{pb}, \qquad (10)$$

  i.e. the *Normalization* block vector is a weighted row sum of the neglected entry and normalization is done using the diagonal entries of the diagonal point blocks of the same row. Note that for the scalar case this definition reduces to the usual ILU_β (cf. [Wi1]).

  With the above definitions we can write
  $$M_{pbilu} = LU. \qquad (11)$$

- **pbilusp:** *damped spectrally shifted point block ILU (incomplete lower-upper decomposition of order 0) smoother*

Let
$$M = (L+D)D^{-1}(U+D) \tag{12}$$
be the standard point block ILU decomposition. Then following Wittum [Wi2] we define the spectrally shifted point block ILU by
$$M_{\text{pbilusp}} = (L+D_\beta)D_\beta^{-1}(U+D_\beta) \tag{13}$$
with
$$D_\beta = WD \tag{14}$$
and
$$\begin{aligned}W &= \text{diag}(W_0, ..., W_{N_{pb}}) \\ W_I &= \text{diag}(1+\beta_i(r_I)_i, i=1,...,n_{pb}), I=1,...,N_{pb}\end{aligned} \tag{15}$$

Depending on the definition of r we distinguish the *globally shifted* method
$$(r_I)_i = \max_{K=1,...,N_{pb}} \sum_{J=1}^{N_{pb}} \text{Normalization}(N_{KJ})_i \tag{16}$$
and the *locally shifted* method
$$(r_I)_i = \sum_{J=1}^{N_{pb}} \text{Normalization}(N_{KJ})_i. \tag{17}$$

2. equation block methods, inner (inexact) solvers in square brackets work on 1×1, i.e. scalar, point blocks:

- **ebgs[pbilu]**: *equation block Gauß-Seidel smoother*

  Let
  $$K = D + E + F \tag{18}$$
  be a splitting of K into an equation block diagonal matrix D and a strictly lower and a strictly upper equation block matrix E, F resp.

  Then the approximate inverse of the ebgs is
  $$M_{\text{ebgs[pbilu]}} = (\tilde{D} + E) \tag{19}$$
  where
  $$\begin{aligned}\bar{D} &= \text{diag}(D_{II}, I=1,...,N_{eq}) \\ \tilde{D}_{II} &= M_{\text{pbilu}}(D_{II}) = \text{the (scalar) pbilu decomposition of } D_{II}\end{aligned} \tag{20}$$

For the following two smoothers we define the (right) transforming iteration by:
Let
$$\tilde{K} = KK \tag{21}$$

with some convenient matrix K and perform the regular splitting on an approximation $\hat{K}$ of $\tilde{K}$:

$$\hat{K} = M - N. \tag{22}$$

Then on step of the transforming iteration reads

$$x^{(i+1)} = x^{(i)} - \hat{\tilde{K}} M^{-1} d^{(i)}. \tag{23}$$

Note that we also replaced K by an approximation $\hat{\tilde{K}}$.

- **simple[pbilu]**: *SIMPLE (semi implicit method for pressure linked equations, Patankar/Spalding, [PS], see also [Wi3]) smoother*

The stiffness matrix for the Navier-Stokes equations plus stabilization term is

$$K = \begin{bmatrix} Q(\mathring{u}_{old}) & 0 & \frac{\partial}{\partial x} \\ 0 & Q(\mathring{u}_{old}) & \frac{\partial}{\partial y} \\ \frac{\partial}{\partial x} & \frac{\partial}{\partial x} & ch^2 \Delta \end{bmatrix}, \tag{24}$$

$Q(\mathring{u}_{old})$ the quasi Newton linearization of viscous plus convective terms (cf. eqn. (1)).

As transforming matrix (using the above notation) we use the Gauß (equation) block elimination matrix

$$K = \begin{bmatrix} 1 & 0 & -(Q(\mathring{u}_{old}))^{-1}\frac{\partial}{\partial x} \\ 0 & 1 & -(Q(\mathring{u}_{old}))^{-1}\frac{\partial}{\partial y} \\ 0 & 0 & 1 \end{bmatrix}. \tag{25}$$

This results in the product matrix

$$\tilde{K} = \begin{bmatrix} Q(\mathring{u}_{old}) & 0 & 0 \\ 0 & Q(\mathring{u}_{old}) & 0 \\ \frac{\partial}{\partial x} & \frac{\partial}{\partial y} & -ch^2\Delta - \vec{\nabla} \cdot (Q(\mathring{u}_{old}))^{-1}\vec{\nabla} \end{bmatrix} \tag{26}$$

Using the approximation $\bar{Q}(\mathring{u}_{old}) = \text{diag}(Q(\mathring{u}_{old}))$ for every occurence of $(Q(\mathring{u}_{old}))^{-1}$ we finally have

$$\hat{\tilde{K}} = \begin{bmatrix} 1 & 0 & -(\bar{Q}(\mathring{u}_{old}))^{-1}\frac{\partial}{\partial x} \\ 0 & 1 & -(\bar{Q}(\mathring{u}_{old}))^{-1}\frac{\partial}{\partial y} \\ 0 & 0 & 1 \end{bmatrix} \tag{27}$$

and

$$\hat{K} = \begin{bmatrix} Q(\mathring{u}_{old}) & 0 & 0 \\ 0 & Q(\mathring{u}_{old}) & 0 \\ \frac{\partial}{\partial x} & \frac{\partial}{\partial y} & -ch^2\Delta - \vec{\nabla} \cdot (\bar{Q}(\mathring{u}_{old}))^{-1}\vec{\nabla} \end{bmatrix}. \quad (28)$$

The terms $-(\bar{Q}(\mathring{u}_{old}))^{-1}\frac{\partial}{\partial x}$ and $-(\bar{Q}(\mathring{u}_{old}))^{-1}\frac{\partial}{\partial y}$ are calculated from the stiffness matrix by matrix multiplication without extending the stencil while $\vec{\nabla} \cdot (\bar{Q}(\mathring{u}_{old}|))^{-1}\vec{\nabla}$ is discretized seperatly. As approximate inverse for $\hat{K}$ we use the ebgs decomposition of $\hat{K}$:

$$M_{simple[pbilu]} = M_{ebgs[pbilu]}(\hat{K}). \quad (29)$$

- **distrib[pbilu]**: *distributive smoother (Brandt/Dinar, [BD], see also [Wi3])*

Here we use a transforming matrix such that the product matrix will contain a commutator of differential operators:

$$\hat{\tilde{K}} = \tilde{K} = \begin{bmatrix} 1 & 0 & \frac{\partial}{\partial x} \\ 0 & 1 & \frac{\partial}{\partial y} \\ 0 & 0 & Q(\mathring{u}_{old}) \end{bmatrix} \quad (30)$$

$$\tilde{K} = \begin{bmatrix} Q(\mathring{u}_{old}) & 0 & \frac{\partial}{\partial x}Q(\mathring{u}_{old}) - Q(\mathring{u}_{old})\frac{\partial}{\partial x} \\ 0 & Q(\mathring{u}_{old}) & \frac{\partial}{\partial y}Q(\mathring{u}_{old}) - Q(\mathring{u}_{old})\frac{\partial}{\partial y} \\ \frac{\partial}{\partial x} & \frac{\partial}{\partial y} & -ch^2\Delta Q(\mathring{u}_{old}) - \Delta \end{bmatrix}. \quad (31)$$

The ladder one is replaced by

$$\hat{K} = \begin{bmatrix} Q(\mathring{u}_{old}) & 0 & 0 \\ 0 & Q(\vec{\mathring{u}_{old}}) & 0 \\ \frac{\partial}{\partial x} & \frac{\partial}{\partial y} & -\Delta \end{bmatrix}. \quad (32)$$

Again the approximate inverse is

$$M_{simple[pbilu]} = M_{ebgs[pbilu]}(\hat{K}). \quad (33)$$

Streamwise Numbering

The model for the following reasoning is a scalar convection equation. Due to the quasi Newton linearization and to the upwind scheme a given node depends only on its upwind neighbours. If we can find a global ordering of the unknowns in a way that the stiffness matrix has nonzero entries only in the lower triangle, then of course we will be able to solve the system of equations in one step even by a Gauß-Seidel method. Unfortunately in most of the relevant cases there are vortices in the flow and therefore cyclic dependencies. But nevertheless we will obtain good results if we introduce arbitrary cuts through the vortices by "removing" just enough of the "cyclic" nodes to get rid of the cylic dependencies. We start the numbering at the inlet going in layers downstream but taking only nodes depending on the already numbered ones (those nodes will form the beginning of our new list). In a similar way we go upstream from the outlet (those nodes will make up the end of our new list). Finally we are left with nodes involved in cycles. We remove some of them and append them to the begin of the new list. If we did that in an efficient way we removed some of the cyclic dependencies such that again we can find nodes depending only on already numbered nodes. So we can just repeat the foregoing steps until all nodes are numbered.

The sets of nodes found in the three subsequent steps will be denoted by FIRST, LAST and CUT resp.

After this rough description we introduce the following algorithm that does the job (the basic ideas can be found in [BW]):

Let $G = \{L_{ij} = (N_i, N_j) : N_i, N_j \in N, N_i \neq N_j, a_{ij} \neq 0 \vee a_{ji} \neq 0\}$ be the graph associated with the stiffness matrix $A = (a_{ij})$ where N is the set of nodes in the grid.

Each Link L_{ij} is assigned a flow character by the upwinding procedure:

$$f(L_{ij}) = \begin{cases} \text{upwind, node } i \text{ is upwinded to node } j \\ \text{downwind, else} \end{cases} \tag{34}$$

(For definition and details of the upwinding cf. [Ra].)

After that we can formulate the algorithm

Algorithm 1: Streamwise numbering

(1) Initialize all nodes $N \in G$ with
$U(N)$ = # of downwind links
$D(N)$ = # of upwind links
$P(N)$ = false

(2) Create two list, containing the leading nodes of the new order L_l and the trailing nodes of the new order L_t in the order the are added to the resp. lists.
Initialize two first-in-first-out stacks F, L with
$F = \emptyset$
$L = \emptyset$
for all nodes $N \in G$
if $(U(N) = 0)$
{
$F = F \cup N$
$P(N)$ = true
append N to L_l
}
else if $(D(N) = 0)$

>   {
> $L = L \cup N$
> $P(N) = 0$
>   append N to L_t
>   }

(3) do
    {
       /* construct next *FIRST*-set: */
       while $(F \neq \emptyset)$
       {
          pop next $N \in F$, $F = F \backslash N$
          decrement all U-counters of downwind neighbours N_d with $P(N_d) =$ false
          if $(U(N_d) = 0)$
          {
             push N_d to the *FIRST*-fifo, $F = F \cup N_d$
             $P(N_d) =$ true
             append N to L_l
          }
       }

       /* construct next *LAST*-set: */
       while $(L \neq \emptyset)$
       {
          pop next $N \in L$, $L = L \backslash N$
          decrement all D counters of upwind neighbours N_u with $P(N_u) =$ false
          if $(D(N_u) = 0)$
          {
             push N_u to the *LAST*-fifo, $L = L \cup N_u$
             $P(N_u) =$ true
             append N to L_t
          }
       }

       /* construct next *CUT*-set: */
       specify a cut subset $C \subset G' \{N \in G : P(N) =$ false$\}$ by any *find-cut-algorithm* such
          that $C \neq \emptyset$ if $G' \neq \emptyset$
       append all nodes $N \in C$ to L_l
       for all nodes from C do
       {
          decrement all D counters of upwind neighbours N_u with $P(N_u) =$ false
          if $(D(N_u) = 0)$
          {
             push N_u to the *LAST*-fifo, $L = L \cup N_u$
             $P(N_u) =$ true
             append N to L_l
          }
       }
    } while $(G' \neq \emptyset)$
(4) Now we only have to join the two lists L_l and L_t where we reverse the order in the lad-

der one, i.e. the first node added to L_t will now be the last node of the new order.

Concerning the *find-cut-algorithm* we have made good experiences in practice with a heuristic procedure which is trying to cut a subdomain with cyclic dependencies from the vortex center to its boundary.

An example of the resulting sparsity pattern of the stiffness matrix could look like this:

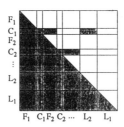

For the Backward Facing Step the result of the reordering could look like:

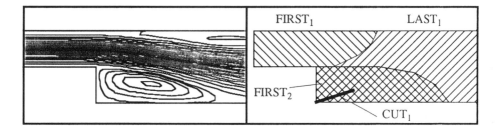

Fig. 1: Streamwise numbering of the Backward Facing Step.

Though the above reasoning and also numerical tests (cf. [BW]) are most convincing for the scalar case it is not *a priori* clear that the same strategy will yield good results for the Navier-Stokes equations where we still have an elliptic part for the pressure (the stabilization term mentioned above) also in the presence of strong convection.

But the results in the next section will show that we basically have the same behaviour as in the scalar case, i.e. the linear convergence rates *decrease* for *increasing* Reynolds number.

One thing that definitly helps is the property of the discretization that the pressure gradient as well as the divergence of the velocity will be shifted to upwind differences as the Reynolds number increases. So also for those terms information propagates along characteristics of the flow field.

Results

In a recent paper ([RW2]) we have compared our smoothers in their behaviour concerning varying aspect ratios for the Stokes equations (the case of dominant diffusion) and their performance for the convection dominated limit employing the streamwise numbering strategy.

In the DFG-benchmark of laminar flow around a cylinder we have a combination of the two regimes in one domain because the recirculation zone and the boundary layers – at least in the limit at the walls – are clearly diffusive dominated while in the inflow and outflow region we have strong convection. Also we have a mixture of nicely shaped elements together with ones having obduce angles and other having bad aspect ratios of up to 1:15 (the intermediate regime of aspect ratios is the worst case as can be seen in our paper [RW2] or, for the scalar case, in [Wi1]).

To see how our solver performs we kept the same nonlinear and linear convergence criteria and tested the smoothers with different kinematic viscosities and different orderings of the nodes in the grid.

To define the benchmark problem have a look at figure Fig. 2 page 11. The lenght of the pipe is 2.2m its width is 0.41m. The cylinder has a diameter of 0.1m and its midpoint has a vertical distance from the lower wall of the pipe of 0.2m and a horizontal distance of 0.2m from the inlet (the left side of the pipe). Boundary conditions are no-slip at the walls, a parabolic profile with a maximum of 0.3m/s and a fixed constant pressure at the outflow to the right.

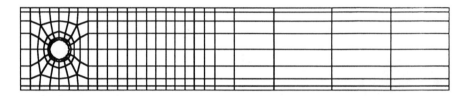

Fig. 2: The computational domain and the coarse grid used

To solve the equations we start with the nested iteration on the coarse grid (level 0) using an exact solver and proceed level by level of the multigrid performing 6 fixed point cycles on each and reducing the linear defect by one order of magnitude per fixed point step.

In the following diagrams we plotted the mean linear convergence rates per level, averaged over the six calls of the linear multigrid in the fixed point point steps, against $\log(1/\nu)$, where ν is the kinematic viscosity. More precise, if at step number i we needed n_i V-cycles to reduce the linear defect by one order of magnitude beginning at a defect $d_0^{(i)}$ and ending with $d_{n_i}^{(i)}$ then we define the mean linear convergence rate κ by

$$\kappa = \left(\prod_{i=1}^{6} \kappa_i\right)^{1/6}, \quad \kappa_i = \left(\frac{d_{n_i}^{(i)}}{d_0^i}\right)^{\frac{1}{n_i}}. \tag{35}$$

The Reynolds number for the problem is defined as $Re = \bar{U}D/\nu$, $\bar{U}$ the effective incoming velocity, i.e. $2U_{max}/3 = 0.1$ m/s and $D = 0.41$m the pipe diameter.

The damping factors for the velocity and pressure components are denoted by ω_u and ω_p resp. and the modification parameters for the different ILU versions are denoted by β_u and β_p.

The parameter settings are found by parameter studies on level 1 and are fixed then for the remaining levels.

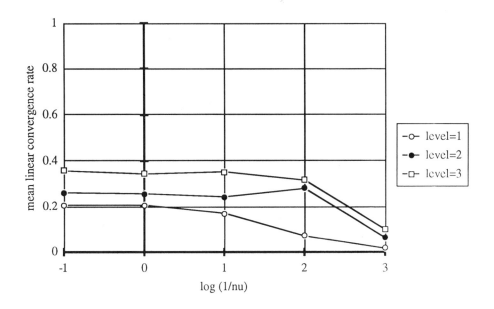

Fig. 3: pbilu ($\omega_u = 1$, $\omega_p = 1$, $\beta_u = 0.25$, $\beta_p = 0$, V(1,1)-cycle)

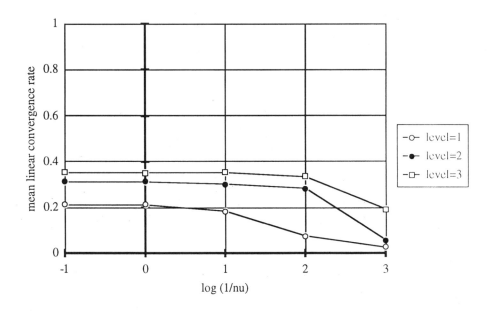

Fig. 4: pbilusp (global, $\omega_u = 1$, $\omega_p = 1$, $\beta_u = 0.125$, $\beta_p = 0$, V(1,1)-cycle)

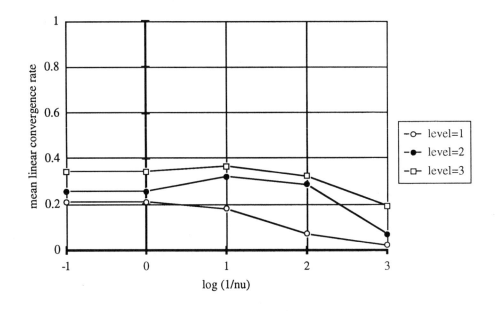

Fig. 5: pbilusp (local, $\omega_u = 1$, $\omega_p = 1$, $\beta_u = 0.5$, $\beta_p = 0$, V(1,1)-cycle)

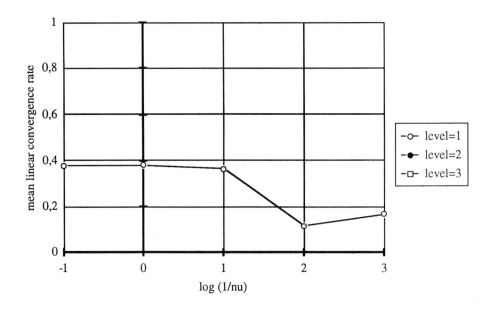

Fig. 6: ebgs[pbgs] ($\omega_u = 1$, $\omega_p = 0.4$, $\beta_u = 0.5$, $\beta_p = 0.5$, V(1,1)-cycle): divergence on level 2 and 3

The above results have been calculated with streamwise ordering.

The diagrams of the pbilu and the pbilusp nicely show the expected decrease of the convergence rate with increasing Reynolds number. The convergence rates are sensitive to the parameter settings of course but in quite a large range the convergence rates deed not exceed values of 0.6 for those smoothers. On the other hand with the ebgs method we found no parameter combination yielding convergence for the whole range of the Renolds numbers we tested. The same comment applies to the pbgs and pbsor whose convergence rates hardly have been below 0.6 even on level 1.

For comparison we give the results for lexicographic ordering and the hierarchical ordering (sort of an elementwise coarse/fine ordering) the refinement module produces in the following figures (for one smoother only). As one can see the streamwise numbering is essential for convergence. With lexicographic ordering the linear solver diverged on level 2 and 3 while the behaviour of the convergence rates on level 1 deteriorate for increasing Reynolds number..

The hierarchical ordering on the other hand failes for low Reynolds number on higher levels.

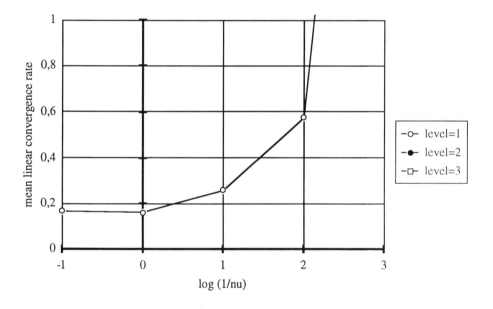

Fig. 7: pbilu ($\omega_u = 1, \omega_p = 1, \beta_u = 0.25, \beta_p = 0.5$) with lexicographic ordering, level 2 and 3 diverged

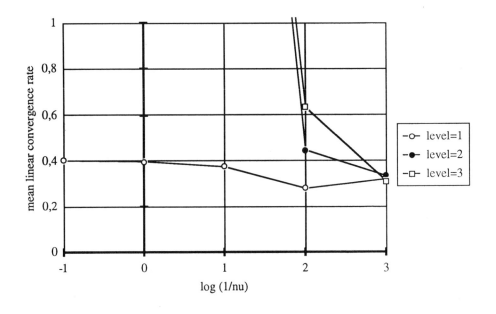

Fig. 8: pbilu ($\omega_u = 1$, $\omega_p = 1$, $\beta_u = 0.25$, $\beta_p = 0.5$) with hierarchical ordering

Conclusions

As far as the above benchmark problem is concerned and also the other test cases we investigated it seems that we found a class of smoothers – namely the pbilu and the pbilusp variants – performing quite well on unstructered locally refined multigrids for a range of Reynolds numbers of at least 4 orders of magnitude. For most of the nonlinear steps with one or two V(1,1)-cycles a sufficient reduction of the defect can bew reached. Also with unfavourable aspect ratios we can maintain a good performance of the linear multigrid solver. The streamwise numbering has prooved to be very satisfactory and can be expected to work very well for all kinds of problems where the flow is not dominated by vortices (which is a quite realistic assumption). But even for the Driven Cavity problem we found an improvement of the convergence rates by a factor of circa 0.6 already at an intermediate Reynolds number of 500.

The future developments will concentrate on the extension to three space dimensions – which is not expected to be a serious problem – and to instationary problems using fully implicite schemes of higher order like the Crank-Nicholson and the fractional step θ-scheme. We also will have to investigate on *a posteriori* error estimators.

Last not least the parallelization of our Navier-Stokes code that is currently being implemented into the older parallel version of ug will be finished and tested. Here emphasis will be laid on the parallel implementation of the numbering strategy.

References

[ABF]   Arnold, D.N., Brezzi, F., Fortin M.:
A stable finite element for the Stokes equations, Calcolo, 21 (1984) 337-344.

[BW]   Bey, J., Wittum, G.:
Downwind Numbering: A Robust Multigrid Method for Convection-Diffusion Problems on Unstructerd Grids, Preprint 1995/2, ICA, Universität Stuttgart.

[BD]   Brandt, A., Dinar, N.:
Multigrid solutions to elliptic flow problems. ICASE Report 79-15 (1979).

[DJRST] Dorok, O., John, V., Risch, U., Schieweck, F., Tobiska, L.:
Parallel Finite Element Methods for the Incompressible Navier-Stokes Equations. In this publication.

[DSW]   Durst, F., Schäfer, M., Wechsler, K.:
Efficient Simulation of Incompressible Viscous Flows on Parallel Computers. In this publication.

[LPS]   Lilek, Z., Peric, M., Seidl, V.:
Development and Application of a Finite Volume Method for the Prediction of Complex Flows. In this publication.

[PS]   Patankar, S.V., Spalding, D.B.:
A calculation procedure for heat and mass transfer in threedimensional parabolic flows. Int. J. Heat Mass Transfer 15 (1972), 1787-1806.

[Ra]   Raw, M.J.:
A New Control-Volume-Based Finite Element Procedure for the Numerical Solution of the Fluid Flow and Scalar Transport Equations, Ph.D. thesis, University of Waterloo, Ontario, Canada, 1985.

[RW1]   Reichert, H. (now Rentz-Reichert), Wittum, G.:
Solving the Navier-Stokes Equations on Unstructured Grids. NNFM, Vol. 38, p. 321-333, Vieweg, Braunschweig 1993.

[RW2]   Reichert, H. (now Rentz-Reichert), Wittum, G.:
Robust Multigrid Methods for the Incompressible Navier-Stokes-Equations. NNFM, Vol. 49, p. 216-228, Vieweg, Braunschweig 1995.

[SR]   Schneider, G.E., Raw, M.J.:
Control volume finite-element method for heat transfer and fluid-flow using co-located variables. Numer.Heat.Transf. 11 (1988) 363.

[Wi1]   Wittum, G.:
On the Robustness of ILU-Smoothing. SISSC 10 (1989) 699-717.

[Wi2]   Wittum, G.:
Spektralverschobene Iterationen, Preprint 1995/4, ICA, Universität Stuttgart.

[Wi3]   Wittum, G.:
Distributive Iterationen für indefinite Systeme. Ph. D. Thesis, Univ. Kiel, 1986.

[VH]   Vilsmeier, R., Hänel, D.:
Computational Aspects of Flow Simulation on 3-D, Unstructured, Adaptive Grids. In this publication.

Simulation of Detailed Chemistry Stationary Diffusion Flames on Parallel Computers

Georg Bader
BTU Cottbus, Institut für Mathematik
Karl Marx Straße 17, 03044 Cottbus

Eckard Gehrke
Universität Heidelberg
Interdisziplinäres Zentrum für Wissenschaftliches Rechnen
Im Neuenheimer Feld 368, 69120 Heidelberg

Summary

We present the numerical simulation of stationary, laminar diffusion flames with detailed chemical reaction mechanism on distributed memory parallel computers. The problem consists in solving the conservation equations of total mass, momentum, energy and species masses for a compressible fluid. For the numerical solution we suggest a hybrid algorithm which is a combination of data and domain decomposition techniques. This approach has been tested on a variety of parallel computers ranging from moderately to massively parallel and has proven uniformly high parallel efficiencies. Results for Parsytec GC/PP, a Cray T3D and an IBM-SP2 computer are presented. Calculations for a model bunsen burner with axisymmetric configuration are presented and discussed.

1. Introduction

The challenge of solving combustion problems with strong heat releases originates from the large number of state variables, the strong nonlinear character of the governing partial differential equations involving drastically different length scales and highly nonlinear, stiff chemical source terms. The mathematical formulation includes the total mass, the velocities, the temperature and typically a large number of species conservation equations for a detailed reaction mechanism. See [17,19] for a description of the model equations. The different length scales in the flame fronts, the steep gradients and the true multidimensional structure of flames require a fine resolution in order to approximate the solution accurately. Nevertheless, there exists a growing demand for simulation of technical combustion processes. Of particular interest is the design of fuel effective and of reduced pollutant productive devices which require a deeper understanding of diffusion flames. This insight can only be gained by true multi-dimensional simulation of detailed chemical reaction models.

As a consequence the numerical simulation of practically relevant combustion devices form an interesting but challenging class of problems. Until recently, the algorithmic difficulties and the sheer computational complexity of combustion problems have severely limited a treatment of detailed chemistry, multi-dimensional formulations. Up to now mostly simplified models, either in one space dimension with detailed chemistry or multi-dimensional with a simplified reaction mechanism were considered. When a more complete problem statement was attempted, it was necessary to resort to the availability of large amounts of expensive computer time on very costly vector computers. The availability of more economic parallel computers nowadays allows to solve complex combustion problems efficiently within an acceptable time horizon. In particular, distributed memory computers with their, in principle, unlimited computing power and storage capacity offer a good perspective for a successful treatment of these computationally demanding problems. It has to be stressed, however, that for an efficient application on these computers, a variety of algorithmic and technical questions have to be answered successfully in order to solve such problems more routinely.

Distributed memory parallel computers are built on the paradigm that the solution of large scale problems decomposes into tasks of almost equal complexity which can be solved with a limited amount of interaction. The times necessary for this interaction, e.g. synchronization and communication are considered as overhead. This implies that parallel computers attain their maximum performance for problems which require, compared to their algebraic complexity, almost no communication. Such problems might be called weakly coupled problems.

Domain decomposition methods try to mimic this basic property of parallel computers. The resulting methods interchange infrequently large chunks of data between neighboring processors. This reduces the communication overhead in favor of the computing performance. The corresponding algorithms are called *coarse grain* parallel methods. In terms of problem solving they are ideally suited for explicit time discretization schemes of transient problems. Hence, the resulting implementation attains high *parallel efficiency*, e.g. a good exploitation of the computing power. Further, they show good *scalability* properties, c.f. [15] for example.

The situation with implicit discretizations of strongly coupled stationary equations is typically more difficult. Here the strong coupling of the differential equations carries over to the discrete problem. Thus, algorithms which reflect this strong coupling have to exchange smaller amounts of data more frequently among the processors. The corresponding parallel algorithms tend to be of *fine granular* nature, implying certain limitations on the *scalability*, e.g. the number of processors which can be exploited efficiently. If we, in order to reduce this difficulty, apply a solution method with weak coupling, e.g. Jacobi's iteration, then the computing performance will increase. However, the experienced convergence speed drops quite drastically, and the overall performance in terms of problem solving is low. This indicates that some kind of compromise between computational overhead and convergence speed has to be found. A successful choice depends typically on the properties of the problem to be solved and the characteristic hard- and software parameters of the parallel computer.

The combustion problems addressed in this paper are stationary, strongly convection dominated nonlinear problems. Hence, they fall naturally into the category of strongly coupled problems. Nevertheless, the resulting parallel algorithm should be efficient over a wide range of distributed parallel computers ranging from moderately parallel with high performance processors to massively parallel computers with moderate performance nodes. We suggest a hybrid method based on a combination of data and domain decomposition. The data decomposition is applied in the dominant convection direction in order to retain the convergence speed and robustness of the sequential algorithm. Domain decomposition is applied in the diffusion dominated direction to allow an efficient exploitation of a larger number of processors.

The paper is organized as follows. In section 2 we describe the problem statement for detailed chemistry combustion problems. The sequential solution approach is outlined in section 3. Especially the various nonlinear aspects of stationary flame problems are considered. In section 4 we present the data and domain decomposition techniques used for the parallel implementation. Finally, in section 5 a selection of numerical results on the model bunsen burner is presented and discussed. Special attention is given to obtained efficiencies on a variety of different parallel computers.

2. The governing equations

Typical flame problems are spatially three dimensional phenomena. A class of practically important flame configurations which allow a two dimensional simulation are axisymmetric devices. As a model diffusion flame we consider the bunsen burner. It can be characterized by a fuel jet discharging into an air stream through two concentric tubes with radii R_{in} and R_{out} as depicted in figure 1. The gases are mixing at the outlet of the inner tube through diffusion and form a thin combustion zone, the flame.

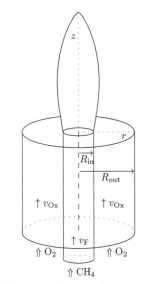
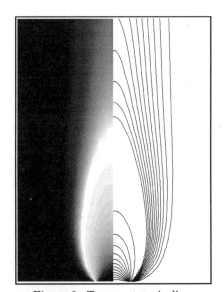

Figure 1: Bunsen burner   Figure 2: Temperature isolines

The constitutive equations for laminar flame models are mathematically formulated by the conservation of total mass, momentum, energy and species masses. A complete description of these equations in primitive variable formulation can be found in [17,19] so that they are just restated in the corresponding streamfunction vorticity formulation, e.g. ψ and ω. Let ρ denote the total mass density and v_r, v_z the radial and axial velocities of the fluid mixture. Further, let $Y_k, 1 \leq k \leq K$ be the mass fraction and $V_k = (V_{k_r}, V_{k_z})$ the diffusion velocity of the species. With these preparations the stationary, streamfunction vorticity formulation is given by:

Streamfunction

$$\frac{\partial}{\partial z}\left(\frac{1}{r\rho}\frac{\partial\psi}{\partial z}\right) + \frac{\partial}{\partial r}\left(\frac{1}{r\rho}\frac{\partial\psi}{\partial r}\right) + \omega = 0, \qquad (2.1)$$

Vorticity

$$r^2\left[\frac{\partial}{\partial z}\left(\frac{\omega}{r}\frac{\partial\psi}{\partial r}\right) - \frac{\partial}{\partial r}\left(\frac{\omega}{r}\frac{\partial\psi}{\partial z}\right)\right] - \frac{\partial}{\partial z}\left(r^3\frac{\partial}{\partial z}\left(\frac{\mu}{r}\omega\right)\right) -$$
$$\frac{\partial}{\partial r}\left(r^3\frac{\partial}{\partial r}\left(\frac{\mu}{r}\omega\right)\right) + r^2 g\frac{\partial\rho}{\partial r} + r^2\left(\frac{\partial\rho}{\partial z}, -\frac{\partial\rho}{\partial r}\right)\cdot\nabla\left(\frac{v_r^2 + v_z^2}{2}\right) = 0, \qquad (2.2)$$

Energy

$$c_p\left[\frac{\partial}{\partial z}\left(T\frac{\partial\psi}{\partial r}\right) - \frac{\partial}{\partial r}\left(T\frac{\partial\psi}{\partial z}\right)\right] - \frac{\partial}{\partial r}\left(r\lambda\frac{\partial T}{\partial r}\right) - \frac{\partial}{\partial z}\left(r\lambda\frac{\partial T}{\partial z}\right)$$
$$+ r\sum_{k=1}^{K}\left\{\rho c_{pk}Y_k\left(V_{k_r}\frac{\partial T}{\partial r} + V_{k_z}\frac{\partial T}{\partial z}\right)\right\} + r\sum_{k=1}^{K}h_k W_k\dot{w}_k = 0. \qquad (2.3)$$

Mass balances, $1 \leq k \leq K$

$$\left[\frac{\partial}{\partial z}\left(Y_k\frac{\partial\psi}{\partial r}\right) - \frac{\partial}{\partial r}\left(Y_k\frac{\partial\psi}{\partial z}\right)\right] + \frac{\partial}{\partial r}(r\rho Y_k V_{k_r}) + \frac{\partial}{\partial z}(r\rho Y_k V_{k_z}) - rW_k\dot{w}_k = 0. \qquad (2.4)$$

The system is closed by the equation of state for an ideal gas

$$\rho = \frac{p\bar{W}}{RT} \qquad (2.5)$$

where $\bar{W}$ denotes the mean molecular weight of the gas mixture. Before completing the problem statement by defining the chemical production rates $\dot{w}_k$ and the boundary conditions, some comments on the system of equations (2.1)-(2.5) are necessary. When summing over all species mass balance equations (2.4) the conservation of total mass for the mixture must appear. This implies, that the sum over the chemical production terms has to vanish identically. Furthermore, the species diffusion velocities must be modeled analytically and numerically such that

$$\sum_{k=1}^{K} Y_k V_k = 0 \qquad (2.6)$$

is satisfied over the complete domain. The conservation of total mass and the species mass conservations are linearly dependent. In order to obtain a uniquely solvable set of equations one of the species balances is usually dropped. The corresponding species mass is calculated from the relation $Y_{N_2} = 1 - \sum_{k \neq N_2} Y_k$ for the mass fractions. The diffusion velocities can be modeled in analogy to Fick's law by $Y_k V_k = -D_k \nabla Y_k$, where D_k denotes the diffusion coefficient of the kth species. See [17] for further details.

Let the reaction mechanism be composed of m reversible elementary reaction steps with K different species. Then by denoting the chemical symbol of species k by S_k and the stoichiometric coefficients for the jth reaction by ν_{kj}^f and ν_{kj}^r we can formally write the reaction mechanism in the general form

$$\sum_{k=1}^{K} \nu_{kj}^f S_k \rightleftharpoons \sum_{k=1}^{K} \nu_{kj}^r S_k, \qquad j = 1, \ldots, m. \tag{2.7}$$

Herewith the molar production rates $\dot{\omega}_k$ can be written as

$$\dot{\omega}_k = \sum_{j=1}^{m} (\nu_{kj}^r - \nu_{kj}^f) \left[k_j^f(T) \prod_{n=1}^{K} \left(\frac{\rho Y_n}{W_n} \right)^{\nu_{nj}^f} - k_j^r(T) \prod_{n=1}^{K} \left(\frac{\rho Y_n}{W_n} \right)^{\nu_{nj}^r} \right], \tag{2.8}$$

where the forward rate constants $k_j^f(T)$ are expressed by the modified Arrhenius law

$$k_j^f = A_j^f T^{\beta_j} \exp(-E_j / \mathcal{R} T), \tag{2.9}$$

with the parameters A_j, β_j and E_j for all reactions. The reverse rate constants are calculated by the forward rates divided by the equilibrium constants which are taken from the JANNAF and CHEMKIN thermochemical data bases [4,12]. For the bunsen burner we consider a C_1-chain reaction mechanism with $m = 46$ elementary reactions and $K = 16$ different species, see [17] for a listing of the mechanism.

The physical configuration of the burner consists of an inner fuel stream with radius R_{in} surrounded by a coflowing air jet with radius R_{out}. Thus, exploiting symmetry of the problem the computational domain is selected as the right upper quadrant of figure 1. It extends from $r = 0$ to $r = R_{out}$ and $z = 0$ to $z = L$, where L is chosen to be sufficiently large. Plug flow is assumed at inflow with $v_z = v_{CH_4}$ and $v_z = v_{air}$ for the fuel and oxidizer jets respectively. Note that the boundary conditions for the bunsen burner are chosen such that the model assumption of laminar, stationary flow are appropriate. They are as follows, c.f. [17].

Axis of symmetry ($r = 0$)

$$\begin{gathered} v_r = 0, \quad \frac{\partial v_z}{\partial r} = 0, \quad \omega = 0, \\ \frac{\partial T}{\partial r} = 0, \quad \frac{\partial Y_k}{\partial r} = 0, k = 1, \ldots, K. \end{gathered} \tag{2.10}$$

Outer boundary ($r = R_{out}$)

$$\frac{\partial v_r}{\partial r} = 0, \quad \frac{\partial v_z}{\partial r} = 0, \quad \omega = \frac{\partial v_r}{\partial z}, \tag{2.11}$$

$$T = 298K, \quad Y_{O_2} = 0.232, \quad Y_{N_2} = 0.768, \quad Y_k = 0, k \neq O_2, N_2.$$

Inlet ($z = 0$)
  fuel jet $r < R_{in}$:

$$v_r = 0, \quad v_z = v_{CH_4}, \quad \omega = \frac{\partial v_r}{\partial z} - \frac{\partial v_z}{\partial r}, \tag{2.12}$$

$$T = 298K, \quad Y_{CH_4} = 1, \quad Y_k = 0, k \neq CH_4.$$

oxidizer jet $R_{in} < r < R_{out}$:

$$v_r = 0, \quad v_z = v_{air}, \quad \omega = \frac{\partial v_r}{\partial z} - \frac{\partial v_z}{\partial r}, \tag{2.13}$$

$$T = 298K, \quad Y_{O_2} = 0.232, \quad Y_{N_2} = 0.768, \quad Y_k = 0, k \neq O_2, N_2.$$

Exit ($z = L$)

$$v_r = 0, \quad \frac{\partial v_z}{\partial z} = 0, \quad \frac{\partial \omega}{\partial z} = 0,$$

$$\frac{\partial T}{\partial z} = 0, \quad \frac{\partial Y_k}{\partial z} = 0, k = 1, \ldots, K. \tag{2.14}$$

3. Sequential solution techniques

In this section we present some of the main algorithmic components used for the numerical solution of laminar diffusion flame models. Of highest importance for the solution of the considered highly nonlinear, convection dominated problems is the choice of linear and nonlinear solution methods. Hence, special attention will be paid to these aspects. See also [3,6,8,9,17] for further comments.

The differential equations on the rectangular domain are discretized on a tensor product grid. In order to ensure unique solubility and an accurate resolution of the conservation equations, a sufficiently fine grid has to be applied. Thus, to resolve the steep gradients around the flame front a statically solution adapted, highly nonuniform grid is used. The discretization of diffusion and source terms is done via symmetric finite differences. Convection terms are treated by a monotonicity preserving upwind differencing of first order. This defines 9-point finite difference equations for the streamfunction vorticity formulation. Let these equations be written in residual form as a large system of nonlinear equations

$$F(u) = 0, \tag{3.1}$$

where u consists of the complete state vector, the streamfunction, the vorticity, the temperature T and the mass fractions Y_k. Note that the domain of definition $F : D \subset X \to Y$ is restricted. The mass fractions, must satisfy the constraints $0 \leq Y_k$ and

$\sum_k Y_k = 1$. If one of these constraints is violated during the iterative solution process the result is unpredictable.

For the solution of (3.1) the damped Newton iteration [7]

$$DF(u^i)\Delta u^i = -F(u^i), \quad u^{i+1} = u^i + \lambda_i \Delta u^i \quad i = 0, 1, \ldots \qquad (3.2)$$

is applied, where $DF(u^i)$ denotes the Jacobian matrix and λ_i a suitably chosen damping parameter, $0 < \lambda_i \leq 1$. This method is only locally convergent so a starting guess u^0 which is sufficiently close to the sought solution u^* has to be provided. Recall, that system (3.1) is highly nonlinear and contains drastically different length scales. A strong interaction of nonlinearity and stiffness is to be expected. The use of natural test functions, [7], reduces the influence of stiffness considerably. Further, the application of a weighted norm compensates for the different scaling of the state vector components, c.f. [6,7].

Approximations for the Jacobian matrices are computed by finite differences. The number of necessary evaluations of the discrete equation (3.1) is drastically reduced by using a coloring algorithm, [3,17]. Due to the complexity of the continuous and discrete systems the evaluation is still very costly. It takes up a considerable part of computing time. For its further reduction a simplified Newton method is employed. In order to guarantee fast convergence the use of backdated Jacobians is monitored.

The linear equations for the Newton corrections (3.2) are solved iteratively with the GMRES [16] or the BiCGstab [18] method. The convergence of Krylov subspace methods is highly dependent on an effective preconditioning for the linear systems. Natural candidates are block variants of the Gauss-Seidel method or ILU-decomposition. The size of detailed chemistry models implies, that block preconditioners are computationally rather expensive. On the other hand the large aspect ratios of the grid, the strong coupling via the Arrhenius law and the convection dominated structure of the underlying differential problem requires a highly effective preconditioning. Therefore, the linear systems, $Ax = f$, with $A = DF(u^i)$ and $f = -F(U)$ are preconditioned by a 9-point block-ILU decomposition. The ordering of the unknowns has a strong impact on the convergence speed of the used iterative methods. Compared with other choices, lexicographic ordering of grid points along the dominant convection direction, e.g. in axial direction, reduces the number of iterations of up to $20 - 30\%$. Practical experience shows, that the discussed solution approach is highly effective and robust for the considered problems.

Newtons method requires a sufficiently good starting approximation for convergence. For stationary models in general and for highly nonlinear flame problems in particular this appears to be a nontrivial task. The difficulty is caused by the exponential nonlinearity introduced through Arrhenius kinetic for the reactive sources. They imply a strong coupling of flow fields with temperature. A natural way to obtain sufficiently good starting estimates for diffusion flames is to solve the flame sheet model, c.f. [3,9,17]. This model considers a single irreversible reaction between fuel and oxidizer in the presence of an inert gas (N_2),

$$CH_4 + 2O_2 \rightarrow CO_2 + 2H_2O. \qquad (3.3)$$

It assumes infinitely fast reaction such that the flame shrinks to an infinitely thin flame sheet. Thus, this flame model amounts to solve the Navier-Stokes equations together with a source free conservation equation of the form

$$\frac{\partial}{\partial z}\left(S\frac{\partial \psi}{\partial r}\right) - \frac{\partial}{\partial r}\left(S\frac{\partial \psi}{\partial z}\right) - \frac{\partial}{\partial r}\left(r\rho D\frac{\partial S}{\partial r}\right) - \frac{\partial}{\partial z}\left(r\rho D\frac{\partial S}{\partial z}\right) = 0. \tag{3.4}$$

Here S denotes a conserved scalar and D the diffusion constant. Approximations for the temperature field and the mass fractions of the species in (3.3) can be recovered from the conserved scalar. The solution of this simplified model provides a structurally correct approximation for the detailed chemistry model. Nevertheless, the quality is generally not sufficient to ensure convergence of the Newton method. An improved starting guess for the stationary model can be found by solving the transient formulation

$$\frac{\partial u}{\partial t} + F(u) = 0, \quad t \geq 0, \tag{3.5}$$

of model (2.1)-(2.5) with appropriate initial-boundary conditions. Note however, that this approach drastically increases the computational complexity of the original stationary problem. Let the temporal derivative be discretized by the implicit Euler method

$$u_n = u_{n-1} - h_n F(u_n), \tag{3.6}$$

with the current time step h_n. Numerical solution of (3.6) implies the solution of the fully nonlinear set of conservation equations in each step. Thus, quasi time stepping does not aim at the resolution of the true transient behavior, but can be seen as a homotopy method for approaching the stationary solution. Consequently, the time steps are restricted only by the condition, that the Newton method for (3.6) converges. This iteration is usually started with a value u_n^0, predicting u_n. Two obvious choices would be either $u_n^0 = u_{n-1}$ or the explicit Euler prediction $u_n^0 = u_{n-1} - h_n F(u_{n-1})$. However, both choices limit the maximal acceptable step size, because they are computationally intense or instable. The so-called semi-implicit Euler predictor

$$u_n^0 = (I + h_n DF(u_{n-1}))^{-1}(u_{n-1} - h_n F(u_{n-1})), \tag{3.7}$$

is a preferable choice. It is stable for stiff initial value problems and provides a higher order approximation for u_n. This approach can be combined with an adaptive stepsize control along the lines, indicated in [7]. The resulting algorithm is highly effective in producing sufficiently close starting approximations for the stationary problem (3.1).

Table 1: Quasi time stepping, CPU-times (sec)

Grid	16×16	32×32	64×64	128×128
transient	72.0	495.0	3000.0	—
homotopy	14.6	73.0	466.3	—
stationary	3.4	13.7	86.8	665.4

Table 1 compares the typical computational costs for a true resolution of the transient problem (3.5) and the described homotopy approach for the flame sheet model with identical initial values u_0. A further, substantial reduction of computing time can be achieved by combining the solution on successively refined grids with the homotopy approach. Here the problem is first solved on a coarse grid. This solution is then taken as a starting guess on the next finer grid. Results are shown in the last row of Table 1.

4. Parallel solution strategies

In order to find efficient parallel algorithms it is mandatory to decompose the algorithmic workload to be performed into equal tasks. The system of conservation equations for the detailed reaction flame model is discretized on a tensor product grid. Hence, the distribution of discrete data is quite straight forward. The data is allocated either on a processor pipe or on a two dimensional array of processors. This induces a decomposition of the nodes into strips or rectangular patches, see Figure 3.

When deciding on the algorithmic part of the parallelization strategy it should be kept in mind that the problem to be solved is highly nonlinear and convection dominated. A change of the computational environment should not imply a failure or even worse, produce wrong results. Therefore we have followed a strategy which keeps a global control of the complete nonlinear and linear subproblems. Hence, parallelization is regarded as a distribution of computational work rather than a decomposition of the original problem into more or less independent tasks. This approach might slightly increase the amount of global communication, but leads to a substantial gain in robustness.

The evaluation of the discrete equations (3.1) and the finite difference approximation of the Jacobian contributes the most intensive calculations to the solution process. Decomposition of nodes into strips or patches allows to perform this computations with local, next neighbor communications only. They can be executed in parallel and are therefore fully scalable. Newton and Krylov subspace methods require the evaluation of so-called global functions like scalar products or norms. This implies global communication and potentially reduces the scalability of the approach for a large number of processors. As the global operations are rather infrequent their impact is quite moderate.

An essential part which has to be performed in parallel is the preconditioning of the linear systems for the Newton method. Let the problem be discretized with $n_r \times n_z$ nodes with a 9-point stencil. Then the Jacobian matrix has a total dimension of $n_r n_z$ block rows and columns. This block-matrix is allocated columnwise according to the corresponding nodes onto the processors. Most effective preconditioning algorithms, in particular ILU-decompositions, are highly recursive. There are two basic approaches to treat this difficulty. Either strictly local preconditioning can be considered [6]. This can however, result in a severe loss of convergence speed and robustness. Alternatively, a global preconditioning can be performed. As a consequence communication is required and produces some kind of overhead.

We have selected a hybrid approach consisting of data and domain decomposition components for parallel preconditioning. In the downstream or convection direction a data decomposition is applied. It is based on a thorough analysis of the data dependence for ILU-decomposition. This reveals the possibility of an implementation as a pipelined algorithm for a pipe of processors. See [3,8,9] for a detailed description of the algorithm.

The most important property of this approach is, that the parallel and the sequential preconditioning are exactly identical. Thus, no loss of convergence speed for the linear and nonlinear iterations has to be expected. This is highly important for convection dominated problems where domain decomposition methods experience a substantial loss of preconditioning quality.

The limitations of this kind of data decomposition are twofold. First, the parallel algorithm exchanges frequently rather small amounts of data. The fine granularity does reduce the parallel efficiency when solving scalar or small systems of differential equations. For the detailed model, the number of algebraic operations per node is sufficiently large in order to keep the parallel efficiency degradations rather moderate. Compare comments in [2,3,8,9] and the following section. Second, the number of efficiently usable processors is essentially limited by the number of nodes $n_z/2$ in axial direction. Thus, while the number of nodes grows quadratically when refining the grid, the number of processors may be increased only linearly. Whether this is a serious limitation depends mainly on the the parallel computer used. As a rule of thumb it is more serious for massively parallel, moderate performance processor machines.

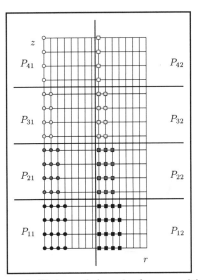

Figure 3: Data and domain decomposition

In order to allow efficient exploitation of more parallel processors the data might be distributed on a two dimensional array, see Figure 3. For each column of processors, $P_{ij}, 1 \leq i \leq p_z$, j fixed, this might be interpreted as a domain decomposition in the radial, rather diffusion dominated direction. To derive a domain decomposition preconditioning let the global Jacobian matrix A be partitioned into blocks according to the decomposition into strips. The remaining coupling blocks between the different subdomains can be eliminated by

$$A = LDU, \quad (4.1)$$

with appropriately shaped block matrices L, U and D. Now D is block diagonal with respect to the domain decomposition. The lower and upper block matrices L and U contain the coupling between neighboring domains. Note, that elimination (4.1) would in general destroy the sparsity structure of the original Jacobian. Therefore it is performed in the sense of an incomplete LU-decomposition such that the original block structure is retained. This step implies a certain loss for the quality of the resulting preconditioning, but still keeps essentially the next neighbor coupling of the linear system consistently. See [8,9] for a more detailed description.

The actual preconditioning is carried out in three steps

$$Lz_1 = b, \quad Dz_2 = z_1, \quad Uy = z_2. \tag{4.2}$$

Noting that the matrix D only consists of local interaction, we can apply a strictly local ILU-decomposition for D and do a standard local preconditioning step herewith. Thus only the remaining factors L and U involve next neighbor communication for this kind of preconditioning. Furthermore, this approach can be interpreted as a minimal overlapping domain decomposition technique.

Finally, data and domain decomposition methods can be combined in a natural way. In each of the p_r strips or domains we can still apply data decomposition with p_z processors. The computing times for various combinations of a fixed number of processors $p = p_z \cdot p_r$ in radial and axial directions on a variety of distributed memory parallel computers will be discussed in the following chapter.

6. Numerical results

In this section we present a number of results on the detailed reaction mechanism model for the bunsen burner, described in section 3. The size of the coflowing input jets is given by $R_{in} = 0.2\,cm$ and $R_{out} = 5.0\,cm$. The inflow velocities are $v_f = 5.0\,cm/s$ for the fuel and $v_{air} = 25.0\,cm/s$ the air jets respectively. The reaction mechanism with 46 chemical reactions and 16 different species is taken from [9]. The most striking property of this model is its extreme complexity. While for the simplified flame sheet model only three equations have to be solved, [2,3,5,6], we now have to deal with a system of 19 strongly coupled equations. Especially the evaluation of the discrete equations imply the highly expensive calculation of the chemical source terms via the CHEMKIN package at all grid points. Moreover, the Jacobian contains large dense blocks per grid point. Thus, the iterative solution of the linear systems becomes rather expensive.

The computation is started on a rather coarse grid. The coarsest grid with successful resolution of intermediate species, such as OH, consists of 32×32 points. On this grid the quasi time stepping, as described in section 3, was applied. The refinement rule used was a simple doubling of gridpoints in axial and radial direction. Note however, that the coarsest grid is locally refined around the flame profile. With this approach no quasi time stepping was needed on finer grid levels. Further, the number of highly damped Newton iterations was kept to a minimum for finer grid levels. The resulting solution profiles for O_2, H_2O, CO and CO_2 are shown in Figures 4-7.

The computation on a Sparc 10-20 workstation took 33h for the coarsest grid. A reasonable accurate resolution of this problem on a single workstation is out of question. Thus, any further computation has been done on various distributed memory parallel computers. Table 2 shows the required computing time and necessary storage requirements for solutions on grids of different sizes. Note, that the solution time on the coarsest grid also contains the time, necessary for the quasi time stepping.

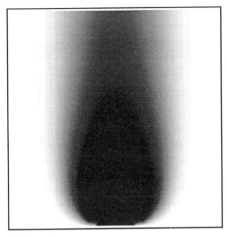

Figure 4: O2 mass fraction

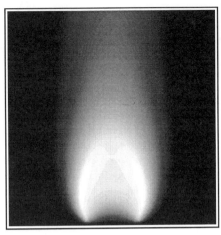

Figure 5: H2O mass fraction

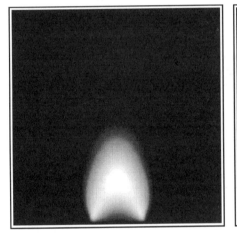

Figure 6: CO mass fraction

Figure 7: CO2 mass fraction

Table 2: Parsytec GC/PP results for different resolutions

Grid $(n_r \times n_z)$	32×32	64×64	128×128
core memory (MByte)	57.4	228.8	827.4
processors $(p_r \times p_z)$	16×4	32×2	64×1
cpu-time (sec)	934.0	416.94	917.80

For a fixed grid size the granularity becomes finer as the number of processors increases. In the limiting case, $p_z = 64$, only two consecutive grid points are located on the same processor, increasing the number of communication steps to an upper limit. If the number of processors is doubled to 64×2, the computing time is reduced by a factor of 1.9. The achieved parallel efficiency of 0.526 is surprisingly high for the data decomposition approach. This is due to the large number of algebraic operations executed per grid point. Hence, even the very fine grain data decomposition shows excellent parallel efficiency.

A systematic analysis of the efficiencies achieved with various number of processors in direction of data and domain decomposition is shown in Table 3. Note that the total number of processors and thus the peak performance as well as the problem size is fixed. The optimal overall performance is gained for a real combination of data and domain decomposition strategies. Even if this optimal choice is not used, the expected loss of efficiency will not be dramatic. This conclusion might however slightly change for different problem sizes and other parallel computers. Similar results hold for the BiCGstab method. Finally, for the detailed chemistry model the amount of storage required by GMRES is strongly dominated by the Jacobian matrix.

Table 3: Parsytec GC results, fixed resolution, cpu-times (sec)

$p = p_z \times p_r$	64×1	32×2	16×4	8×8	4×16
cpu-time(GMRES)	917.8	742.6	686.5	918.3	1028.8
cpu-time(BiCGstab)	974.3	757.1	906.2	749.01	881.7

We compare the achieved performance on the Parsytec machine with results on two other popular parallel computers. First, consider the performance of Cray T3D. For small numbers of processors this computer shows essentially the same processing power as the Parsytec Power/PC. If however, the full number of processors are employed, the computing performance and the efficient interaction with the powerful communication network is demonstrated quite impressively, see Table 4.

Table 4: Cray T3D results

Grid $(n_r \times n_z)$	32×32	64×64	128×128
$p = p_z \times p_r$	16×4	32×2	64×1
cpu-time (sec)	1120.7	504.4	798.1
$p = p_z \times p_r$	–	32×8	64×4
cpu-time (sec)	–	331.9	285.6

The IBM-SP2 parallel computer has quite different architectural parameters. It is based on high performance processing nodes, with 256 MFlops each. They are connected via a high performance network. Thus, high overall performance is already achieved for a rather moderate number of nodes. Table 5 shows the computing times on a 16 node IBM-SP2 in direction of data decomposition. The performance is comparable to the application of a Parsytec PowerPC or a Cray T3D machine with 64 processors each.

For parallel computers with moderate performance nodes a considerably finer granularity of the problem is necessary. For typical engineering applications this might imply more effort for parallelization strategies, to gain sufficient scalability. The domain decomposition has been developed for exploitation up to 256 processors with reasonable efficiency. The results show, that this would have not been necessary for the IBM-SP2.

Table 5: IBM-SP2 results, cpu-times for $p = p_z p_r = 16 \times 1$

Grid $(n_r \times n_z)$	32×32	64×64	128×128
cpu-time (sec)	1443.01	655.31	984.94

Hence, we conclude that the parallelization strategies presented, provide the necessary flexibility to achieve high parallel efficiencies on quite different distributed memory parallel computers. Further, they provide an interesting and robust alternative to pure domain decomposition methods for highly convection dominated and strongly nonlinear problems.

Acknowledgment: This research is supported by the Deutsche Forschungsgemeinschaft within the *Schwerpunktprogramm "Strömungssimulation mit Hochleistungsrechnern"*. The numerical results on the IBM-SP2 were achieved at the "Gesellschaft für Mathematik und Datenverarbeitung" in Bonn, St. Augustin and at the BTU Cottbus. Computations on the Cray T3D/MCN320-8 were performed at the Edinburgh Parallel Computing Centre, University of Edinburgh during a visit under the TRACS program by the Human Capital and Mobility Program of the European Community. We also thank D. Eggers for helpful comments on an early draft of this paper.

References

[1] D. A. ANDERSON, J. C. TANNEHILL, R. H. PLETCHER, *Computational Fluid Mechanics and Heat Transfer*, New York, Hemisphere Publishing Corporation (1984).

[2] G. BADER, E. GEHRKE, *On the performance of transputer networks for solving linear systems of equations*, Parallel Computing, Vol. 17 (1991) 1397–1407.

[3] G. BADER, E. GEHRKE, *Solution of Flame Sheet Models on Transputer Systems*, in Flow Simulation with High-Performance Computers I, DFG Priority Research Programme Results 1989-1992, E. H. Hirschel (Hrsg.), Notes on Numerical Fluid Mechanics, Vol. 38, Vieweg (1993) 31–40.

[4] M. W. CHASE JR., C. A. DAVIES, J. R. DOWNEY JR., D. J. FRURIP, R. A. MCDONALD AND A. N. SYVERUND, *Jannaf thermochemical tables, third edition*, J. Phys. Chem. Ref. data Suppl. 1, 14 (1985) 1–1856.

[5] A. R. CURTIS, M. J. POWELL, J. K. REID, *On the Estimation of Sparse Jacobian Matrices*, J. Inst. Math. Appl. 13 (1974) 117–119.

[6] A. ERN, C. C. DOUGLAS, M. D. SMOOKE, *Detailed chemistry Modelling of laminar Diffusion Flames on parallel Computers*, Yale University, (1994)

[7] P. DEUFLHARD, *A modified Newton method for the solution of ill-conditioned systems of nonlinear equations with applications to multiple shooting,* Numer. Math., 22 (1974) 289–315.

[8] E. GEHRKE, *Parallele Methoden zur numerischen Simulation reaktiver Strömungen am Beispiel laminarer Diffusionsflammen,* PhD thesis, Institute of applied mathematics, University of Heidelberg (1994).

[9] E. GEHRKE, G. BADER, *Parallel Methods for the Numerical Simulation of Stationary Diffusion Flames,* in Proceedings of the 2nd Summer Conference Numerical Modelling in Continuum Mechanics, Academy of Sciences of the Czech Republic, Vol. 40, Prague, (1995).

[10] W. G. GROPP, D. E. KEYES, *Domain Decomposition with local Mesh Refinement,* Techn. Report 91-19, NASA Langley Research Center (1991).

[11] W. GROPP, E. LUSK AND A. SKJELLUM, *Using MPI: Portable Parallel Programming with the Message Passing,* MIT Press (1994).

[12] R. J. KEE, J. A. MILLER, T. H. JEFFERSON, *CHEMKIN: A General-Purpose, Transportable Fortran Chemical Kinetics Code Package,* Sandia National Laboratories Report, SAND80-8003.

[13] D. E. KEYES, M. D. SMOOKE, *Flame Sheet Starting Estimates for Counterflow Diffusion Flame Problems,* J. of Comp. Physics, vol. 73 (1986) 267–288.

[14] D. E. KEYES, M. D. SMOOKE, *A Parallelized Elliptic Solver for Reacting Flows,* in Parallel Computations and Their Impact on Mechanics. A.K. Noor (Ed.), Numerical Society of Mechanical Engineers (1989) 375–402.

[15] M. MEINKE, E. ORTNER, *Implementation of Explicit Navier-Stokes Solvers on Massively Parallel Systems,* in Flow Simulation with High-Performance Computers I, DFG Priority Research Programme Results 1989-1992, E. H. Hirschel (Hrsg.), Notes on Numerical Fluid Mechanics, Vol. 38, Vieweg (1993) 138–151.

[16] Y. SAAD, M. H. SCHULTZ, *GMRES: a generalized minimal residual algorithm for solving nonsymmetric linear systems,* SIAM J. Sci. Stat. Comput. 7 (1986) 856–869.

[17] M. SMOOKE, R. MITCHELL AND D. KEYES, *Numerical Solution of Two-Dimensional Axisymmetric Laminar Diffusion Flames,* Combust. Sci. and Tech., 67 (1989) 85–122.

[18] H. A. VAN DER VORST, *BI-CGSTAB: A FAST and Smoothly Converging Variant of BI-CG for the Solution of Nonsymmetric Linear Systems,* SIAM J. Sci. Stat. Comput. 13 (1992) 631–644.

[19] F. A. WILLIAMS, *Combustion Theory,* Addison-Wesley Publishing Company Inc, second edition (1988).

An Adaptive Operator Technique for Hypersonic Flow Simulation on Parallel Computers

Ioannis St. Doltsinis, Jürgen Urban

University of Stuttgart
Institute for Computer Applications
Pfaffenwaldring 27, 70569 Stuttgart, Germany

Abstract

The aerodynamics of reentry vehicles raised to the authors a number of tasks related to the physical modeling and the numerical simulation of chemically reacting hypersonic viscous flows [1, 2, 3]. An incorporation of real gas effects into the numerical simulation necessitates enormous computer capacity and demands an enhanced efficiency of the numerical computation as well as the introduction of parallel processing. For this purpose, a specific adaptive method is presented aiming at a reduction of the number of numerical operations, and a parallel algorithm is described. In connection with parallel processing, the adaptive method is seen to cause an uneven distribution of numerical operations among the processors. In order to avoid idle-times on any processing unit, a dynamic re-distribution technique has been conceived and investigated.

Introduction

Hypersonic flows are characterised by various physical phenomena which occur in certain regions within the flow field and influence each other. The specific phenomena have been studied intensively for various reentry configurations by different resarch groups utilising high performance computing [4, 5, 6]. In the hypersonic regime, the convective transport of mass, momentum and energy has a determining influence on the development of the flow. However, in certain restricted regions where the convective transport is low, e.g. near solid walls, diffusive transport is of importance. Chemical reactions affect markedly the flow field only in certain high-temperature zones behind shock waves. Consequently, species transport has to be considered merely in the flow field downstream of the reactive zone.
Apart from the convective transport of mass, momentum and energy, all other physical mechanisms appear in regions of limited extent within the flow field. Therefore, the computation of these effects may be neglected outside these regions and may thus lead to a significant reduction of the numerical operations. To this end, the system of differential equations is represented by a number of flux and source operators which may be activated or neglected in confined regions. The application of the individual operators is controlled by a frequent analysis of the evolving solution for the flow field. The proposed adaptive

operator technique has been studied extensively for the two-dimensional case of a flow around a double ellipse [7]. In the present account, the method is considerably advanced, especially towards large-scale three-dimensional applications.

The adaptive technique has been also implemented in an integrated parallel software being developed at the ICA as based on the Finite Element Programming System FEPS of the Institute [8, 9]. This parallel software comprises a parallel mesh generator performing a block-level decomposition for arbitrary two- and three-dimensional domains and subsequent generation of the mesh. The decomposed treatment of the computational domain necessitates data exchange on the interfaces between adjected subdomains. For that purpose, a particular library has been developed which provides a set of functions for communication tasks within the parallel finite element algorithm. Finally, visualisation is again fully integrated into a parallel environment [10].

Application of the adaptive operator technique in the parallel execution mode causes an uneven distribution of numerical operations among the processing units, disturbing the initial balance of the computational load. In order to avoid idle–times on any participating processor and thus a loss of efficiency of the parallel computation, a re-balancing of the computational load is mandatory. For this purpose, a dynamic re-distribution technique has been conceived and investigated. In this context, parts of the numerical operations are evacuated from overloaded to underloaded processors. For applications which are characterised by coarse granularity, the communication time required for the balancing process is low enough and the re-distribution technique is seen to increase the efficiency of the overall procedure.

Method of adaptive operators

A mathematical description of viscous hypersonic flows is provided by the well-known compressible Navier–Stokes equations. The occurrence of chemical reactions in a multi-component gas mixture necessitates additional considerations. In order to evaluate convective and viscous fluxes, the caloric and thermal properties of the gas mixture have to be determined. The determination of these properties requires the knowledge of the local gas composition which has to be deduced from the solution of the equations for the conservation of species. As the species equations are strongly coupled with the flow equations via the velocity field, they have to be solved simultaneously with the latter. The system of differential equations governing hypersonic viscous flow with chemical reactions may then be written in terms of distinct physical operators as follows

$$\mathcal{G} : \frac{\partial}{\partial t} \begin{pmatrix} \boldsymbol{u} \\ \boldsymbol{w} \end{pmatrix} + \begin{pmatrix} \mathcal{B} \\ \mathcal{B}^S \end{pmatrix} + \begin{pmatrix} \mathcal{D} \\ \mathcal{D}^S \end{pmatrix} = \begin{pmatrix} \boldsymbol{0} \\ \mathcal{P}^S \end{pmatrix} \tag{1}$$

where the array $\boldsymbol{u} = \{\varrho, \varrho\vec{v}, \varrho\varepsilon\}$ comprises the density ϱ, velocity $\vec{v}$ and energy ε of the mixture governed by the Navier-Stokes equations, and $\boldsymbol{w} = \{\varrho_1 \ldots \varrho_{n_s}\}$ the densities of the species governed by the individual mass conservation equations.

Definition of physical operators

Physical flux operators A restriction of the modelling to convective transport of mass, momentum and energy defines the basic Euler operator $\mathcal{B} = \{\mathcal{B}^\mathcal{K}, \mathcal{B}^{\vec{\mathcal{I}}}, \mathcal{B}^\mathcal{E}\}$. The constituents of the operator are defined by

$$\mathcal{B}^\mathcal{K} = \vec{\nabla} \cdot (\varrho \vec{v}), \qquad \mathcal{B}^{\vec{\mathcal{I}}} = \vec{\nabla} \cdot \left(\varrho \vec{v} \vec{v} + p \overline{\overline{I}}\right), \qquad \mathcal{B}^\mathcal{E} = \vec{\nabla} \cdot (\varrho \vec{v} \varepsilon + p \vec{v}). \tag{2}$$

This basic operator refers to an inviscid and non-reacting gas flow where energy and pressure do not depend on the species distribution, thus $e = e(T)$ and $p = p(\varrho, T)$.

In regions where viscous flux significantly influences the flow field, a diffusive operator $\mathcal{D} = \{\mathcal{D}^\mathcal{K}, \mathcal{D}^{\vec{\mathcal{I}}}, \mathcal{D}^\mathcal{E}\}$ associated with heat conduction and shear stresses is added to the basic operator. The constituents of the diffusive operator read

$$\mathcal{D}^\mathcal{K} = 0, \qquad \mathcal{D}^{\vec{\mathcal{I}}} = -\vec{\nabla} \cdot (\overline{\overline{\tau}}), \qquad \mathcal{D}^\mathcal{E} = -\vec{\nabla} \cdot \left(k \vec{\nabla} T + \overline{\overline{\tau}} \cdot \vec{v}\right) \tag{3}$$

and contain molecular transport coefficients which may be introduced by simple expressions like the Sutherland–Fick law where the thermal conductivity k and the viscosity μ are functions of the temperature solely. The coefficient μ defines the viscous stress $\overline{\overline{\tau}}$.

The consideration of convective transport of the n_s species requires an additional set of differential equations governing the unknown vector $\boldsymbol{w} = \{\varrho_1 \ldots \varrho_{n_s}\}$ by the operator $\mathcal{B}^\mathcal{S} = \{\mathcal{B}_1^{\mathcal{K}_r} \ldots \mathcal{B}_{n_s}^{\mathcal{K}_r}\}$ with constituents

$$\mathcal{B}_r^{\mathcal{K}_r} = \vec{\nabla} \cdot (\varrho_r \vec{v}) \qquad \text{for} \quad r = 1 \ldots n_s. \tag{4}$$

Within the flow regimes, where convective transport of species is considered, energy and pressure depend on all species mass fractions ξ_r, thus $e = e(T, \xi_r)$ and $p = p(\varrho, T, \xi_r)$.

Diffusive transport of the species in boundary layers may be treated as molecular mass diffusion. Therefore, the operator $\mathcal{D}^\mathcal{S} = \{\mathcal{D}_1^{\mathcal{K}_r} \ldots \mathcal{D}_{n_s}^{\mathcal{K}_r}\}$ has to be considered. The constituents of this diffusive operator read

$$\mathcal{D}_r^{\mathcal{K}_r} = \vec{\nabla} \cdot \left(\vec{j}_r\right) \qquad \text{for} \quad r = 1 \ldots n_s. \tag{5}$$

Mass diffusion causes additional transport of dissipative energy, considered in the operator $\mathcal{D}^\mathcal{E}$ in eq. (3) by the term

$$\mathcal{D}^{\mathcal{E},\mathcal{S}} = -\vec{\nabla} \cdot \left(\sum_r \varepsilon_r \vec{j}_r\right). \tag{6}$$

In regions where the flow of a multi-component mixture has to be considered, the molecular transport coefficients in the diffusive operators $\mathcal{D}$ and $\mathcal{D}^\mathcal{S}$ may be defined by explicit expressions which have been introduced by the Straub model [11, 2]. In contrast to the simple Sutherland–Fick formula, these expressions also depend on the mass fraction and pressure, thus $\mu = \mu(p, T, \xi_r)$, $k = k(p, T, \xi_r)$, $D_{ij} = D_{ij}(p, T, \xi_r)$.

Physical source operator The production and reduction of species by chemical reactions is considered by source terms in the continuity equations for the species. They are represented in eq. (1) by the production operator $\mathcal{P}^S = \{\mathcal{P}_1^{\mathcal{K}_r} \ldots \mathcal{P}_{n_s}^{\mathcal{K}_r}\}$ which may be considered as a horizontal enlargement of the operator $\mathcal{B}^S$. The entities of the production operator are

$$\mathcal{P}_r^{\mathcal{K}_r} = \dot{\omega}_r = \dot{\omega}_r(\boldsymbol{w}) \quad \text{for} \quad r = 1 \ldots n_s. \tag{7}$$

Definition of numerical operators

In order to integrate eq. (1) within the small interval in time $\tau = {}^b t - {}^a t$, it appears advantageous to distinguish between flow and chemical phenomena because they occur as a rule on significantly different time scales. In this connection, the chemical reactions are first assumed frozen ($\mathcal{P}^S = \boldsymbol{o}$). The numerical integration is then performed by an explicit two-step finite element procedure. Here, the governing equation eq. (1) is first discretised with respect to time by a Taylor–series expansion and is subsequently approximated in the space domain via the Galerkin technique [1]. The change in the gas composition due to chemical reactions is then determined separately by integration of the rate expression eq. (7) within the same interval of time and is added to the homogeneous part of eq. (1).

A set of numerical operators is defined which represent the contributions of the individual physical operators to the volume and boundary integrals of the discretised system. Thus, the discretisation of the homogeneous system $\mathcal{G}_h$ in the time and space domain leads to a system of linear equations which may be written in terms of numerical operators analogous to those in eq. (1)

$$\boldsymbol{G}_h : \begin{pmatrix} \boldsymbol{M}\delta \boldsymbol{U}^n \\ \boldsymbol{M}\delta \boldsymbol{W}^n \end{pmatrix} + \begin{pmatrix} \boldsymbol{B} \\ \boldsymbol{B}^S \end{pmatrix} + \begin{pmatrix} \boldsymbol{D} \\ \boldsymbol{D}^S \end{pmatrix} = \begin{pmatrix} \boldsymbol{o} \\ \boldsymbol{o} \end{pmatrix}. \tag{8}$$

In order to solve eq. (8), the consistent mass matrix $\boldsymbol{M}$ is first replaced by the diagonalised lumped mass matrix $\boldsymbol{M}_L$. The incremental solutions $\delta \boldsymbol{U} = \{\delta \boldsymbol{U}_{\mathcal{K}}, \delta \boldsymbol{U}_{\vec{\mathcal{I}}}, \delta \boldsymbol{U}_{\mathcal{E}}\}$ and $\delta \boldsymbol{W} = \{\delta \boldsymbol{W}_{\mathcal{K}_1} \ldots \delta \boldsymbol{W}_{\mathcal{K}_{n_s}}\}$ of the conservative variables are then obtained iteratively. A further improvement of the numerical solution is achieved via artificial damping as described in [1].

The change of the gas composition due to chemical reactions is then determined separately by integration of the production rate and provides

$$\delta \boldsymbol{W}_{qV}^n = \boldsymbol{P}^S. \tag{9}$$

Discrete flux operators The numerical fluxes arising from the integration of the discretised Euler equations are represented by the numerical basic operator $\boldsymbol{B} = \{\boldsymbol{B}^{\mathcal{K}}, \boldsymbol{B}^{\vec{\mathcal{I}}}, \boldsymbol{B}^{\mathcal{E}}\}$ which consists of contributions from the volume and surface integrals, $\boldsymbol{R}$ and $\boldsymbol{S}$.

$$\boldsymbol{B}^{\mathcal{K}} = \boldsymbol{S}^{\mathcal{K}} - \boldsymbol{R}_C^{\mathcal{K}}, \qquad \boldsymbol{B}^{\vec{\mathcal{I}}} = \boldsymbol{S}^{\vec{\mathcal{I}}} - \boldsymbol{R}_C^{\vec{\mathcal{I}}}, \qquad \boldsymbol{B}^{\mathcal{E}} = \boldsymbol{S}^{\mathcal{E}} - \boldsymbol{R}_C^{\mathcal{E}}. \tag{10}$$

An additionally activated diffusive operator $\boldsymbol{D} = \{\boldsymbol{D}^{\mathcal{K}}, \boldsymbol{D}^{\vec{I}}, \boldsymbol{D}^{\varepsilon}\}$ accounts for diffusive effects in the flow field. The contributions from the diffusive fluxes from the volume and surface integrals read

$$\boldsymbol{D}^{\mathcal{K}} = \boldsymbol{0}, \qquad \boldsymbol{D}^{\vec{I}} = \boldsymbol{S}_D^{\vec{I}} - \boldsymbol{R}_D^{\vec{I}}, \qquad \boldsymbol{D}^{\varepsilon} = \boldsymbol{S}_D^{\varepsilon} - \boldsymbol{R}_D^{\varepsilon} \; . \tag{11}$$

The numerical fluxes arising from the integration of the discretised species equations are represented by the operator $\boldsymbol{B}^S = \{\boldsymbol{B}_1^{\mathcal{K}_r} \ldots \boldsymbol{B}_N^{\mathcal{K}_r}\}$, which reads

$$\boldsymbol{B}_r^{\mathcal{K}_r} = \boldsymbol{S}^{\mathcal{K}_r} - \boldsymbol{R}_C^{\mathcal{K}_r} \qquad \text{for} \quad r = 1 \ldots n_s. \tag{12}$$

Diffusive fluxes from the species equations are represented by the operator $\boldsymbol{D}^S = \{\boldsymbol{D}_1^{\mathcal{K}_r} \ldots \boldsymbol{D}_{n_s}^{\mathcal{K}_r}\}$ which again contains contributions from volume and surface integrals

$$\boldsymbol{D}_r^{\mathcal{K}_r} = \boldsymbol{S}_D^{\mathcal{K}_r} - \boldsymbol{R}_D^{\mathcal{K}_r} \qquad \text{for} \quad r = 1 \ldots n_s. \tag{13}$$

Discrete source operator In order to account for changes in the gas composition caused by chemical reactions, the nodal operator $\boldsymbol{P}^S = \{\boldsymbol{P}_1^{\mathcal{K}_r} \ldots \boldsymbol{P}_{n_s}^{\mathcal{K}_r}\}$ has to be introduced

$$\boldsymbol{P}_r^{\mathcal{K}_r} = \int_{0_t}^{b_t} \mathcal{P}_r^{\mathcal{K}_r} dt \qquad \text{for} \quad r = 1 \ldots n_s. \tag{14}$$

Activation of numerical operators

Within the incremental scheme, the numerical operators $\boldsymbol{B}, \boldsymbol{D}, \boldsymbol{B}^S, \boldsymbol{D}^S$ and $\boldsymbol{P}^S$ may now be activated in restricted regions of the flow field in compliance with the actual physics of the problem. To this end, the discretised flow domain Ω and its boundary Γ are divided on the element level in non-overlapping subdomains Ω_i and Γ_i respectively, where

$$\Omega = \bigcup_j \Omega_j, \quad \text{and} \quad \Gamma = \bigcup_j \Gamma_j \qquad \text{for} \quad j = B, D, B^S, D^S.$$

Here, the index j marks the flux operator which may be activated in addition to the basic operator. The operator $\boldsymbol{D}^S$ however may only be activated if both $\boldsymbol{B}^S$ and $\boldsymbol{D}$ are operative, thus

$$\Omega_{D^S} = \Omega_{B^S} \cap \Omega_D \qquad \text{and} \qquad \Gamma_{D^S} = \Gamma_{B^S} \cap \Gamma_D.$$

A division of the flow field is performed also on the nodal point level. To this end, non-overlapping subdomains Ω_k are defined, where the production operator $\boldsymbol{P}^S$ or a virtual operator $\boldsymbol{O}$ with components 0 is activated

$$\Omega = \bigcup_k \Omega_k \qquad \text{for} \quad k = P^S, O \qquad \text{and} \quad \Omega_{P^S} \subset (\Omega_{B^S} \cap \Omega_{D^S}).$$

As a restriction, the production operator can only be activated within regions where at least convective transport of species is considered.

A local activation of numerical operators is performed on the basis of a frequently repeated analysis of the development of the numerical solution for the viscous multi-component flow field. In the course of the procedure, the contributions of the individual numerical operators are normalised in order to provide suitable indicators which may be compared with specified threshold values. Thus, the regions where the diffusive operator D is activated are characterised by the fact that the value of the indicator $\zeta_D = ||D||/||B||$ exceeds the threshold value $\overline{\zeta}_D$. Accordingly, the region where convective transport of species is considered by activation of the operator B^S is restricted by the condition that the value of the indicator $\zeta_{B^S} = ||B^S||/||B^K||$ must exceed the threshold value $\overline{\zeta}_{B^S}$. Finally, the production operator P^S which represents non-equilibrium chemistry is activated if the value of the indicator $\zeta_{P^S} = ||P^S||/||B^S||$ exceeds a threshold value $\overline{\zeta}_{P^S}$ defined a priori.

Application of the adaptive method

The proposed adaptive technique has been studied extensively for the two-dimensional case of a flow around a double ellipse [7]. The consequences of the restricted employment of the numerical operators have been explored for the inviscid as well as for the viscous case. At the same time, the effect of variation of the threshold values on the convergence behaviour was investigated with respect to increasing numerical efficiency.

As a result of our investigations, the application of the adaptive operator technique achieves significant reductions of the computational effort. The confinement of the viscous operator to regions with low convective transport has no pronounced effect on the convergence behaviour or on the quality of the numerical solution if the threshold value does not exceed $\overline{\zeta}_D = 10^{-2}$. Also, the restriction of the convective and diffusive species transport to the regions detected by the identifier does not affect the numerical solution although the convergence behaviour shows some deviations from the regular non-adaptive algorithm. For threshold values $\overline{\zeta}_{B^S} = 10^{-4}$ and $\overline{\zeta}_{P^S} = 10^{-1}$, a satisfactory reproduction of the regular numerical solution is achieved. In all studies, the required cpu-time was reduced by at least 15% by an application of the adaptive method.

For a large-scale application of the adaptive operator technique, the numerical simulation of the symmetric re-entry manoeuvre of a complete three-dimensional configuration of the Hermes space-shuttle has been considered. The manoeuvre conditions are specified by the Mach number $M_\infty = 25$, the angle of incidence of the longitudinal axis of the shuttle $\alpha = 30°$ and the atmospheric conditions at the altitude $H = 75$ km. The three-dimensional space around the orbiter has been discretised by 110,488 hexahedral finite elements. The discretisation mesh has a total of 117,973 nodal points corresponding to 1,178,730 unknown field variables in the discretised Euler equations including chemical reactions and convective species transport. For the purpose of comparison, the solution for the non-reacting flow and two-dimensional solutions for the symmetry plane were computed as well.

As a result, fig. 1.a shows the element fraction for the region Ω_{B^S} with species transport and the nodal point fraction for Ω_{P^S} with chemical production in the course of the incremental procedure for the three-dimensional as well as for the two-dimensional reacting case. In the simulations, the regions Ω_{B^S} reached their maximum extension already after 1000, resp. 2000 time steps. In the stationary state, 89% of the elements were activated for the two-dimensional and 68% in the three-dimensional case. The extent of Ω_{PS} still rises even near stationary state, but with no effect on the solution itself. This is explained by the fact that the indicator $\zeta_{PS} = ||\boldsymbol{P}^S||/||\boldsymbol{B}^S||$ may exceed the threshold value, even if the absolute values of the components $||\boldsymbol{P}^S||$ and $||\boldsymbol{B}^S||$ are insignificantly small. At the end of the computations, 71% of the nodal points were activated for the two-dimensional and 55% in the three-dimensional case.

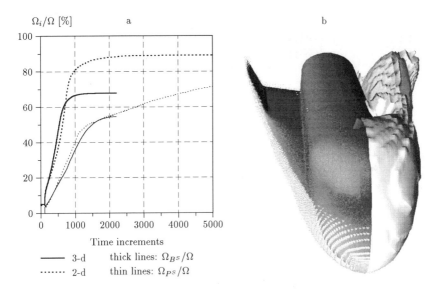

Figure 1: Fraction of activated operators for species transport and chemical production (a). Extent of the activated operators at time step 500 (b). Left: chemical reactions, right: species transport.

A snapshot after 500 time steps indicating the extent of the operators activated in addition to the basic one is presented in fig. 1.b. The mesh nodal points with active production operator $\boldsymbol{P}^S$ are marked at the left-hand side of the symmetry plane of the orbiter. The region where convective species transport has to be taken into account via the operator $\boldsymbol{B}^S$ is contained by the surface on the right-hand side of the symmetry plane. The three-dimensional result was obtained with the Cray C94 D computer. Here, the application of the adaptive technique leads to a reduction of the cpu–time of at least 20%.

An assessment of the quality of the numerical solution may be based on the development of the density residuum in fig. 2.a. For the two-dimensional reacting flow, the density residuum shows a behaviour characteristic of the adaptive technique. If the threshold value $\overline{\zeta}_{B^S}$ exceeds the density residuum, the frequently repeated analysis of the reacting flow field causes an oscillating density residuum, with amplitude related to the threshold value. However, the solution is not markedly affected since the threshold value is small.

As a result of the computation, fig. 2.b indicates the variation of the temperature along the stagnation stream line. It is seen that three-dimensional effects considerably reduce the shock stand-off distance in the chemically frozen and the reacting flow, whilst the maximum temperature is less sensitive. An enormous reduction of the temperature in the flow field and a further decrease of the shock stand-off distances is obtained by the introduction of chemical kinetics. Thus, the stand-off distance is reduced from 34 cm to 15 cm in the two-dimensional case and from 11 cm to 7 cm in the three-dimensional case.

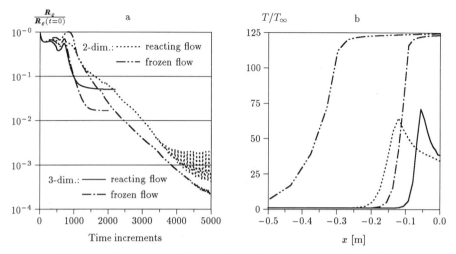

Figure 2: Convergence history for adaptive computations (a). Temperature variation along stagnation stream line (b).

Integrated parallelisation concept

Parallel mesh generator

The parallelisation of the finite element algorithm is based on domain decomposition. Thereby, the computational domain is decomposed into non-overlapping subdomains. This necessitates data exchange at the interface between adjacent subdomains. The parallel algorithm has to be combined with an automatic procedure that subdivides a given mesh into a number of sub-meshes equal to the number of processors available.

The majority of the existing decomposition techniques works on a generated mesh and on the element-level which has a number of drawbacks. The mesh generation and decomposition has to be performed sequentially and can impose a significant overhead on the parallel computation, especially if complex three-dimensional domains have to be decomposed among a large number of processors. Moreover, the length of the interior boundary between subdomains, which for the applications described here is the most important parameter determining the communication overhead and thus the efficiency of the parallel flow computations, cannot be controlled with this method.

To overcome the problems associated with the element-level algorithm a block-level algorithm has been developed at the ICA. Domain decomposition and mesh generation are based on a structured multi-block algorithm for arbitrary two- or three-dimensional domains that allows for user-defined areas of refinement. For the parallel computations, the number of nodes on the faces and edges of the two- or three-dimensional blocks defining the computational domain are calculated. The blocks are then distributed in a balanced manner onto all available processors by recursively splitting the domain in half, resulting in a binary tree structure depicted in fig. 3. During this process, a processor, after splitting the domain at any level, retains the first half and sends the other half-domain to the processor connected to it on the next lower level of the tree. After the block structure is distributed onto the processors in this way, the meshes on each sub-domain can be generated in parallel with no need for further inter-processor communication.

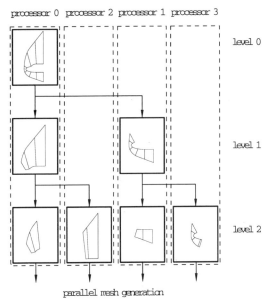

Figure 3: Recursive domain decomposition.

Parallel finite element algorithm

The parallel finite element computation is performed on the same processors where the sub-domain meshes were generated. The inter-processor communication required during the computation is performed via high-level communication primitives specifically designed for finite element algorithms. These are based on a virtual processor topology defined by the tree structure of fig. 3. In particular the exchange of internal boundary data can thus be handled in a quite general and optimised manner. This approach also facilitates the portability of the code considerably by limiting machine-dependent portions to a small kernel, performing mainly a mapping from the virtual topology onto the actual hardware topology. The procedure has been implemented under PARIX on a transputer-based Parsytec MC-3 DE with 44 T805 transputer and a SC with 256 T800 transputer. Recently, it has also been implemented under OSF/1 on an intel Paragon XP/S-5 with 72 i860XP

173

processors.

As an example for the inter-processor communication, the summation of the flux at the interior boundaries will be explained the following. In a first step, the volume and surface integrals for the flux balance are computed on the element-level and accumulated on the nodes of every sub-domain. For the description of the communication process, the flux balance according to eq. (8) may now be written in a simplified manner as

$$M_i(\delta U^n)_i = R_i \quad \text{for} \quad i = 0 \ldots n_p - 1. \tag{15}$$

In order to determine the flux balance at the nodes of the interior boundaries, contributions from both neighbouring subdomains have to be accumulated. This is done by passing twice through the binary tree. First, the binary tree is passed from the highest level n_l ($n_l=2$ in fig. 3) to level 0 and the contributions are accumulated in global boundary vectors R_Γ. The accumulated data are distributed into the subdomains while passing the binary tree back from level 0 to the highest level n_l.

With reference to fig. 3, starting at level n_l of the tree structure, the contributions $R_{\Gamma_{j,\nu}}$ from the interior boundaries generated at this decomposition level (ν) are communicated from the left processor j to the right processor k and accumulated in the global boundary vector $R^\star_{\Gamma_{k,\nu}}$ of processor k:

$$R_{\Gamma_{j,\nu}} \longrightarrow R^\star_{\Gamma_{k,\nu}} = R_{\Gamma_{k,\nu}} + R_{\Gamma_{j,\nu}} \tag{16}$$

where $l = n_l \ldots 1, \quad j = 0 \ldots 2^{(l-1)} - 1, \quad k = j + 2^{(l-1)}$.

Afterwards, the contributions $R_{\Gamma_{k,o}}$ from the interior boundaries generated at a lower decomposition level (o) are communicated from the right processor to its left neighbour. The contributions are then accumulated in the boundary vector $R^\star_{\Gamma_{j,o}}$ of processor j.

$$R^\star_{\Gamma_{j,o}} = R_{\Gamma_{j,o}} + R_{\Gamma_{k,o}} \longleftarrow R_{\Gamma_{k,o}} \tag{17}$$

where $l = n_l \ldots 1, \quad j = 0 \ldots 2^{(l-1)} - 1, \quad k = j + 2^{(l-1)}$.

This procedure is repeated recursively at every level of the tree structure until level 0 is reached and all contributions at the internal boundaries are accumulated. Then, the accumulated data have to be distributed again recursively among the processors and are finally copied into the local arrays.

$$R^\star_{\Gamma_{j,\nu}} = R^\star_{\Gamma_{k,\nu}} \longleftarrow R^\star_{\Gamma_{k,\nu}} \tag{18}$$

$$R^\star_{\Gamma_{j,o}} \longrightarrow R^\star_{\Gamma_{k,o}} = R^\star_{\Gamma_{j,o}} \tag{19}$$

where $l = 1 \ldots n_l, \quad j = 0 \ldots 2^{(l-1)} - 1, \quad k = j + 2^{(l-1)}$.

Benchmark tests

In order to compare the performance of the parallel computers available at ICA, the cpu and communication rates have been measured for simple test cases. The processor performance may be assessed by means of the execution time for a scalar product of two arrays. As a result of the measurements, a rate of 0.43 MFlops (2.3μs/byte) has been achieved with the T805 transputer, a rate of 0.36 MFlops (2.7μs/byte) for the T800 transputer.

The rate of intel's i86XP processor was measured to be 6.1 MFlops (0.16μs/byte) which is more than one order of magnitude higher compared to the Parsytec transputers.

The communication performance is characterised by the latency time and the communication rate. The dependence of these characteristic values on the distance between the communicating processors as well as on the size of the message may be measured for a processor pipe configuration. The results of the measurements reflect the characteristics of the data transfer on the parallel platforms.

For the communication between neighbouring processors, a minimum latency time of 70μs was measured for MC-3 and 80μs for the SC. The communication rate was measured on both transputer networks to less than 4μs/byte even for large data packages. With an increasing number of processors which have to transmit the data package from the sender to the receiver processor, the latency time in the networks increased by approximately 12μs per interim step. This behaviour is explained by the fact that in every interim step test and copy tasks on the transmitted data are required. In contrast to this, the communication rate is nearly independent of the number of interim steps (5.2μs/byte for MC-3, 6.3μs/byte for SC), which is achieved by a pipelining technique. The communication performance of intel's Paragon is characterised by a latency time of 45μs and communication rate of 0.2μs/byte. Both values are found almost independent of the distance between the communicating processors in the network and the package size. As a result of the investigations, the ratio of compute to communicate performance is approximately two for the Parsytec transputer and one for intel's Paragon.

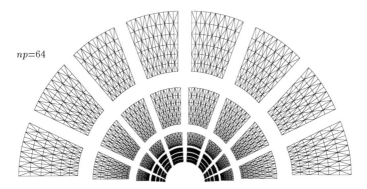

Figure 4: Finite element mesh decomposed into 64 subdomains.

Hypersonic flow past a circular cylinder In the following, a set of benchmark tests will investigate the scalability of the finite element algorithm on the available parallel platforms. As an example, the inviscid hypersonic flow over a circular cylinder is computed [12]. The nitrogen flow over the cylinder with a radius of 2.54 cm is characterised by the free stream Mach number M_∞=6.14. The discretised flow field is covered by 8256 triangular elements and was decomposed into 2^n subdomains for n=0,1,2... as shown in fig. 4 for 64 subdomains.

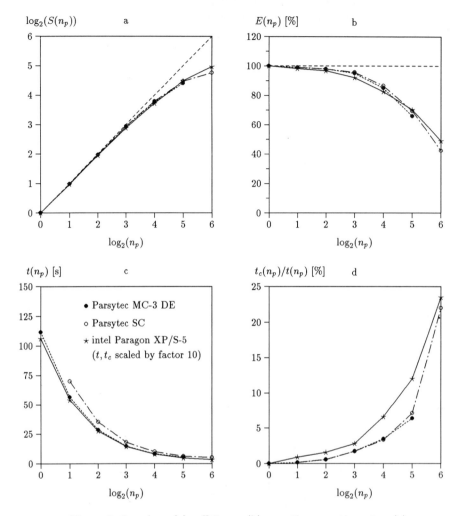

Figure 5: Speed-up (a), efficiency (b), cpu-time per time step (c) and relative communication time (d).

The results of the benchmark tests are presented in fig. 5. In the figure, the speed-up (a) and the efficiency (b) are depicted together with the execution time per time step (c) and the normalised communication time (d). In order to compare the execution and the communication times of the Parsytec transputers with the intel machine, all time measurements from the intel Paragon have been scaled by a factor 10. This scaling approximately leads to a congruence of the data from MC-3 and Paragon. Accordingly, the communication time for MC-3 and Paragon may be normalised by the execution time from MC-3 and thus a direct comparison of the systems is possible. The differences between the transputer networks MC-3 and SC are merely explained by the distinct efficiency of the processors. For this reason, the MC-3 apparently achieves a slightly better performance.

The comparison of the computational time per time step demonstrates the competitiveness of parallel computing. Considering a network of 32 processors, the proposed numerical problem has been solved in less than 5% of the necessary sequential processing time. However, the decomposition of a problem is limited by the effort for the inter-processor communication. Because of this, the efficiency of the parallel computation decreases rapidly for higher processor numbers.

The efficiency of the parallel algorithm is reflected by the dependence of the speed-up on the size of the processor network. For a balanced load distribution, the deviation from linearity is mainly caused by the increasing communication. As an example from a network consisting of 32 processors, the fraction of the execution time required for the data exchange is 12% for intel Paragon, 6.4% for Parsytec MC-3 and 7.1% for Parsytec SC. The high communication fraction on the intel Paragon is due to the fact that the ratio of communication to computation performance is lower than for the transputers. This observation conflicts with the performance investigations for the simple test cases shown before. The reason might lie in the difference of the performance of the transputers between the sequential mode as in the test case and the present network application where the execution is interrupted for the data transfer.

Further deviations from a linear increase of the speed-up are caused by an uneven distribution of the load and by the multiple computation of the source terms at the interior boundaries. The importance of these effects increases for a finer granularity of the distributed processing.

Execution of the adaptive method on parallel computers

The incorporation of the adaptive technique into the parallel algorithm causes an uneven distribution of numerical operations among the processing units. The initial balance of the computational load is disturbed when the adaptive modeling affects the number of numerical operations and thus the execution time. An uneven load distribution causes idle-times whenever synchronous data exchange between neighboring processors is required. This situation arises whenever data computed on the element-level as for the flux balance, the mass matrix correction and the artificial viscosity have to be accumulated on the nodal point level.

The idle-times arising in a programme module can be determined explicitly on the basis of the execution time for a single flux or source operator and the number of operations of a certain category in every subdomain. For this purpose, each one of the n_p subdomains may be decomposed into subregions distinguished by the activated operators:

$$\Omega_i = \bigcup_j \Omega_{i,j} \quad \text{and}$$

$$\Gamma_i = \bigcup_j \Gamma_{i,j} \quad \text{for} \quad i = 0 \ldots n_p - 1 \quad \text{and} \quad j = B, D, B^S, D^S \quad \text{or}$$

$$\Omega_i = \bigcup_j \Omega_{i,j} \quad \text{for} \quad i = 0 \ldots n_p - 1 \quad \text{and} \quad j = P^S, N \ .$$

The execution time t_{c_i} of any processor may be determined by summation of the execution times for the operators t_j multiplied by the number of operators n_j from the category j

$$t_{c_i} = \sum_j n_{i,j} \cdot t_j \quad \text{for} \quad i = 0 \ldots n_p - 1.$$

The idle-time of the processor in any programme module may then be determined by comparison of the individual execution time with the maximum execution time

$$t_{w_i} = \max_i \{t_{c_i}\} - t_{c_i} \quad \text{for} \quad i = 0 \ldots n_p - 1.$$

The effective idle-time t_w which represents the loss of efficiency caused by unbalanced load is defined as the difference between the maximum and the mean execution time t_p,

$$t_w = \max_i \{t_{c_i}\} - t_p \quad \text{with} \quad t_p = n_p^{-1} \sum_i t_{c_i}.$$

As the effective idle-time reduces the efficiency of the overall parallel procedure, it has to be reduced by appropriate load-balancing techniques.

Re-distribution of adaptive operators

In order to avoid idle-times on any participating processor, the computational load has to be re-balanced within every programme module of the parallel algorithm. For this purpose, an adaptive re-distribution technique has been developed such that partitions of the numerical operations are evacuated from overloaded to underloaded processors, as is discussed in the following.

The sender processor p_s with the maximum and a receiver processor p_r with the minimum amount of computational load have to be identified first

$$p_s = p\left(\max_i \{t_{c_i}\}\right) \quad \text{and} \quad p_r = p\left(\min_i \{t_{c_i}\}\right) \quad \text{for} \quad i = 0 \ldots n_p - 1.$$

The number of operations $n_{j,s}$ to be evacuated from the overloaded to the underloaded processor has to determined as the difference of the execution time on the sender processor $t_{c_{p_s}}$ and the mean processing time t_p divided by the execution time t_j for a single operation of category j. The category is chosen such that the execution time for the operation is high compared to the data which have to be communicated between the exchanging processors. In addition, the number of operations $n_{j,s}$ is restricted by two different limitations. First, the number of evacuated operations is limited by the number of available operations of this category on the sender processor. Furthermore, the receiver processor should not be overloaded on its part. Thus, the number of transferable operations is given by the statement

$$n_{j,s} = \min\left\{\frac{t_{c_{p_s}} - t_p}{t_j}, n_{p_s,j}, \frac{t_p - t_{c_{p_r}}}{t_j}\right\}.$$

The number of evacuated operations modifies the execution time on the sender and receiver processor in accordance with

$$t_{c_{p_s}} = \left(\sum_j n_{p_s,j} \cdot t_j\right) - n_{j,s} t_j \quad \text{and} \quad t_{c_{p_r}} = \left(\sum_j n_{p_r,j} \cdot t_j\right) + n_{j,s} t_j$$

and aims at a balancing of the computation load in n_p-1 re-distribution steps.

Application of the load balancing technique

The parallel execution mode of the adaptive operator method will now be studied for the cylinder flow decomposed on four processors. The partition of the regions Ω_{i,B^S} for convective species transport and Ω_{i,P^S} for chemical production demonstrates an uneven evolution within the four subdomains as depicted in fig. 6.

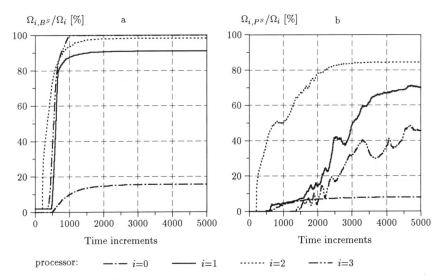

Figure 6: Fraction of activated operators for convective transport of species (a) and chemical production rates (b).

Snapshots after 500, 1000 and 5000 time steps indicating the extent of the operators activated in addition to the basic one are presented in fig. 7. The mesh nodal points with active production operators $\boldsymbol{P}^S$ (chemical reactions) are marked by black dots and the region where convective species transport has to be taken into account via the operator $\boldsymbol{B}^S$ is shaded with grey color.

Obviously processor $i=0$ is underloaded during the entire incremental process. In the stationary state, the area Ω_{i,B^S} has an extension of 16% in subdomain $i=0$ compared to 91...100% in the subdomains $i=1...3$. The unevenness of the load distribution is partly amplified by the extension of the areas Ω_{i,P^S}.

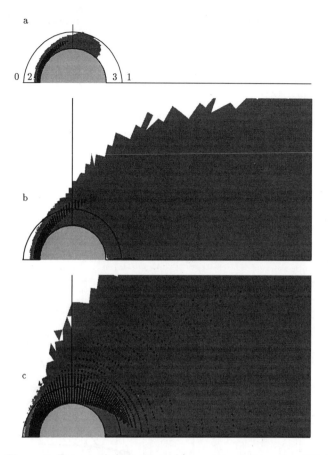

Figure 7: Extent of the activated operators at time step 500 (a), 1000 (b) and 5000 (c). Black dots: chemical reactions, grey shading: convective species transport.

Idle-times and execution times within the individual programme modules may be discussed with the averaged values in fig. 8. In the figure, averaged idle and execution times are depicted for 200 and 800 time steps. Figs. 8.a.1 and 8.b.1 show the distribution of the execution time among the programme modules for the regular parallel algorithm. Accordingly, the computation of the flux balance requires 40.6%, the mass matrix correction requires 10.2%, the incorporation of artificial viscosity requires 31.6% and the integration of the chemical source terms requires 17.6%. The application of the adaptive method results in a mean reduction of 36.9% within 200 time steps (a.2) and 21.4% within 800 time steps (b.2) as compared to the regular algorithm. The average effective idle-time caused by the unbalanced load is 9.7% within 200 time steps and 14.7% within 800 time steps. These idle-times are reduced by an adaptive re-distribution of the operators from overloaded to underloaded processors. This may result to a maximum reduction of 46.6% within 200 time steps (a.3) and 36.1% within 800 time steps (b.3). It has to be mentioned that the re-balancing of source operators has already been implemented in the parallel adaptive algorithm whilst the re-balancing of flux operators has still to be implemented.

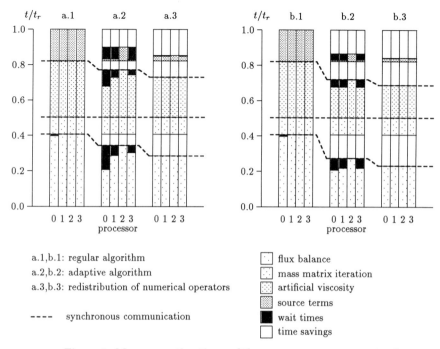

Figure 8: Mean execution times of the programme components after 200 (a) and 800 (b) time steps.

The scalability of the load balancing technique has been investigated on the Parsytec MC-3 DE. All time measurements are based on the execution time for 200 time steps. The load balancing is limited to the node-level source operations. As a result, fig. 9 shows the execution time including communication related to the execution time for the regular sequential algorithm (a) and the communication time related to the individual execution times (b).

On closer examination of the execution times, the benefit of the adaptive method decreases rapidly with increasing number of subdomains. For the sequential problem, the reduction of execution time is 47% compared to the regular algorithm. In the parallel execution mode with 32 subdomains, the reduction of execution time decreases to 19%. This is explained by the fine granularity of the parallel computation, where the number of operations which might be reduced by application of the adaptive method becomes small compared to the communication times required for the regular data exchange. The load balancing on the node level effects a small reduction of the overall execution time which is compensated for high processor numbers by the additional communication time required for the evacuation of the numerical operations.

181

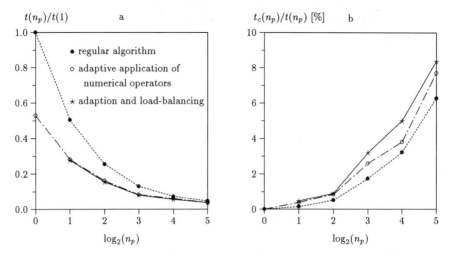

Figure 9: Normalised execution time (a), communication time (b).

Acknowledgement

The authors wish to express their gratitude to Professor J. Argyris, F.R.S., for the support he provided as the then Director of the Institute for Computer Applications towards the realisation of the research project.

References

[1] J. Argyris, I.St. Doltsinis, H. Friz. Hermes Space Shuttle: Exploration of reentry aerodynamics. *Comp. Meth. Appl. Mech. Eng.*, 73:1–51, 1989.

[2] J. Argyris, I.St. Doltsinis, H. Friz, J. Urban. An exploration of chemically reacting viscous hypersonic flows. *Comp. Meth. Appl. Mech. Eng.*, 89, 85–128, 1991.

[3] J. Argyris, I.St. Doltsinis, H. Friz, J. Urban. Physical and computational aspects of chemically reacting hypersonic flows. *Comp. Meth. Appl. Mech. Eng.*, 111, 1–35, 1994.

[4] J. Argyris, H. Friz, F. Off. Domain decomposition and operator splitting for parallel finite element computations of viscous real gas flows. *Notes on Numerical Fluid Mechanics*, Vol. 38, Vieweg, Braunschweig, 1993.

[5] E. Laurin, J. Wiesbaum. Three-dimensional numerical simulation of the aerothermodynamic reentry. *Notes on Numerical Fluid Mechanics*, in this publication.

[6] S. Brück, B. Brenner, D. Rues, D. Schwamborn. Investigations of hypersonic imulation of the aerothermodynamic reentry. *Notes on Numerical Fluid Mechanics*, in this publication.

[7] J. Argyris, I.St. Doltsinis, J. Urban. Adaptive application of physical operators in hypersonic flow computation. *ERCOFTAC bulletin*, 21:28-29, 1994.

[8] I.St. Doltsinis, S. Nölting. Generation and decomposition of finite element models for parallel computations. *Computing Systems in Engineering*, 2:427-449, 1992.

[9] H. Friz, S. Nölting. Towards an integrated parallelization concept in computational fluid dynamics. *ERCOFTAC bulletin*, 18:8-11, 1993.

[10] J. Argyris, H.U. Schlageter. A parallel interactive and integrated visualization system. *Notes on Numerical Fluid Mechanics*, in this publication.

[11] D. Straub. Exakte Gleichungen für die Transportkoeffizienten eines Fünfkomponentengemisches als Modellgas dissoziierter Luft. *DLR-FB* 72-34, 1972.

[12] H.G. Hornung. Non-equilibrium dissociating nitrogen flow over spheres and circular cylinders. *J. Fluid Mechanics*, Vol.53, Part 1, 1972.

Parallel Interactive and Integrated Visualization

John Argyris, Hans-Ulrich Schlageter
Institute for Computer Applications,
Pfaffenwaldring 27, 70569 Stuttgart, Germany

Summary

An integrated visualization system comprising both parallel and interactive components is presented. The system allows numerous independent software entities to be individually grouped and linked together to user-specific applications. The module package runs on a powerful graphics workstation under AVS and permits the connection of parallel machines as the back end in order to distribute computational effort and to receive visual output during the run of a parallelized CFD simulation. A focal point of discussion is the exchange of data and instructions over the network between main components of the system which is effected directly by newly developed transmitter software able to offer point-to-point connections which need be neither specified nor set up in advance.

1 Introduction

The wide use of computers with distributed memory in CFD caused a bottleneck in interpreting vast quantities of data, since usually visualization is still effected on sequential workstations despite this evolution.
One solution which will be proposed here is to let parallelized analysis [3, 4] and visualization form a hybrid system with a strongly interactive character. The aim is achieved by integrating time-intensive graphics calculations into the analysis and transporting thus obtained polygon data over the network to an associated graphics workstation for rendering purposes [2].
A further parallel application sequentially post-processes data which are already stored on a hard disk, but performs time-consuming calculations in parallel, e.g. if a bundle of streamlines is to be computed. Besides, the graphics system is not limited to parallel applications but features many other requirements [1], since the majority of the users will use graphics in the conventional way as a pre- and post-processing tool.
A very important point in many applications is the management of the data flow between the main components of the system and different machines. Point-to-point connections need to be established interactively whenever requested by the user. As soon as not needed any further, connections are dismissed, together with the termination of a parallel computation. Thus, in parallel multiuser systems, for instance, processors and endpoints of communication likewise are at other user's disposal again.
The paper is subdivided as follows: section 2 gives a brief overview of the main components and subparts of the system and section 3 presents some often needed sequential

applications. In section 4 the use of parallel computers is described guided by two general parallel applications. In section 5 techniques to transfer data over the network are discussed. Results obtained by a parallel simulation [8] are presented in section 6, followed by concluding remarks.

2 System Components

The newly developed graphical system, whose sequential parts are bedded in the modular and networking structure of AVS, comprises two major blocks: graphics preparation and graphics rendering (Fig. 1). While graphics preparation encompasses facilities which manage the input, output, import and export of data and the computation of the graphics results, either single-threaded or in parallel, the task of the sequential graphics renderer is polygon building and polygon rendering. Applications generally are arranged into three subsequent phases: the provision of data, the graphics computation and building of graphical objects (polygon building) and the rendering of these objects on to screen. Owing to the structure of AVS, the user interface is application dependent and integrated into the modules, allowing stepwise user interactions with every part and subpart of the system.

All modules and algorithms presented in this work are newly developed except for the rendering module *geometry viewer* which belongs to AVS. The graphics computation algorithms of AVS could not be applied because they own a different data structure, their source code is not available, they neither can be parallelized nor integrated into a finite element programming system.

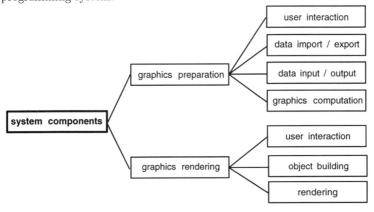

Figure 1: System components overview.

Provision of the Analysis Data

To handle the data interface between analysis and graphics, modules for conversion, transmission and reduction of data are needed. Their purpose is to standardize analysis data of various formats to a common base for the next level of downstream modules. While modules of the group *input* read stored data from a hard disk, modules belonging to the group *import* are able to receive analysis data directly from a running simulation. Some additional modules reduce data simply by selection or, e.g., by ignoring those which reside outside a user-specified control volume.

Visualization of the Analysis Data

The graphics computation algorithms are adjusted to the data structure of the analysis programming system [3] which yields a number of benefits. Data, for instance, that are derived from elements or nodes can be processed more easily, more consistently and independently of any kind of grid, structured or unstructured. Since the majority of the graphics algorithms is orientated to the element structure of the grid, advantages can be gained for parallelization of these algorithms. Moreover, the algorithms can be permanently built into the analysis packet as graphics pre-processors. In a word, the algorithms were designed flexibly to fit into sequential and parallel applications, in a standalone mode or integrated into an analysis, then operating on the given data structure.

Various algorithms and procedures newly developed are implemented into visualization modules [1], programming entities, which process analysis results data into graphical results data. The superposition of visualized results is supported. The modules are provided to compute both nodal and element attributes, isolines, contour planes, vectors, isosurfaces, cutting planes, volume cutouts, stationary and transient streamlines, a simulation of a particle movement, the movement of predefined surfaces, e.g., as shown in Figure 9, and more. Some modules present two- and three-dimensional grids and optionally element groups, others process displacements of grid structures and are reserved above all for structural mechanical applications.

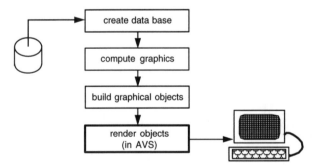

Figure 2: General scheme of sequential graphics applications.

3 Sequential Graphics Applications

Figure 2 illustrates the general scheme of a sequential graphics application, starting with the creation of a common database by either reading or receiving analysis results, followed by the computation of the graphics results, the building of graphical objects by means of a graphics library and finally the rendering of the objects.

In former releases of this graphics system, graphics computation and building of graphical objects were effected in a single step [1]. The use of parallel computers, however, made it necessary to perform these steps separated from each other. Thus, adaptations to graphics libraries other than AVS are also simplified. In addition, generalized polygon building modules have been designed to operate on graphical results and obeying to various user options.

The course of the sequential visualization will now be demonstrated guided by two example applications. Figure 3 shows the first demonstration module network, creating contour planes or vectors mapped on predefined arbitrary cutting planes. In this example and also the second example, the database is formed by reading in both of the analysis data files, configuration and nodal results, and generating either surface or point data.

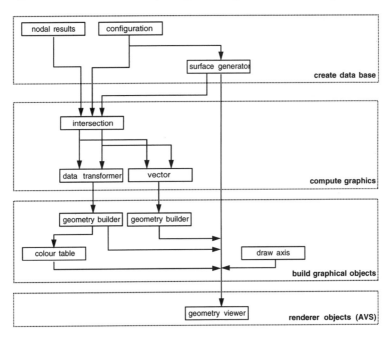

Figure 3: Module network: first example application.

The surfaces with which the three-dimensional problem is to be intersected are to be defined in the module *surface generator*. Surfaces can either be generated in single mode and added to a surface description one by one or computed as a sequence by giving an increment in the normal direction and the number of surfaces. The module *intersection* needs the grid data and surface description data as input to compute cutting planes with the values of a physical quantity data mapped on them. Optionally, the remaining geometry cut-off can be drawn in wireframe description or shaded, either by using a selected rgb colour or the variation of a mapping quantity. For each cutting plane and/or remaining geometry in a sequence of surfaces, an object number may be added for animation purposes later. Thus, the downstream module *geometry builder* is able to compute exactly one graphical object to each object number and add it to an animation cycle.

The module *data transformer* can be inserted between the modules intersection and geometry builder in order to compute either isolines or contour planes in the cutting planes. In the same way, vectors can be computed by simply connecting the module *vector*.

187

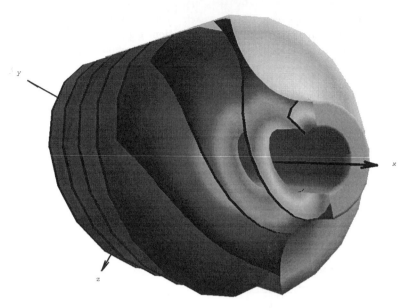

Figure 4: Sequence of cutting planes with velocity data intersecting a nozzle flow.

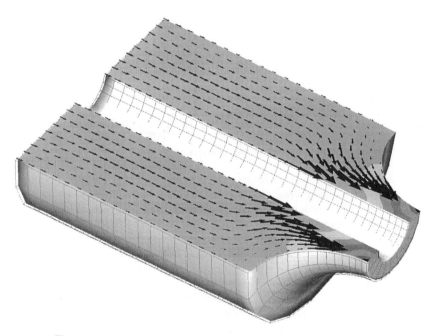

Figure 5: Velocity vectors and density contour planes, both mapped on one cutting plane through a nozzle.

For the sake of completeness, it should be mentioned that the data transformer is multifunctional and could be also connected directly with modules *nodal results* and *configuration*, in order to create isosurfaces of a selected physical quantity. A sequence of isosurfaces can be animated and shaded to represent another mapping quantity.

Figures 4 and 5 both present graphical results of the same three-dimensional FE problem. The case of a laminar isothermal flow through an axisymmetric nozzle is considered. Figure 4 shows a set of parallel cutting planes shaded with velocity values. In Figure 5 the results of an intersection were post-processed twice: into velocity vectors and density contour planes.

The second demonstration module network shown in Figure 6 generates a special data structure containing a space-time description of particles released in a vector field, which forms the base for several applications: streamlines, particle tracing and the tracing of points which in the beginning reside on a two-dimensional mesh.

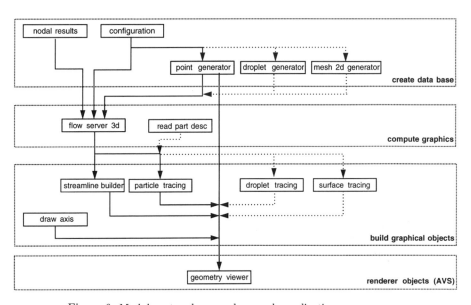

Figure 6: Module network: second example application.

First, a space-time description is computed by the module *flow server 3d* by integrating along a field of velocity vectors using an adaptive shoot procedure. The computation is based on the input data and the initial coordinates of the released particles, which have been produced in either one of the modules *point generator*, *droplet generator* and *mesh 2d generator*. While the *point generator* is able to create starting points in single mode and also in an arbitrary rectangle, in a bounding box or by using any combination of these modes, the droplet generator produces starting points which are at the same time nodes of a triangular mesh approximating a sphere in a user specified grade of fineness. The *mesh 2d generator* finally is applied to produce segments as demonstrated in Figure 7.

Second, an animation cycle of images can be obtained by the corresponding module tracing the original point set or mesh. Here, module *droplet tracing* corresponds with *droplet generator* and *surface tracing* with *mesh 2d generator*. The modules *streamline builder* and *particle tracing*, however, can be applied together with any of the three generator modules. For an animation, the user selects the number of images to be computed and a time maximum less than the absolute time maximum of the computation. An extract will be taken at each time-step by calculating the location of each particle out of the space-time description. Moreover, a space-time description can be stored on hard disk to be read in later by the module *read part desc*.

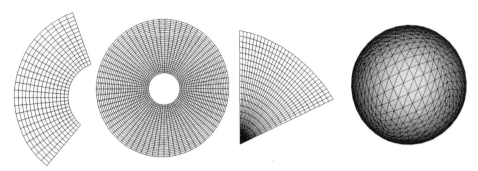

Figure 7: Generation of different surface segments.

Figures 8 and 9 both show results of the same laminar isothermal flow through a axisymmetric nozzle. Figure 8 illustrates the movements of three droplets of different diameter $d_0 = 2 * d_1 = 5 * d_2$ and Figure 9 shows the results when a circular mesh was released.

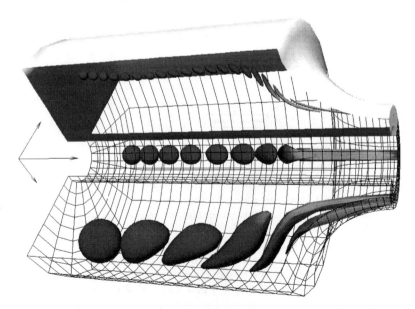

Figure 8: Three droplets of different initial radius moving through a nozzle. Scalar velocity values are mapped on the surfaces.

Here, the effects due to the boundary layer can be seen, which the vector plot in Figure 5 was hardly able to resolve. In the examples, the droplet surface was approximated by 1200 triangles and 602 nodes, the surface segment by 864 rectangles and 900 nodes. In order to accelerate the process, the computation of the space-time description was performed on an Intel Paragon on 32 processors, using the parallel version of the module *flow server 3d*.

Figure 9: Tracing of a surface segment (left) through a nozzle (three extracts).

4 Parallel Graphics Applications

Two general parallel graphics applications will be distinguished: in the first case a parallel computer is used just to support graphics, while in the second example the simulation process as a whole except for the rendering is performed in parallel, the graphics computation integrated into the analysis allowing the extraction of visual results during the run.

Parallel Supported Graphics

The purpose of this application is to minimize the computational effort in ordinary interactive graphics post-processing, where large-scale calculations may lead to long waiting periods. Therefore, graphics computations will be effected in parallel and interactively steered, whenever it is requested by the user.

The example module network in Figure 10 permits both an interactive streamline calculation in the conventional way and in conjunction with an associated parallel computer. For parallel use modules for parallel handling (centre) are needed in addition to the modules of the sequential application (left). Parallel handling modules govern the parallel side (right), which remains hidden from the user.

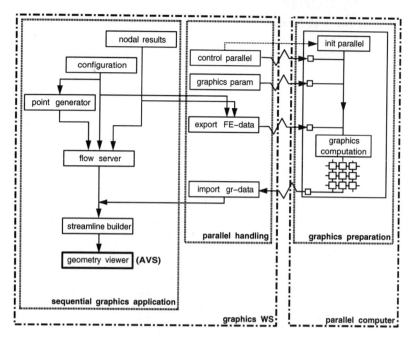

Figure 10: Streamline calculation supported by a parallel computer.

In this example, the modules of the first level read in the analysis results: *configuration* and *nodal results*. Startpoints will be produced in the grid by the module *point generator*. The streamline data will be computed either sequentially using the module *flow server* or in parallel. In the latter case, the parallel program is directly invoked interactively from the module *control parallel* on a specified number of processors. If the file system of the parallel machine is not mounted the user is requested to initialize the parallel program manually.

Graphics parameters set by the user and steering the computation are transmitted next and the analysis data are exported over the network. On finishing the parallel computation, graphics streamline results are re-imported and integrated into sequential fields by the receiving module *import gr-data*. In both applications, sequential and parallel, the results will be polygonized in the module *streamline builder* and rendered on to the screen by the AVS geometry viewer.

Parallelism is achieved by dividing up the particles and the starting points and distributing them on the processors. In this example, every processor operates on the same grid, in contrast to other cases where the elements of the grid are distributed on the processors if a sequence of isosurfaces, for instance, is to be computed in a three-dimensional mesh. In order to balance the load, each processor receives approximately the same number of elements as a base to compute parts of the isosurfaces. It is the module *export FE-data*, in which the grid data are decomposed and sent to the parallel program in a processor loop. The processorwise reception of the data is explained in Figure 13 in section 6.

In addition, distributed analysis results files outputed by a parallel CFD analysis can also be processed by reading them in by a parallel graphics application running on a corresponding number of processors.

Quasi-Real-Time Simulation

The visualization is usually preceded by a mesh generation and a run through the analysis. The system FEPS [3] functions as a parallel analysis programming system and is coupled with a parallel mesh generator PAGE [4]. Both systems are owned and designed by the Institute of Computer Applications in Stuttgart. The abbreviation PAGE stands for Parallel Generator and FEPS for Finite Element Programming System. The latter is a general FE-development system for complex nonlinear applications. Instead of PAGE the programming system PATRAN can also be employed for nonparallel mesh generation and communicates with other systems via a post-processor called the neutral system. The analysis data are specified in the same neutral format for an onward procession.

The parallel application presented next considers the case that a visualization of the analysis results is being requested during the course of a parallel simulation. One possibility offered here is to transfer the analysis results directly into a module ready to receive. The graphics computation is then effected sequentially on the Graphics workstation. In the second possibility, graphics is computed in parallel via the analysis executing the appropriate graphics functions. Graphics results are then transferred to the workstation in order to be polygonized and rendered. In order to explore the data during the simulation, a graphics flag is read by the analysis master processor to a certain time, e.g. after each time integration step. Only if this flag bit is set to on, are additional graphics parameters read or imported and the appropriate graphics routines are then executed by the analysis. Provided that the file systems of both graphics workstation and parallel computer are mounted, graphics flag and graphics parameters can be set directly by an AVS module, using a remote procedure. Otherwise, the parameter file stored by the module must be transferred, e.g. via File Transfer Protocol.

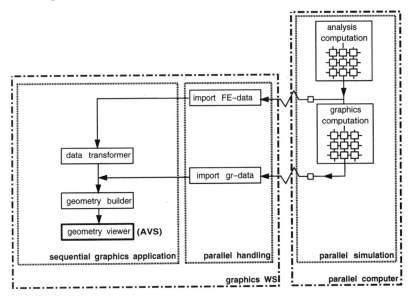

Figure 11: Taking extracts out of a parallel simulation.

At the end of the graphics calculation, the system tries to connect with the graphics workstation to transmit the graphics results. If it happens that connection fails, the simulation process is resumed immediately as well as after the successful data transmission. On the graphics workstation the results are built to polygonal objects and rendered by AVS (Figure 11). Moreover, a generalized scheme for both parallel applications discussed can incidentally be drawn from Figures 16 and 17.

5 Data Transfer Using Transmitters

In a hybrid system comprising several independent components, the management of the data flow becomes very important. Point-to-point connections must be established at the user's request, and only then. Because commercial software such as P4 and PVM could not meet the demands because their communication network must be established in advance and remains unchangeable throughout the application, a set of transmitter routines has been developed to avoid this problem [5].

The Transmitter Concept

Transmitters are mediatory processes based on the sockets of the inter process communication IPC which act on the transport level according to the OSI reference model.
Sockets are endpoints of communication and can be classed in two socket families: AF-UNIX sockets are used to communicate between processes on the same machine and AF-INET sockets with different machines. A transmitter is a stand-alone process, which is initiated by either of the parties willing to exchange data. The data flow is enabled as soon as both sides are connected to a transmitter (Figure 12). Transmitters terminate automatically after their task is finished. A task is always defined by a data header preceding the data to be transferred.

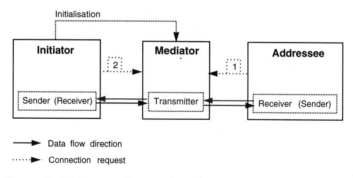

Figure 12: Initiator, mediator and sequential addressee.

The data header interpreted by the transmitter contains field-defining data, e.g. number of fields to be sent, byte sizes and type of data. Thus, the receiving party knows what will be accepted and is able to allocate enough memory, provided the data are bound for it.

Once activated, the transmitter opens two passive sockets, one listening towards the addressee and the other one listening for a connection with the initiator. Since passive sockets do not stop listening until they accepted a connection, it is wise to have no listening processes opened by either of the applications in order to avoid blocking, e.g. in case of connection failure or transmission error occurrence. A blocking transmitter, however, can easily be killed without, e.g., disturbing the simulation process. Initiator and addressee both open active sockets, which cannot block the run of the calculation.

From the parallel side, data fields are sent processorwise (token ring) including pointers to build up sequential fields on the sequential side (Figure 13).

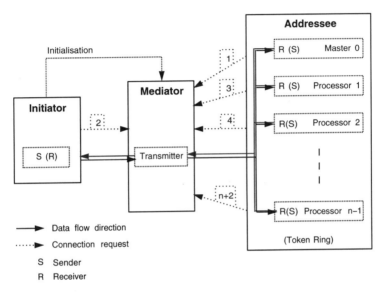

Figure 13: Initiator, mediator and parallel addressee.

Transmitter applications

Several transmitter programs have been established depending on the machine architecture and the needs of the user. Figure 14 illustrates the transmitter scheme when a graphics application requests graphics results from a parallel CFD simulation. Here the situation (c) refers to the case where an Intel Paragon is used as parallel computer, while situations (a) and (b) both elucidate the interconnection with a Parsytec transputer network. The following is related to the illustration (a), where the host machine and graphics workstation are not identical. A more detailed data flow scheme for this constellation is explained in Figure 15 by means of the integrated parallel simulation discussed in the previous sections.

Graphics preparation routines and data import facilities are integrated into the numerical analysis, in fact into the time-increment loop where they are to be executed right after the step of the numerical solution accompanied by actualization of the analysis data. Since a potential graphics request is generally expected, the following question arises after each time step: will data be exported at all and, if this is the case, will analysis data be further processed to graphics results before the transmission is taken up? These questions are handled by some graphics flags read in by the master processor as default from a record file. Default parameters may be altered interactively if the file system is mounted.

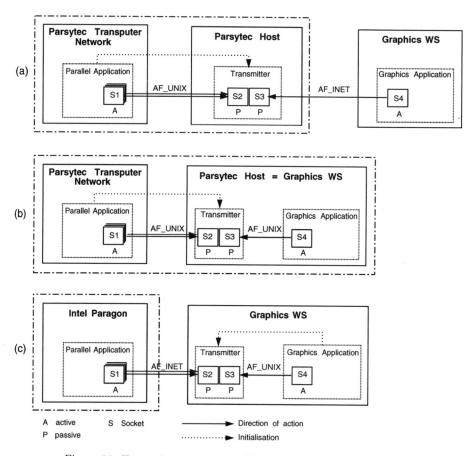

Figure 14: How to interconnect parallel computer and graphics workstation: three different example configurations.

As can be seen from Figure 15, the transmitter is initialized on the host machine by a system call from the function *start transmitter*. As soon as initialized, the transmitter subsequently opens two passive sockets with the first one (AF_INET) listening towards the graphics workstation. Only if a connection is interactively established by the user does the transmitter open the second socket (AF_UNIX) listening towards the transputer network. If not, the parallel side will fail in trying to connect within a certain default time interval and therefore will resume the numerical computation.

Assuming the connection was correctly established, the data can now be transferred. In the illustration, transmitter 1 will transmit processorwise graphical results which, once received by the sequential side, will be integrated into sequential fields with the aid of pointers delivered by the transmitter. Thus, overall graphics results are received and overall analysis results likewise, where analysis results were passed through transmitter 2. Finally, overall graphics results are polygonized and rendered. Overall analysis results, on the other hand, need to be post-processed before by appropriate graphics computation routines.

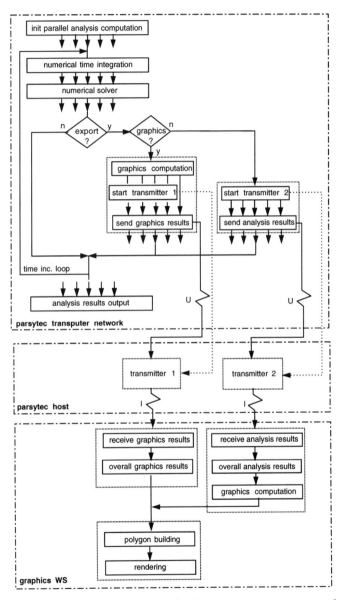

Figure 15: Data and instruction flow between a transputer network, its host machine and a remote workstation.

197

For a better understanding, a summarizing overview of the two parallel graphics applications, including its variations and transmitter processes, is presented in Figure 16 and 17.

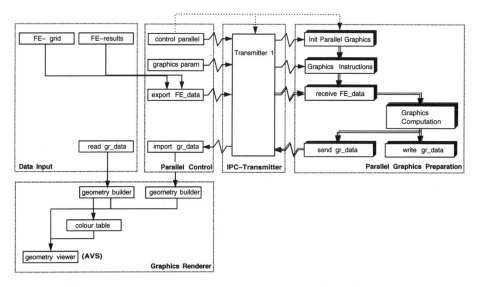

Figure 16: Parallel applications 1: parallel supported graphics.

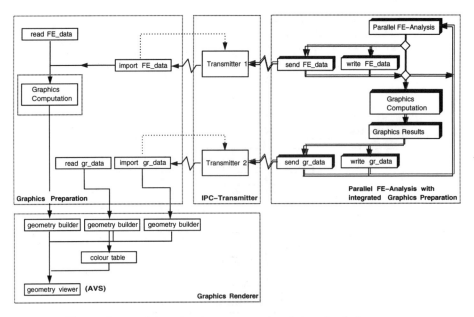

Figure 17: Parallel applications 2: a quasi-real-time simulation.

6 Parallel Results

In this section results for the two general applications discussed in section 4 will be presented.

In the first example, a parallel computer is used to accelerate the graphical post-processing of results obtained by a sequential finite element calculation of a Bénard convection in a rectangular box with 25872 tetrahedrons [7]. Initially, only the top surface of the box was hot, while the medium and other sides were kept cold, except for the left and right side, where a linear temperature profile was given. The analysis results reflect the quasi-stationary state. Sixty-four streamlines were calculated in parallel on 2 – 32 processors by integrating along a field of velocity vectors using an adaptive shoot procedure with a maximum of 2500 shoot iterations. The sequential analysis results were transmitted to the master processor, which broadcasted the data so that every processor could work on the same data. The start points were distributed on the processors, so every processor of a considered network of 32 processors only had to compute 2 streamlines.

Figure 18 shows the results obtained on both a Parsytec and an Intel Paragon with scalar temperature values mapped on the streamlines.

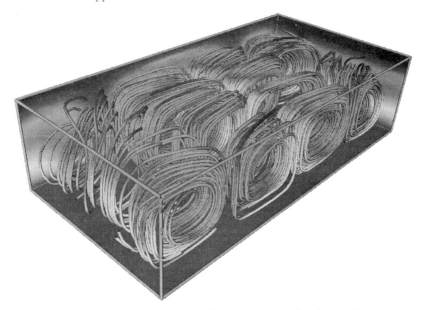

Figure 18: Quasi-stationary state of a Bénard convection: hot top layer and cold bottom layer.

The results of the benchmark tests are presented in Figure 19. Speed-ups (a) and efficiencies (b) for both parallel machines are compared together with the overall computing time (c) and the normalized communication time (d). Not considered was the time needed for a data transmission from and to an associated graphics workstation because here the speed was nearly the same for both machines, due to the slower data reception on the side of the workstation which happened to be a Stardent GS 1000. The transmission time was in the range of the communication time for a Parsytec.

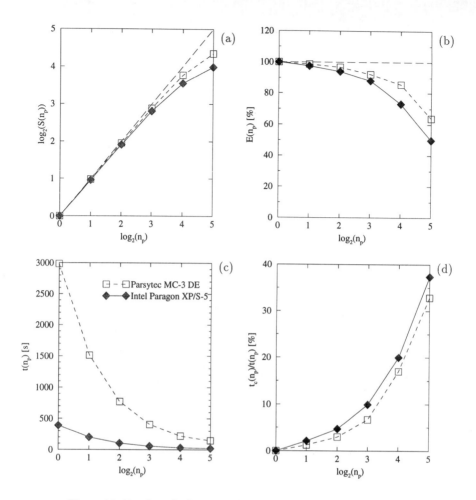

Figure 19: Benchmarks for a distributed streamline computation on 2 to 32 parallel processors.

The second example refers to quasi-real-time simulation, the scheme of which has been presented in Figure 11. In the problem illustrated in Figure 21, the evolution of the incremental solution for a inviscid flow around a two-dimensional circular cylinder in dissociating nitrogen is considered. The calculations, which are related to the experiment done by Hornung [6] under free stream conditions, Ma = 6.14, T = 1833, were performed on an Intel Paragon using 32 processors. Three extracts were taken during the run of a parallel simulation after 500, 1500 and 4000 time-steps showing contour planes and isolines of pressure [8]. Whenever requested, the graphics results were computed in parallel on the processors of the analysis. The result data were transmitted directly over the network to be rendered on a graphics workstation. The graphics computation and the data transmission took less than 20 percent of the time needed to complete a time-step of the analysis. This time is only mentioned to give an idea, because it is of course dependent on the algorithm and, e.g., the number of contour planes selected.

For this example, the decomposed mesh with 8000 triangles is shown in Figure 20.

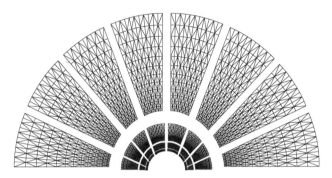

Figure 20: Decomposed mesh for 32 processors.

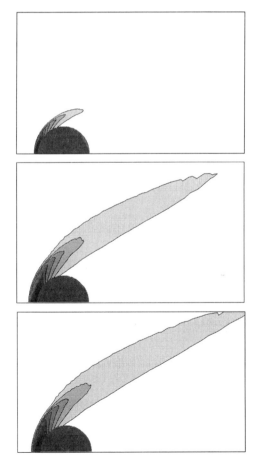

Figure 21: Pressure distribution around a circular cylinder in dissociating nitrogen after 500, 1500 and 4000 time-steps.

7 Conclusion

The integrated interactive and parallel graphics system presented here can be used as a conventional single-threaded tool on a graphics workstation as well as in connection with a parallel computer.

A parallel computer is attached to support two major tasks: either to distribute the computational effort concerning time-consuming graphics computations or to perform a quasi-real-time simulation of the numerical analysis. In the latter case, a graphics software block consisting of a set of parallelized graphics algorithms and data import and export facilities is implemented together with a parallelized numerical analysis.

Parallel computer actions and data flow are steered interactively by modules running on the graphics workstation, which can be linked individually together, in order to build a user specific application. The exchange of data between different processes is effected by transmitter processes developed to this end and enabling direct point-to-point access whenever needed. The graphics system was developed to run under AVS on a graphics workstation. Parallel computers of MIMD architecture are supported.

References

[1] Argyris, J. and Schlageter, H.-U. (1993) Visualization of analysis data with reference to CFD. In: Flow Simulation with High-Performance Computers, 1. DFG Priority Research Programme, Results 1989-1992 (E.H.Hirschel, ed.) NNFM, Vol 38. Vieweg, Braunschweig.

[2] Argyris, J., Friz, H., Nölting, S., Schlageter, H.-U. (1993) An integrated concept for the parallelization of finite element simulations of compressible real gas flows and realtime visualization. Parallel Link.

[3] FEPS 3.3, (1986) Finite Element Programming System, User's Guide, ICA, Universität Stuttgart.

[4] Nölting, S. (1995) Parallelisierung komplexer Finite Element Systeme, Dissertation, ICA, Universität Stuttgart.

[5] Schlageter, H.-U. (1995) An Integrated Visualization System with Support of Parallel Computers, Proceedings, Workshop on Visualization, D. Kröner and R. Rautman (eds.).

[6] Hornung, H.G. (1972) Non-Equilibrium Dissociating Nitrogen Flow Over Spheres and Cylinders, Journal of Fluid Mechanics, Vol. 53, Pt. 1, May 1972, pp. 149-176.

[7] Beddies, G. and Szimmat, J. (1993) Petrov-Galerkin Finite Element Approximation dreidimensionaler Konvektionsströmungen, ICA-Bericht Nr. 42, Stuttgart.

[8] Doltsinis, I.S. and Urban, J. (1996) An Adaptive Operator Technique for Hypersonic Flow Simulation on Parallel Computers, In this publication.

Interactive Visualization:
On the way to a virtual wind tunnel

Carsten Meiselbach and Ralph Bruckschen
Universität-GH Paderborn, Fachbereich Mathematik-Informatik
Warburger Str. 100, D-33098 Paderborn, Germany
E-Mail: meisel@uni-paderborn.de, rwb@uni-paderborn.de

Summary

In this paper we will give you a short introduction to a visualization program for three dimensional flow data, generated by Navier Stokes solvers (W. Borchers(1992)). First, we will give you a short introduction to the problem. The 2nd section discusses one main problem for simulating flows: the lack of memory. After that we describe the used hardware and the features of our implemented tool for visualizing the flow (sections 3 and 4). At last we show some software details of transposing our software to OpenGL/X and Motif. Please refer to two other very interesting works in this volume of Schlageter and Wierse.

1. Introduction to the problem

The calculation example is a 3D instationary flow past a ball. Our solver is working for the largest problem on a rectangular grid of 128 times 128 times 1024 nodes. The results are usually stored as 3D single precision real vectors, so we get about 192 megabytes of data for each timestep. New approaches in storage and rendering of flow data were used to find an acceptable solution to this problem (for more details about this problem refer to [1], for the solver [2] and [3]). The developed visualization tool has the following features:

- Handling the instationary flow data in short time.

- Portability on massive parallel systems.

- Flexibility for different geometries.

- Real time rendering.

- Interactiv selection of starting points.

2. 3D flow data, 3D user interface, 3D interaction ?

With the current technology, it is very expensive, or even impossible to store and handle time dependent vector-field data in real time. The throughput and the seeking time of Harddisks are not sufficient for retrieving that large amount of data. Usually, the whole main-memory of a computer is used to calculate one vectorfield, so a visualization machine might have far more memory than the supercomputer - that's neither common and nor economic. Currently it is possible to visualize one vector field interactively, as long as it fits into the main memory of the graphic workstation. The question is now, how to visualize large vector fields quasi-interactive? Our solution is, to generate as much graphic primitives as possible, to fit them into the main-memory of a workstation, and select them interactively, to get a close approximation to "real" interaction. In our example, we computed about 200 timesteps of the fluid flow, and strewed in particles in a raster of 18 times 18 times 18 around the ball. Using the here described tool, the user is capable of selecting the particle entry points interactively, hence interactivly visualizing the flow. To develop this tool, we needed a 3D input device, a sort of 3D user interface and some data-compression technique to fit as much data as possible into the memory. The second problem is the portability of this software onto other machines, cheaper than the used graphics-supercomputer.

3. Some specifications of the Polhemus Fasttrak and usability

Figure 1: User with stylus, visible the small receiver fixed on a pen

One advantage of our program is the usage of a real 3d input device, the Polhemus Fastrakker. With this device is it possible to determine the exact position of a small receiver at a resolution of about 0.0002 inches exactly within a range of 30 inches around the normal transmitter, and up to 33 feet using a long range antenna. The transmitter was positioned beneath the users monitor, the receiver was fixed on a stylus, so that the user can use the trakker like a 'magic wand'. It is now very easy to move an virtual cursor object in the simulation simultaneously with the stylus. If the user manipulates his viewing point of the simulation with the mouse (or space-mouse), the 3D-cursor will stay on its position, as we inverted the rotation matrix for the cursor object. So the input device can be handled intuitively, the cursor seems to stick on the stylus. At every time it is possible to recalibrate the cursor, to obtain a comfortable position for the user in front of the display.

4. Features of the prototype

Using the described input device, it is possible to move a 3D-cursor on the screen, to select parts of the visualization. Usually, the cursor is a ball or a cube. Its size is user-defined. The cursor has the function as an object, from which particles are released into the flow. Actually, it is checked, wether an injection point is within this cursor-object. As the particles are already precalculated and injected in a raster around the object, this is more a selection of particle injector, then a real injection. To get a "natural" behavior of the so selected particles, we show only particles which are injected during the timestep, at the point wheere the cursor touches the injection point. If the cursor is moved away, these particles are kept visible until they leave. This mode is analogous to an injection of real, fluorescent particles into a fluid flow. In the other mode, all particles from the selected injection points are switched on, as long as the cursor touches. To aid the user in finding positions of injection points, we applied a simple optical hint. If the cursor touches injection points, it turns blue. If it's not, it turns red. So the user knows, wether he is capable of injecting particles. The user can now interactively explore regions of the flow.

To visualize larger regions of the flow, it is possible to switch injection points on and off, and to give them individual colors. It is possible to change the rendering mode of each injection point, eg. instead of pixels, small balls are rendered. An eraser and a full reset aids the user in modeling the structure of the injection points. Individual particles can be selected and their pathlines can be generated. Those pathlines can be interactively colored, deleted or recalculated.

We have seen that depthqueues are not sufficient to aid the user in navigating through the 3D scene of the flow. We used LCD-shutter glasses to generate a stereskopic view of the scene. Now, the user has a 3D view and is able to move the 3D cursor around in a real 3D scenario.

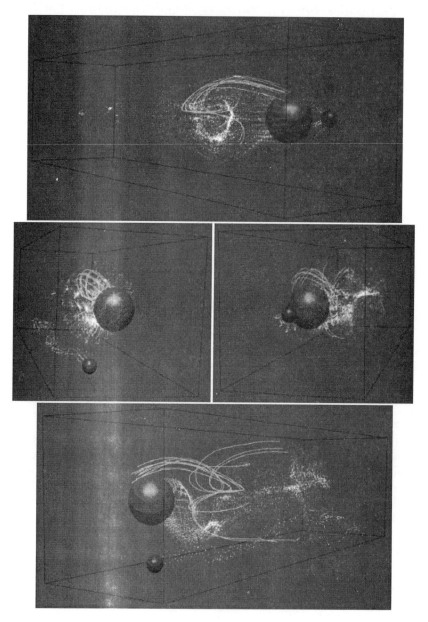

Figure 2: Four timesteps of our flow past a ball, captured from different viewpoints. Visible are colored particles and some injected streamlines above our ball. The small ball is the particle injector, from which particles are released into the flow.

To summarize the main features of our visualization tool:

- Pseudo-interactive injection of particles via a 3d input device
- Interactive selection of particle injectors
- Interactive generation of streamlines
- Usage of different colors for the particle visualization by the user
- Stop, rewind and forward functions like a Videorecorder
- Stereoscopic view using shutter glasses
- Navigation with mouse or spacemouse

5. From IRIS GL to OpenGL/X

A next step in the history of our flow visualization tool is the transposing from SGI's old graphics library (GL) to OpenGL. OpenGL and especially Open Inventor are SGI's new strategic interfaces for 3D computer graphics. The old IRIS GL is being maintained and bugs will be fixed, but SGI will no longer add enhancements. After other implementors had experience trying to port the IRIS GL to their own machines, it was learned that the IRIS GL was too tied to a specific window system or hardware. OpenGL code is neither binary nor source code compatible with IRIS GL code, so it is impossible to mix your old IRIS GL code with OpenGL. Based upon consultations with several implementors, OpenGL is much more independent on several hardware platforms. Today on most platforms you can find an installation of OpenGL. It is supported by many hardware and software vendors like IBM, HP, Digital Equipment and Sun (optimized for the SPARC/Solaris enviroment). Most of this systems are UNIX and use under the X Window enviroment graphical frontend-librarys like OSF/Motif.

There are two approaches to using OpenGL with Motif. One is to render into a standard Motif drawing area widget, but this requires each application window to use a single visual for its window hierachy. A better approach was to use the special OpenGL drawing area widget allowing windows used for OpenGL rendering to pick freely an appropriate visual without affecting the visual choice for other widgets. We use the IRIS IM widget set as a basic class for OpenGL rendering. IRIS IM is SGI port of the industry standard OSF/Motif software enviroment and is commonly used in mixed-model programs. It will be easy to replace the Motif rendering area by an different, OSF/Motif-like widget class. We have only to re-implement the callback functions for the rendering type we use. In our mixed-model program, the X part of the code manages the window initialization, window control and event handling. Some librarys are available for OpenGL that enchance the functionality. Very useful are the OpenGL Utility Library Toolkit (GLUT) and the auxiliary library for OpenGL. The GLUT is a portable toolkit which performs window and event operations to support OpenGL rendering. The auxiliary library provides several simple routines for windowing, event handling and drawing three-dimensional objects.

Still another way to integrate OpenGL rendering with widgets is the Open Inventor object-oriented 3D graphics toolkit which renders using OpenGL and integrates with X Toolkit widgets. Open Inventor allows you to specify 3D scenes in an object-oriented fashion instead of low-level OpenGL rendering primitives. But the higher functionality has the disadvantage in performance. The mixed-model OpenGL with OSF/Motif is as fast as the old GL.

There is a fair amount of work to convert an IRIS GL program to OpenGL. Most of it was in substituting the window management and input handling routines, for which the equivalents are not OpenGL, but the local window system, like the X Windows System. More about the programming details you will find in [4] and [5], especially in [6].

References

[1] R. Bruckschen: "3D Particle Tracing for Real Time Visualization", Workshop on Visualization, D. Kröner and R. Rautmann (Editors) (1995).

[2] W. Borchers: "Zur Stabilität und Faktorisierungsmethode für die Navier-Stokes Gleichungen inkompressibler viskoser Flüssigkeiten", Habilitationsschrift für das Fach Mathematik im Fachbereich Mathematik-Informatik der Universität-GH Paderborn (1992).

[3] S. Blazy, W. Borchers, U. Dralle: "Parallelization Methods for a Characteristic's Pressure Correction Scheme".

[4] "Graphics Library Programming Guide", Silicon Graphics Inc., Mountain View, California (1992).

[5] "OpenGL Programming Guide", Addison-Wesley Publishing Company (1993).

[6] "OpenGL Porting Guide", Silicon Graphics Inc., Mountain View, California (1993).

[7] Foley, van Dam, Feiner, Hughes: "Computer Graphics, Principles and Practice", Addison-Wesley Publishing Company (1994).

[8] Frits H. Post, Theo van Walsum: "Fluid Flow Visualization", H.Hagen, H.Müller, G.M. Nielson (Eds.), Focus on Scientific Visualization, Springer Verlag (1993).

II. DIRECT AND LARGE-EDDY SIMULATION OF TURBULENCE

INTRODUCTION TO PART II

by

C. Härtel

Institute of Fluid Dynamics
Swiss Federal Institute of Technology
CH-8092 Zürich, Switzerland

The papers presented in this section deal with the study of turbulent fluid flows by direct numerical simulation (DNS) and large-eddy simulation (LES). In the past decade the development and improvement of these two closely related flow-simulation techniques have been strongly stimulated by the rapidly increasing performance of modern supercomputers.

In the DNS approach the evolution of all significant scales of motion is fully resolved in space and time and consequently no turbulence model is needed. Since turbulence typically contains a very broad range of eddies, the resolution requirements for DNS are immense. In fact, significant supercomputing resources are needed for the DNS of all but the simplest turbulent flows. Therefore the application of this method is restricted to flows in rather simple geometries and at low Reynolds numbers at present time.

Large-eddy simulation, a combination of DNS with statistical turbulence modeling, is more widely applicable than DNS. In LES only the large inhomogeneous scales are directly simulated while the effect of the missing small scales is supplied by a so-called subgrid model. Although LES in general is much more expensive than statistical simulations of the same flow, the method is expected to play a major role for the analysis and prediction of complex turbulent flows in the future.

In the paper of Güntsch and Friedrich results are reported from a DNS study of turbulence subject to a compression in a cylinder. This study was conducted in order to clarify some issues related to the air flow in a piston engine. The numerical scheme used by the authors is based on a second-order accurate finite-volume discretization in a coordinate system which follows the compression. A key finding of this study is that the growth of the turbulent kinetic energy during the compression is not affected by the compression rate, but depends on the compression ratio only.

The paper of Maaß and Schumann is concerned with the effect of a train of surface waves on a turbulent flow. The authors present results from direct numerical simulations of the flow in a channel with a wavy lower wall which they compute using a finite-difference scheme in fitted coordinates. The influence of the numerical resolution, the Reynolds number and the geometrical shape of the surface is investigated and a comparison of numerical results with existing experimental data is made.

Härtel and Kleiser address the issue of subgrid modeling in the near-wall region of a wall-bounded turbulent flow. In this flow regime pronounced inhomogeneities occur even in the small scales which renders their modeling very difficult. Analyzing DNS data of turbulent channel and pipe flow, the authors find that within the buffer layer a reversed flux of energy from small to large scales exist which cannot be accounted for by currently applied models. An improved near-wall formulation of the eddy viscosity is proposed which shows promising results in *a priori* tests.

Breuer and Rodi use LES to study a number of turbulent flows in more complex geometries using a finite-volume method in curvilinear body-fitted coordinates. Curvilinear and non-orthogonal grids are required in many practical engineering applications, but little experience has been gained yet with such grids in large-eddy simulation. The flow problems studied by Breuer an Rodi include the flow in a 180 degree bend and the flow around surface-mounted obstacles. A discussion of the influence of the subgrid model, the numerical resolution and the boundary conditions is provided along with a comparison of LES results with available experimental data.

The flow over a surface-mounted obstacle, namely a hemisphere, is also considered in the paper of Manhart and Wengle. Contrary to Breuer and Rodi the authors employ a cartesian grid where the hemisphere is blocked out. Special emphasis is laid on realistic inflow data which are obtained from auxiliary large-eddy simulations of boundary-layer flows downstream of a barrier fence and a set of vorticity generators. For the analysis of the simulation results Manhart and Wengle utilize the proper orthogonal decomposition (POD) technique.

Direct numerical simulation of turbulence compressed in a cylinder

Eberhard Güntsch and R. Friedrich
Lehrstuhl für Fluidmechanik,
Technische Universität München,
Arcisstraße 21, 80333 München, Germany

SUMMARY

The behaviour of initially isotropic turbulence during compression in a cylinder is investigated by means of direct numerical simulation (DNS). The flow is governed by the 3D time-dependent Navier-Stokes equations, which have been simplified assuming an adiabatic compression process in which sound waves are unimportant and density, temperature and shear viscosity are functions of time alone. A finite volume technique is used to integrate these equations in a cylindrical coordinate system which is axially compressed while the piston moves. The numerical scheme is central and essentially second-order accurate in space and time. Initial conditions with well correlated velocity and pressure fluctuations according to decaying isotropic turbulence are generated from a stochastic divergence-free velocity field which is made to satisfy a prescribed energy spectrum. Two flow cases with different evolution of the compression rate are considered in order to study its influence on the growth rate of the turbulent kinetic energy. It is found that only the compression ratio (not the rate) has an effect. Furthermore, the simulations show that turbulence cannot be sustained in zones close to the piston surface and the cylinder head.

INTRODUCTION

The air flow in a piston engine involves a complicated system of turbulent shear layers and recirculation regions. The fact that the flow generated during the intake stroke can have a great number of different global and local characteristics (e.g. swirling or tumbling motion) may explain why the behaviour of turbulence during the compression stroke is not yet fully understood. The literature contains contradictory conclusions concerning the development of turbulence intensities. They range from rapid decay through no change to an increase. On the other hand, simulations of homogeneous, compressed turbulence [1] clearly demonstrate the increase in turbulent kinetic energy. The aim of the present investigation is to contribute to a better understanding of turbulence during compression in a cylinder, considering a simplified flow situation which is closer to reality than homogeneous compressed turbulence.

BASIC EQUATIONS
Concept of uniform density flow

It is widely assumed [1], [9] and has been shown experimentally [10] that the gas density during the compression stroke of a combustion engine is homogeneous in space and only a function of the crank angle (or time), provided the cylinder walls are not hot. This means that the effect of sound waves can be neglected. Furthermore, for high compression rates the compression process can be considered as adiabatic, so that density, temperature and thus viscosity are functions of time alone. Under these conditions the energy equation is irrelevant and the Navier-Stokes equations take the simplified form

213

$$\frac{1}{\rho}\frac{\partial \rho}{\partial t} + \nabla \cdot \vec{u} = 0, \quad (1)$$

$$\frac{\partial}{\partial t}(\rho\vec{u}) + \nabla \cdot (\rho\vec{u}\vec{u}) = -\nabla p + \nabla \cdot \bar{\bar{\tau}}. \quad (2)$$

Since the divergence of the velocity field is homogeneous in space, the shear stress term in Eqn. (2) is free of dilatational effects, viz

$$\nabla \cdot \bar{\bar{\tau}} = \mu \nabla \cdot (\nabla \vec{u} + \nabla \vec{u}^T). \quad (3)$$

The Navier-Stokes equations (1) and (2) are solved in an (r,φ,z) cylindrical coordinate system using a finite volume technique. The computational domain has the shape of a straight circular cylinder which is compressed in the z-direction. In order to conserve the number of mesh cells during the compression stroke, the axial coordinate is transformed, according to

$$z = \zeta L(t), \quad t = t'. \quad (4)$$

where L(t) is the instantaneous cylinder (or clearance) height. Using Eqn. (4), time and space derivatives become

$$\frac{\partial}{\partial t} = \frac{\partial}{\partial t'} - \frac{\zeta}{L(t)}U_p\frac{\partial}{\partial \zeta}, \quad \frac{\partial}{\partial z} = \frac{1}{L(t)}\frac{\partial}{\partial \zeta}. \quad (5)$$

where U_p represents the piston velocity $U_p(t) = dL/dt$. L(t) can be used to define an instantaneous compression ratio:

$$K_c(t) = L(0)/L(t) \quad (6)$$

which should be distinguished from the compression rate S:

$$S = |U_p|/L(t) = \frac{1}{\rho}\frac{\partial \rho}{\partial t}. \quad (7)$$

Numerical method and boundary conditions

The Navier-Stokes equations transformed according to Eqns. (4) and (5) are integrated over a mesh volume $\Delta V = r\Delta\varphi \Delta r \Delta\zeta$, using Schumann's [2] volume averaging procedure. This leads to an essentially second-order accurate central space discretization. The same technique is used in the contribution [5], [6] to this volume. In non-dimensional form these mesh-averaged equations are

$$Sr\Delta V \frac{1}{\rho}\frac{\partial \rho}{\partial t} + \sum_\alpha (\Delta A_\alpha \overline{u_\alpha}|_+ - \Delta A_\alpha \overline{u_\alpha}|_-)C_\alpha = 0 \quad (8)$$

$$Sr\Delta V \frac{\partial \overline{u_\alpha}}{\partial t} + \frac{U_p}{L}\left(\Delta A_\zeta \overline{\zeta u_\alpha}|_{\zeta+} - \Delta A_\zeta \overline{\zeta u_\alpha}|_{\zeta-} - \Delta V \overline{u_\alpha}\right)$$
$$+ \sum_\beta (\Delta A_\beta (\overline{u_\alpha}\overline{u_\beta} + \overline{p}/\rho \delta_{\alpha\beta} - \overline{\tau_{\alpha\beta}})|_+$$
$$- \Delta A_\beta (\overline{u_\alpha}\overline{u_\beta} + \overline{p}/\rho \delta_{\alpha\beta} - \overline{\tau_{\alpha\beta}})|_-)C_\beta$$
$$- (Term)_\alpha = 0 \quad (9)$$

$$C_\beta = 1 + (1/L(t) - 1)\delta_{\zeta\beta} \quad , \quad \alpha, \beta = r, \varphi, \zeta$$
$$Sr = l_{ref}/(u_{ref}t_{ref}) \quad , \quad \overline{\tau_{\alpha\beta}} = \frac{1}{Re}\overline{D_{\alpha\beta}} \tag{10}$$

$(Term)_\alpha$ contains the curvature terms:

$$(Term)_\zeta = 0,$$
$$(Term)_\varphi = \Delta\varphi \Delta A_\varphi(-\overline{u}_\varphi \overline{u}_r + \overline{\tau_{\varphi r}}),$$
$$(Term)_r = \Delta\varphi \Delta A_\varphi(\overline{u}_\varphi^2 + \overline{p} - \overline{\tau_{\varphi\varphi}}). \tag{11}$$

$\overline{D_{\alpha\beta}}$ is the discrete form of the deformation tensor in the present moving cylindrical coordinate system. The overbar over variables (except the pressure) denotes surface averages; $\overline{p}$ is a volume average. The mesh surfaces $r\Delta\varphi\Delta r$, etc., are chosen of the order of the initial Kolmogoroff length scale squared. The momentum equations in their finite volume formulation are integrated in time on staggered grids. This is done using a semi-implicit scheme, in which all (convection and diffusion) terms containing derivatives in the φ-direction are treated implicitly while the remaining terms are integrated explicitly in a leapfrog step. A fractional-step algorithm is applied to achieve the proper coupling between the pressure and velocity fields. The resulting Poisson equation for the pressure contains the compression rate S as a known function of time. Using FFT in the φ-direction it is converted into a set of 2D Helmholtz equations which are treated with a cyclic reduction technique.

Wall boundary conditions are implemented without any approximation. The wall normal velocity component is zero at the cylinder wall and equals U_p at the piston surface. Since the tangential velocity components are not defined at the walls in this scheme, the tangential shear stresses are specified instead. The algorithms described here have all been implemented in the code FLOWSI, which has been developed in our group and is used in the work of [7].

The shape of the combustion chamber has been chosen as simple as possible. The flow conditions at the start of the compression stroke (when the piston is at BDC) are well defined, repeatable and known in all details. Turbulence initially is isotropic with skewness factors of the velocity gradients between -0.4 and -0.5. Several realizations of these initial conditions which are equal in the mean, but different in their details, allow for a number of realizations of the compression stroke and provide acceptable statistics.

FLOW and SIMULATION PARAMETERS

Parameters of the simulation will now be specified. We begin with the initial values of mean dimensional quantities which are obtained from a separate DNS of decaying isotropic turbulence, see [8]:

- velocity scale $q = (2K)^{1/2} = 0.2806 m/s$

- Taylor microscale $\lambda = (\langle u'^2 \rangle / \langle (\partial u'/\partial x)^2 \rangle)^{1/2} = 9.3 \cdot 10^{-3} m$

- integral length scale $\Lambda = \frac{3\pi \int k^{-1} E(k) dk}{4 \int E(k) dk} = 35.7 \cdot 10^{-3} m$

- Kolmogoroff length scale $\eta = (\nu^3/\epsilon)^{1/4} = 0.92 \cdot 10^{-3} m$. \hfill (12)

Taking η as the length scale to be resolved in a DNS, the computational domain $R^2\pi L(0)$ is divided into $64 \times 64 \times 128$ cells in the (r, φ, z) directions, providing a resolution of

$$\Delta z = 1.64 \cdot 10^{-3} m,$$
$$\Delta r = 0.58 \cdot 10^{-3} m,$$
$$r\Delta\varphi = (0.057 \div 3.68) \cdot 10^{-3} m. \quad (13)$$

Grid and geometry are illustrated in Figure 1. The spanwise resolution near the wall would

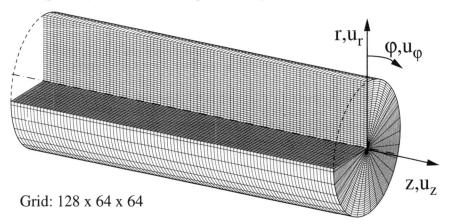

Grid: 128 x 64 x 64

Figure 1: Grid and geometry of the simulated flow (only each second grid line is plotted).

certainly not be sufficient for fully developed shear flow dominated by streaky structures. In the present case, however, the flow is not fully developed in this sense. It separates or reattaches at the cylinder wall during compression so that the streaky structures cannot form. The flow parameters used for non-dimensionalization are the pipe diameter D and the maximum piston velocity U_0. A few features of the simulation are with respect to a Cray YMP 8/864:

- Code performance: 140 MFLOPS
- Core memory: 20 MW
- Time step: $1.6 \cdot 10^{-4}$
- CPU time per timestep per point: $3.97 \cdot 10^{-6}$ s
- total CPU-time for 1 realisation (180° crank angle): 11.1 h
- CPU-time for the whole simulation: 1000 h

Important non-dimensional parameters of the present problem are

$$Sk/\epsilon = 3.02, \quad Re_t = \frac{\Lambda q}{\nu} = 183, \quad Re_\lambda = \frac{\lambda q}{\nu} = 47. \quad (14)$$

The ratio 3.02 of the turbulent time-scale K/ϵ to the mean time-scale $\overline{S}^{-1}$ states that the turbulence dynamics is roughly three times slower than the compression process. Effects of compression will therefore outweigh effects of turbulence decay. It is worthwhile considering the initial turbulent field before discussing its change during compression. Figure 2 shows contour lines of the initial axial and radial velocity components. The whole cylinder is filled with eddies of fairly large scale. The wall has not yet had time to affect the turbulent field.

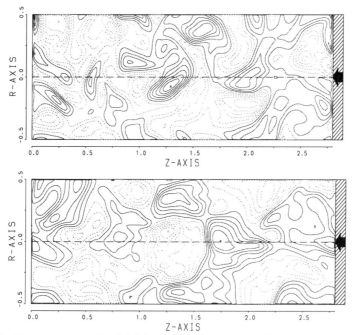

Figure 2: Contour lines of the initial axial (top) and radial (bottom) velocity components, u'_z, u'_r.

Two cases have been studied with the following characteristics:

- Case I : the compression rate is constant (S=20),
- Case II: the compression rate changes with time according to a sinusoidally varying piston speed U_p.

In Case I, U_p decreases exponentially, viz

$$U_p(t) = U_0 exp(-St). \qquad (15)$$

and in Case II it varies as

$$U_p(t) = U_0 sin(\pi n t/30) \quad , \quad n = 1000 \ min^{-1}. \qquad (16)$$

Accordingly, the compression times t_c, namely the time intervals the piston needs to reach the TDC, are different. Figure 3 shows the time variation of S along with compression times.

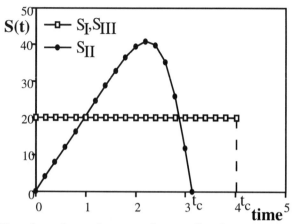

Figure 3: Time dependence of compression rate S and compression time t_c for cases I and II.

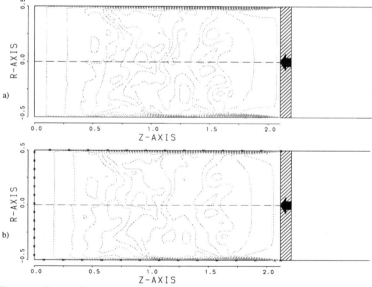

Figure 4: Contour lines of instantaneous axial velocity component u_z. a) $t = 0.8R/U_0$ (Case I); b) $t = 1.2R/U_0$ (Case II).

Figures 4a and b show contour lines of the total instantaneous axial velocity for Cases I and II. The velocity field reflects the generation of a boundary layer along the cylinder wall. All contour lines indicate a flow towards the cylinder head (negative u_z values). Lines close to the piston surface and the cylinder head are straight, showing that turbulence has decayed in these regions. At this time the piston positions and thus the instantaneous compression ratios $K_c(t)$ of Cases I and II coincide. There are strong similarities even in the instantaneous flow states between the two cases which indicate that it is not the history of the compression rate which matters, but the compression ratio. This fact is also supported by the profiles of the rms velocity fluctuations (see [3], [4] for more details). Therefore, we will concentrate on the discussion of Case II alone.

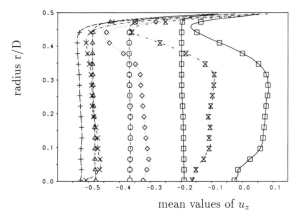

Figure 5: Mean axial velocity profiles taken at half the clearance height (Case II). (□) 12% stroke; (○) 25% stroke; (△) 38% stroke, (+) 50% stroke; (×) 63% stroke; (×) 76% stroke; (◇) 89% stroke; (□) 100% stroke.

RESULTS

Statistical quantities

Statistical data are obtained by averaging in the circumferential direction and over a certain number of realizations of the compression process with different initial conditions. Since averaging in the φ direction is of no use close to the cylinder axis, the data there must be considered with some care. The number of realizations differs for the presented piston positions (Case II). For positions up to 50% of the piston stroke we have 56 and for the remaining ones we have 36 realizations. We first discuss the mean axial velocity profiles. Figure 5 presents $\overline{u_z}$ profiles taken at half the clearance height ($\zeta=1/2$) for eight different piston positions (crank angles $\pi/8, ..., \pi$). They clearly reflect the piston movement which is accelerating in the beginning and then slowing. With the execption of profiles for 89% and 100% piston stroke, the sampling seems to provide sufficiently stable data. In Figure 6 profiles of the turbulent kinetic energy are presented at the same locations and instants in time. A clear increase in turbulent kinetic energy can be observed, with maximum values being reached at 76% of the total piston stroke. This supports observations that turbulence is produced during the compression process. In a previous paper [3], we had argued that turbulence cannot be maintained close to the piston surface and the cylinder head. We will now consider the axial distribution of the radial mean velocity at different

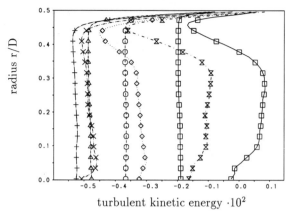

Figure 6: Turbulent kinetic energy profiles taken at half the clearance height (Case II). (□) 12% stroke; (○) 25% stroke; (△) 38% stroke, (+) 50% stroke; (×) 63% stroke; (×) 76% stroke; (◇) 89% stroke; (□) 100% stroke.

locations r/D and for two different piston positions. Figure 7a shows a typical feature of the flow field, namely the strong displacement of the fluid near the piston surface due to the inhomogeneity of the boundary conditions in the corner region. It causes the fluid to leave the corner with radial velocities that are comparatively high (a similar effect can be observed at the leading edge of a flate plate where the incoming horizontal flow is quickly decelerated, while the vertical flow is accelerated). Comparable effects are found in Figure 7b, but with higher amplitudes. It seems as if the flow displacement near the piston surface induces a successive thickening and thinning of the 'boundary layer' along the cylinder wall. The question of what makes the turbulence fluctuations die out close to the piston surface remains open. A primary source for this could be sought in the impermeability constraint which has a much stronger damping effect than viscosity.

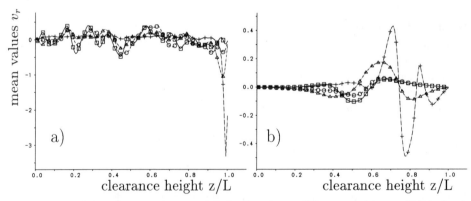

Figure 7: Axial distribution of mean radial velocities at different positions r/D(k). (□) k=16; (○) k=32; (△) k=48; (+) k=62. a) 12.5% stroke; b) 89% stroke.

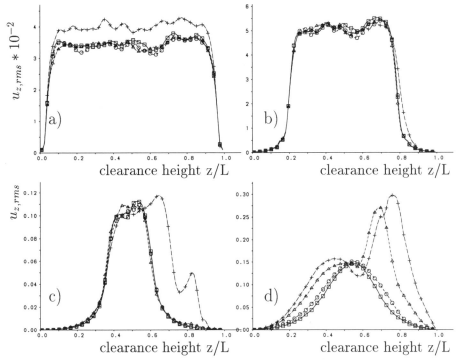

Figure 8: Comparison of axial rms velocities at different radial positions r/D(k). (□) k=16; (○) k=32; (△) k=48; (+) k=62. a) 12.5% stroke; b) 38% stroke; c) 63% stroke; d) 100% stroke.

Figure 8 shows the axial distribution of the axial rms velocity fluctuations at different radial locations and for different piston strokes. Near the piston surface and the cylinder head the rms values decrease. The fact that the values increase during the compression process and that the plateaux are reaching higher levels again indicates that turbulence is produced at the earlier stages of compression. The figure also reflects an axial variation of the rms values which is induced by the structure of the mean velocity field.

It is very instructive to discuss the effect of compression on the integral length scales. Here we concentrate on discussing the integral length scale in the z-direction. To avoid difficulties during the calculation of the integral length scale L_{zz}, which might occur owing to the change of sign of the correlation function along the z-axis, we calculate it as shown in Figure 9. The integration is performed up to that point where L_{zz} first reaches 10% of its maximum value. In Figure 10a we plot the axial distribution of L_{zz} over the clearance height z/L at four different radial positions and at a piston stroke of 50%. We note that the smallest values of L_{zz} are found near half the clearance height. From there the length scale increases again towards the piston surface and the cylinder head because most of the turbulence in those regions has already decayed. Another feature of the flow field is documented by the distribution of L_{zz} over the crank angle, which is shown in Figure 10b. L_{zz} is computed at half the clearance height and for three different radial positions. At crank angles from 50 up to 76% of the piston stroke the integral length scale is minimal.

These angles coincide with those where the turbulent kinetic energy is maximal. Next

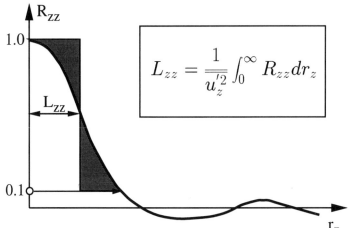

Figure 9: Calculation of the integral length scale L_{zz}.

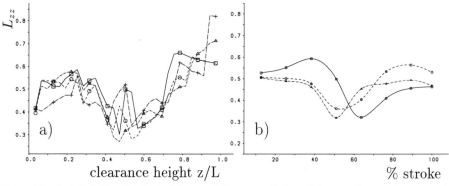

Figure 10: a) Axial distribution of L_{zz} at different radial positions and at a piston stroke of 50%. b) Distribution of L_{zz} over the crank angle at different radial positions. (□) k=16; (○) k=32; (△) k=48.

we inspect the production process of turbulent kinetic energy. To do so we write the total production as the sum of shear and dilatational production. Eqn. (17) is the general expression in cartesian coordinates for simplicity reasons

$$\underbrace{\overline{u'_i u'_j} \frac{\partial \overline{u_i}}{\partial x_j}}_{P_{Tot}} = \underbrace{\overline{u'_i u'_j} \left(\frac{\partial \overline{u_i}}{\partial x_j} - \frac{1}{3} \delta_{ij} \frac{\partial \overline{u_k}}{\partial x_k} \right)}_{P_{Shear}} + \underbrace{\frac{1}{3} \overline{u'_i u'_j} \delta_{ij} \frac{\partial \overline{u_k}}{\partial x_k}}_{P_{Dil}}. \qquad (17)$$

In cylindrical coordinates we find, e.g. for the total production P_{Tot},

$$P_{Tot} = \underbrace{\frac{1}{L} \overline{u'^2_z} \frac{\partial \overline{u_z}}{\partial \zeta}}_{P_{Tot1}} + \underbrace{\overline{u'_r u'_z} \frac{\partial \overline{u_z}}{\partial r}}_{P_{Tot2}} + \underbrace{\overline{u'^2_r} \frac{\partial \overline{u_r}}{\partial r}}_{P_{Tot3}} + \underbrace{\frac{1}{L} \overline{u'_r u'_z} \frac{\partial \overline{u_r}}{\partial \zeta}}_{P_{Tot4}} + \underbrace{\frac{\overline{u_r}}{r} \overline{u'^2_\varphi}}_{P_{Tot5}}. \qquad (18)$$

By plotting the axial distribution of the different production terms in Eqn. (17) we are able to tell which is of greater influence. In Figure 10a we show the axial distribution of the total production, the production due to shear and the production due to dilation over the clearance height z/L at a piston position of 12% stroke. We can see that nearly all of the production of turbulent kinetic energy is coming from dilatational production, while the production due to shear is not contributing to it. Figure 11b displays the axial distribution of the three dilatational production terms. It is clearly visible that there is

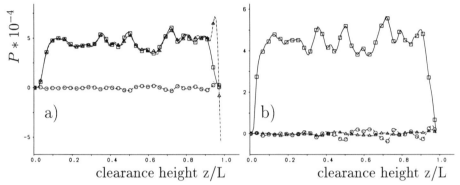

Figure 11: Axial distribution of the different production terms at 12% stroke. a) (□) P_{Tot}; (○) P_{Shear}; (△) P_{Dil}. b) (□) P_{Dil1}; (○) P_{Dil2}; (△) P_{Dil3}.

only one dominating term, namely that containing the mean axial strain rate,

$$P_{Dil1} = -\frac{2}{3}k\frac{1}{L}\frac{\partial \overline{u_z}}{\partial \zeta}. \tag{19}$$

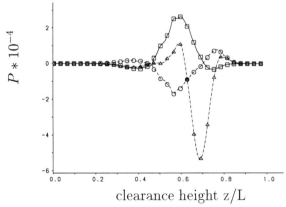

Figure 12: Axial distribution of the different production terms at 100% stroke. (□) P_{Tot}; (○) P_{Shear}; (△) P_{Dil}.

A look at the same distribution at 100% stroke shows a different behaviour (see Figure 12). Now the total production of turbulent kinetic energy is mostly coming from shear production while dilatational production plays only a minor role. The dominating term hereby is $-\overline{u'_r u'_z} \frac{\partial \overline{u_z}}{\partial r}$. It is also important to note that turbulence is produced only in the core of the cylinder.

Instantaneous flow quantities

Figures 13a and b show contour lines of the instantaneous axial velocity component in

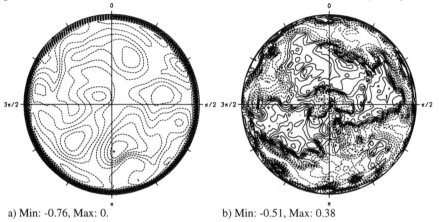

a) Min: -0.76, Max: 0. b) Min: -0.51, Max: 0.38

Figure 13: Contour lines of the total instantaneous axial velocity at different piston positions. a) 50% stroke, b) 100% stroke.

a plane ζ=constant at half the clearance height for two different piston positions (solid and dashed lines in correspond to positive and negative values, respectively. The u_z field reflects the generation of a boundary layer along the cylinder wall. In Figure 13a all contour lines indicate a flow towards the cylinder head (negative u_z values). This phase of the compression process seems to be mainly dominated by the piston motion. Figure 13b displays a different behaviour. Although still visible, the wall layer is thinner and the contour lines indicate spots of reversed flow. The piston motion has lost its dominance. The difference between minimum and maximum values of u_z also increases during the compression process. Figure 14 presents streamlines of the instantaneous flow field for four different piston positions. One can see how the flow generates a small recirculation region near the wall, then strongly separates and forms into recirculation zones which then induce others. Similar flow phenomena were already observed in slow compression of air which was initially at rest [3]. Figure 14d gives insight into the complex turbulent flow situation at TDC.

CONCLUSIONS

The DNS technique has been used to investigate the effects of compression on initially isotropic turbulence. Only mean compression effects have been taken into account assuming the density to be homogeneous in space, but time-dependent. Thus, acoustic effects were completely discarded. The following findings can be summarized:

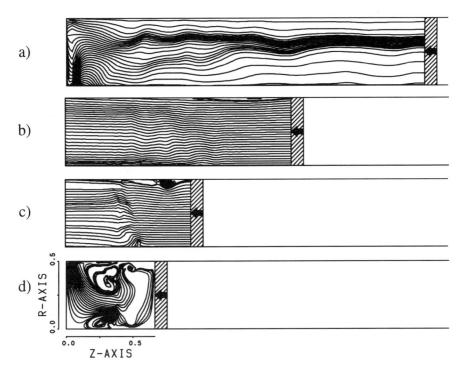

Figure 14: Streamlines of the instantaneous velocities u_z, u_r. a) 12% stroke, b) 50% stroke, c) 76% stroke, d) 100% stroke.

- During compression (Cases I and II) an increase in turbulent kinetic energy is found in situations where $SK/\epsilon \geq 1$. It is due to dilatational production ($-\frac{1}{3}\langle u_i' u_i' \rangle \partial \langle u_k \rangle / \partial x_k$).

- The compression history ($S(t)$) does not seem to affect the amplification rate of turbulence (as long as the temporal mean of S is nearly the same in both cases), but the compression ratio $K_c(t)$ does.

- Turbulence cannot be maintained close to the piston surface and the cylinder head.

- Integral length scales show the strongest decrease at strokes between 50 and 76% where the turbulent kinetic energy is maximal.

- In the earlier stages of compression turbulent kinetic energy is mainly produced by dilatation. At TDC production due to shear outweighs dilatational production.

- The structure of the mean flow field has a strong impact on the spatial distributions of turbulence intensities. This must be taken into account in comparisons with experimental data.

Acknowledgement

The authors thank the German Research Association (DFG) for supporting these studies in its Priority Research Programme Strömungssimulation mit Hochleistungsrechnern.

REFERENCES

[1] C.T. Wu, J.H. Ferziger and D.R. Chapman, 1985: Simulation and modeling of homogeneous, compressed turbulence. *Thermosciences Div. Rept. TF-21*, Mech. Engg. Dept., Stanford University.

[2] U. Schumann, 1975: Subgrid scale model for finite difference simulations of turbulent flows in plane channels and annuli., *J. Comp. Physics*, **18**, 376–404.

[3] E. Güntsch and R. Friedrich, 1994: On the influence of compression and rotation on initially isotropic turbulence in a cylinder., *Proc. 2nd Europ. Comp. Fluid Dyn. Conf. 5-8 Sept. 1994*, Stuttgart, S. Wagner et al. (Eds), Wiley, New York, pp. 525–534.

[4] E. Güntsch and R. Friedrich, 1995: Compression of initially isotropic turbulence in a cylinder at low Reynolds number., *Proc. 10th Symp. on Turb. Shear Flows, Pennstate University, 14-16 August, 1995*.

[5] C. Maaß and U. Schumann, 1996: Direct numerical simulation of separated turbulent flow over a wavy boundary. This volume.

[6] M. Manhart and H. Wengle, 1996: Large eddy simulation and eigenmode decomposition of turbulent boundary layer over a hemisphere. This volume.

[7] M. Griebel, W. Huber and C. Zenger, 1996: Numerical turbulence simulation on a parallel computer using the combination method. This volume.

[8] E. Güntsch, 1996: Kompression isotroper Turbulenz in einem Zylinder. - Eine numerische Studie. Doctoral Dissertation. TU München.

[9] W.C. Reynolds, 1980: Modeling of fluid motions in engines - An introductory overview. In: J.N. Mattavi and C.A. Amann (Eds.), *Combustion Modeling in Reciporcating Engines*, pp. 11-20

[10] G. Serre, 1994: Etude Experimental Et Modelisation De La Turbulence Homogene Compressee. Doctoral Dissertation. L'Ecole Central de Lyon.

Direct Numerical Simulation of Separated Turbulent Flow over a Wavy Boundary

Carsten Maaß and Ulrich Schumann

DLR, Institut für Physik der Atmosphäre
D-82230 Oberpfaffenhofen, Germany

Summary

The impact of a wavy surface on turbulent flow is investigated by direct numerical simulation. By means of finite differences in terrain following coordinates, the method treats the flow in a plane channel with wavy lower and flat top surfaces. Both surfaces are smooth. The lower surface wave amplitude is 0.05 and the wavelength is 1 in units of the mean channel height. The Reynolds number in terms of mean velocity and mean channel height is 6760. Parameter studies are performed with different resolution, Reynolds number and geometrical shape of the surface wave. If the vertical resolution is fine enough to resolve the viscous surface layer, a recirculation zone develops as expected for this surface geometry and Reynolds number. The comparison with existing experimental data shows good agreement when the precise details of the surface wave geometry, which deviates slightly from a sinusoidal profile, is taken into account.

Introduction

Turbulent boundary-layer flows in complex geometries are typical for environmental and technical flow problems. Traditional research on turbulent boundary layers considers flat plane channels and walls (Andrén et al. [2], Härtel and Kleiser [11], Schumann [24]) and pipes (Eggels et al. [9]). In this sense considering a plane channel with one wavy wall is a small step to more complex geometries. If one neglects friction and turbulence, i. e. considers a potential flow with constant mean streamwise velocity, the pressure variation at the surface is 180° out of phase with the wave and causes zero wave drag. Therefore disturbances decay exponentially with distance from the wall (Lamb [18]). Benjamin [3], using a curvilinear coordinate system, included friction and a mean boundary-layer velocity profile. He showed that the shear layer causes a phase shift of the velocity distribution that produces a drag. Thorsness et al. [26] found a quasi-laminar model sufficient to describe the wall shear stress for wavenumbers $2\pi\nu/(u_\tau\lambda) > 0.01$, but for smaller wavenumbers, the prediction of the phase shift is rather poor. Turbulence models using the van Driest function produce results which agree better with experiments (Kuzan et. al. [17]). For small amplitudes δ and large wavelength λ the flow responds linearly to sinusoidal disturbances. For large enough amplitudes the positive pressure gradient behind the wave is sufficient to cause separation. The flow over a train of waves differs from other separated flows [6] because the separation which may occur behind each wave crest affects the flow over the following waves.

All these models can describe some properties of the flow but no details of the separation. Recently, direct numerical simulations (DNS) of channel flow without separation above mild waviness were performed [12, 14], but in computational domains covering only one surface wave. Hino and Okumura [12] found stronger coherent eddy motion on the wavy wall than on the flat wall. The shear-stress distribution showed quasi-stationary streaky patterns corresponding to longitudinal vortices. The waviness has little effect on the total bottom drag. Simulations of flows over two- and three-dimensional rough hills have been made by Wood and Mason [28] using a numerical model which employs a $1\frac{1}{2}$-order turbulence closure model.

Zilker and Hanratty [30] gave an overview on experimental work, from which the bulk of our understanding of such flows is mainly derived. The experiments concentrated on the measurement of surface pressure, wall shear stress, and velocity components. Motzfeld [21] pointed to the difference between roughness and waviness. Zilker et al. [29] showed that the wall shear stress is a better measure for the linearity of the flow than the velocity.

Buckles et al. [5] found a thin turbulent boundary layer which forms behind the reattachment point and extends to the next separation point. There, a free shear layer develops away from the wall behind the next reattachment point, so that there are now two layers above each other. Gong et al. [10] examined a flow above a wavy wall at two different surface roughnesses. The measurements exhibited an approximately two-dimensional flow with separation over the relatively rough surface. Over the smoother surface they observed a three dimensional secondary flow. The findings can be well modeled by large-eddy simulations (LES). The experiments of Kuzan et al. [17] with $\delta/\lambda = 0.1$ show a decreasing separation with increasing Reynolds number.

Recently, Hudson [13] made measurements in a water channel for two different amplitudes and several flow rates. For a case with separation in the mean flow field, he investigated how the Reynolds stresses and turbulence production differ from what would be observed over a flat wall.

Most theoretical studies were two-dimensional and limited to cases with weak nonlinear effects. Experiments were carried out for a wide set of parameters but not directly comparable and sufficient for turbulence modelling. There is little information about the three-dimensional structures. Considering turbulent separated flow over a wavy wall [1] as a test case for advanced statistical turbulence models, it has been shown recently that these models are still not able to really predict the flow [22]. There is a need for well defined experiments as well as for results from direct numerical simulations. The data presented here could be helpful for improving statistical models as well as LES models, e. g. Breuer and Rodi [4].

In a previous article [20], results of DNS at $Re = 4780$ have been shown and compared to experimental data. There we did not reach complete agreement with the laboratory measurements. In the mean time it turned out that the Reynolds number in the experiment was 6760 instead of 4780 as reported initially. Also the shape of the surface wave differed slightly from the purely sinusoidal shape assumed before. In this study we investigate the three-dimensional nonlinear flow at $Re = 6760$ over a wave which approximates the measured surface shape by means of DNS and compare the results to revised experimental data of Hudson [13].

Method

The method is explained in detail by Krettenauer and Schumann [16]. In this study the model uses grid refinement near the walls, contains different boundary conditions, and is applied to a flow with higher requirements on the accuracy of the method because of thinner shear layers at the surface. For these reasons it still remains to validate the code although the present numerical method has been used for DNS and LES of turbulent convection over wavy boundaries without [16, 23] and with mean flow [7, 8, 19].

The basic equations describe the conservation of mass and momentum for incompressible flows,

$$\frac{\partial u_j}{\partial x_j} = 0, \tag{1}$$

and

$$\frac{\partial u_i}{\partial t} + \frac{\partial}{\partial x_j}(u_j u_i) = -\frac{\partial p}{\partial x_i} + \frac{1}{Re}\frac{\partial^2 u_i}{\partial x_j^2} + P_x \delta_{1i}, \qquad i = 1, 2, 3, \tag{2}$$

where u_i denotes the velocity components, x_i the Cartesian coordinates, t the time, p the deviation from a reference pressure p_0, P_x the driving pressure gradient in x-direction and $Re = UH/\nu$ the Reynolds number for unit density. All quantities are made dimensionless with the channel height H, with the mean streamwise velocity U, with the time scale $t/t_{\text{ref}} = H/U$ and density $\rho = \text{const}$. The equations of motion are formulated for the Cartesian velocity components (u, v, w) as a function of curvilinear coordinates $(\bar{x}, \bar{y}, \bar{z})$ which are related to the Cartesian coordinates according to the transformation $\bar{x} = x$, $\bar{y} = y$, $\bar{z} = \eta(x, y, z)$. Here,

$$\eta = H\frac{z-h}{H-h} \tag{3}$$

maps the domain above the wavy surface at height $h(x,y) = \delta \cos(2\pi x/\lambda)$ and below a plane top surface at $z = H$ onto a rectangular transformed domain. The geometrical parameters are H, the lateral lengths L_x and L_y, the wave amplitude δ, and the wavelength λ, see Figure 1. To get higher resolution at the rigid top and bottom surfaces we use an additional hyperbolical transformation function of a coordinate ξ which is discretized equidistantly,

$$\eta = \frac{H}{2}\left(1 + \frac{\tanh(c_g \xi)}{\tanh c_g}\right), \qquad -1 \leq \xi \leq 1. \tag{4}$$

The parameter c_g is choosen so that two adjacent grid spacings $\Delta \eta$ differ by less than 11 %.

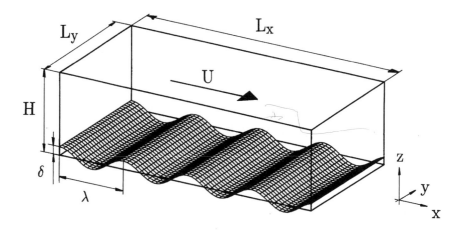

Figure 1: Perspective view of the computational domain in three dimensions showing the sinusoidal surface-wave in x-direction; the surface height is constant in y-direction. In this example, the wavelength is $\lambda = 1$, the wave-amplitude is $\delta = 0.1$, and the domain-size is $H = 1$, $L_x = 4$, and $L_y = 2$.

The differential equations are approximated by finite differences in a spatial staggered grid. The first discrete grid point for the horizontal velocity components is located $\Delta \eta/2$ above the wall. The momentum equation is integrated in time using the Adams-Bashforth scheme. The mean velocity U is defined as the average velocity in the x-direction across a y-z plane at a position with vanishing wave amplitude. After each time step Δt, the actual mean velocity is tested and a mean pressure gradient P_x in the axial direction is determined such that U remains constant. The pressure is determined iteratively. After five iterations the divergence in the first grid level above the bottom surface is reduced to $10^{-3}U/H$. At other grid levels it is lower by the order of 3 magnitudes. Periodicity conditions are used at the lateral boundaries. At the bottom and top surfaces we implement the no-slip condition.

The initial conditions prescribe uniform velocity at every grid point. Random perturbations are added to the velocity field which eventually becomes turbulent. This situation is comparable to turning on the pump in a laboratory set-up. The details of this initialisation are unimportant for the final statistics at late times of the simulation.

The simulations are performed on the DLR high performance computers Cray-YMP and NEC SX-3. All cases are at least run until the final dimensionless time $t_{\text{ref}} = H/U = 70$, which required 15 h CPU time on the NEC SX-3 using $256 \cdot 128 \cdot 96$ grid points. The obtained perfomance on the NEC SX-3 is 20% of the peak performance and at least 7 times higher as on the Cray-YMP. The internal data management, which is carried along although it is not needed, devides the data in vertical x-z slices. This embarrasses optimal vectorisation for the used number of grid points. For the future, we plan to parallelize the code by using parallel algorithms as described in Schumann and Strietzel [25].

Results and Discussion

We report below results obtained from a set of direct numerical simulations with different Reynolds numbers, bottom shapes, and resolutions, see Table 1, which were performed during the approach to an experimental set-up. In all cases the longitudinal domain size extends over 4 surface wavelengths. The lateral domain size should be large enough to cover the largest turbulent structures. The wave amplitude $2\delta/\lambda$ is 0.1 in all cases. The horizontal grid spacings are equal ($\Delta x = \Delta y$). The Reynolds number in DNS 5 (see [20]) is smaller by a factor of $\sqrt{2}$ than in all other cases. The parameters of DNS 1, which corresponds to our best knowledge to the experiment wb3 in the thesis of Hudson [13], differ from DNS 2 only in the shape of the wavy bottom (see below). DNS 2 to 4 differ only in resolution. DNS 4 is the only run with equidistant vertical grid spacing. All other cases use variable grid spacings with maximum resolution near the top and bottom surfaces. Using the friction velocity at the wavy bottom $u_{\tau,\text{wa}}$, the finest resolution in terms of viscous scale is $(\Delta x^+, \Delta \eta^+) = (10.2, 1.6)$ for case 1. This is still at the limit of what is required to resolve the viscous sublayer, but of the same order as in other successfull DNS calculations [15, 27].

From the map of Zilker and Hanratty [30] one expects turbulent flow with mean separation for all cases in Table 1.

Table 1: Model parameters of various DNS. In all cases the computational domain extends over $4H \times 2H \times 1H$; $\Delta x^+ = \Delta y^+$ denotes the horizontal mesh size, $\Delta \eta^+$ the mesh height, c_g the parameter in equation (4), $\Delta(\Delta\eta) = \left(\frac{\Delta\eta_{k+1}-\Delta\eta_k}{\Delta\eta_k}\right)_{\max}$ the difference between two adjacent grid spacings, Δt the nondimensional time step, $u_{\tau,\text{fl}}$ and $u_{\tau,\text{wa}}$ the nondimensional friction velocity at the flat and wavy wall, respectively. Values marked with $^+$ are made nondimensional with wall units. DNS 1 is performed with the experimental wave profile, all other simulations with a sinusoidal bottom.

DNS	Re	$N_x \cdot N_y \cdot N_z$	c_g	Δt	Δx^+	$\Delta(\Delta\eta)$	$\Delta\eta^+_{\min}$	$\Delta\eta^+_{\max}$	$u_{\tau,\text{fl}}$	$u_{\tau,\text{wa}}$
1	6760	$256 \cdot 128 \cdot 96$	1.7	0.003	10.2	0.068	1.6	12.4	0.070	0.097
2	6760	$256 \cdot 128 \cdot 96$	1.7	0.003	10.9	0.068	1.8	13.3	0.070	0.104
3	6760	$96 \cdot 48 \cdot 48$	1.4	0.009	30.1	0.107	5.4	23.8	0.071	0.107
4	6760	$80 \cdot 40 \cdot 40$	1.0	0.010	33.8	0.000	16.9	16.9	0.074	0.100
5	4780	$160 \cdot 80 \cdot 64$	1.4	0.006	12.4	0.079	2.7	12.2	0.071	0.104

By monitoring the kinetic energy E_{kin} of the whole domain we made sure that the flow is sufficiently stationary at the end of the simulation (Figure 2). The evolution of E_{kin} for DNS 1 – 3 is very similar and distinct from DNS 4, the only case without separated region. It can be seen already from this picture, that the mean separated flow is not influenced very much by resolution and by details of the surface wave. Due to the random perturbations, adjustment to the boundary conditions, and realisation of the continuity equation, all curves start slightly above the theoretical value of 0.5. Before $t/t_{\text{ref}} = 5$ a relative maximum is passed which is caused by the growth of initial vorticies in the trough region. From studying instantaneous flow fields at early times of the simulations, we know that first one almost laminar vortex in spanwise direction appears at the downslope side of the wave. While this vortex moves in upstream direction through the trough region it gets stronger and forces the mean flow towards the center of the channel. After that, more but weaker vorticies appear at the downslope side of the wave while the primary vortex disappears under the action of strong shear at the upslope side of the wave. Finally a relatively small separated region remains which is nearly unchanged for $t/t_{\text{ref}} > 4.5$. The following time evolution is mainly caused by processes connected with the development of shear layers near the walls, as in a flat channel. For the coarsest grid and for equidistant vertical resolution (DNS 4) the results do not show a separation. The separation arises only in a transient initial phase but disappears at later times. For all DNS it takes about $50\,t/t_{\text{ref}}$ to reach stationarity. For statistical evaluation the three-dimensional flow fields are averaged over the last $10\,t/t_{\text{ref}}$ with an interval of $0.5\,t/t_{\text{ref}}$.

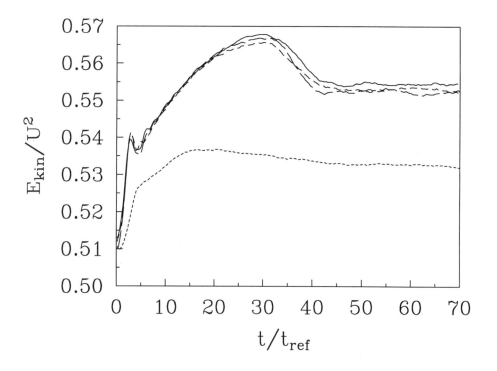

Figure 2: Volume averaged kinetic energy E_{kin} versus time t. —— DNS 1, — — DNS 2, - - - DNS 3, - - - - - DNS 4.

Of main interest for applications is the vertical flux of downstream momentum τ. It is composed of advective contributions due to the mean flow field and frictional parts due to turbulent fluctuations. Both contributions are modified by the wavy wall. In addition, at wavy surfaces the pressure causes a further contribution $\tau_{pres} = \overline{p(\partial z/\partial x)_\eta}$, where the bar denotes the horizontal average over coordinate planes $\eta = \text{const}$. The pressure

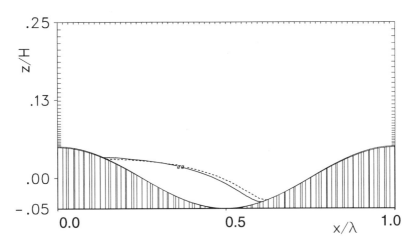

Figure 3: Mean streamline $\Phi = 0$ of DNS 2 (—) and DNS 5 (- - -).

contribution is strongly influenced by the shape of the bottom surface and of the separated region. Because it exceeds the sum of advectional and frictional momentum transport it is mainly responsible for the total drag at the lower boundary. The friction velocities in Table 1 are computed from the sum of the three contributions to the momentum flux.

The friction velocity at the flat wall varies only by about 1.5% for all cases with recirculation, so the influence of resolution, Reynolds number, and bottom surface on the upper boundary is very small. On the contrary, comparing the friction velocity at the lower boundary from DNS 1 and 3 we find a difference of about 10%. Comparing DNS 1 and 2 the difference of about 7% is due to the small differences in the shape of the bottom surface. This shows for the considered flows, that the shape of the surface wave has relatively large influence. Mainly because of the additional pressure forces, the effective friction velocity is up to 50% (DNS 3) larger at the wavy surface than at the flat surface.

Hino & Okumura [12] found for flow without recirculation that the total shear stress on the upper flat wall decreases a little compared with that over the flat parallel bottom. In all of our simulations the total shear stress at the flat wall is significantly higher as proposed e.g. from Blasius' law.

The mean streamline $\Phi = 0$ bounds the separated region. In Figure 3 the separated region is shown for two simulations with different Reynolds numbers. For $Re = 6760$ and 4780 the influence of Reynolds number is small. Corresponding to other observations [17] the size of the separated region decreases slightly with increasing Reynolds number.

As already shown for the lower Reynolds number [20], the flow structure near the wavy surface shows a streaky pattern with downstream elongated vortices at the upstream slope but less regular patterns on the downstream side. The spanwise spacing of the regular patterns is about $0.3\,H$.

Now we compare the results of DNS 1 with experimental data from Hudson [13]. The parameters of the DNS correspond to a water channel with channel height $H = 50.8$ mm,

Table 2: Profile of Hudson's wave. Accuracy of locating surface height h: $0.2 \cdot 10^{-3} H$.

x/λ	0.0	0.1	0.2	0.3	0.4	0.5	0.6	0.7	0.8	0.9
h/H	0.1	0.089	0.066	0.036	0.0095	0.0015	0.0195	0.0505	0.0785	0.095

Table 3: Fourier coefficients a_l and b_l of the harmonic analysis of Hudson's wave model.

Mode l	a_l	b_l	$\sum \delta^2$
0	$.109E+00$	$.000E+00$	$.418E-01$
1	$.484E-01$	$-.702E-02$	$.846E-04$
2	$-.391E-02$	$.996E-03$	$.312E-05$
3	$.736E-03$	$.131E-03$	$.329E-06$

wave amplitude $2\delta/\lambda = 5.08$ mm, wave length $\lambda = H$, the mean velocity $U = 122.2$ mm/s, and $Re = UH/\nu = 6760$. In the simulation we used the Fourier series

$$z_s/H = \sum_{l=1}^{3} \left(a_l \cos\left(2\pi \frac{lx}{\lambda}\right) + b_l \sin\left(2\pi \frac{lx}{\lambda}\right) \right), \tag{5}$$

with the coefficients shown in Table 3 to represent the experimental bottom surface from Table 2. Note that in equation (5) $a_0/2$ is neglegted in order to obtain vanishing mean wave height. This curve mainly differs at the upslope side from the ideal cosine profile with wave amplitude $\delta = 0.05 H$ and wavelength $\lambda = 1 H$ which has been used for the other simulations. To our best knowledge the only difference to Hudson's case wb3 concerns the mean velocity. In the simulation we used the mean over the whole channel U to build the Reynolds number while Hudson used the mean velocity U_{ref}, which is an average solely in the lower half of the channel. Because of the asymmetric mean profile U_{ref} is smaller than the bulk velocity U of the whole channel. For DNS 1 we get $U_{\text{ref}} = 0.912 U$, so the Reynolds number may be in effect about 9 % smaller in the computation than in the experiment. For the purpose of comparing simulation and experiment, the velocities are made dimensionless with U_{ref}. From the picture in Figure 3 we assume that the influence from changing the Reynolds number by 10% should be negligible.

Figure 4 depicts profiles of the computed and measured velocities and turbulence quantities. The numerical results are ytp-mean values, i. e. averages over the y coordinate, time ($60 < t/t_{\text{ref}} \leq 70$), and over the four surface positions of equal phase angle. The streamwise velocity profiles (Figure 4 a) are characterised by very strong vertical gradients at the upslope side of the wave. Even with that high resolution of DNS 1 the largest differences between simulation and experiment occur in this region. In the simulation the u-velocity averaged over planes $\eta = const.$ has a maximum of $1.245 U$ at a height $\eta/H = 0.639$, so the maximum is shifted by more than $2\delta/H$ in upward direction. The position of the experimental maximum is unknown. The separated region, bounded by the streamline $\Psi = 0$, extends from $x/\lambda = 0.15$ to $x/\lambda = 0.59$ in the DNS compared to

$x/\lambda = 0.3$ and $x/\lambda = 0.5$ in the experiment. One has to remind the coarse experimental streamwise resolution of $\Delta x/\lambda \leq 0.1$.

Because the vertical velocity component (Figure 4 b) is approximately coupled with the streamwise velocity component the greatest differences between DNS and experiment occur at the same positions, namely at the upslope side of the wave.

The component $-\overline{u'w'}$ of the Reynolds stress tensor is shown in Figure 4 c). At the upslope side $-\overline{u'w'}$ becomes negative near the wall. This is an artefact of calculating the Reynolds stress in Cartesian coordinates. Here the measurements are in good agreement with the numerical results. At the downslope surface $-\overline{u'w'}$ should be positive inside as well as outside the separated region, but the experiment has negative values at the first measurement points from the wall.

The rms-value of u (Figure 4 d) in the DNS develops a relative maximum near the surface at $x/\lambda \approx 0.8$ and reaches an absolute maximum over the next trough. If one defines a shear layer through an intensity maximum of the turbulent velocity fluctuations [5], two shear layers can be discerned for positions $0.8 \leq x/\lambda \leq 1.3$. The agreement with the experiment is very good. The intensity of the v-velocity (Figure 4 e) reaches a maximum of $\sqrt{\overline{v'^2}}/U_{\text{ref}} = 0.2$ at a streamwise position $x/\lambda = 0.69$ which is about $0.1\,x/\lambda$ behind the reattachment point. In this region the v-component gives a large contribution to the turbulence energy near the surface. The rms-value of the w-velocity (Figure 4 f) has its absolute maximum near $x/\lambda = 0.55$. The agreement with the experiment is very good at the upslope side of the wave whereas at the downslope side the experiment shows higher intensities near the bottom at $0.1 \leq x/\lambda \leq 0.4$.

In general the experimental values seem to deviate systematically from the expected and computed results near the center of the channel. Because the spreading near the surfaces is very high we assume that the measurements suffer from some disturbing influences from the wall. Although the number of experimental samples is three times the number of DNS samples the measurements show individual data points far outside the expected statistical spread.

The pressure field is responsible for the form drag of the wavy wall. The figure assumed from a potential flow is disturbed by the separated region and the stagnation point at the location of reattachment (Figure 5). In the simulation the isolines $p/(\rho U^2) = 0.0$ and 0.1 show the effect of the recirculation, decreasing the pressure gradient in x-direction along $\Psi = 0$. In the experimental pressure field, which was obtained indirectly by calculating it from the measured velocity field using a reference pressure measured near the channel center, such effect is not seen. The location of the absolute maximum of the surface pressure is located at $x/\lambda = 0.65$ for the DNS which is $0.05\,x/\lambda$ behind the reattachment point. In the experiment this shift is $0.1\,x/\lambda$. The values of the surface pressure maxima are $2(p-p_{\text{ref}})/(\rho\,U_{\text{ref}}^2) = 0.38$ and 0.81 for the experiment and the simulation, respectively. Because the numerical results agree very well with values from similar experiments [5, 10] we assume that the experimental results from Hudson are too high by a factor of 2. The maximum of pressure intensities (not shown) coincides with the surface pressure maximum. From this comparison the indirect experimental determination of the pressure field from the measured velocity field seems not to be able to substitute direct pressure measurements.

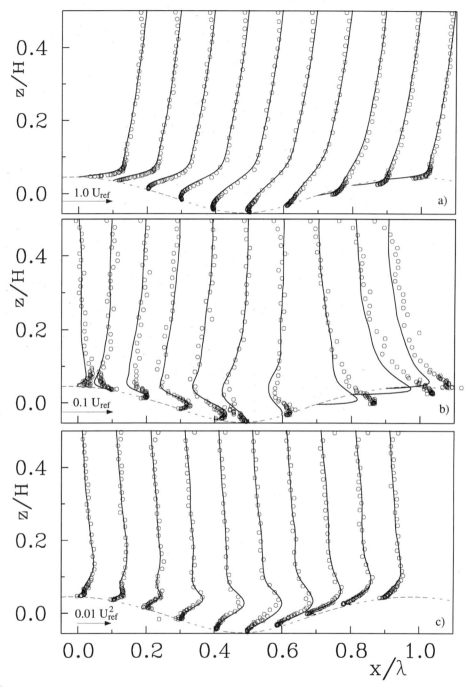

Figure 4: Comparison of DNS 1 and experiment by means of vertical flow profiles at positions $x/\lambda = 0.0, 0.1\ldots, 0.9$, referenced with $U_{\text{ref}} = 0.912\,U$ (The respective origin is on the bottom ordinate). —— ytp-averaged results of DNS 1, ○ LDA measurements [13], a) $\overline{u}/U_{\text{ref}}$ b) $\overline{w}/U_{\text{ref}}$ c) $-\overline{u'w'}/U_{\text{ref}}^2$ d) $\sqrt{\overline{u'^2}}/U_{\text{ref}}$ e) $\sqrt{\overline{v'^2}}/U_{\text{ref}}$ f) $\sqrt{\overline{w'^2}}/U_{\text{ref}}$.

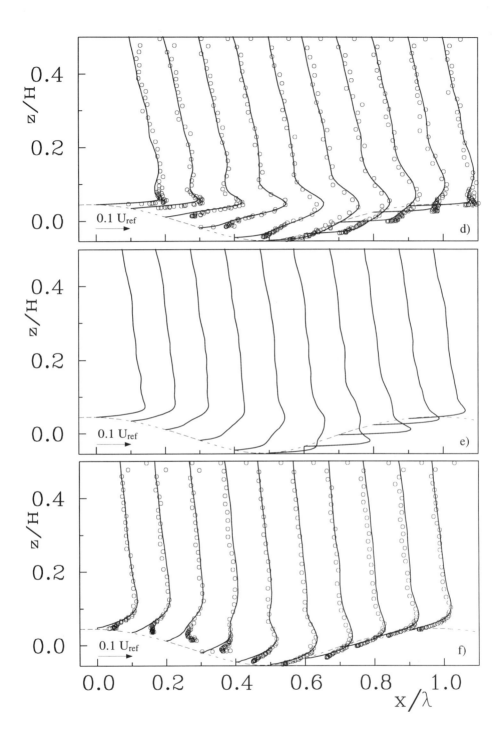

Figure 4: Continuation.

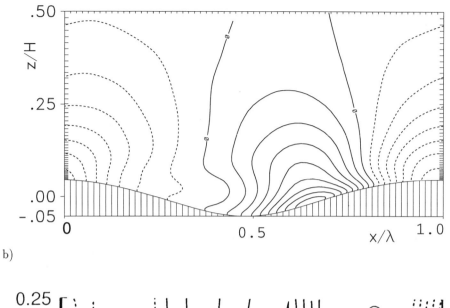

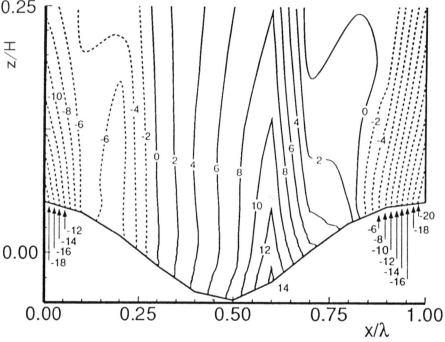

Figure 5: Mean pressure field (— positive values, - - - negative values). a) ytp-averaged pressure field $\bar{p}/(\rho U^2)$ of DNS 1 (mean pressure gradient P_x subtracted, increment 0.01), b) time averaged pressure field $\bar{p}/(\rho u_{\tau,\mathrm{wa}}^2)$ deduced from velocity measurements of the experiment [13].

Conclusion

In this study, direct numerical simulations of turbulent flow over a wavy boundary have been carried out and compared to measurements by Hudson [13]. Our method is able to predict the main features of separated flow and the flow structure can be described. In particular we find that a recirculation zone develops. At $Re = 6760$ this recirculation zone could be simulated with a rather coarse grid but only if a variable vertical grid spacing is used with finer resolution of the order of ν/u_τ near the walls. Mainly because of the additional pressure drag, the effective friction velocity is about 50% larger at the wavy lower surface than at the flat upper surface. The computed local mean velocity profiles and mean turbulence profiles agree rather well with the measurements. The size of the separation zone in the simulation is larger than in the experiment, which yields a smaller pressure drag. We have shown that the influence of the Reynolds number at such low turbulence levels is rather small whereas the shape of the bottom surface is of great importance.

From this comparison between simulation and experiment we conclude that our method is at least as precise as laboratory measurements.

Acknowledgement

The authors thank the Deutsche Forschungsgemeinschaft (DFG) for supporting this study in its Priority Research Program 'Strömungssimulation mit Hochleistungsrechnern'. We also thank Dr. J. D. Hudson and Professor T. J. Hanratty for providing the experimental data and for recalculating some results.

References

[1] G. P. Almeida, D. F. G. Durão, and M. V. Heitor. Wake flows behind two-dimensional model hills. *Experimental Thermal and Fluid Science*, 7:87–101, 1993.

[2] A. Andrén, A. R. Brown, J. Graf, P. J. Mason, C.-H. Moeng, F. T. M. Nieuwstadt, and U. Schumann. Large-eddy simulation of a neutrally stratified boundary layer: A comparison of four computer codes. *Q. J. R. Meteorol. Soc.*, 120:1457–1484, 1994.

[3] T. B. Benjamin. Shearing flow over a wavy boundary. *J. Fluid Mech.*, 6:161–205, 1959.

[4] M. Breuer and W. Rodi. Large-eddy simulation of complex flows of practical interest. In this publication.

[5] J. J. Buckles, T. J. Hanratty, and R. J. Adrian. Turbulent flow over large-amplitude wavy surfaces. *J. Fluid Mech.*, 140:27–44, 1984.

[6] M. Dianat and I. P. Castro. Turbulence in a separated boundary layer. *J. Fluid Mech.*, 226:91–123, 1991.

[7] A. Dörnbrack, T. Gerz, and U. Schumann. Turbulent breaking and overturning gravity waves below a critical level. *Appl. Sci. Res.*, 54:163–176, 1995.

[8] A. Dörnbrack and U. Schumann. Numerical simulation of turbulent convective flow over wavy terrain. *Boundary-Layer Meteorol.*, 65:323–355, 1993.

[9] J. G. M. Eggels, F. Unger, M. H. Weiss, J. Westerweel, R. J. Adrian, R. Friedrich, and F. T. M. Nieuwstadt. Fully developed turbulent pipe flow: a comparison between direct numerical simulation and experiment. *J. Fluid Mech.*, 268:175–209, 1994.

[10] W. Gong, P. A. Taylor, and A. Dörnbrack. Turbulent boundary-layer flow over fixed, aerodynamically rough, 2-d sinusoidal waves. *J. Fluid Mech.*, 1996. in press.

[11] C. Härtel and L. Kleiser. Large-eddy simulation of near-wall turbulence. In this publication.

[12] M. Hino and T. Okumura. Coherent structure of turbulent flow over wavy walls. In *Proc. 9th Turbulent Shear Flow Symp. Kyoto, Aug. 16-18*, pages 14.3.1–4, 1993.

[13] J. D. Hudson. *The effect of a wavy boundary on turbulent flow*. PhD thesis, Dept. Chemical Engineering, University of Illinois, Urbana, 1993.

[14] T. Kajishima, Y. Miyake, and T. Ohta. Direct numerical simulation of turbulent flow in a wavy channel. In *Proc. of The International Symp. on Mathematical Modelling of Turbulent Flows. Tokyo, Dec. 18-20*, pages 176–180, 1995.

[15] J. Kim, P. Moin, and R. Moser. Turbulence statistics in fully developed channel flow at low reynolds number. *J. Fluid Mech.*, 177:133–166, 1987.

[16] K. Krettenauer and U. Schumann. Numerical simulation of turbulent convection over wavy terrain. *J. Fluid Mech.*, 237:261–299, 1992.

[17] J. D. Kuzan, T. J. Hanratty, and R. J. Adrian. Turbulent flows with incipient separation over solid waves. *Exper. in Fluids*, 7:88–98, 1989.

[18] H. Lamb. *Hydrodynamics*. Dover, New York, 1945.

[19] C. Maaß, A. Dörnbrack, and U. Schumann. Grobstruktursimulationen turbulenter Strömungen über welligem Untergrund. Deutsche Meteorologen-Tagung, 16.-20.3.1992, Berlin. In *Ann. Meteorol.*, volume 27, pages 306–307, 1992.

[20] C. Maaß and U. Schumann. Numerical simulation of turbulent flow over a wavy boundary. In P. R. Voke, L. Kleiser, and J.-P. Chollet, editors, *Direct and large-eddy simulation I: selected papers from the First ERCOFTAC Workshop on Direct and Large-Eddy Simulation*, pages 287–297. Kluwer Academic Press, Dordrecht, 1994.

[21] H. Motzfeld. Die turbulente Strömung an welligen Wänden. *Z. angew. Math. Mech.*, 17:193–212, 1937.

[22] W. Rodi, editor. *ERCOFTAC Workshop on Data Bases and Testing of Calculation Methods for Turbulent Flows*, Karlsruhe, April 3-7 1995.

[23] U. Schumann. Large-eddy simulation of turbulent convection over flat and wavy terrain. In B. Galperin and S. A. Orszag, editors, *Large-Eddy Simulation of Complex Engineering and Geophysical Flows*, pages 399–421. Cambridge Univ. Press, 1993.

[24] U. Schumann. Stochastic backscatter of turbulence energy and scalar variances by random subgrid-scale fluxes. *Proc. Roy. Soc. London A*, 451:293–318 and 811, 1995.

[25] U. Schumann and M. Strietzel. Parallel solution of tridiagonal systems for the Poisson equation. *J. Sci. Comput.*, 10:181–190, 1995.

[26] C. B. Thorsness, P. E. Morrisroe, and T. J. Hanratty. A comparison of linear theory with measurements of the variation of shear stress along a solid wave. *Chem. Eng. Sci.*, 33:579–592, 1978.

[27] F. Unger. *Numerische Simulation turbulenter Rohrströmungen*. PhD thesis, Technische Universität, München, 1994.

[28] N. Wood and P. Mason. The pressure force induced by neutral, turbulent flow over hills. *Q. J. R. Meteorol. Soc.*, 119:1233–1267, 1993.

[29] D. P. Zilker, G. W. Cook, and T. J. Hanratty. Influence of the amplitude of a solid wavy wall on a turbulent flow. Part 1. Non-separated flows. *J. Fluid Mech.*, 82:29–51, 1977.

[30] D. P. Zilker and T. J. Hanratty. Influence of the amplitude of a solid wavy wall on a turbulent flow. Part 2. Separated flows. *J. Fluid Mech.*, 90:257–271, 1979.

LARGE-EDDY SIMULATION OF NEAR-WALL TURBULENCE[1]

C. HÄRTEL and L. KLEISER

Institute of Fluid Dynamics, Swiss Federal Institute of Technology

CH-8092 Zürich, Switzerland

SUMMARY

The present paper summarizes results from a numerical study conducted in order to clarify the requirements for the large-eddy simulation of near-wall turbulence. A key finding of the study is that an inverse cascade of turbulent kinetic energy from small to large scales exists within the buffer layer which cannot be accounted for by currently applied subgrid models. This may lead to significant errors in large-eddy simulations when the wall layer is resolved. Therefore a new model for the eddy viscosity is proposed which is more consistent with the near-wall physics than are the models commonly employed. *A priori* tests of this new model show promising results. The present investigation is primarily based on direct numerical simulation data of turbulent channel flow at various low Reynolds numbers. However, no significant Reynolds-number effects were observed which suggests that the findings may essentially be generalized to flows at higher Reynolds numbers.

1 INTRODUCTION

During the past two decades the large-eddy simulation (LES) technique has established its role as a powerful and efficient research tool for the investigation of turbulent flows in geophysics and engineering. In LES the flow-geometry-dependent large-scale (grid-scale, GS) motions of the flow are explicitly computed, while the small-scale (subgrid-scale, SGS) motions, which are expected to be more universal, are accounted for by a model. In the engineering field LES primarily was employed for fundamental studies in the past, but has recently started to become more extensively utilized for more complex flow problems (see e.g. the publications [1, 2] printed in this volume).

The key difficulties with the large-eddy simulation of wall-bounded flows arise from the presence of a wall layer where viscosity plays a major role and where locally significant turbulence structures exist which become extremely small at high Reynolds numbers. Essentially two different approaches exist to treat the near-wall flow in a LES. In the first approach, which in practice is more common, one circumvents the very costly resolution of the near-wall layer by bridging the wall region with the aid of empirical boundary conditions for the outer layer [3, 4, 5]. In the second approach a refined mesh is employed in the vicinity of solid boundaries in order to resolve at least coarsely the

---

[1]Work performed while the authors were affiliated with DLR Göttingen

most dominant near-wall structures. In general no further empirical information about the near-wall flow is required in this approach. For high-Reynolds-number flows approximate boundary conditions appear to be the only feasible way to treat the wall layer in a LES; it must be emphasized, however, that it is desirable in principle to treat the near-wall turbulence structures accurately whenever possible, since they may strongly influence the entire flow field.

The development of subgrid models suitable for application in the highly inhomogeneous wall layer requires detailed knowledge about the prevailing physical mechanisms in this flow regime. In this respect the analysis of direct numerical simulation (DNS) data has led to considerable advances in recent years [5, 6, 13, 16]. For example, in the study of Härtel et al. [13] it was found that within the buffer layer the energetic interactions between large and small turbulent scales are characterized by an inverse transfer of energy where large eddies are fed with kinetic energy from the small scales. Clearly, a consistent subgrid modeling should take proper account of this important feature of the near-wall flow.

Within the present project we examined in detail the requirements for an efficient and reliable subgrid modeling in near-wall turbulence. The study focussed on the turbulent flow in a plane channel and a circular pipe. Part of the work was done in close cooperation with Prof. R. Friedrich and Dr. F. Unger within the present DFG Priority Research Program. Among other things the energetic interactions between GS and SGS motions were analyzed and *a priori* and *a posteriori* tests of various subgrid models were conducted. Some key findings of this study are summarized in the following. For more detailed discussions the reader is referred to the earlier publications from this project [7, 8, 9, 10, 11, 12, 13, 14, 15, 16].

2 GOVERNING EQUATIONS

In LES any dependent flow variable f is divided into a GS part $\overline{f}$ and a SGS part f', i.e. $f = \overline{f} + f'$. The definition of $\overline{f}$ can be based on a spatial filtering performed by convolving f with a filter function H [17, 18]. Several suitable filter functions have been suggested in the literature, the most widely used ones being the Gaussian filter, the box filter, and the cutoff filter in spectral space. In our investigations we used primarily the cutoff filter H_i^c which can conveniently be described by its Fourier transform $\hat{H}_i^c$

$$\hat{H}_i^c(k_i) = \begin{cases} 1 & \text{for } |k_i| \leq K_i^c = 2\pi/\Delta_i \\ 0 & \text{otherwise} \end{cases} \quad (1)$$

where k_i denotes the wavenumber, Δ_i the width of the filter in the i-th direction and K_i^c the cutoff wavenumber. However, a comparison of results obtained with different filters gave evidence that the shape of the filter plays only a minor role for the characteristics of the interactions between GS and SGS motions [10, 14].

In this paper also the common Reynolds decomposition will be employed, where a quantity g is divided into a statistical mean value $\langle g \rangle$ and a fluctuation $\tilde{g}$, i.e.

$$g = \langle g \rangle + \tilde{g} \quad . \quad (2)$$

In the case of turbulent channel flow the operator $\langle \cdot \rangle$ symbolizes an averaging over the wall-parallel planes, the two channel halves and time. Such averages are referred to as global averages in the remainder.

Applying the filtering operation to the continuity and Navier-Stokes equations yields the evolution equations for the large energy-carrying scales. For an incompressible fluid they read

$$\frac{\partial \overline{u}_k}{\partial x_k} = 0 \qquad (3)$$

$$\frac{\partial \overline{u}_i}{\partial t} + \frac{\partial}{\partial x_k}(\overline{\overline{u}_i \overline{u}_k}) = -\frac{\partial \overline{p}}{\partial x_i} + \frac{\partial Q_{ik}}{\partial x_k} + \nu \frac{\partial^2 \overline{u}_i}{\partial x_k \partial x_k} \quad . \qquad (4)$$

In (3) and (4) u_i denotes the velocity component in the direction x_i, p the pressure and ν the kinematic viscosity of the fluid. The effect of the unresolved scales appears in the SGS stress tensor Q_{ij} which consists of two components C_{ij} and R_{ij} usually termed "SGS cross stresses" and "SGS Reynolds stresses", respectively

$$Q_{ij} = C_{ij} + R_{ij} \quad \text{where} \quad C_{ij} = -(\overline{\overline{u}_i u'_j} + \overline{u'_i \overline{u}_j}) \ , \quad R_{ij} = -\overline{u'_i u'_j} \ . \qquad (5)$$

For the analysis of the energetic interactions between resolved and unresolved motions one has to consider the budget equation for the GS kinetic energy. In this equation the effect of the unresolved motions on the energy balance of the resolved flow field is represented by the term

$$\overline{u}_k \frac{\partial Q_{kl}}{\partial x_l} = \frac{\partial}{\partial x_l}(\overline{u}_k Q_{kl}) - Q_{kl}\overline{S}_{kl} \ , \qquad (6)$$

where $\overline{S}_{ij} = (\partial \overline{u}_i/\partial x_j + \partial \overline{u}_j/\partial x_i)/2$ denotes the GS rate-of-strain tensor. The second term $Q_{kl}\overline{S}_{kl}$ on the right-hand-side of (6) is a source term which governs the exchange of energy between GS and SGS turbulence and which is usually called "SGS dissipation". Positive values of the SGS dissipation indicate a flow of kinetic energy from large to small scales, whereas the transfer is reverse whenever it takes negative sign. The global average of the SGS dissipation will be designated ε here.

By application of (2) ε may be decomposed in two parts, one of these being due to mean and the other due to fluctuating rates of strain, respectively [8, 13]

$$\varepsilon = \varepsilon^{MS} + \varepsilon^{FS} \quad \text{where} \quad \varepsilon^{MS} = \langle Q_{kl}\rangle\langle \overline{S}_{kl}\rangle \ , \quad \varepsilon^{FS} = \langle \tilde{Q}_{kl}\tilde{\overline{S}}_{kl}\rangle \quad . \qquad (7)$$

In (7) the term ε^{MS} means an enhancement of SGS turbulence in the presence of mean-flow gradients. The second term ε^{FS} accounts for a redistribution of energy within the turbulence spectrum without affecting the mean flow directly. A separate analysis of ε^{MS} and ε^{FS} provides a much more detailed insight into the energy budget of the flow than can be obtained from considering their sum only.

Most of the results presented in the following sections will be given in common wall units, i.e. normalized by the friction velocity u_τ and the kinematic viscosity ν which are the appropriate reference quantities for wall turbulence. Any quantity scaled in these wall units is indicated by the usual superscript "+".

3 DNS DATABASES

The present analysis is primarily based on a direct numerical simulation database of turbulent channel flow at a wall Reynolds number of $Re_\tau = 210$ (based on friction velocity u_τ and channel half width h). The mass-flux Reynolds number Re_Q (based on the bulk velocity u_Q and h) of this simulation is $Re_Q = 3333$. The simulation was performed by Gilbert & Kleiser [19] and the results have been validated carefully by comparison with other simulations and recent experimental results. This simulation, denoted as simulation II here, falls within the range of low Reynolds numbers where a universal scaling of flow quantities in wall units cannot be expected [20]. Therefore, part of the analysis was repeated using two further DNS databases, one of them having a lower and the other one a higher Reynolds number than simulation II. These DNS will subsequently be referred to as simulation I and III, and the respective Reynolds numbers are $Re_\tau = 115$ ($Re_Q = 1667$) and $Re_\tau = 300$ ($Re_Q = 5000$). A comparison of results obtained from the three databases should reveal whether significant Reynolds-number effects exist.

Besides the influence of the Reynolds number, the influence of some changes in the flow geometry is generally of interest. Therefore we also examined a DNS database of turbulent pipe flow at a wall Reynolds number of $Re_\tau = 180$ which was computed by Unger [21]. No further details of this study will be shown here, but we remark that an excellent agreement between the respective results for channel and pipe flow was obtained [9, 13].

4 NUMERICAL METHOD

A detailed discussion of the numerical method and the relevant physical and numerical parameters of the three direct simulations of turbulent channel flow is given in [10]. The geometry of the flow domain and the coordinate system used in these simulations are illustrated in figure 1, where the streamwise, spanwise and normal directions are denoted by x_1, x_2 and x_3, respectively. The extents L_1 and L_2 of the computational domain differ for the three DNS and were chosen such that the dominant correlation lengths of the turbulence fields can be accommodated. In all of the simulations the respective mass flux Q was kept constant during the computation.

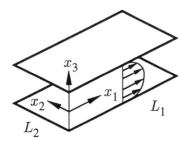

Figure 1: Computational domain and coordinate system.

The numerical scheme employed is based on a spectral discretization in space together with finite differences in time. The temporal discretization is done in a semi-implicit manner where an explicit method is employed for the non-linear terms together with a Crank-Nicolson scheme for the diffusive terms and the pressure.

The resulting mean-velocity profiles from the three DNS are displayed in a semi-logarithmic scale in figure 2, where the wall distance y is used rather than the wall-normal coordinate x_3. All profiles exhibit the characteristic linear behavior within the viscous sublayer, but a marked logarithmic regime can be discerned for the two higher Reynolds numbers only. It is seen from the figure that the profile of simulation III is already very close to the experimentally established law-of-the-wall for high-Reynolds-number flows as given by Dean [22]. In contrast to simulation II, no detailed experimental reference data are available for simulation I and III, but some further validation of these simulations can be achieved by comparison of global characteristics. For example, from a formula given by Dean [22] one can derive the following relation between the wall Reynolds number Re_τ and the mass-flux Reynolds number Re_Q of the flow

$$Q^{A,B} = \frac{\ln(Re_\tau^A/Re_\tau^B)}{\ln(Re_Q^A/Re_Q^B)} = \frac{7}{8} \approx 0.88 \quad , \tag{8}$$

where "A" and "B" designate two channel flows with different Reynolds numbers. The present DNS results are in excellent agreement with this relation

$$Q^{I,II} \approx 0.88 \quad , \quad Q^{II,III} \approx 0.87 \quad , \quad Q^{I,III} \approx 0.87 \quad . \tag{9}$$

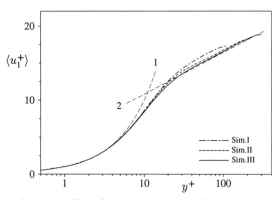

Figure 2: Mean-velocity profiles of turbulent channel flow in wall units. Results of three DNS at different Reynolds numbers. $Re_\tau = 115$ (Sim. I), 210 (Sim. II), and 300 (Sim. III). Curve 1: $\langle u_1^+ \rangle = y^+$, 2: $\langle u_1^+ \rangle = 2.44 \cdot \ln y^+ + 5.17$ (Dean, 1978).

5 ANALYSIS OF THE ENERGY BUDGET

The subgrid-scale effects in a turbulent flow depend qualitatively and quantitatively on the width of the filter function applied. Generally speaking, the filter width should be as large as possible in order to minimize the computational needs of a simulation. However, larger filter widths give rise to a more complex SGS turbulence which puts higher demands on the subgrid models. In our analysis the following cutoff wavenumbers (see equation (1)) were employed in streamwise and spanwise direction

$$(K_1^c)^+ \approx 0.04 \quad , \quad (K_2^c)^+ \approx 0.08 \quad . \tag{10}$$

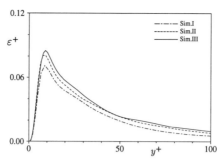

Figure 3: SGS dissipation ε. Comparison of results for different Reynolds numbers.

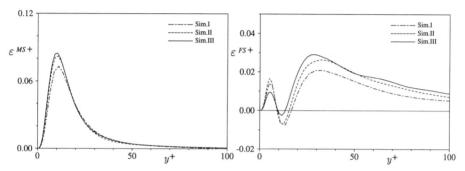

Figure 4: SGS dissipation ε^{MS} due to mean rates of strain (left) and ε^{FS} due to fluctuating rates of strain. Comparison of results for different Reynolds numbers.

These cutoff wavenumbers correspond approximately to the minimum grid spacings for the LES of wall turbulence suggested by Zang [23]. No filtering was performed in the wall-normal direction x_3, since in this direction a rather fine resolution is required anyway in order to resolve the steep mean-flow gradients.

Figure 3 shows the global averages of the SGS dissipation as computed from the three DNS databases. The results are displayed for the near-wall region $y^+ \leq 100$ which corresponds to one third of the channel half width in simulation III, but essentially to the full half width h in simulation I. In all cases the SGS dissipation is a non-negative function throughout the channel, meaning that on the average the net transfer of kinetic energy is from GS to SGS motions. However, the individual components ε^{MS} and ε^{FS}, which are depicted in figure 4, reveal that the energetic interactions in the near-wall flow are more complex than suggested by the results for ε. While ε^{MS}, like ε, is non-negative throughout, the component ε^{FS} exhibits a pronounced negative minimum within the buffer layer which gives rise to a reverse transfer of kinetic energy within the turbulence spectrum. It is seen that this important characteristic of the wall layer is only little affected by the Reynolds number of the flow.

An analysis of the contributions of the individual stress components Q_{ij} to ε^{FS} (termed ε_{ij}^{FS} hereafter) shows that the inverse cascade of kinetic energy is primarily caused by the stress component Q_{13}, i.e. the subgrid stress aligned with the mean rate of strain [7, 13]. For ε_{13}^{FS} an evolution equation may be derived and an examination of the individual terms in this equation can provide a much better understanding of the underlying physical mechanisms which cause the inverse transfer of energy. We performed such an analysis and found that the pronounced negative minimum in ε_{13}^{FS} is related to a systematic phase shift between the GS shear $\partial \overline{u}_1/\partial x_3$ and the wall-normal SGS stress Q_{33} [10, 12]. This finding may be exploited in order to derive an improved definition of the characteristic subgrid velocity and length scales on which the SGS eddy viscosity should be based in the near-wall flow. This new model for the eddy viscosity will briefly be discussed in Section 7.

6 ASSESSMENT OF SUBGRID MODELS

Whether or not currently applied subgrid models are capable of taking proper account of the near-wall physics can be investigated with the aid of an *a priori* test, where exact and modeled SGS quantities are compared for the same filtered flow. We conducted *a priori* tests for several more widely applied SGS models which were all of the eddy-viscosity type. In eddy-viscosity models the deviatoric part τ_{ij} of the SGS stress tensor Q_{ij} is set proportional to the rate-of-strain tensor $\overline{S}_{ij}$ of the resolved flow field, i.e.

$$\tau_{ij} = Q_{ij} - \delta_{ij} Q_{kk}/3 = 2 \nu_t \overline{S}_{ij} \quad . \tag{11}$$

The closure problem for the tensor Q_{ij} is thus reduced to finding a (scalar) eddy viscosity ν_t as a function of the resolved flow variables. As an example, the respective results for ε^{MS} and ε^{FS} as obtained from three different SGS models are shown in figure 5 along with the correct DNS data. The first of these models is the classical Smagorinsky [24] model (abbreviated as "Smago." in the figure), where the SGS eddy viscosity is derived under the assumption that the small-scale turbulence is locally in equilibrium of production and dissipation of kinetic energy. This leads to

$$\nu_t = (C_S \Delta x)^2 \, \|\overline{S}\| \quad \text{where} \quad \|\overline{S}\| = \sqrt{2 \overline{S}_{kl} \overline{S}_{kl}} \quad . \tag{12}$$

In (12) the grid size of the computational mesh is denoted by Δx and C_S is a yet undetermined model constant, named Smagorinsky constant. For homogeneous isotropic turbulence Lilly [25] analytically derived a value of $C_S = 0.17$ which, however, was found to be too large in practice. Therefore the more common value $C_S = 0.1$ was used in our analysis [3]. Δx was computed as the geometric mean of the mesh sizes in the three coordinate directions [26] and the eddy viscosity was supplied with an additional Van Driest-type damping function (see [27]) to ensure the proper near-wall behavior $\nu_t \propto (y^+)^3$. This yields

$$\nu_t = C_S^2 \left[1 - \exp\left(-(y^+/A^+)^3\right)\right] (\Delta x_1 \Delta x_2 \Delta x_3)^{2/3} \, \|\overline{S}\| \quad \text{where} \quad A^+ = 25 \quad . \tag{13}$$

The second model is the structure-function model ("Str.Fu.") proposed by Métais & Lesieur [28]. Following Chollet & Lesieur [29] the authors assumed that the proper characteristic SGS velocity, on which the eddy viscosity should be based, is given by the square root of the kinetic energy residing at the cutoff wavenumber K^c. In [28] this velocity scale was evaluated with the aid of the second-order velocity-structure function. We employed a two-dimensional version of this model, where the structure function is computed from velocity differences taken in the homogeneous wall-parallel planes of the

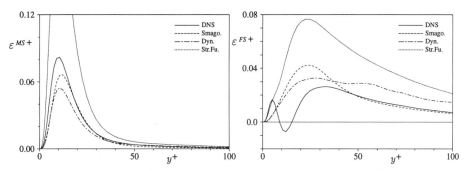

Figure 5: SGS dissipation ε^{MS} due to mean rates of strain (left) and ε^{FS} due to fluctuating rates of strain. Comparison of DNS results with various SGS models. Results for simulation II.

channel [30]. To achieve an improved near-wall behavior, the resulting eddy-viscosity was supplied with an additional wall-damping function $D(y^+) = (1 - \exp(-y^+/25))^2$ (see Zang et al. [31]).

The third model is the dynamic Smagorinsky model ("Dyn.") suggested by Germano et al. [32] and improved by Lilly [33]. In the dynamic Smagorinsky model C_S is considered as a function of space and time which is motivated by the common experience that the optimum value of the Smagorinsky constant may greatly vary from flow to flow [34]. It is desirable hence to avoid an uncertain ad hoc specification of the Smagorinsky constant by somehow adjusting C_S to the actual state of the flow. In the dynamic model this is achieved by introducing a second filter (the so-called "test filter") in addition to the original LES filter. Assuming that the SGS stresses due to both the original filter and the test filter are similar and can be modeled using the identical ansatz, a relation for the evaluation of C_S can be derived. In the present study the model is employed in an averaged version, where C_S is taken as a function of time and the inhomogeneous coordinate x_3 only, while being constant within the homogeneous wall-parallel planes.

From figure 5 it is seen that the results for the structure-function model are significantly in excess of the DNS data which confirms conclusions already arrived at by Comte et al. [30]. While the models give qualitatively satisfactory results with respect to ε^{MS}, the curves for ε^{FS} reveal that none of them can capture the inverse cascade of energy. Rather, the modeled transfer of energy within the turbulence spectrum is from GS to SGS motions throughout the channel. For all models ε^{FS} monotonically increases from the wall to a global maximum located at about $y^+ = 25$ which gives rise to significant differences in the modeled and correct dissipation. Concerning the adjustment of C_S in the dynamic model an interesting consequence can be drawn from the curves in figure 5. Within the buffer layer the dynamically obtained values for C_S are clearly too large concerning ε^{FS}, but at the same time slightly too small concerning ε^{MS}, which reveals that the constant cannot be adjusted such that a satisfactory modeling of both ε^{MS} and ε^{FS} is achieved.

If one of the above models is employed in a large-eddy simulation, the excessive values of the modeled dissipation ε^{FS} may degrade the LES results within the buffer layer. This is particularly severe because in this flow region the maximum production of turbulent kinetic energy is encountered while simultaneously the transport of turbulence energy takes its negative minimum. Consequently, an excessive damping of turbulence cannot be compensated by additional wall-normal transport. Since the near-wall flow and the core flow are essentially decoupled by the logarithmic regime, a good performance of LES models in the outer flow will generally have little effect on the quality of the simulation results in the near-wall region.

A comprehensive assessment of subgrid models has to include an *a posteriori* test, i.e. large-eddy simulations where the subgrid model is employed. We performed a series of LES of turbulent channel flow for three bulk Reynolds numbers Re_Q identical to those of the direct simulations [10]. The grid sizes used in spanwise and streamwise direction were chosen such that they correspond to a filtering with the cutoff wavenumbers given in equation (10). Since the previous *a priori* tests did not reveal significant differences between the various subgrid models, only the Smagorinsky model was employed. The numerical method used for the LES is essentially the same as the one used for the DNS and is described in more detail in [10].

In an LES where the Smagorinsky model is applied the results may considerably depend on the model constant, and therefore a proper choice of C_S is of primary importance. A somehow optimized value of C_S may be obtained from the requirement that the overall exchange of kinetic energy between GS and SGS motions is represented correctly by the model, i.e.

$$\left\langle \tau_{kl}^{mod} \overline{S}_{kl} \right\rangle_{D,t} = \left\langle \tau_{kl}^{DNS} \overline{S}_{kl} \right\rangle_{D,t} , \qquad (14)$$

where the operator $\langle \cdot \rangle_{D,t}$ indicates an averaging over the whole computational domain and time. The superscripts *mod* and *DNS* in equation (14) denote the modeled SGS stresses and the stresses computed directly from the DNS data. Table 1 gives the respective values of this optimized Smagorinsky constant, termed C_S^* hereafter, for different bulk Reynolds numbers. The significance of C_S^* will become evident from the results presented below.

Table 1: Optimized Smagorinsky constant C_S^* for various Reynolds numbers Re_Q.

	$Re_Q = 1667$	$Re_Q = 3333$	$Re_Q = 5000$
C_S^*	0.078	0.100	0.114

In figure 6 the mean-velocity profiles of the three direct simulations are displayed together with velocity profiles obtained from large-eddy simulations. For the two lower Reynolds numbers C_S was set to C_S^*, while for the highest Reynolds number C_S was set to a value slightly lower than the optimized one. The reason is that the latter large-eddy simulation was conducted with a guess for C_S^*, since the full DNS results were not yet available when the LES was performed. Note that the velocity profiles in figure 6 are not given in wall units, but were non-dimensionalized by the channel half width h and the bulk velocity u_Q of simulation I. The respective mass flux Q is identical for DNS and LES.

Figure 6 gives evidence that away from the wall the quality of the LES results improves with increasing Reynolds number. The relative error in the center-line velocity decreases from about 5 % for the lowest Reynolds number to approximately 1 % for

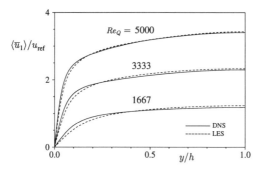

Figure 6: Mean-velocity profiles of turbulent channel flow. Filtered DNS data and LES results at various bulk Reynolds numbers Re_Q. Reference velocity u_{ref} is the bulk velocity u_Q of simulation I. Model constants: $Re_Q = 1667, 3333$: $C_S = C_S^*$, $Re_Q = 5000$: $C_S = 0.96\, C_S^*$.

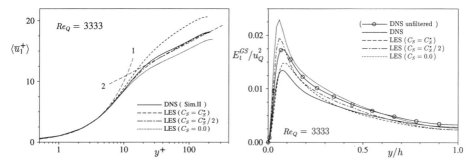

Figure 7: Comparison of DNS results with LES results for various values of the Smagorinsky constant C_S. Left: Mean velocity profiles in wall units (Curve 1: $\langle u_1^+ \rangle = y^+$, 2: $\langle u_1^+ \rangle = 2.44 \cdot \ln y^+ + 5.17$), right: Averaged turbulence energy E_t^{GS} of the grid-scale motions normalized by the bulk velocity (circles mark unfiltered DNS data).

the highest one. It should be emphasized that this result is not surprising, since both theoretical considerations and practical experience suggest that within the outer flow the Smagorinsky model is a more efficient subgrid model at higher than at lower Reynolds numbers.

The errors in the LES results are generally larger in the near-wall region than in the core flow, in accordance with the findings from the *a priori* test. To illustrate this more clearly the velocity profiles for $Re_Q = 3333$ from figure 6 are displayed in wall units in figure 7. Clearly, both DNS and LES give the correct linear profile within the viscous sublayer, but above a wall distance of approximately $y^+ = 10$ the curves for the direct simulation and the large-eddy simulation diverge significantly. Much larger values are obtained from the LES than from the DNS, as a consequence of the discrepancies in the mean-velocity gradients at the wall.

In figure 7 also LES results for $C_S = C_S^*/2$ and $C_S = 0.0$ are included for comparison. In the latter case the simulation is performed with no subgrid model at all. With respect to the mean-velocity profile the best agreement between LES and DNS

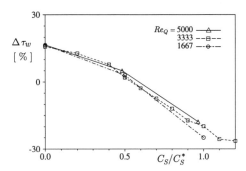

Figure 8: Relative error $\Delta \tau_w / \tau_{w,DNS}$ in the resulting wall shear stress of the LES as a function of the normalized Smagorinsky constant C_S/C_S^*. Symbols identify the individual simulations.

is achieved for $C_S = C_S^*/2$ from which one might speculate that this choice of the Smagorinsky constant is more appropriate than the optimized value C_S^*. Note, however, that for $C_S = C_S^*/2$ the integrated dissipation due to the SGS model accounts for merely a quarter of the correct one, owing to the fact that the modeled subgrid stresses (13) are proportional to C_S^2. For a conclusive assessment of the influence of C_S it is necessary to consider the resulting GS turbulent fluctuations in addition to the mean-velocity profiles. In the second graph in figure 7 the averaged GS turbulent kinetic energy E_t^{GS}

$$E_t^{GS} = \frac{1}{2} \langle \tilde{\tilde{u}}_k \tilde{\tilde{u}}_k \rangle \; , \tag{15}$$

is shown for the large-eddy simulations together with filtered and unfiltered DNS results. To allow for a direct quantitative comparison, the bulk velocity u_Q (being the same for LES and DNS) has been chosen as the reference velocity here. It is observed that for $C_S = 0.0$ and $C_S = C_S^*/2$ the GS turbulent kinetic energy is considerably larger than the filtered DNS results which illustrates the insufficient damping of the large-scale structures in these cases.

In figure 7 the LES with $C_S = C_S^*$ agrees best with the filtered DNS, but it must be emphasized that no direct relation between this agreement and the optimization (14) can be inferred. Since the production of turbulent kinetic energy directly depends on the mean-flow gradients, the significant errors in the mean-velocity profile strongly affect the entire energy budget of the resolved turbulence. The satisfactory performance of the Smagorinsky model for $C_S = C_S^*$ may hence be a mere coincidence. The two graphs in figure 7 make clear that the Smagorinsky constant cannot be optimized in such a way that good results are achieved with respect to both mean-velocity profiles and turbulent fluctuations.

At this place we wish to make a remark concerning the comparison of LES results with unfiltered and filtered DNS data. The unfiltered turbulent kinetic energy of simulation II is included in figure 7 and it is seen that the LES results for $C_S = C_S^*/2$ are in much better agreement with these data than with the filtered ones. If a comparison is based on unfiltered data from simulations or experiments, $C_S^*/2$ may hence appear to be the optimum choice for the Smagorinsky constant. However, from figure 7 it is seen that a considerable amount of energy resides within the subgrid scales in the near-wall flow. This makes clear that conclusive results can only be obtained if a comparison is based on filtered reference results. The use of unfiltered data may lead to erroneous conjectures about the model performance in the near-wall flow.

The above findings for $Re_Q = 3333$ hold equally for the other two Reynolds numbers. This is illustrated by figure 8, where the relative error $\Delta \tau_w$ in the computed wall stress

$$\Delta \tau_w = \frac{\tau_{w,LES} - \tau_{w,DNS}}{\tau_{w,DNS}} = \frac{Re_{\tau,LES}^2 - Re_{\tau,DNS}^2}{Re_{\tau,DNS}^2} \quad , \tag{16}$$

is plotted as a function of the Smagorinsky constant for all large-eddy simulations we performed. In the figure the Smagorinsky constant is given in normalized form C_S/C_S^* on the abscissa. The appropriateness of this normalization, and hence the significance of C_S^*, is impressively demonstrated by the fact that the curves for the different Reynolds numbers almost collapse. From the *a priori* test it was already presumed that the near-wall errors of the LES should essentially be independent of Reynolds number, and this presumption is fully confirmed by the figure.

7 AN IMPROVED SGS EDDY VISCOSITY

The eddy-viscosity ansatz is a particularly convenient approach to model the unknown turbulent stresses and has become most common in LES. Therefore it is necessary to clarify whether the shortcomings of current SGS models in the near-wall flow are inherent in the eddy-viscosity ansatz. This issue may be examined by comparing various models for the eddy viscosity which differ with respect to the characteristic scales on which they are based. For the evaluation of the eddy viscosity ν_t in (11) the specification of two characteristic subgrid scales is required, for example a time scale T and a (quadratic) subgrid velocity scale E

$$\nu_t \propto T \cdot E \quad . \tag{17}$$

In general the selection of the most appropriate characteristic scales it no trivial task, and it may strongly depend on the actual flow. In LES it is common to relate E to the kinetic energy of the subgrid motions by setting $E = \langle u_k' u_k' \rangle / 2$, while the grid spacing Δx is used as the characteristic subgrid length scale L (which corresponds to setting the subgrid time scale to $T = \Delta x / E^{1/2}$). In this case an additional damping function will be required in the near-wall region in order to ensure that the proper near-wall limiting behavior $\nu_t \propto (y^+)^3$ is achieved. The resulting eddy viscosity, which we will refer to as "model **A**" hereafter, reads

$$\nu_t = D(y^+) \, C_A \, \Delta x \, \sqrt{u_k' u_k'} \quad , \quad D(y^+) = 1 - \exp(-(y^+/25)^2) \quad , \tag{18}$$

where $\Delta x = (\Delta x_1 \Delta x_2 \Delta x_3)^{1/3}$. In (18) C_A denotes a model constant. For homogeneous isotropic turbulence C_A can be derived analytically [10], but in more general cases the constant needs to be determined empirically according to some additional condition. Applying the optimization (14) yields a value of $C_A = 0.073$.

Horiuti [6] suggested an alternative definition of the eddy viscosity, where E is evaluated using the wall-normal SGS velocity only, i.e. where $\langle u_3' u_3' \rangle / 2$ is employed to evaluate E rather than $\langle u_k' u_k' \rangle / 2$. This model, named "model **B**" hereafter, utilizes the

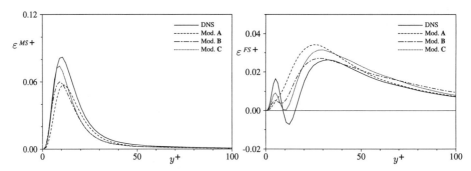

Figure 9: SGS dissipation ε^{MS} due to mean rates of strain (left) and ε^{FS} due to fluctuating rates of strain. Comparison of DNS results with three SGS models. Results for simulation II.

preferential damping of the normal velocity close to the wall. The complete definition of the eddy viscosity reads

$$\nu_t = C_B \, \Delta x \, \frac{u'_3 u'_3}{\sqrt{u'_k u'_k}} \quad . \tag{19}$$

Like for model **A**, Δx is computed as $\Delta x = (\Delta x_1 \Delta x_2 \Delta x_3)^{1/3}$. The constant C_B in (19) can be determined from the normalization (14) which gives a value of $C_B = 0.197$.

Our analysis of the turbulent energy budget has confirmed that in the vicinity of the wall the normal component of the SGS kinetic energy is a more adequate velocity scale than the total SGS kinetic energy [12]. Moreover, it could be inferred from this analysis that the proper subgrid time scale T is given by the viscous time scale $t_w = \nu/u_\tau^2$ of the wall layer rather than by E^{SGS} and Δx. Consequently, the eddy viscosity should be computed as

$$\nu_t = C_C \, \frac{-Q_{33}}{u_\tau^2} \, \nu \, , \tag{20}$$

which we will call "model **C**". From (14) a value of $C_C = 6.42$ is computed for the constant. In model **C** the limiting behavior of ν_t is $\propto (y^+)^4$ rather than $\propto (y^+)^3$, but we made no attempt to correct this slight discrepancy.

At first sight it might be surprising that, contrary to the other two models, the grid size of the numerical discretization does not enter the eddy viscosity in model **C**. Thus ν_t becomes a function of a single SGS scale (Q_{33}) only rather than a function of two. This, however, is not inconsistent with the known physical properties of near-wall turbulence. The wall layer, according to the law-of-the-wall, tends to establish a universal behavior if properly scaled in viscous wall units which implies that all flow quantities, like averaged velocity profiles, fluctuation intensities or energy spectra, collapse in this nondimensional form. Consequently, the dimensionless SGS stress Q_{33}^+ is uniquely determined for a given dimensionless grid size Δx^+ and vice versa. This means that only one degree of freedom is left on which the SGS eddy viscosity may depend if the problem is recast in wall units. As a consequence (20) can only be physically meaningful in the near-wall flow, since the universal scaling using ν and u_τ has no significance within the outer layer of a wall-bounded flow, i.e. above the logarithmic regime.

The three eddy-viscosity models outlined above were examined in an *a priori* test using the database of simulation II. Figure 9 gives a comparison of the results for ε^{MS} and ε^{FS} obtained from the DNS data and the models **A**, **B** and **C**, respectively. Note that the model coefficients have been adjusted such that $\varepsilon^{FS} + \varepsilon^{MS}$, integrated over the whole channel width, is identical with the DNS value as a consequence of (14). From the results shown in figure 9 it is seen that **C** performs best in the near-wall region. Although a kink in the curve for ε^{FS} can also be found for model **B**, it is much more pronounced for model **C**. In the curve for model **A** such a kink is not discernible at all.

The above results illustrate that the inverse cascade of kinetic energy in the buffer layer may in principle be accounted for by an eddy-viscosity ansatz, but the proper choice of the characteristic scales is of crucial importance. The most common approach to compute the eddy viscosity from $(u'_k u'_k)/2$ and Δx clearly is inappropriate in the near-wall flow. The definition of the eddy viscosity given by model **C**, on the other hand, can be valid in the near-wall region only and is not applicable in the outer flow. This suggests that subgrid models for complex wall-bounded flows may have to be adjusted to the various flow regimes in a very substantial way.

8 CONCLUDING REMARKS

In the present paper results of a detailed numerical study have been summarized which was conducted in order to clarify several issues relevant to the large-eddy simulation of wall-bounded turbulence. Among other things, the turbulent energy budget in the near-wall flow was thoroughly analyzed using direct numerical simulation data of turbulent channel flow and turbulent pipe flow. A key finding of this study is that in the buffer layer an inverse transfer of turbulent kinetic energy exists.

The performance of currently applied SGS models was examined in an *a priori* test where exact and modeled SGS quantities are compared for the same filtered flow field. It turned out that none of the investigated models was able to account for the inverse energy cascade. This deficiency may result in significant errors in LES results in the near-wall region which was illustrated by a series of large-eddy simulations at different Reynolds numbers. Consistent with what was found in the *a priori* test and in the analysis of the energy budget, no significant Reynolds-number effects appeared in these LES results. It is conjectured hence that the present findings can essentially be generalized to flows at higher Reynolds numbers of more practical interest.

Motivated by the observed deficiencies of the eddy-viscosity models, the eddy-viscosity ansatz was examined in a more fundamental manner. Three models for the eddy viscosity were compared which differ with respect to the characteristic subgrid scales employed. One of these was directly derived from our analysis of the turbulent kinetic energy budget. Satisfactory results could be obtained with this new model only which made clear that the proper choice of the characteristic scales is of utmost importance for the subgrid modeling in the near-wall flow.

The authors wish to acknowledge the financial support provided by the Deutsche Forschungsgemeinschaft.

REFERENCES

[1] M. Breuer and W. Rodi: In this publication.

[2] M. Manhart and H. Wengle: In this publication.
[3] J. W. Deardorff: J. Fluid Mech. **41**, 453-480, 1970.
[4] U. Schumann: J. Comput. Phys. **18**, 376-404, 1975.
[5] U. Piomelli, J. H. Ferziger and P. Moin: Phys. Fluids A **1**, 1061-1068, 1989.
[6] K. Horiuti: Phys. Fluids A **5**, 146-157, 1993.
[7] C. Härtel and L. Kleiser: Computational Fluid Dynamics '92, Vol. 1, eds. Ch. Hirsch et al., Elsevier, Amsterdam, 1992.
[8] C. Härtel and L. Kleiser: Engineering Applications of Large-Eddy Simulations, eds. S. A. Ragab and U. Piomelli, FED-Vol. 162, ASME, 1993.
[9] C. Härtel and L. Kleiser: Flow Simulation With High-Performance Computers I, ed. E. H. Hirschel, NNFM 38, Vieweg, Braunschweig, 1993.
[10] C. Härtel: Doctoral Dissertation, Technical University of Munich, Germany, 1994.
[11] C. Härtel: Transition, Turbulence and Combustion, Vol. II, eds. M. Y. Hussaini et al., Kluwer Academic Publishers, Dordrecht, 1994.
[12] C. Härtel and L. Kleiser: Direct and Large-Eddy Simulation I, eds. P. Voke et al., Kluwer Academic Publishers, Dordrecht, 1994.
[13] C. Härtel, L. Kleiser, F. Unger and R. Friedrich: Phys. Fluids **6**, 3130-3143, 1994.
[14] C. Härtel and L. Kleiser: submitted to Physics of Fluids, 1995.
[15] C. Härtel: Handbook of Computational Fluid Mechanics, ed. R. Peyret, Academic Press, to appear 1996.
[16] J. A. Domaradzki, W. Liu, C. Härtel and L. Kleiser: Phys. Fluids **6**, 1583-1599, 1994.
[17] A. Leonard: Adv. Geophys. **18** A, 237-248, 1974.
[18] A. A. Aldama: Lecture Notes in Engineering 56, Springer Verlag, Berlin, 1990.
[19] N. Gilbert and L. Kleiser: Proc. 8th Symp. on Turbulent Shear Flows, Munich, Germany, September 9-11, 1991.
[20] R. A. Antonia and J. Kim: J. Fluid Mech. **276**, 61-80, 1994.
[21] F. Unger: Doctoral Dissertation, Technical University of Munich, Germany, 1994.
[22] R. B. Dean: J. of Fluids Eng. **100**, 215-223,1978.
[23] T. A. Zang: Phil. Trans. R. Soc. Lond. A **336**, 95-102, 1991.
[24] J. Smagorinsky: Monthly Weather Review **91**, 99-164, 1963.
[25] D. K. Lilly: Proc. IBM Scientific Computing Symposium on Environmental Sciences, Yorktown Heights, N.Y., 1967.
[26] A. Scotti, Ch. Meneveau and D. K. Lilly: Phys. Fluids A **5**, 2306-2308, 1993.
[27] E. R. Van Driest: J. Aero. Sci. **23**, 1007-1011, 1956.
[28] O. Métais and M. Lesieur: J. Fluid Mech. **239**, 157-194, 1992.
[29] J.-P. Chollet and M. Lesieur: J. Atmos. Sci. **38**, 2747-2757, 1981.
[30] P. Comte, S. Lee and W. H. Cabot: Proc. 1990 Summer Program, Center for Turbulence Research, 31-45.

[31] T. A. Zang, C.-L. Chang and L. L. Ng: Proc. Fifth Symposium on Numerical and Physical Aspects of Aerodynamic Flows, Long Beach, January 13-15, 1992.

[32] M. Germano, U. Piomelli, P. Moin and W. H. Cabot: Phys. Fluids A **3**, 1760-1765, 1991.

[33] D. K. Lilly: Phys. Fluids A **4**, 633-635, 1992.

[34] R. S. Rogallo and P. Moin: Ann. Rev. Fluid Mech. **16**, 99-137, 1984.

LARGE EDDY SIMULATION FOR COMPLEX TURBULENT FLOWS OF PRACTICAL INTEREST

M. Breuer [*] and W. Rodi

Institut für Hydromechanik, Universität Karlsruhe (TH),
Kaiserstr. 12, D-76128 Karlsruhe

SUMMARY

LES is expected to become a powerful tool for turbulent flow calculations of practical interest in the near future. The paper is concerned with the development and first applications of a LES technique based on an explicit finite–volume method for curvilinear body–fitted grids suitable to simulate flows in or around complex geometries. Two different subgrid–scale models are implemented and tested. For the near-wall treatment different wall function approaches as well as no–slip boundary conditions can be applied. This technique has been used to simulate six different internal and external flow configurations: plane channel flow, flow through a straight square duct and a 180° bend, flow around a surface–mounted cubical obstacle and the flow past a long, square as well as a circular cylinder. For four test cases results are discussed in detail and compared with experimental data available. The influence of different aspects of a LES method (e.g. resolution, subgrid–scale modelling, wall boundary conditions) are investigated.

INTRODUCTION

Turbulent flows of practical interest are in general very complex including phenomena such as separation, reattachment and vortex shedding. An appropriate description by Reynolds–Averaged Navier–Stokes (RANS) equations combined with statistical turbulence models is difficult to achieve. This is due to the necessity of modelling the whole spectrum of turbulent scales. The method of direct numerical simulation requires no model assumptions but will not be applicable to engineering flows in the foreseeable future because of extremely high computing costs. The concept of large–eddy simulation (LES) seems to be a promising way of solving such flow problems. In LES the large eddies that depend strongly on the special flow configuration are resolved numerically whereas only the fine–scale turbulence has to be modelled by a subgrid–scale model.

The goal of the work reported here is the development of a LES technique for calculating practically relevant flows with complex boundaries. In contrast to many previous LES studies which were mainly concerned with geometrically very simple flow configurations, this investigation requires the solution of the Navier–Stokes equations on curvilinear boundary–fitted grids which in general are non–orthogonal. A finite–volume approach based on the pressure correction method is used to solve the governing equations on a non–staggered grid. Two different wall function approaches were implemented in order to

---

[*] Present address: Lehrstuhl für Strömungsmechanik, Universität Erlangen–Nürnberg, Cauerstr. 4, D–91058 Erlangen

enable high–Reynolds–number LES at reasonable computing costs. Fine scale–turbulence is modelled by either the Smagorinsky [29] or Germano et al.'s [14] dynamic model. For the first time the dynamic model is applied to a fully inhomogeneous flow. After extensive tests for plane channel flow (not presented here), an investigation of the flow through a straight square duct at different Reynolds numbers was carried out. Because of small turbulence–induced secondary motions in the duct the flow field significantly differs from plane channel flow. The first test case requiring a curvilinear body–fitted grid is the flow through a 180° bend with square cross–section at Re=56,690. The flow around a surface mounted cubical obstacle placed in a plane channel was chosen as a typical bluff–body flow. Detailed experimental data ($Re = U_B H/\nu = 40,000$, $U_B =$ bulk velocity, $H =$ obstacle height) have been provided by Martinuzzi et al. [23]. In addition, the flow at a Reynolds number of 3,000 is simulated. These flows have been test cases for the International 'LES Workshop of Flows past Bluff Bodies' held at Rottach Egern (Germany) on June 26–28, 1995 [27]. LES results using both subgrid–scale models and different near-wall approaches are shown and compared with experimental data available. Finally first results for the flow around a circular cylinder at Re=3,900 are presented. More results from this project are published in [3, 4, 5, 6, 7, 27]. Related topics can be found in this publication [16, 21].

DESCRIPTION OF THE LES METHOD

The LES code developed in Karlsruhe [3, 4, 5, 6, 7] is based on a finite–volume method for solving the incompressible (filtered) Navier–Stokes equations on general body–fitted, curvilinear grids (**LESOCC** = **L**arge **E**ddy **S**imulation **O**n **C**urvilinear **C**oordinates). A non–staggered, cell–centred grid arrangement is used. Both convective and viscous fluxes are approximated by central differences of second order accuracy. It is well–known for LES now that the numerical dissipation produced by a scheme for the convective fluxes is of much greater relevance than its order of accuracy in space itself. Preliminary calculations for the LES standard test case, the plane channel flow, applying third order upwind schemes have clearly demonstrated this, leading to highly damped velocity fluctuations and therefore rather poor turbulence statistics, especially for the higher order moments. The time integration is performed with a predictor–corrector scheme, where the explicit predictor step for the momentum equations is either an Adams–Bashforth scheme or a low–storage multi–stage Runge–Kutta method (3 or 5 sub–steps). The corrector step covers the implicit solution of the Poisson equation for the pressure correction (SIMPLE). Both schemes are of second order accuracy in time because the Poisson equation for the pressure correction is not solved during the sub–steps of the Runge–Kutta algorithm in order to save cpu–time. Due to the higher stability limit of the Runge–Kutta scheme, much larger time steps ($CFL = O(1)$) can be used which leads to a reduction in computing time by a factor of about 2 compared with the Adams-Bashforth scheme. However, the time steps applied still guarantee good resolution in time necessary for LES calculations. The linear system of equations is solved by an incomplete LU decomposition method which can be accelerated by a FAS multigrid technique. In order to avoid decoupling of pressure and velocity on the non–staggered grid, the momentum interpolation proposed by Rhie and Chow [25] is applied. Two different subgrid–scale models are implemented, namely the standard Smagorinsky model [29] with Van Driest damping ($l = C_s \Delta (1 - exp(-y^+/25)^3)^{0.5}$) near solid walls as well as the promising dynamic model of Germano et al. [14]. One of the major drawbacks of the Smagorinsky model is the now well–known fact that the Smagorinsky constant C_s was found to depend on the flow problem considered. Secondly, in an inhomogeneous flow, the optimum choice for C_s may be different for different points in the flow. Furthermore, the Smagorinsky model needs some additional assumptions to describe flows undergoing transition or near solid walls. The dynamic approach eliminates some of these disadvantages by calculating a 'Smagorinsky constant' as a function of time and position. Following a suggestion of Lilly [20], a least–squares approach is used to obtain values for $C_s{}^2$. Depending on the flow problem considered, different kinds of averaging procedures can be applied in the dynamic approach. If the flow is homogeneous in a certain direction, averaging can be done over this direction. For fully inhomogeneous flows, like the 3–D obstacle flow, only an averaging procedure in time is applicable. In order not to restrict the values of C_s to a fully time–independent function and to allow variations with low frequencies, a special form of time averaging is chosen, which is well–known as a recursive lowpass digital filter. With an appropriate value of the free parameter of the filter function [3], all high frequency oscillations are damped out and only the low frequency variations remain. This seems to be better than fully freezing C_s. Additionally, negative eddy viscosities are clipped.

Another basic element of a LES technique is an appropriate formulation of the boundary conditions. Solid walls as well as inflow and outflow boundaries are difficult to model. Only if the investigated flow configuration has one or more homogeneous directions, simple periodic boundary conditions can be applied, choosing an integration domain which guarantees nearly uncorrelated inflow and outflow values. In the case of plane channel flow, periodic conditions can be used in the streamwise and spanwise directions whereas for the developed square duct flow these conditions can only be applied in the mean flow direction. However, the curved duct flow as well as the flow around a cubical obstacle, which do not have any homogeneous direction, require real inflow and outflow boundary conditions.

The difficulty of prescribing appropriate outflow conditions is well–known from laminar flow computations. Especially for unsteady vortical flows these formulations have to ensure that vortices can approach and pass the outflow boundary without significant disturbances or reflections. An outflow boundary condition which satisfies these requirements is the convective boundary condition given by $\frac{\partial u_i}{\partial t} + c\frac{\partial u_i}{\partial x} = 0$ where c is the mean exit velocity. For all simulations described in this paper this condition works very well.

At the inlet section three velocity components have to be prescribed. In contrast to laminar or Reynolds–averaged flow computations, for LES it is not sufficient to specify mean values at the inlet, but the time dependent velocity field in the whole cross–section must be given. Experimental results for this are generally not available. Another possibility is the use of numerical data produced by a previous LES (or DNS) with periodic boundary conditions. This often used technique is applied for the bend and obstacle flow. The calculated and stored results for one cross–section of the square duct/plane channel flow simulation are reused as inflow data for the fully inhomogeneous flow computations. This technique is rather expensive concerning consumed cpu–time and required storage devices, but up to now other approaches are not practically applicable.

In principle, Stokes' no–slip boundary condition on solid walls is valid for LES of turbulent flows. However, the realization of this condition demands an extremely high resolution of the flow field in the vicinity of the wall to account for the viscous sublayer. LES for complex flows of practical interest which are normally high–Reynolds–number flows cannot afford the no–slip condition. Special wall functions are therefore necessary to model the near-wall region rather than to resolve it directly. Schumann's wall boundary conditions

[28] are well-known. They assume a phase coincidence of instantaneous wall shear stress and tangential velocity component in the computational cell nearest to the wall. The relation between the average wall shear stress and the average velocities is given by the law of the wall split into three regions, the viscous sublayer, the buffer layer, and the logarithmic layer. Werner and Wengle's approach [31] also assumes that the instantaneous tangential velocity component inside the first cell is in phase with the instantaneous wall shear stress. However, they propose a linear law of the wall ($y^+ \leq 11.81$) or a 1/7 power law distribution ($y^+ > 11.81$) directly for the instantaneous velocity distribution. This procedure offers the advantages that no average values are required, numerical problems in reattachment regions are avoided, and no iteration method is necessary to determine the wall shear stress because it can be obtained analytically. Both approaches are implemented in **LESOCC**.

RESULTS AND DISCUSSION

Square duct flow at Re = 4410

The turbulent flow through a straight square duct is chosen as one of the test cases for the LES code. Turbulent flow in ducts can be accompanied by two different kinds of secondary motions in the plane perpendicular to the mean flow direction. In curved ducts (laminar or turbulent flow), centrifugal forces cause pressure–induced motions (Prandtl's first kind) which can reach velocities of the order of 20–30 % of the mean velocity. Prandtl's second kind of secondary motion however is much weaker (2–3 %) and can only be observed in turbulent flows through straight or curved ducts with non–circular cross–sections. Although these turbulence–induced motions are quite small, they have important consequences for the flow, e.g. they cause a bulging of the mean velocity contours towards the corners.

The computational domain has the dimensions ($4\pi D \times D \times D$), where D is the hydraulic diameter of the duct. The grid consists of 62 grid points in streamwise direction and 41×41 grid points in the cross–section. The Reynolds number based on the hydraulic diameter D and the bulk mean velocity u_B is $Re = 4410$. Before starting the calculation of the mean values and statistics of the turbulent flow, the simulation is carried out for several turnover times to ensure independence from the randomized initial condition. The computation is then continued for nearly 100000 time steps. Every fiftieth time step a sample is taken. Furthermore, the flow can be averaged in the homogeneous direction. Fig. 1 shows the axial mean velocity profiles along the wall bisector. Two slightly different dynamic subgrid–scale models were applied. Model 1 has a test filter which filters the variables in all directions in space, while in model 2 the test filter is only used in the homogeneous direction. The results of three LES are compared with the DNS data of Gavrilakis [13] for the same Reynolds number and the LES data of Madabhushi & Vanka [30] for $Re = 5810$. Compared with the DNS data all LES show slightly lower mean velocities near the center of the duct. However, the overall agreement is quite good and only negligible differences appear due to the different subgrid–scale models. In Figs. 2 – 4 the distributions of the root mean square (RMS) velocity fluctuations are plotted. For all LES (including Madabhushi's) the velocity fluctuations in the streamwise direction are overpredicted compared with the DNS results, while the fluctuations in the cross flow directions are in general somewhat lower than Gavrilakis'. However, the overall agreement is still satisfactory. It is worth mentioning that the type of test filter for the dynamic model seems to play an important role. The secondary, turbulence–induced motion is shown in

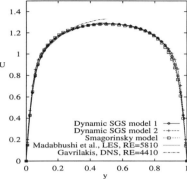

Fig. 1: Mean streamwise velocity distribution along wall bisector using different subgrid–scale models ($Re = 4410$) and comparison with DNS [13] ($Re = 4410$) and LES [30] ($Re = 5810$).

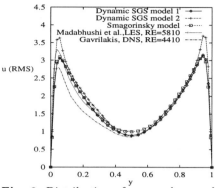

Fig. 2: Distribution of u_{RMS} along wall bisector using different subgrid–scale models ($Re = 4410$) and comparison with DNS [13] ($Re = 4410$) and LES [30] ($Re = 5810$).

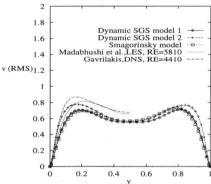

Fig. 3: Distribution of v_{RMS} (for key to symbols see fig. 2).

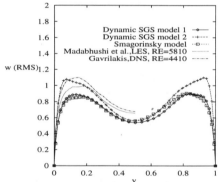

Fig. 4: Distribution of w_{RMS} (for key to symbols see fig. 2).

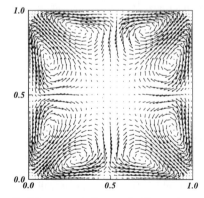

Fig. 5: Mean secondary velocity vectors, $Re = 4410$.

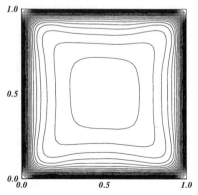

Fig. 6: Contours of mean streamwise velocity, $Re = 4410$.

Fig. 5 by velocity vectors, and its influence on the axial velocity distribution in Fig. 6 by isolines. The flow field is symmetric with respect to the y– and z–axes as well as the diagonals forming a system of eight vortices. Fluid approaches the corner regions along the diagonals and leaves it along the wall and the wall bisector. The distortion of the mean velocity contours towards the corner is due to the transport of high–momentum fluid into the corner. The maximum secondary velocity is 2% of the bulk velocity whereas Gavrilakis' DNS gives a maximum of 1.9% nearly at the same position.

Square duct flow at Re = 56,690

In order to examine the performance of a LES at higher Reynolds numbers, a square duct flow at $Re = 56,690$ (value in bend flow experiments) was chosen which at the same time was used for generating the inflow data for the bend flow. Two different equidistant grids with 63^3 ($4D \times D \times D$) and 101^3 ($2\pi D \times D \times D$) grid points were employed. In Fig. 7 the axial mean velocity profiles along the wall bisector are plotted in comparison with available experimental data for different Reynolds numbers. Because the differences in axial velocities using the Smagorinsky model and the dynamic model (on both grids) are very small, only one LES result is presented. Although the experimental data are somewhat scattered, they show generally significant differences from the results obtained by LES. The calculation overpredicts the axial velocity in the vicinity of the wall which, for a fixed mass flow rate, results in a velocity profile that is too flat. However, the average friction velocity $\overline{u_\tau}$ agrees fairly well (error $< 0.5\%$) with the theoretical value given by Blasius' law $\lambda = 0.3164/Re^{0.25}$. The application of both types of wall functions mentioned above yields virtually the same results. LES for plane channel flow at comparable Reynolds numbers, however, agree very well with theoretical and measured data. Therefore the discrepancies in the square duct case arise due to the small secondary motion in this flow configuration. A collection of measured and calculated (RANS) secondary velocities along the wall and corner bisectors at different Reynolds numbers is shown in [12]. In spite of normalizing the secondary velocities with the friction velocity, considerable scatter of the data can be observed indicating difficulties and uncertainties involved in the determination of this small secondary motion. As described in [3] LES applying wall functions near solid walls underpredicts the secondary velocity similar to the algebraic Reynolds stress model of Demuren & Rodi [12]. In order to investigate the influence of the wall treatment a LES on a finer grid in the cross–section ($151 \times 151 \times 65$; $2\pi D \times D \times D$) was carried out. The grid points were clustered in the vicinity of the walls ($\Delta y_{min} = \Delta z_{min} = 5*10^{-4}D$; $\Delta y^+{}_{min} = \Delta z^+{}_{min} \approx 1.4$) in order to allow the application of the no–slip boundary condition instead of any wall function. (Compared with the equidistant grids mentioned above the distance between the wall and the first grid point is about 10 times smaller for the stretched fine grid.) Fig. 8 shows a comparison of the results on both grids (stretched & equidistant). The upper part of the figure displays the secondary velocities along the wall bisector. Here the influence of the wall treatment is clearly visible. The strength of the secondary motion is nearly doubled by the use of no–slip boundary conditions compared with the calculation using wall functions. As expected the influence of the wall treatment is much lower for the secondary velocities along the corner bisector (lower part of Fig. 8). This figure clearly demonstrates that the wall functions commonly applied in LES today are not suitable to capture small turbulence induced motions with the right strength.

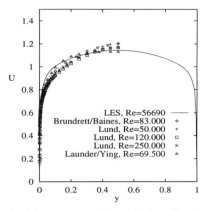 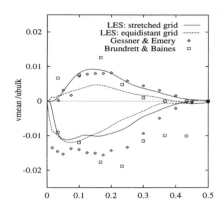

Fig. 7: Mean streamwise velocity distribution along the wall bisector ($Re = 56,690$) and comparison with measurements.

Fig. 8: Secondary velocities along wall & corner bisectors (LES: $Re = 56,690$) and comparison with experimental data.

180° bend flow at Re = 56,690

The fully inhomogeneous flow through a 180° bend ($R_m/D = 3.375$) is the first test case which cannot be simulated by spectral or Cartesian finite–difference methods most commonly employed in LES and DNS to date. The curvilinear grid consists of $121 \times 63 \times 63$ grid points, while the cross–section is the same as in the corresponding straight duct flow. The computational domain includes an upstream inlet and a downstream outlet tangent of three hydraulic diameters. For $Re = 56,690$ experimental data are available from Choi et al.[10] and Chang et al.[11]. Without any averaging procedure for the C value the dynamic model turned out to destabilize strongly the numerical method by producing very high eddy viscosity values. After implementing the lowpass filter the dynamic model determines reasonable values for C_S in the whole flow field with a maximum of the order of 1 and an average slightly higher than the commonly used Smagorinsky constant ($C_S = 0.1$). Figs. 9 ($\theta = 90^o$) and 10 ($\theta = 135^o$) show profiles of the mean streamwise velocity along lines parallel to the flow symmetry plane ($2z/d = 0.25, 1.0$, see Fig. 11 for explanation). At $\theta = 90^o$ the velocity along the symmetry line exhibits a small trough, which however is not strong enough and too close to the inner wall of the bend compared with the experimental data. Further downstream at $\theta = 135^o$ in contrast to the experiments no dip can be observed in the velocity profile at the symmetry plane but at the plane closer to the wall ($2z/d = 0.25$). Nevertheless a strong curvature–induced secondary motion, responsible for the dips, exists. In Fig. 11 secondary velocity vectors and streamlines are plotted for the plane $\theta = 135^o$. LES produces a multi–cellular secondary flow pattern similar to Reynolds–averaged computations with an algebraic stress model (ASM) [19]. An extra vortex near the symmetry plane, which cannot be resolved by k–ϵ turbulence modelling, can be observed in the LES results similar to the ASM predictions. However, the overall agreement with the experimental data is not satisfactory. Comparisons of velocity profiles in the inlet tangent already show deviations from the measurements, indicating that the inflow data produced by the preceding periodic duct flow computation are at least partly responsible for the discrepancies between computed and measured velocity distributions

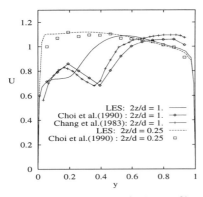

Fig. 9: Mean streamwise velocity profiles at station 90 deg., ($Re = 56,690$).

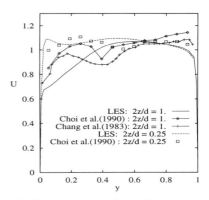

Fig. 10: Mean streamwise velocity profiles at station 135 deg., ($Re = 56,690$).

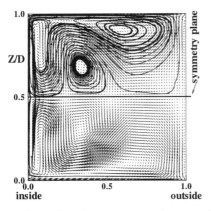

Fig. 11: Secondary flow vectors and streamlines at 135 deg. station.

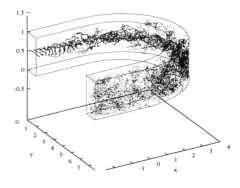

Fig. 12: Streaklines for the bend flow ($Re = 56,690$).

in the bend. Another reason may be the insufficient resolution of the flow field especially in the vicinity of the walls. As found out later for the square duct flow (mentioned in the previous chapter) the grid used in the cross-section (63×63) was not sufficient to resolve all important flow features. Therefore it could not be sufficient for the much more complex bend flow. Simulations with refined grids are necessary to prove this. Fig. 12 finally shows computed streaklines in the bend flow. The injection of massless particles (200 sources) takes place every twentieth time step in the plane of symmetry near the inlet section. The streaklines break up already at the entrance of the curved section. Further downstream they are totally mixed by turbulence fluctuations, filling the whole cross-section at the outlet.

Flow around a surface–mounted obstacle at Re=3,000/40,000

A first test case for bluff body aerodynamics is the three–dimensional flow around a surface–mounted cubical obstacle (height $= H$) placed in a plane channel. This is a fully turbulent, complex flow with multiple separation regions and vortices. It has been the test case **B** of the previously mentioned LES workshop [27]. Test case **A**, the flow past a long square cylinder at Re=22,000, has also been calculated with **LESOCC** by M. Pourquié. Results for this case can be found in [7]. For the obstacle flow two different Reynolds numbers (**B1**: Re=3,000, **B2**: Re=40,000) have been investigated; measurements are available for the case **B2** with Re=40,000 [23].

For all simulations a computational domain with an upstream length $x_1/H = 3$ and a downstream length of $x_2/H = 6$ is used. The width is set to $b/H = 7$ and the height of the channel is $2H$ as in the experiment. All computations are performed on a stretched grid with $165 \times 65 \times 97$ grid points for the x, y and z directions. In the streamwise direction 70 grid points are distributed in the region in front of the obstacle. On the surface of the obstacle 31 grid points are applied in all directions. The smallest cell volume near solid walls has an extension of $(0.0125\,H)^3$. Due to the non–staggered cell–centred grid arrangement the smallest distance between the wall and the first grid point is $(0.0125\,H)/2$.

The following table gives an overview of the models and parameters used for the 5 different simulations. The inflow is fully developed turbulent channel flow and was generated by LES of plane channel flow (same grid in the cross–sectional plane). For the lateral boundaries (x–y plane) at $\pm z/H = 3.5$, periodic boundary conditions are chosen. At solid walls the no–slip condition is used for $Re = 3,000$ whereas for $Re = 40,000$ Werner and Wengle's wall function approach is applied, except in simulation UKAHY5. A convective boundary condition is used at the outflow boundary where the convective velocity is assumed to be the mean bulk velocity U_{bulk}.

Table 1: Parameters for the 3-D obstacle flow

Key	Test case	Re	SGS mod.	C_s	Δt	$T_{averaging}$	Wall b.c.
UKAHY1	B1	3,000	Smago.	0.1	10^{-3}	≈ 103.5	no–slip
UKAHY2	B1	3,000	Dynamic	–	10^{-3}	≈ 200.5	no–slip
UKAHY3	B2	40,000	Smago.	0.1	10^{-3}	≈ 150.1	wall function
UKAHY4	B2	40,000	Dynamic	–	10^{-3}	≈ 108.6	wall function
UKAHY5	B2	40,000	Smago.	0.1	10^{-3}	≈ 151.2	no–slip

Fig. 13 shows a first qualitative comparison of the computed results for both Reynolds numbers and the measurements [23] at Re=40,000. The streamlines in the plane of symmetry as well as the surface streamlines of the bottom channel wall are plotted. In all LES the velocities are averaged over a long period of more than 100 dimensionless time units (H/U_B) to achieve good statistics. Regarding the streamlines in the symmetry plane the influence of the Reynolds number on this flow is quite small. For both Re numbers the choice of the subgrid–scale model seems to be more important. Table 2 gives an overview of the size of the different separation/reattachment regions. The reattachment length behind the obstacle is only slightly overpredicted by the LES with the Smagorinsky model and underpredicted somewhat by the LES with the dynamic model. All results are in much closer agreement with the experiment than typical RANS/2–equation model predictions.

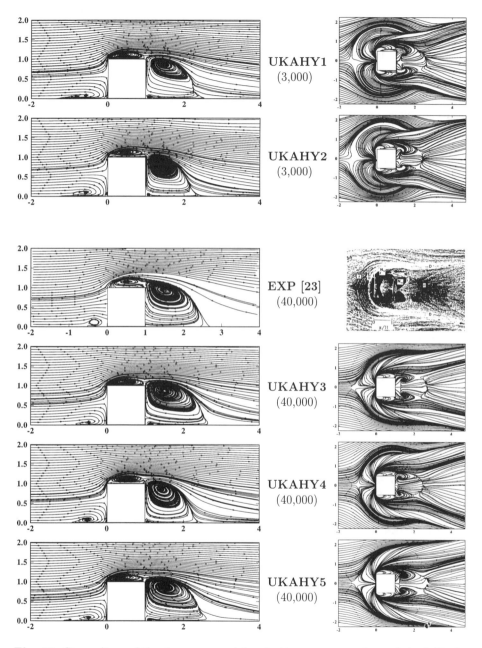

Fig. 13: Streamlines of the time–averaged flow in the symmetry plane of the 3–D obstacle and surface streamlines on the bottom wall of the channel, (see Table 1 for legend).

In [6] a detailed comparison between LES and RANS results is given for test case **B2** applying various versions of the $k-\varepsilon$ model: the standard model and the Kato–Launder (KL) modification using wall functions, and the standard model in a two–layer approach using a one–equation model near the wall. The computed length scales predicted with these RANS models are included in Table 2. It can be seen that all the RANS models overpredict the length of the separation zone behind the cube significantly. The standard $k-\varepsilon$ model predicts early reattachment of the flow on the roof. Due to suppression of turbulence production in front of the cube, this situation is improved somewhat using the KL modification; the most realistic prediction is, however, obtained with the two–layer approach which better resolves the flow on the roof and near the ground wall. The extension of the separation bubble on the roof computed by LES agrees fairly well with the measurements, too. As observed in the experiments, no reattachment of the time–averaged flow takes place on the roof of the obstacle in all simulations. In order to capture this flow feature a fine grid near the obstacle walls is very significant. This was one of the conclusions at the LES Workshop [27], where some other contributors with coarser resolution near the walls achieved (incorrect) reattachment on the roof.

Table 2: Separation / reattachment length for the 3–D obstacle flow

Method	Key	Re	X_{front}	X_{roof}	X_{rear}
LES	UKAHY1	3,000	1.39	–	1.60
LES	UKAHY2	3,000	1.29	–	1.42
EXP [23]	EXP	40,000	1.040	–	1.612
LES	UKAHY3	40,000	1.287	–	1.696
LES	UKAHY4	40,000	0.998	–	1.432
LES	UKAHY5	40,000	1.228	–	1.700
RANS [6]	Standard $k-\varepsilon$	40,000	0.651	0.432	2.182
	KL modification	40,000	0.650	–	2.728
	Two–layer $k-\varepsilon$	40,000	0.950	–	2.731

Fig. 13 displays also a comparison of the experimental versus the calculated surface streamlines. The primary separation line, the recirculation region behind the obstacle as well as the outer border of the wake region are quite similar to the oil flow patterns of the experiment. However, here the influence of the Re number seems to be more significant. Hence a comparison with the experiment has to be restricted to the high Re number computations. The figure shows that the horseshoe vortex generated between the primary and the secondary separation lines (B) is fairly well predicted by LES. The flow patterns suggest also that the structure of the outer limit of the wake region formed by the lateral arms of the horseshoe vortex (line D), varies between the different subgrid–scale models. In the experiment, the width of this wake decreases up to approximately the reattachment point; then it increases again. This behavior is well described by LES with the dynamic model (UKAHY4). In the simulation using the Smagorinsky model combined with wall functions (UKAHY3) this tendency is much weaker. If no–slip boundary condition are applied at solid walls (UKAHY5) instead of the wall functions this characteristic behavior of the outer wake totally disappears. Here the influence of the wall modelling becomes visible for the first time. All LES approaches seem to predict correctly the corner vortices generated downstream of the vertical leading edges of the cube at the channel–body junction. The overall agreement between the experiment and the time–averaged flow field

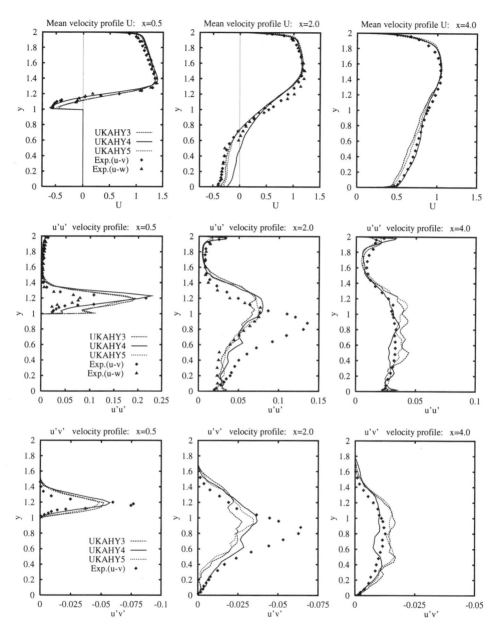

Fig. 14: Comparison of mean velocity profiles $\overline{U}$ and Reynolds stresses $\overline{u'u'}$ and $\overline{u'v'}$ at three different locations of the symmetry plane of the 3–D obstacle, time–averaged flow, Re = 40,000 (Note different scales of the plots; see Table 1 for legend)

calculated by LES is satisfactory.

Fig. 14 compares calculated and measured streamwise velocity profiles $\overline{U}$ at three different locations in the symmetry plane. One obstacle height in front of the cube all streamwise velocity profiles agree fairly well with the measurements (not shown here). However, small differences can be observed for the profile in the middle of the roof ($x/H = 0.5$). Here the best result compared with experiment is provided by the LES with the dynamic model (UKAHY4). The size of the separation bubble and the magnitude of the reversed flow velocity is well reproduced. The Smagorinsky model with (UKAHY3) and without wall functions (UKAHY5) gives nearly the same results. Moving further downstream, the effect of the variations in the computed length of the recirculation region (Tab. 2) is clearly visible. In the wake region ($x/H = 2.0$), the computed velocity magnitudes are globally underestimated; here LES results using the Smagorinsky model are closest to experiment. Far away from the reattachment point at $x/H = 4.$, again LES with the dynamic model (UKAHY4) gives a slightly better representation of the flow than UKAHY3/5. Fig. 14 displays also profiles of the Reynolds stresses $\overline{u'u'}$ and $\overline{u'v'}$ at the same locations as before. It should be noted that for LES only the resolved part of the turbulent stress is included. At $x/H = 0.5$ all simulations give similar peak values for $\overline{u'u'}$ and $\overline{u'v'}$. Further downstream ($x/H = 2.$) some scatter in the experimental data for $\overline{u'u'}$ can be detected. This component has been measured twice, one time using the laser doppler–anemometry in the u–v–plane ($\diamond$) of the experimental setup, next time in the u–w–plane ($\triangle$). Both measurements differ by about 100 % in the middle of the channel. All LES agree better with the measurements in the u–w–plane. Due to this scatter it is not clear whether the $\overline{u'v'}$ component at the same location is trustworthy. At the third position ($x/H = 4.$) in the recovery region of the flow the profiles of the Reynolds stress spread around the experimental values. No trend is recognizable as to which subgrid–scale model performs best. Again, the influence of the wall modelling (compare UKAHY3/5) in this region seems to be nearly insignificant and smaller than that of the subgrid–scale model. All profiles are not very smooth which may be an indication of too short averaging time for these higher order moments.

Flow around a circular cylinder at Re=3,900

The last test case shown here is the flow around a long, circular cylinder at the subcritical Re=3,900, based on the cylinder diameter D and free–stream velocity U_∞. This flow field has already been simulated by LES and was analyzed in great detail by Beaudan and Moin [2]. It was found to be highly three–dimensional. For instance, counter-rotating vortices in the main stream direction were found, which cannot be represented in two–dimensional calculations, showing the necessity of three–dimensional calculations. A curvilinear, O–type grid was generated which consists of 165×165 control volumes in the cross–sectional plane but only 32 computational cells in the spanwise direction. The grid points were clustered in the vicinity of the cylinder and in the wake region. The whole integration domain has a radial extension of $15D$ in the cross–section and πD in the direction of the cylinder axis. For this low Reynolds number case no–slip boundary conditions can be applied at the cylinder surface. At the inlet of the computational domain a constant, uniform velocity distribution (no fluctuations) is prescribed. Again, the convective boundary condition is used at the outlet. In the spanwise direction of the circular cylinder periodicity of the flow is assumed. The calculations were performed at first without any subgrid–scale model, because the influence of modelled subgrid–scale stress is known to be small at low Re numbers [2]. Statistics are compiled over about $90D/U_\infty$ time units, or approximately 20 vortex shedding cycles and in the spanwise direction. In addition to the 3–D LES, a 2–D simulation on the same cross–sectional grid has been performed to examine the impact of three–dimensionality of the flow. Fig. 15 shows the streamlines of the time–averaged flow for both simulations. Totally different streamline pattern can be observed. The largest difference is given by the absence of an attached recirculation region behind the cylinder in the 2–D case. In spite of nearly the same averaging time for the 2–D simulation the 2–D flow field is more asymmetric than the one of the 3–D calculation because no averaging is done in the spanwise direction. In contrast to experimental observations and the 3–D LES the 2–D simulation shows a positive time–averaged streamwise velocity at all centerline stations downstream of the cylinder. In the 3–D case the flow field consists of a large recirculation region behind the cylinder and two additional, small separation bubbles attached to the downstream face of the cylinder. Table 3 gives an overview of some computed bulk parameters compared with various experimental data and the LES results of Beaudan & Moin [2]. They used an O–type grid with $144 \times 136 \times 48$ grid points which was optimized concerning the distribution of the grid points by extensive refinement studies.

Strouhal number St and primary separation angle $\overline{\Theta_1}$ agree fairly well with the other LES results and are within experimental uncertainty. However, the recirculation length behind the cylinder is about 20% too small compared with the mean experimental value [9], whereas the simulation of Beaudan et al. [2] show in general too long recirculation bubbles (up to 30%). For this quantity the influence of the subgrid–scale model seems to be larger than for other bulk quantities. The computed drag coefficient $\overline{C_D}$ of LESOCC is also about 10% higher than the experimental value. This is consistent with the computed back–pressure coefficient ($C_{P_b} = -1.07$) which is about 10% too high. The reason for these discrepancies may be at least partly the coarse resolution especially in the spanwise direction because the 2–D simulation has shown similar effects: too large drag and back–pressure coefficients. More detailed investigations applying refined grids and both subgrid–scale models are planned for the future.

Table 3: Bulk numbers for the cylinder flow

	present no SGS model	LES [2] no SGS model	LES [2] Smago. model	LES [2] Dyn. model	Experiment
St Strouhal	0.22	0.216	0.209	0.203	0.215 ± 0.005 (Cardell [9])
$\overline{C_D}$ Drag	1.13	0.96	0.92	1.00	0.98 ± 0.05 (Norberg [24])
$\overline{\Theta_1}$ Separat.	±87	±85.3	±84.8	±85.8	±85 ± 2 (Son et al. [26])
$\overline{\Theta_2}$ Separat.	±111	±109.7	±110.5	±110.6	
$\overline{\Theta_3}$ Separat.	±148	±154.1	±146.2	±158.3	
L/D Length	1.04	1.56	1.74	1.36	1.33±0.2 (Cardell [9])

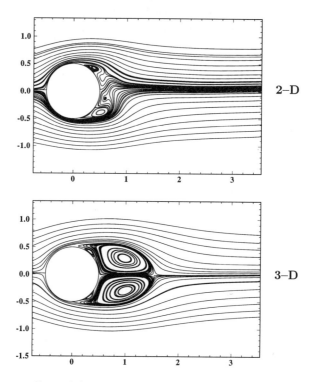

Fig. 15: Mean streamlines of the time–averaged flow around a circular cylinder, Re = 3,900, upper: 2–D simulation, lower: 3–D simulation.

Conclusions

Six different turbulent flows have been simulated by the method of LES. The finite–volume code employed was written for general curvilinear grids in order to enable the computation of flows in or around complex geometries. The recently proposed dynamic subgrid–scale model was implemented as well as the Smagorinsky model. The dynamic model was successfully adjusted and applied to fully inhomogeneous flows.

The results may be categorized as follows: For low Re numbers the agreement with experimental/DNS data is in general quite good (square duct case, obstacle and cylinder flow). Two reasons are mainly responsible for this; the influence of the subgrid–scale model is quite small and the wall modelling can be done without any empirical wall functions. When moving to higher Re numbers, both items become important. A large portion of the turbulence spectrum has to be modelled and the no–slip boundary condition at solid walls is no longer applicable. Now different kinds of flow have to be distinguished. For bluff body flows like the flow around the 3–D obstacle, which are characterized by strong large–scale motions, the influence of wall modelling seems to be small as demonstrated above. The agreement with measurements is still satisfactory for high Re numbers. Such flows are particularly suitable for the LES technique which yields generally much better results than any RANS approach, albeit at a considerably higher price. On the contrary, for flows with small turbulence–driven secondary motions like the flow through the square duct the wall treatment seems to be extremely important. Especially for this kind of flows investigations with refined near–wall models have to be carried out.

Acknowledgements

The work reported here was sponsored by the Deutsche Forschungsgemeinschaft. The calculations were carried out on the SNI S600/20 vector computer of the University of Karlsruhe (Computer Center). The authors should also like to acknowledge helpful discussions with Prof. J.H. Ferziger from Stanford University.

References

[1] Akselvoll, K., Moin, P.: *Large Eddy Simulation of Backward Facing Step Flow,* Eng. Turbulence Modelling & Experiments 2, (Rodi et al.,eds.), Elsevier Science Publishers B.V., pp. 303–323, (1993).

[2] Beaudan, P., Moin, P.: *Numerical Experiments on the Flow past a circular cylinder at a Sub-Critical Reynolds Number,* Report No. TF–62, Thermosciences Division, Dept. of Mech. Engineering, Stanford University, (1994).

[3] Breuer, M., Rodi, W.: *Large-Eddy Simulation of Turbulent Flow through a Straight Square Duct and a 180° Bend,* Fluid Mech. and its Appl., vol. **26**, Direct & LES I, Sel. papers f. the First ERCOFTAC Workshop on Direct & LES, Guildford, Surrey, U.K., 27–30 March 1994, ed. Voke, Kleiser & Chollet, Kluwer Acad. pub., (1994).

[4] Breuer, M., Rodi, W.: *Large-Eddy Simulation of Turbulent Flow through Straight and Curved Ducts,* ERCOFTAC Bulletin, vol. **22**, Sept. (1994).

[5] Breuer, M., Pourquie, M., Rodi, W.: *Large Eddy Simulation of Internal and External Flows,* 3rd International Congress on Industrial and Applied Mathematics, Hamburg, 3–7 July, (1995), to be published in ZAMM, (1996).

[6] Breuer, M., Lakehal, D., Rodi, W.: *Flow around a Surface Mounted Cubical Obstacle: Comparison of LES and RANS–Results,* accepted for the IMACS–COST Conference on Computational Fluid Dynamics, Three–Dimensional Complex Flows, Lausanne, Switzerland, Sept. 13–15, (1995), to be publ. by Vieweg Verlag.

[7] Breuer, M., Pourquie, M.: *First Experiences with LES of Flows past Bluff Bodies,* accepted for the 3rd Intern. Symposium of Engineering Turbulence Modeling and Measurements, Crete, Greece, May 27–29, (1996), to be publ. by Elsevier Science B.V..

[8] Brundrett, E., Baines, W.D.: *The production and diffusion of vorticity in duct flow,* J. Fluid Mech., vol. **19**, pp. 375–394, (1964).

[9] Cardell, G.S.: *Flow past a circular cylinder with a permeable splitter plate,* Ph.D. Thesis, Graduate Aeronautical Lab. California Inst. of Techn. (1993).

[10] Choi, Y.D., Moon, C., Yang, S.H.: *Measurements of turbulent flow characteristics of square duct with a 180° bend by hot wire anemometer,* Intern. Symposium on Engineering Turbulence Modelling and Measurement, (1990).

[11] Chang, S.M., Humphrey, J.A.C., Modavi, A.: *Turbulent flow in a strongly curved U-bend and downstream tangent of square cross–sections,* Physico Chemical Hydrodynamics, vol. **4**, no. 3, pp. 243–269, (1983).

[12] Demuren, A.O., Rodi, W.: *Calculation of turbulence-driven secondary motion in non-circular ducts,* J. Fluid Mech., vol. **140**, pp. 189 –222, (1984).

[13] Gavrilakis, S.: *Numerical simulation of low–Reynolds–number turbulent flow through a straight square duct*, J. Fluid Mech., vol. **244**, pp. 101–129, (1992).

[14] Germano, M.; Piomelli, U.; Moin, P.; Cabot, W. H.: *A dynamic subgrid–scale eddy viscosity model*, Phys. Fluids A. **3** (7), pp. 1760–1765, (1991).

[15] Gessner, F.B., Emery, A.F.: *The numerical prediction of developing turbulent flow in rectangular ducts*, Trans. ASME I: J. Fluids Engng. **103**, pp. 445–455, (1981).

[16] Härtel, C., Kleiser, L.: *Large-Eddy Simulation of Near-Wall Turbulence*, In this publication, (1996).

[17] Kato, M., Launder, B.E.: *The Modeling of Turbulent Flow around Stationary and Vibrating Square Cylinders*, Proc. 9th Symp. Turb. Shear Flows, Kyoto, 10-4-1, (1993).

[18] Launder, B.E., Ying, W.M.: *Secondary flows in ducts of square cross-section*, J. Fluid Mech., vol. **54**, pp. 289–295, (1972).

[19] Launder, B.E.: *Second-moment closure: present .... and future*, Lecture Series, Introd. to the Mod. of Turb., von Karman Inst. for Fluid Dyn., Belgium, pp. 282–300, (1991).

[20] Lilly, D.K.: *A proposed modification of the Germano subgrid–scale closure method*, Phys. Fluids A **4** (3), pp. 633–635, (1992).

[21] Manhart, M., Wengle, H.: *Large-Eddy Simulation and Eigenmode Decomposition of Turbulent Boundary Layer Flow Over a Hemisphere*, In this publication, (1996).

[22] Martinuzzi, R.: *Experimentelle Untersuchung der Umströmung wandgebundener, rechtiger, prismatisher Hindernisse*, Dissertation, Univ. Erlangen, (1992).

[23] Martinuzzi, R. and Tropea, C.: *The Flow around surface-mounted, prismatic obstacle placed in a Fully Developed Channel Flow*, J. of Fluids Engineering, vol. 115, (1993).

[24] Norberg, C.: it Effects of Reynolds number and low–intensity free–stream turbulence on the flow around a circular cylinder, Publ. No. 87/2, Dept. of Applied Thermosc. and Fluid Mech. Chalmer Univ. of Techn. Gothenburg, Sweden, (1987).

[25] Rhie, C.M., Chow, W.L.: *A numerical study of the turbulent flow past an isolated airfoil with trailing edge separation*, AIAA–J., Vol. 21, pp. 1225–1532, (1983).

[26] Son, J., Hanratty, T.J.: *Velocity gradients at the wall for flow around a cylinder at Reynolds numbers from 5×10^3 to 10^5*, J. Fluid Mech. **35**, 353–368, (1969).

[27] Rodi, W., Ferziger, J.H., Breuer, M., Pourquié, M.: *Status of Large Eddy Simulation: Results of a Workshop*, to be published soon, Workshop on LES of Flows past Bluff Bodies, Rottach–Egern, Tegernsee, Germany, June 26–28, (1995).

[28] Schumann, U.: *Subgrid–scale model for finite–difference simulations of turbulent flows in plane channels and annuli*, J. Comput. Phys., **18**, pp. 376–404, (1975).

[29] Smagorinsky, J.: *General circulation experiments with the primitive equations, I, The basic experiment*, Mon. Weather Rev. **91**, pp. 99–165, (1963).

[30] Madabhushi, R.K., Vanka, S.P.: *Large eddy simulation of turbulence-driven secondary flow in a square duct*, Phys. Fluids A **3** (11), pp. 27–34, (1991).

[31] Werner, H., Wengle, H.: *Large-Eddy Simulation of Turbulent Flow over and around a Cube in a plate Channel*, 8th Symp. on Turb. Shear Flows, (Schumann et al., eds.), Springer Verlag, (1993).

Large-Eddy Simulation and Eigenmode Decomposition of Turbulent Boundary Layer Flow over a Hemisphere

MICHAEL MANHART and HANS WENGLE

Institut für Strömungsmechanik und Aerodynamik, LRT/WE7,
Universität der Bundeswehr München, D-85577 Neubiberg, Germany

Summary

Large-eddy simulation (LES) has been applied to turbulent boundary layer flow over a hemisphere with a rough surface (at $Re_D = 150000$). The shape of the surface-mounted hemisphere is blocked-out within a cartesian non-equidistant grid. The time-dependent inflow condition is provided from a separate LES of a boundary layer developing behind a barrier fence and a set of vorticity generators. Two results from LES with different grid resolution are compared with experimental data from a corresponding wind tunnel experiment. In addition, the data sets from LES have been analysed using proper orthogonal decomposition (POD). From this analysis it can be concluded that the first three fluctuating eigenmodes (representing about 24 % of the turbulent kinetic energy) can be related to the major events in the separated flow behind the flow obstacle.

Introduction

There are *two basic simulation concepts* available to calculate the three-dimensional and time-dependent structure of a turbulent flow. On the one hand, there is the so-called *Direct Numerical Simulation (DNS)* which requires to resolve *all* the relevant scales in a turbulent flow, but its range of application is limited to relatively small Reynolds numbers (often too small from an engineering point of view). On the other hand, there is the so-called *Large-Eddy Simulation* (LES) which is capable to predict directly the spatial and temporal behaviour of at least the large-scale structures of *high* Reynolds number flows, and only the effects of the small-scale motions which cannot be resolved on a given computational mesh need to be modelled with a so-called subgrid-scale model.

In recent years LES has been applied to more complex flow fields, including cases with separation regions with reattaching free shear layers. The case of a fully turbulent flow over a backward facing step has been considered by Arnal and Friedrich [2] for Re=155000 and by Silveira-Neto *et al.* [19] for Re=38000. Akselvoll and Moin [1] also performed an LES for the backward facing step flow, for Re=5100, and their results can be compared with results from a direct numerical simulation (DNS) of Le *et al.* [5]. Yang and Ferziger [24] performed an LES over a periodic arrangement of two-dimensional ribs in a channel for Re=3200 and compared the LES results with a DNS at the same Reynolds number. Werner and Wengle [22] considered the case of a *single* two-dimensional square rib in fully developed channel flow at Re=40000. Murakami *et al.* (1991) [13], Werner and Wengle [23] and Breuer and Rodi [4] used LES to calculate the turbulent flow field over a *single* surface-mounted cube (for Re=84000 and Re=40000, respectively). In the case of a single flow obstacle, a time-dependent inflow condition must be provided, usually from a separate LES of a turbulent boundary layer flow or a turbulent channel flow. In the case of a fully three-dimensional *mean* flow field (no homogeneous directions), long integration times are needed to evaluate the second-order statistics. All the papers mentioned above deal with sharp-edged flow obstacles. In Manhart and Wengle [10] a first attempt was made to extend the range of application of LES to more general geometries, with the curved surfaces still approximated within a cartesian grid. The cartesian grid approach is attractive because it can be implemented efficiently and the extension in the order of accuracy of the discretization to higher than second order is straightforward. The problems with such an approach are linked to the undesirable effects of the stepwise approximation of curved geometry on, for example, the separation angle, or the vortex shedding process. The geometry of the blocked-out shape of a curved flow obstacle introduces an artificial roughness on the curved surface. The effects of surface roughness on the flow around surface-mounted hemispheres have been investigated by Savory and Toy [16, 17, 18] and Toy *et al.* [21] in wind tunnel experiments, at a Reynolds number of $Re_D = U_\infty D/\nu = 150000$ (D is the diameter of the hemisphere, U_∞ is the undisturbed free-stream velocity). From these experiments first-and second-order statistics of the flow field in the near wake is available for comparison with results from LES. The surface of some of the models was roughened using a random coat of spherical glass beads to give a roughness-to-diameter ratio of about 0.01. This leads to flow separation with a so-called supercritical flow behaviour (i.e. the pressure distribution becomes Reynolds number independent above Re=120000). Measurements in three different boundary layers revealed that the wake region is sensitive to the turbulence level and the momentum in the incoming flow. In the experiment, the case 'thin' was a boundary layer which naturally developed along a smooth wall. At the position where (later) the hemisphere was mounted the mean velocity profile and the vertical profile of the longitudinal Reynolds normal stress have been measured (with the model absent). At that location the ratio of boundary layer thickness to obstacle diameter was $\delta/D = 0.5$. We tried to reproduce this case using numerically simulated flow over vorticity generators (as it is usually done in wind tunnel experiments to shorten the length of development of the turbulent boundary layer until a fully turbulend developed boundary layer is attained).

The dynamics of separating flow fields considered in this paper is governed by large scale structures (coherent structures) which behave in a non-universal way. For a deeper understanding of the physics of such flows it is useful to study the spatio-temporal bevaviour of these coherent structures. Lumley [6] proposed the so-called *Proper Orthogonal Decomposition* (POD) method for the identification of coherent structures in turbulent flow fields. In a classical POD, the three-dimensional and time-dependent flow field is expanded into deterministic spatial basis functions and random coefficients in time, and a related eigenvalue problem is solved for the spatial eigenfunctions. From a more general point of view, Aubry et al. [3] derived a deterministic space-time symmetric version of POD to be used as a systematic mathematical tool to study the space and time evolution of a complex system simultaneously. A similar view has been taken by Sirovich [20], leading to alternative formulations of the related basic eigenvalue problems for either the orthogonal spatial modes or the orthogonal temporal modes (see also Manhart and Wengle [9]).

An ideal data base for POD evaluation is the three-dimensional and time-dependent flow field from DNS. For example, Rempfer [14] analysed in his dissertation and in a successor paper [15] the DNS data of a transitional flat-plate boundary layer flow. The

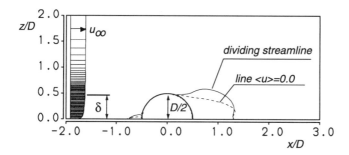

Figure 1: Schematic view of the geometry of boundary layer flow approaching a hemisphere

three-dimensional and time-dependent results from LES can also be analysed by POD (although there are questions to be answered concerning the effects of subgrid-scale modelling on the eigenmodes of POD). Manhart et. al. [7, 8] carried out a POD analysis of the turbulent shear layer above and behind a square rib (at Re=50000), Manhart and Wengle [9] analysed the fully inhomogeneous turbulent flow field over a surface-mounted cube (at Re=50000), and first results on coherent structures in turbulent boundary layer flow over a hemisphere (Re=150000) have been presented by Manhart and Wengle [11].

The main objectives of this paper are: (a) to carry out LES and POD for two significantly different grid resolutions, (b) to analyse the effects of representing a rough and curved surface by simply blocking out the geometry within a cartesian grid, (c) to compare the results with a wind tunnel experiment, and (d) to analyse the vortex-shedding process from the viewpoint of conventional statistics and of the statistics of an eigenmode decomposition.

LES of turbulent flow over a hemisphere

The Navier Stokes equations (top-hat filtered) are solved numerically on a staggered and non-uniform cartesian grid, using second-order finite-differencing in time and space. The problem of pressure-velocity coupling is solved iteratively using a multigrid solver for the Poisson equation. The subgrid scale stresses, arising from the nonlinear convection terms, have been evaluated by the Smagorinsky model (with $c_1 = 0.1$). The geometry of the hemisphere is approximated by simply blocking out the 'body-filled' grid cells within a cartesian grid. For further details of the numerical scheme, see [9, 23].

The geometry of the flow obstacle is evident from figure 1. The origin of the coordinate system is located in the center of the hemisphere. Two cases with different spatial resolutions have been evaluated (in this paper referenced as HEMI2 and HEMI3, respectively). Measured in units of the reference length D of the flow problem (D is the diameter of the hemisphere) the dimensions of the computational domain are for the case HEMI2: (X,Y,Z)=(15.2,4.8,4.0) with (N_X, N_Y, N_Z)=(224,128,65) grid points and for the case HEMI3: (X,Y,Z)=(21.0,4.4,3.3) with (N_X, N_Y, N_Z)=(168,80,48) grid points, i.e. for HEMI2 we used three times as many grid points (1.86 Mio. grid points) in comparison to

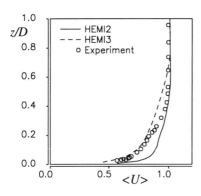
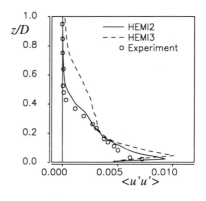

Figure 2: Mean streamwise velocity profile of the undisturbed boundary layer. Lines: LES; symbols: Savory und Toy [16]

Figure 3: Streamwise normal Reynolds stress of the undisturbed boundary layer. Lines: LES; symbols: Savory und Toy [16]

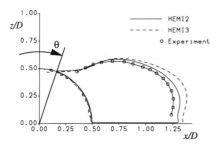
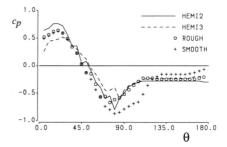

Figure 4: Dividing streamline of the time mean velocity field in the wake of the hemisphere ($y = 0.0$). Lines: LES; symbols: Savory und Toy [16]

Figure 5: Pressure distribution along the surface of the hemisphere ($y = 0.0$). Lines: LES; symbols: Savory und Toy [16]

HEMI3 with 0.65 Mio. grid points. This leads to a minimum grid spacing of $\Delta x_i = 0.015$ in case HEMI2 ($\Delta x_i = 0.03$ in case HEMI3, respectively).

As a time-dependent inflow condition we used the instantaneous velocity profiles from an LES of a turbulent boundary layer ($\delta/D = 0.5$) developing behind a barrier fence and a set of vorticity generators. The quality of this inflow condition has been checked by comparing the *mean* streamwise velocity profile and the streamwise normal Reynolds stress of the separate LES with the results of Savory and Toy [16, 18]. The location of the vertical profiles, shown in figures 2 and 3 (experimental and numerical results), corresponds to the later position of the center of the hemisphere. The difference in the shape of the vertical mean velocity profile (figure 2) can be explained with the presence of a slight pressure gradient in the case HEMI2, which is caused by the no-slip boundary condition at the upper wall. However, in the experiment a zero pressure gradient is provided by tilting the upper wall in the streamwise direction. The mean streamwise velocity in the case HEMI3, where a slip boundary condition has been applied at the

Figure 6: Pressure fluctuations. Top view of isosurfaces $p' = 0.04$ (light grey) and $p' = -0.04$ (dark grey).

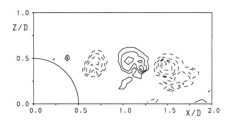

Figure 7: Pressure fluctuations. Isolines in the symmetry plane $y = 0.0$. ——— : 0.04, 0.06, ..., 0.2; — — — : -0.04, -0.06, ..., -0.2.

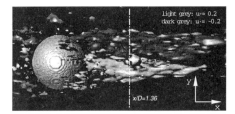

Figure 8: Streamwise velocity fluctuations. Top view of isosurfaces $u' = 0.2$ (light grey) and $u' = -0.2$ (dark grey).

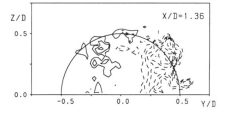

Figure 9: Streamwise velocity fluctuations. Isolines in the plane $x/D = 1.36$ (marked in figure 8). ——— : 0.2, 0.3, ..., 0.8; — — — : -0.2, -0.3, ..., -0.8.

upper wall, coincides better with the measured profile if normalized with the boundary layer thickness. The vertical profile of the normal stress figure 3 is satisfying, with the exception of the peak value close to the wall, which is too large in comparison to the experiment. In the case HEMI3 the boundary layer thickness is about 50% larger than in the corresponding experiment.

A more detailed description of the simulations is given in [10], [12]. The time-mean flow field as well as the second-order moments in the near-wake region of the hemisphere show good agreement between LES and experiment [10]. The separation angle ϕ_s evaluated by the LES ($\phi_s = 110°$) compares well with the measured value ($\phi_s = 105°...110°$). The shape of the dividing streamline and the reattachment length X_R (see figure 4) is predicted within 10% accuracy (LES: $X_R = 1.31$ for HEMI2, $X_R = 1.36$ for HEMI3, experiment: $X_R = 1.25$). This result clearly shows the improvement in simulating the correct shape of the recirculation zone by increasing the number of grid points. Savory and Toy also measured the pressure distributions on the surfaces of smooth and rough hemispheres. From figure 5 it can be concluded that our stepwise approximation of the surface of the hemisphere had the same effect as the roughening of the surface of the wind tunnel model by spherical glass beads. Again, increasing the number of grid points significantly improves the results.

Looking at pressure and velocity fluctuations at the same instant (figures 6, 7, 8 and 9), two different physical processes can be detected in the wake region of the hemisphere.

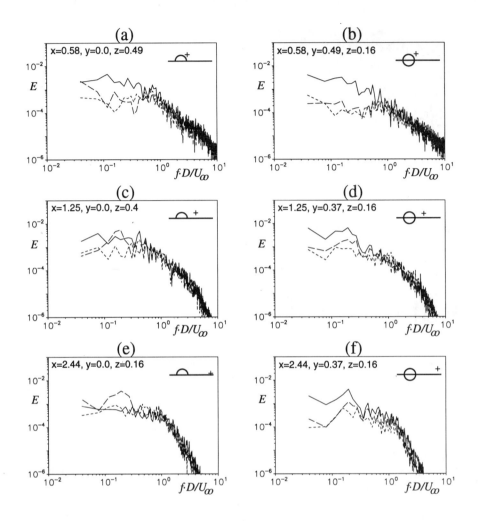

Figure 10: Velocity spectra in the wake of a hemisphere evaluated of simulation HEMI2.
———— : u; — — — : v; - - - - - - : w

Regions of negative pressure fluctuation correspond to vortex cores whereas regions of positive pressure fluctuation indicate the position of stagnation points in the turbulent velocity field. Therefore, the regions of alternating positive and negative pressure fluctuations (figures 6 and 7), being nearly symmetric to the plane $y/D = 0.0$, indicate the presence of symmetric vortex shedding from the top region of the hemisphere. These spanwise vortices would lead to symmetric (relative to the plane $y/D = 0.0$) velocity fluctuations in the streamwise and vertical velocity components. However, the streamwise velocity fluctuations seem to be strongly connected to the alternating vortex shedding from the sides of the hemisphere, and therefore we get the asymmetric view of figure 8.

Figure 10 shows energy spectra from time records (about 3000 samples collected over a dimensionless period of time of $150 \cdot t \cdot U_\infty / D$) at different locations in the recirculation zone (the positions of these spectra are indicated by a cross-mark in each diagram of figure 10). In the immediate neighbourhood of the hemisphere (x/D=0.58) the spectrum of the u-fluctuations shows a significantly higher energy content than the other fluctuating velocity components (up to a Strouhal number $S = f \cdot D/U_\infty = 1.0$). The turbulent energy seems to be produced mainly in the streamwise component, from which it can be concluded that (in a cylindrical coordinate system where the axis points in the streamwise direction) the shear layer exhibits a 2D structure in the region immediately after separation. However, further downstream (x/D=1.25 and 2.44) the lateral v-component in the symmetry plane (y=0.0) has taken up an energy content comparable to (figure 10c) or even higher than (figure 10e) the longitudinal u-component. A frequency peak at $S \approx 0.2$ appears in the symmetry plane for the lateral v-component (figure 10c and 10e) and outside of the symmetry plane (y=0.47) in the longitudinal u-component (figure 10d and 10f). These peaks could be related to the time-periodic separation events at the lateral side surfaces of the hemisphere. A higher frequency peak at $S \approx 0.5$ is observed in the u-fluctuations in the symmetry plane at x/D=1.25 (figure 10c). This frequency peak seems to be related to the roller-like separation events at the top surface region of the hemisphere. At frequencies higher than about S=1.0 an energy decay proportional to $S^{-5/3}$ can be observed within the separation region, whereas in the reattachment zone (around x/D=2.4) a much stronger energy decay is to be observed (figure 10e and 10f). Whether this stronger energy decay is caused or affected by the grid stretching in streamwise direction is not clear yet.

The detailed results show the significant effects of the two dominating separation processes: the separation and roll-up of the shear layer at the top region of the hemisphere (with $S \approx 0.5$) and the alternating (approximately time-periodic) separation of turbulent structures from the side regions of the hemisphere (with $S \approx 0.2$). The strong lateral v-fluctuations in the reattachment zone observed in [10] seem to be caused by the latter process (S=0.2, see figure 10e).

POD of the flow field

The two different LES data sets mentioned above (HEMI2 and HEMI3, respectively) have been analysed by POD. They differed by the spatial resolution within a test volume smaller than the original computational domain. This test volume just covered the regime of increased turbulence energy around the flow obstacle. Within that test volume the flow fields have been filtered (with a top hat filter and a filter width of $2\Delta x_i, i = 1, 2, 3$), thereby decreasing the number of collected data by a factor of eight. In the case HEMI2 about 80000 sampling points in space have been chosen (case HEMI3: 20000 sample points) and in both cases about 3000 samples in time have been collected, e.g. in case HEMI2 the dimensionless time interval was 0.05 units and the total time covered was about 150 units (HEMi3: $\Delta t = 0.1$; $T = 300$).

In our application of POD we follow the treatment proposed by Sirovich [20] and Aubry et al. [3], see also [9]. The turbulent space-time velocity signal $u_i(\vec{x}, t)(i = 1, 2, 3)$ is characterized by N_M spatio-temporal modes which form a complete orthogonal basis for the following expansion:

$$u_i(\vec{x}, t) = \sum_{n=1}^{N_M} a^n(t) \phi_i^n(\vec{x}) \quad . \tag{1}$$

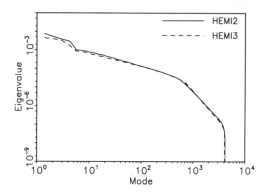

Figure 11: Eigenvalue-spectra of the Karhunen-Loève decomposition. ———— : HEMI2; — — — — : HEMI3

The temporal modes $a^n(t)$ are the eigenvectors of the temporal correlation tensor $C(t,t')$:

$$\int_T C(t,t')a^n(t')dt' = \lambda^n a^n(t) \quad ; \quad C(t,t') = \frac{1}{T}\int_V u_i(\vec{x},t)u_i(\vec{x},t')d\vec{x} \quad . \tag{2}$$

The spatial modes $\phi_i^n(\vec{x})$ are calculated by projecting the velocity fields onto the temporal modes:

$$\phi_i^n(\vec{x}) = \frac{1}{T}(\lambda^n)^{-1}\int_T a^n(t)u_i(\vec{x},t)dt \quad . \tag{3}$$

The POD gives an optimal basis in the sense that, with a given number of modes n, the expansion (1) converges optimally fast. The time-averaged energy content of a spatio-temporal mode n is given by the corresponding eigenvalue λ^n. Therefore, each eigenvalue represents the energy contribution of the corresponding mode, and the sum of the eigenvalues is equal to the total kinetic energy of the flow field. In a statistically stationary turbulent flow field the first mode represents the *mean* flow field, in our case HEMI2 about 96% of the total kinetic energy, and the remaining 4% represent the turbulent kinetic energy (within the test volume selected).

Table 1 shows the separate energy contributions (eigenvalues) of the fluctuating modes (n > 1) to the turbulent kinetic energy and the accumulated contributions for the two data sets available, i.e. case HEMI2 with a fine spatial resolution and case HEMI3 with a coarse spatial resolution. In figure 11, the corresponding eigenvalue spectra are shown, from which it can be concluded that the first three fluctuating modes (n=2,3, and 4) are the (energetically) dominating modes of the flow. They together represent 24% (case HEMI2) and 18% (case HEMI3), respectively, of the turbulent energy. The energy of the higher modes (n = 5, ..., 500) is decaying with a power-law (exponent about -0.75), and for wave numbers above n=500 there is an energy decay corresponding to a power-law with an exponent of about -3.0. These exponents are neither dependent on the grid resolution in the LES nor dependent on the number of time samples taken for calculating the temporal correlation tensor (eq. 2). But they are strongly affected if the POD is performed in a test volume which is too small to capture all regions of the flow field with

Table 1: Contribution of different modes to the turbulent energy.

n	$100 \cdot \lambda^n / \sum_{i=2}^{NM} \lambda^i$		$100 \cdot \sum_{i=2}^{n} \lambda^i / \sum_{i=2}^{NM} \lambda^i$	
	HEMI2	HEMI3	HEMI2	HEMI3
2	12.6	8.6	12.6	8.6
3	6.6	6.6	19.2	15.2
4	4.9	3.3	24.1	18.5
5	1.7	1.6	25.8	20.1
6	1.6	1.6	27.4	21.7
28	0.50	0.47	45.2	38.6
200	0.095	0.11	77.2	73.2
1000	0.0037	0.0041	98.4	98.4
2000	0.00046	0.00045	99.8	99.8

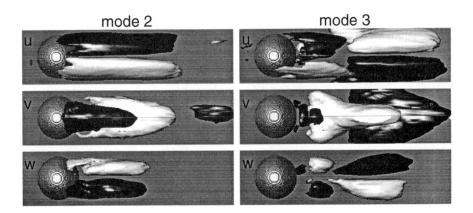

Figure 12: Spatial modes 2 and 3 in the wake of the hemisphere (HEMI2). Top view of isosurfaces ϕ_i(i.e.: u,v,w), light grey: +0.25, dark grey: −0.25

significantly enhanced turbulent energy [12]. However, the mode number at which the exponent of the eigenvalue spectrum changes is determined by the number of samples in time. With 3000 time samples used in this investigation about 1500 modes are required to capture 99% of the turbulent energy. This should be interpreted as a lower bound for the *total* number of degrees of freedom of the flow case considered. Note, that an upper bound for the *total* number of degrees of freedom in a fully turbulent flow field can be estimated as a number proportional to $Re^{9/4}$.

Each term in the expansion (1) can be interpreted as a spatio-temporal mode of the flow (containing energy which corresponds to its eigenvalue) which is a combination of a three-dimensional spatial mode (describing the instantaneous spatial form of a coherent structure) and a temporal mode (describing its temporal evolution).

Spatial Modes. The spatial eigenfunctions $\phi_i^n(\vec{x})$, i=1,2,3 are linked to the three

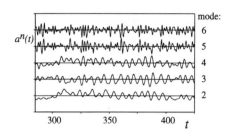
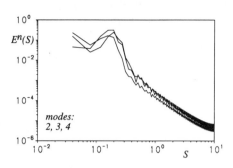

Figure 13: Temporal modes $a^n(t)$, $2 \leq n \leq 6$ in case HEMI2 with fine grid resolution

Figure 14: Spectra of the temporal modes $E^n(S)$, $n = 2, 3, 4$ in case HEMI2 with fine grid resolution

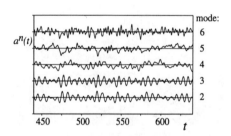
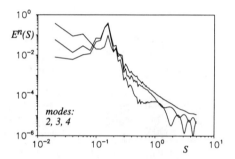

Figure 15: Temporal modes $a^n(t)$, $2 \leq n \leq 6$ in case HEMI3 with coarse grid resolution

Figure 16: Spectra of the temporal modes $E^n(S)$, $n = 2, 3, 4$ in case HEMI3 with coarse grid resolution

velocity components $u_i(\vec{x}, t)$, i=1,2,3 and represent the separate contribution of each mode to the spatial structure of the turbulent flow. Figure 12 shows the three-dimensional isosurfaces of mode 2 and mode 3. Mode 4 is similar to mode 3 but with a phase shift in x-direction. These spatial modes are anti-symmetric for the longitudinal u- and the vertical w-component and symmetric for the lateral v-component of the flow. The spatial structure of the first three fluctuating modes and their dominating energy content indicate their relation to the separation events at the lateral sides of the flow obstacle. Each of them represents a large-scale vortex similar to the 'von Kármán' vortices known from laminar flow.

Temporal Modes. Figure 13 shows the temporal modes $a^n(t)$, $2 \leq n \leq 6$ of the case HEMI2 in a time window between 290 and 420 dimensionless time units $t \cdot U_\infty/D$. The first three fluctuating modes (n=2,3, and 4) exhibit (for the time greater than 350 time units) a quasi-periodic behaviour with a period of about 5 units and with a temporal phase shift between them. The corresponding frequency spectra $E^n(S)$ of these three modes in figure 14 have a peak at a Strouhal number $S = fD/U_\infty \approx 0.2$. It is an obvious advantage

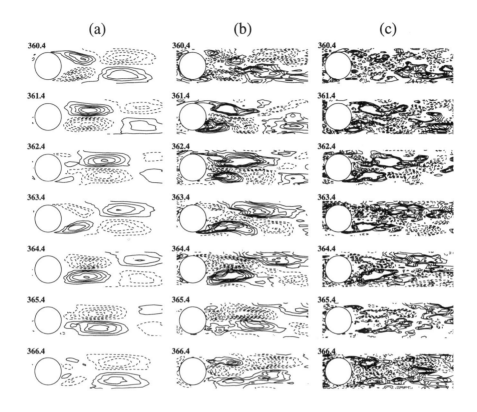

Figure 17: Approximation of streamwise fluctuations with different numbers of modes during the quasi-periodic phase: a) $\sum_2^4 \phi_1^n$ b) $\sum_2^{28} \phi_1^n$ c) $\sum_2^{200} \phi_1^n$

of POD that this typical frequency which has been observed in the time spectra of the velocities (not present in the spectra for the higher modes) can now be directly related to the corresponding basic eigenflows and their common (dominating) contribution to the separation processes in the turbulent flow field analysed.

The quasi-periodic behaviour just described is interrupted by shorter periods of time during which the intensity of the time variations (the amplitude) is significantly lowered, and no positive/negative periodic behaviour can be observed. We call this different phase of activity in the flow field a quasi-stationary phase. The total times of these quasi-stationary phases vary between 10 and 20 reference time units, whereas the total times of the quasi-periodic phases lasts about 100 time units. In our case HEMI2, 100 time units correspond to about 40000 time steps of the LES and, of course, it would be desirable to cover a larger number of these intervals of quasi-periodic and quasi-stationary behaviour to evaluate a more accurate statistics.

Spatio-temporal Modes. Looking at spatial and temporal modes (with the same eigenvalue) *simultaneously*, we consider just *one* complete term in the expansion (1), or

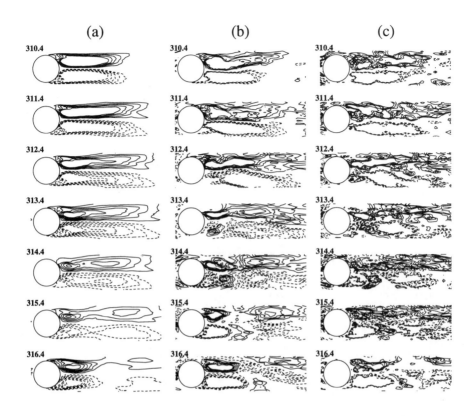

Figure 18: Approximation of streamwise fluctuations with different numbers of modes during the quasi-stationary phase: a) $\sum_2^4 \phi_1^n$ b) $\sum_2^{28} \phi_1^n$ c) $\sum_2^{200} \phi_1^n$

we consider the sum over a selected range of mode numbers n. In the figures 17 and 18 we present three series of isolines of the streamwise velocity fluctuations as a result of accumulating the effects of (a) the three dominating fluctuating modes n=2,3, and 4, (b) the first 27 fluctuating modes, and (c) the first 199 fluctuating modes. The more modes we include in our expansion (1) the better the small details of the original field from LES are reconstructed. If we compare the details of the isolines from 199 modes (representing 77% of the turbulent energy) with the corresponding result from LES (equivalent to 3000 modes) a significant difference cannot be observed.

The results of figure 17 are typical for the quasi-periodic phase of the separation process, and in figure 18 the corresponding results for the quasi-stationary phase are presented. The typical behaviour of the dominating flow field in the quasi-stationary phase is totally different from that in the quasi-periodic phase: during the (longer lasting) quasi-periodic phase we observe the classical separation of counter-rotating vortices from the flow obstacle, whereas during the quasi-stationary we note the creation and separation of larger vortices not changing the sign of rotation. However, it is possible that the sign of rotation changes and a new series of equally rotating vortices is created (see, e.g. in figure 13 the change in sign of $a_2(t)$ at the dimensionless time 305).

Effects of the grid resolution. As mentioned above the grid resolution of the LES has a negligible effect on the decay of the eigenvalue spectrum at higher mode numbers. However, comparing the energy content of the first three fluctuating modes, which represent the 'von Kármán' vortex shedding, one can see a significant difference between the two simulations (see table 1). As shown in [12] the spatial modes ϕ_i^n, $n = 2, 3, 4$ form in both simulations the same subspace, i.e. they represent the same vortices. Comparing in the figures 13 and 15 the behaviour of the temporal modes, the most significant conclusion is that in the case HEMI3 with coarse spatial resolution the so-called quasi-stationary behaviour of the dominating modes (n=2,3, and 4) cannot be observed. This leads to a more pronounced peak in the time spectrum of the case HEMI3 (compare figure 16 with figure 14). Looking again at the time series of the streamwise velocity component (figures 17 and 18) it is obvious that the large scale structures in the near wake originate from very small structures in the immediate neighbourhood of the separation line on the surface of the hemisphere.

Conclusions

From a comparison of two results from LES, using different grid resolutions and comparing both results with experimental data, the following conclusions can be drawn:

(a) Using different grid resolutions, the finer spatial resolution usually gives better agreement with the experiment. Therefore, a further improvement of the results can be achieved by further increasing the number of grid points.

(b) The blocking out of the curved surface of the flow obstacle (a hemisphere) within a cartesian grid allows to simulate a roughened surface with a desired roughness-to-diameter ratio. However, the numerical simulation of a very small roughness (important in many technical applications) still remains an unsolved problem.

(c) The results from LES show good agreement with experimental data for the first and second order statistics. From the instantaneous flow and pressure fields, and from time spectra at different locations in the recirculation zone, two significant mechanisms in the separation process can be identified: the separation and roll-up of the curved shear layer in the top region of the hemisphere (goverened by a Strouhal number $S \approx 0.5$) and the alternating (approximately time-periodic) separation from the side regions (with $S \approx 0.2$).

(d) With the aid of the Karhunen-Loève decomposition (POD) of the turbulent flow field it is possible to identify these two different mechanisms, to describe precisely their spatial and temporal bevaviour, and to evaluate their energetic significance. The first three fluctuating modes only (representing about 24% of the turbulent energy) are sufficient to characterize the major events in the separation process and in the total recirculation zone.

(e) Within the framework of POD it comes out that longer phases (about 100 time units) of quasi-periodic behaviour are interrupted by shorter phases (about 10 to 20 time units) of quasi-stationary behaviour. This typical behaviour can also be represented by the first three fluctuating modes. However, in the coarse grid LES, the quasi-stationary behaviour can hardly be detected.

(f) The interaction between quasi-periodic and quasi-stationary behaviour of the flow is influenced by smaller scale (and less energetic) processes.

(g) Within the framework of POD, it is possible to identify and to quantitatively evaluate: the interactions between the different Karhunen-Loève modes, the production of turbulent energy, the convective transport by the mean flow field, the interaction with the pressure field, the diffusive processes caused by the *molecular* viscosity and by the *subgrid-scale* viscosity. Therefore, in principle, it will be possible to answer still open questions about the effects of the subgrid-scale modelling on the dynamics of the large coherent structures which are directly represented in an LES.

Acknowledgements: This work has been supported by the German Research Society (DFG), Priority Research Program 'Flow Simulation With High-Performance Computers', Project No. We 705/3 (Wengle/Römer). We also gratefully acknowledge the support by the computing center of the Universität der Bundeswehr München (UniBwM) and by the Leibniz Computing Center (LRZ) of the Bavarian Academy of Sciences.

References

[1] Akselvoll, K. and Moin, P. (1993): Large-eddy simulation of a backward facing step flow, in: W. Rodi and F. Martelli (eds.) *Engineering Turbulence Modelling and Experiments 2*, Elsevier Science Publ., Amsterdam.

[2] Arnal, M. and Friedrich, R. (1992): Large eddy simulation of a turbulent flow with separation, 8th Symp.Turb.Shear Flows, September 9-11, 1991, Munich, Germany, in: F. Durst et al.(eds.) *Turbulent Shear Flows 8*, Springer-Verlag, 1993.

[3] Aubry, N., Guyonnet, R. and Lima, R. (1991): Spatio-temporal analysis of complex signals: theory and applications, J. Statist. Phys. 64(3/4), pp. 683-739.

[4] Breuer, M. and Rodi, W. (1996): Large eddy simulation for complex turbulent flows of practical interest, In this publication.

[5] Le, H., Moin, P. and Kim. J. (1993): Direct numerical simulation of turbulent flow over a backward facing step, 9th Symp.Turb.Shear Flows, August 16-18, 1993, Kyoto.

[6] Lumley, J.L. (1970): *Stochastic Tools in Turbulence*, Academic Press, New York.

[7] Manhart, M., Wengle, H., Schmid, P. and Werner, H. (1993): Eigenmode decomposition of the turbulent shear layer above a square rib, Applied Scientific Research 51, pp. 359-364, and in: F.T.M. Nieuwstadt (ed.) *Advances in Turbulence IV*, Kluwer Academic Publishers, Dordrecht, The Netherlands, 1993.

[8] Manhart, M. and Wengle H. (1993): Eigenmode decomposition of the turbulent velocity and vorticity fields above a square rib, in: E.H. Hirschel (ed.) *Flow Simulation with High-Performance computers I*, pp. 186-200, Vieweg Verlag, Braunschweig, 1993.

[9] Manhart, M. and Wengle, H. (1993): A spatio-temporal decomposition of a fully inhomogeneous turbulent flow field, Theoret.Comput.Fluid Dynamics 5, pp. 223-242.

[10] Manhart, M. and Wengle, H. (1994): Large-edy simulation of turbulent boundary layer flow over a hemisphere, First ERCOFTAC Workshop on 'Direct and Large-Eddy Simulation', March 27-30, Guildford, England, in: P.R. Voke et al. (eds.) *Direct and Large-Eddy Simulation I*, Kluwer Academic Publs. , Dordrecht, The Netherlands, 1994.

[11] Manhart, M. and Wengle (1994): Coherent structures in turbulent boundary layer flow over a hemisphere, Fifth European Turbulence Conference, July 5-8, 1994, Siena, Italy. in: Benzi (ed.) *Advances in Turbulence V*, Kluwer Academic Publishers, Dordrecht, The Netherlands, 1993.

[12] Manhart, M (1995): Umströmung einer Halbkugel in turbulenter Grenzschicht - Grobstruktursimulation und Eigenmodeanalyse der Ablöseprozesse, Dissertation (in german), Universität der Bundeswehr München.

[13] Murakami, S., Mochida, A. and Hayashi, Y (1991): Scrutinizing $k - \epsilon$ EVM and ASM by means of LES and wind tunnel for flowfield around cube, 8th Symp.Turb.Shear Flows, September 9-11,1991, Munich, Germany.

[14] Rempfer, D. (1991): Kohärente Strukturen und Chaos beim laminar-turbulenten Umschlag, Dissertation (in german), Universität Stuttgart.

[15] Rempfer, D. and Fasel, H. (1994): Evolution of three-dimensional coherent structures in a flat-plate boundary layer, J.Fluid Mech. 260, pp. 351-375.

[16] Savory, E. and Toy, N. (1986a): Hemispheres and hemisphere-cylinders in turbulent boundary layers, J. Wind Eng. and Ind. Aerodyn. 23, pp. 345-364.

[17] Savory, E. and Toy, N. (1986b): The flow regime in the turbulent near wake of a hemisphere, Experiments in Fluids 4, pp. 181-188.

[18] Savory, E. and Toy, N. (1988): The separated shear layers associated with hemispherical bodies in turbulent boundary layers, J.Wind Eng. and Ind. Aerodyn. 28, pp. 291-300.

[19] Silveira-Neto, A., Grand, D., Metais, O. and Lesieur, M. (1991): Large eddy simulation of the turbulent flow in the downstream region of a backward-facing step, Phys.Rev. Letters 66, No. 18, pp. 2320-2323.

[20] Sirovich. L. (1987): Turbulence and the dynamics of coherent structures: I, II, III, Quart. Appl. Math. 5, pp. 561-590.

[21] Toy, N., Moss, W. D. and Savory, E. (1983): Wind tunnel studies on a dome in turbulent boundary layers, J. Wind Eng. and Ind. Aerodyn. 11, pp. 201-212.

[22] Werner, H. and Wengle, H. (1989): Large-eddy simulation of turbulent flow over a square rib in a channel, 7th Symp.Turb.Shear Flows, August 21-23, 1989, Stanford, USA, in: H. H. Fernholz and H. E. Fiedler (eds.) *Advances in Turbulence 2*, Springer, Berlin, 1989.

[23] Werner, H. and Wengle, H. (1991): Large-eddy simulation of turbulent flow over and around a cube in a plate channel, 8th Symp.Turb.Shear Flows, September 9-11,1991, Munich, Germany, in: F. Durst et al. (edis.) *Turbulent Shear Flows 8*, Springer, Berlin, 1993.

[24] Yang, K.-S. and Ferziger, J. H. (1993): Large-eddy simulation of turbulent obstacle flow using a dynamic subgrid-scale model, AIAA J., Vol.31, No. 8, pp. 1406-1413.

III. MATHEMATICAL FOUNDATIONS, GENERAL SOLUTION TECHNIQUES AND APPLICATIONS

INTRODUCTION TO PART III

by

M. Meinke

Aerodynamisches Institut
RWTH-Aachen
Wüllnerstraße zw. 5 und 7, 52062 Aachen

The development of efficient and accurate numerical methods for the solution of the Navier-Stokes equations is, and always has been, of the same importance and success as the development of computer hardware. The numerical solution consists of the algorithmic core, which usually solves a linear or linearized system of equations, the grid arrangement, discretization schemes and the physical modelling of transport phenomena. Papers which are concerned with these topics in a general sense are presented in this part. The main emphasis of these papers varies from mathematical aspects – e. g. the formulation of equations, boundary conditions and consistent and stable discretization schemes – the development of solution techniques – e. g. adaptive grid concepts, solution algorithms and physical modelling – to the application to certain physical problems. In the following a few remarks are made about the contents of some papers.
One of the problems to be solved in the near future will be the accurate prediction of chemically reacting flows, e. g. in piston engines or burning chambers. In such flows the Mach number varies from the incompressible limit to supersonic conditions, which renders methods for compressible flows inefficient. Geratz et. al. used an asymptotic analysis of the Euler equations in order to adapt an explicit algorithm for moderate and large Mach numbers also for an application to weakly compressible flows. The solution process may be accelerated by this approach, nevertheless the storage requirements for an application in three space dimensions still remain high, due to additional unknowns arising in the modelling of chemistry and turbulence. Solution adaptive grids are therefore an indispensable tool for flows with small spatial scales, like reaction zones, shock waves, and also vortex structures in unsteady flows. They allow a reduction of the total number of grid points without diminishing the accuracy of the solution. Hentschel and Hirschel propose a hierarchically ordered system for block-structured grids, while Vilsmeier and Hänel developed adaptation concepts for unstructured meshes composed of tetrahedrons, even for unsteady wakes. The question whether an unstructured or structured grid arrangement is better suited for flow simulations cannot be answered in general. When unstructured grids are applied, it is, for example, more difficult to formulate efficient solution schemes. This is one topic of the investigations of Roehl and Simon, who present an implicit method, and Greza et. al., who implemented a multigrid method and a domain splitting technique for unsteady flows. Another disadvantage of unstructured grids

is the inaccurate evaluation of the viscous stresses in highly anisotropic regions of the mesh, e. g. in boundary layers. In both papers hybrid, structured-unstructured grids are used to alleviate this problem for the simulation of flows in turbomachines. An application of unstructured meshes with moving boundaries is given by Wierse. She developed an upwind scheme for the simulation of compressible flow in a two-stroke engine with a solution of the Euler equations.

Other applications presented are the computation of the aeroelastic deformations of a Fokker wing with a coupling of a solution of the Euler equations with a finite element method for the wing structure by Nellessen et. al., and the simulation of hypersonic flows for reentry vehicles by Brück et. al. and Laurin and Wiesbaum. The physical modelling of ionized gases at the first point of a re-entry trajectory has been investigated by Steiner. He proposes modifications to a weighted particle method, which lead to a conservation of the relevant physical quantities and a stable and convergent scheme. Lilek et. al. applied different discretization schemes to the simulation of laminar and turbulent compressible and incompressible flows.

The difficulties in the solution of the Navier-Stokes equations for incompressible flows originate in the decoupling of the pressure from the velocity field. Several approaches exist how to solve the equations for this case. Xue et. al. solved the Navier-Stokes equations in a two stream function Euler Potential formulation with an incomplete lower-upper decomposition. Blazy et. al. propose a combination of a pressure correction method with the method of characteristics, while Weimer et. al. investigated different solution schemes for the artificial compressibility concept. A comparison of the accuracy and efficiency of the different approaches can be found in the benchmark computations of part IV in this book.

NAVIER-STOKES APPROXIMATIONS IN HIGH ORDER NORMS

R. Rautmann
Universität-GH Paderborn, Fachbereich Mathematik-Informatik
Warburger Str. 100, D-33098 Paderborn, Germany

SUMMARY

This note presents recent results on the convergence in high order norms of approximate solutions to the Navier-Stokes initial-boundary value problem: For a semi-discret splitting scheme using a 2-grid approach with Lagrangean transport stepping, H^2-convergence has been proved in [32]. To Rothe's semi-discret approximation scheme, H^2-convergence has been proved in [29], and $H^{2,p}$-convergence in [34] with explicit convergence rates established in [35].

For parallelization methods to related splitting schemes, see Blazy, Borchers and Dralle's contribution in this volume, for time splitting in kinetic gas equations, see Steiner's contribution.

1. INTRODUCTORY REMARKS ON THE NAVIER-STOKES APPROXIMATION PROBLEM

The task, numerically to approximate Navier-Stokes flows, leads us to several exciting mathematical questions: On the one side, the regularity of the real flow should be as high as possible, not only in order to get high convergence rates, but also to ensure convergence of the approximations as well as the convergence even of the boundary values of their (spatial) gradients. Namely, the forces (which the engineers are mainly interested in) acting on a rigid body in a viscous flow are caused by the production of vorticity (i.e. of velocity gradients) at the body's surface. Therefore in any case the boundary values of the gradients of the approximations should approximate the boundary values of the gradient of the real flow velocity. By the trace theorems, H^2-convergence of the approximations would be sufficient (i.e. L^2-convergence of the approximations together with their first and second spatial partial derivatives).

Beyond it the main problem of computational flow analysis is the loss of stability in any known approximation scheme with increasing Reynolds number. Naturally this will be unavoidable as far as it reflects the transition to turbulence of the real flow. Since artificial viscosity would decrease the actual Reynolds number of the calculated flow, the damping influence of the natural flow viscosity (which stems from the diffusion part $-\Delta u$ in the Navier-Stokes equations) should be used as far as possible in order to overcome the destabilizing effect of the transport part (expressed by the nonlinear convection term).

Taking this aim, Lighthill [18] and later Chorin and many other authors [8, 9, 22, 1, 40, 5, 6, 16, 26, 3, 4, 32] proposed to split the evolution in time of the flow into alternating diffusion- and convection steps, the latter using the particle paths following the flow velocity field. Thus a Lipschitz condition for the flow velocity (or a L^∞-bound for its gradient) would be needed for the wellposedness of the numerical approach. By Sobolev's imbedding theorem, norm bounds in the Sobolev space $H^{2,q}$ would be sufficient with $q > 3$.

On the other side, if we assume more than a critical degree of the flow's regularity (specified in [25, 47]), the initial value of the flow velocity has to fulfil a non-realistic compatibility condition [17 p. 277, 281-282]. Therefore the optimum regularity of Navier-Stokes solutions and with it the strongest norms in which they can be approximated uniformly on a compact time interval are well defined.

In the following we will present recent results on the convergence of Navier-Stokes approximations in high order norms which just allow to avoid the non-realistic compatibility condition mentioned above. Our main issue will be to find out strongest norms in which uniform convergence of Navier-Stokes approximations on a compact time interval can be achieved realistically. For the question of optimum convergence rates see [7, 37]

2. H^2-CONVERGENT PRODUCT FORMULA SCHEME WITH LAGRANGEAN TRANSPORT STEPPING(A 2-GRID APPROACH FROM [32])

2.1 NOTATIONS

The velocity $u(t,x) = (u_1, u_2, u_3)$ and the pressure $p(t,x) \geq 0$ of a viscous incompressible flow at time $t \geq 0$ in an open bounded set $\Omega \subset \mathbb{R}^3$, having a sufficiently smooth boundary $\partial\Omega$, solve the Navier-Stokes initial-boundary value problem

$$\begin{aligned} \tfrac{\partial}{\partial t}u - \Delta u + u \cdot \nabla u &= -\nabla \tilde{p}, & \nabla \cdot u &= 0 & \text{in } \Omega, \quad t > 0, \\ u_{|\partial\Omega} &= 0, & u_{|t=0} &= u_0, \end{aligned} \quad (2.1)$$

if the outer forces have a potential $\bar{p} = \tilde{p} - p$ and if we put the mass density and the viscosity constant equal to 1. In the (complex) Hilbert-spaces H^m which contain all measurable vector functions in Ω having square integrable spatial partial derivatives of all orders smaller or equal to m (i.e. in $L^2(\Omega) = H^0$), the divergence freeness $\nabla \cdot u = 0$ is weekly formulated in the subspace

$$H = \text{ closure of } D \subset L^2(\Omega) \text{ in the } L^2 - \text{norm},$$

D denoting the space of test functions φ having compact support in Ω and spatial partial derivatives of all orders, φ being divergence free.

The condition of adherence $u_{|\partial\Omega} = 0$ is weakly expressed in the subspace

$$V = \text{ closure of } D \subset H^1 \text{ in the } H^1 - \text{norm}.$$

By

$$P : L^2(\Omega) \to H$$

we denote H. Weyl's orthogonal projection which sends into zero exactly the generalized gradients. The Stokes operator $A = -P\Delta$ is the closure in H of the projected Laplacean $-P\Delta$. Its domain is $D_A = H^2 \cap V$. The operator A being selfadjoint and positiv definite, its fractional powers A^α with domains $D_{A^\alpha} \subset H$ are defined for any real $\alpha \geq 0$. The compact imbedding

$$D_{A^\beta} \hookrightarrow D_{A^\alpha}$$

holds, if $\alpha < \beta$. Applying P formally on (2.1) we find the Navier-Stokes evolution equation

$$\frac{\partial}{\partial t}u + Au + Pu \cdot \nabla u = 0, \qquad u(0) = u_0 \qquad (2.2)$$

for the vector function $u(t,\cdot)$ taking its values in D_A. Finally for any Banach space B and any real interval J we denote by $C^0(J, B)$ the linear space of uniformly bounded continuous functions $f : J \to B$.

In any convection - diffusion splitting scheme to the initial value problem (2.2) we have to give special attention to the first step modelling the momentum transport along the particle paths $X(t, t_0, x_0), t \geq 0$. We calculate them from the system of ordinary differential equations

$$\frac{d}{dt}X = u(t, X), \qquad X(t_0, t_0, x_0) = x_0$$

where $x_0 \in \bar{\Omega}$. The Lipschitz condition for u (which gives the uniqueness of the solution X) would follow from the assumption $u \in C^0([0,T], H^3)$ which, however, would require the non-realistic compatibility condition for the initial value $u(0, \cdot)$ mentioned above. In order to avoid this difficulty, in our approximation scheme we will use a mollified version $P(u^* \cdot \nabla u)$ of the convective term, containing the Yosida approximation

$$u^* = (1 + rA)^{-2}u, \qquad r > 0.$$

2.2 THE PRODUCT FORMULA SCHEME

On a grid of time points $t_k = k \cdot \epsilon, \epsilon = \frac{T}{K}, k = 0, 1, ..., K$, (the "rough grid"), we will work with the semi-implicit linearization scheme

$$\begin{aligned}
\partial_t u^\epsilon + Au^\epsilon + Pu_k^{\epsilon*} \cdot \nabla u^\epsilon &= 0, \qquad t \in J_k = (t_k, t_{k+1}] \\
u^\epsilon(t_k) &= u_k^\epsilon, \\
u_k^{\epsilon*} &= (1 + hA)^{-2}u_k^\epsilon, \\
u_{k+1}^\epsilon &= u^\epsilon(t_{k+1}).
\end{aligned} \qquad (2.3)$$

To solve numerically each linear problem (2.3) is not a trivial task, because it contains the non constant term $u_k^{\epsilon*}$. We will solve (2.3) by a transport-diffusion splitting on a fine grid with step length $h = \frac{\epsilon}{N}$:

$$t_{k,n} = t_{k,0} + nh, t_{k,0} = t_k, \quad n = 0, 1, ..., N, \; k = 0, \ldots, K-1,$$

$$t_{k+1,0} = t_{k,N}.$$

The transport- or Euler step

$$(E_h) \qquad E_{k,h} v_{k,n} = P(v_{k,n} \circ X_{k,h}) \qquad (2.4)$$

with

$$X_{k,h} : \bar{\Omega} \to \bar{\Omega}, \quad X_{k,h}(x) = x - hv_{k,0}^*(x), \quad x \in \bar{\Omega} \qquad (2.5)$$

gives approximate solutions of the (linearized) initial value problem

$$\frac{\partial}{\partial t}v + Pv_{k,0}^* \cdot \nabla v = 0, \quad v(t_{k,n}, \cdot) = v_{k,n}$$

which describes the transport of v along the particle lines $X(t, t_{k,n}, x)$ from

$$\frac{d}{dt}X = v_{k,0}^*(X), \qquad (2.6)$$

$$X(t_{k,n}, t_{k,n}, x) = x.$$

From the strong regularity of $v_{k,0}^*$ (including its Lipschitz continuity) and the vanishing of $v_{k,0}^*$ at the boundary $\partial\Omega$ we conclude that also $X_{k,h}$ (like any $X(t, t_{k,n}, \cdot)$) represents a continuously differentiable bijective map of Ω and $\bar{\Omega}$ on itself, respectively, for all $h \in (0, h_0]$ and sufficiently small $h_0 > 0$. Therefore $E_{k,h}v_{k,n} \in H$ is well defined. Then by means of the subsequent diffusion- or Stokes step

$$(S_h) \qquad S_h v = (1 + hA)^{-1} v \qquad (2.7)$$

we define the product formula scheme

$$(P) \qquad \begin{aligned} v_{k,n+1} &= S_h E_{k,h} v_{k,n} \\ &= (S_h E_{k,h})^{n+1} (S_h E_{k-1,h})^N ... (S_h E_{0,h})^N v_{0,0}, \\ n &= 0, ..., N-1, \text{ where } v_{0,0} = u_0. \end{aligned} \qquad (2.8)$$

Due to the divergence freeness of $v_{k,0}^*$, the map $X_{k,h}$ is measure preserving up to $O(h^2)$. Therefore the m-acretivity of the operator A shows that the composed mappings $S_h E_{k,h}$ are quasi-contractive in $L^2(\Omega)$.

For proving the H^2-convergence of the scheme (P), in [32] we have estimated the error of the local linearization scheme (2.3), and the error of Rothe's scheme with respect to the scheme (2.3). Then by comparison of the product formula scheme (P) with Rothe's scheme we have proved

THEOREM 2.1: Assume $u \in C^0([0,T], D_{A^{1+\varsigma}})$ denotes a solution of the Navier-Stokes initial-boundary value problem (2.2). Then all approximations $v_{k,n} \in D_{A^{1+\varsigma}}$ from (P) exist for all sufficiently large values of K, N and are converging to $u(t_{k,n})$ in H^2 uniformly in $k = 1, ..., K$, $n = 1, .., N$, with $K, N \to \infty$.

Also explicit convergence rates in H, H^1 and H^2 have been established in [32]. An important tool in the mathematical proofs are the fractional powers $A^\alpha, \alpha \in \mathbb{R}$, of the Stokes operator.

3. EXPLICIT CONVERGENCE RATES OF ROTHE'S SCHEME TO THE NAVIER-STOKES EQUATIONS IN $H^{2,q}$

By definition (S_h) of the Stokes step, the product formula scheme (P) is equivalent to

$$(1+hA)v_{k,n+1} = Pv_{k,n} \circ X_{k,h}, \qquad (3.1)$$

$$v_{k+1,0} = v_{k,N}, n = 0, ..., N-1, k = 0, ..., K-1.$$

The argument $X_{k,h}$ on the right hand side represents the first order approximation (2.5) of the particle paths from (2.6). Replacing it by a semi-implicit approximation to the convective term in the Navier-Stokes equation we find the more straightforward first order scheme of Rothe

$$(1+hA)u_{k,n+1} = u_{k,n} - hPu_{k,0}^* \cdot \nabla u_{k,n+1}, \qquad (3.2)$$

$$u_{k+1,0} = u_{k,N}$$

which is useful to prove the convergence of (3.1), [32].

The Yosida approximation $u_{k,0}^*$ of $u_{k,0}$ was needed in [32] only in order to have well defined particle paths in the direction field $u_{k,0}^*$ even in case $u_{k,0} \in H$ (which does not imply any Lipschitz condition or pointwise continuity). As we have shown in [29, 31], convergence of Rothe's scheme holds in high order norms also without use of the additional smoothing by the Yosida approximation.

Next we will approximate a given solution $u \in C^0([0,T], D_{A^{1+\varsigma}})$ of (2.2) on the grid

$$t_k = k \cdot h, \quad k = 1, ..., K$$

with time step length $h = \frac{T}{K}$ by means of Rothe's scheme

$$(1+hA)v_k^h = v_{k-1}^h - Pv_{k-1}^h \cdot \nabla v_k^h. \qquad (3.3)$$

Using energy estimates and tools from semigroup theory, in [31] we have proved

THEOREM 3.1: *Let $u \in C^0([0,T]), D_A)$ denote a solution of the Navier-Stokes initial value problem (2.2) with right hand side $Pf = 0, v_k^h$ the approximations from (3.3). Assume the initial values $v_0^h \in D_A$ are uniformly bounded in H^2 and satisfy*

$$||v_0^h - u_0|| \le ch$$

with a constant $c \ge 0$. Then all $v_k^h \in D_A$ exist, and for any $h \in (0, h_0], h_0$ being sufficiently small, the error estimates

$$||v_k^h - u(t_k)|| \le b_0 h,$$
$$||A^\beta(v_k^h - u(t_k))|| \le b_\beta h^{1-\beta}$$

for $\beta \in [0,1]$ hold uniformly in $k = 1, ..., K \ge \frac{T}{h_0}$. The constants b_0, b_β depend on $\sup_{t \in [0,T]} ||Av(t)||, c, T$, the constant b_β additionally on β, too.

For Navier-Stokes solutions $u \in C([0,T], D_{A^{1+\varsigma}}), \varsigma \in (0, \frac{1}{4})$, even convergence rates in H^2 have been established in [29, Theorem II].

As we have pointed out in the beginning, uniform bounds for the boundary values of the gradients ∇v_k^h and their norms $||\nabla v_k^h||_{L^\infty(\Omega)}$ would be highly desirable, but (in the L^2-framework which we used until now) unfortunately they would lead again to the non-realistic compatibility condition just mentioned. Therefore now we will study (2.2) in the framework of the Lebesgue spaces $L^q = L^q(\Omega), q \geq 2$. By the well known results of Fujiwara-Morimoto [12] the space $L^q = X^q \oplus G^q$ is the direct sum of the spaces

$$X^q = \text{closure of } D \subset L^q(\Omega) \text{ in the norm of } L^q(\Omega),$$

$$G^q = \text{closure of } \{\nabla \varphi | \varphi \in C^1(\bar{\Omega})\} \subset L^q(\Omega) \text{ in the norm of } L^q(\Omega).$$

Let
$$P_q : L^q \to X^q$$
denote the projection along the space G^q of generalized gradients.

Since Ω is bounded, for the Sobolev spaces $H^{m,q}(\Omega)$ with norms

$$||g||_{H^{m,q}} = \{ \sum_{|\alpha| \leq m} \int_\Omega |\partial^\alpha g(x)|^q dx \}^{\frac{1}{q}},$$

where $\alpha = (\alpha_1, \alpha_2, \alpha_3), \alpha_j = 0, ..., m, |\alpha| = \sum_{j=1}^3 \alpha_j, m \in \mathbb{N}, 1 \leq q < \infty$, the imbedding $H^{m,r} \hookrightarrow H^{m,q}$ holds if $q \leq r$. We will write $H^{m,2} = H^2, H^{0,q} = L^q$. Thus the construction of $P_r u = u - \nabla \varphi$ in [12, p.694] shows that we have

$$P_r = P_{q|L^r} = P_{2|L^r}$$

if $2 \leq q \leq r, P_{2|L^r}$ denoting the restriction of P_2 to L^r. Since we will restrict us to spaces L^q with $q \geq 2$, we can write $P = P_2$. The projection P is bounded on each Sobolev space $H^{m,q}, m \in \mathbb{N}, q \in [2, \infty), [47, \text{p. XXIII}]$. As usual for $s \in (0, m)$ we denote by $H^{s,q}$ the complex interpolation space between $H^{m,q}$ and L^q. $H^{s,q}$ is independent of $m > s$, for details see [45]. In addition $\overset{\circ}{H}{}^{s,q}$ stands for the closure in $H^{s,q}$ of the subspace C_c^∞ of C^∞ - vector functions which have compact support in Ω.

A formal application of the projection P on (2.1) again leads us to a Navier-Stokes evolution problem:

$$\partial_t u + A_q u + P u \cdot \nabla u = P f, \quad t > 0, \quad u(0) = u_0 \qquad (3.4)$$

for a strong solution $u(t) \in D_{A_q}$, where now the Stokes operator A_q is the closure of $-P\Delta$ in the space X^q (with $A_2 = A$, see above). Its domain is $D_{A_q} = H^{2,q} \cap \overset{\circ}{H}{}^{1,q} \cap X^q$. The resolvent estimates in [13, 14, 39, 47] show that - A_q generates the bounded holomorphic semigroup $e^{-tA_q}, t > 0$, acting on X^q. In addition, the fractional powers A_q^α with dense domains $D_{A_q^\alpha} \hookrightarrow X^q$ are defined for all real $\alpha > 0$. In the case $\alpha < \beta$, the domain $D_{A_q^\beta}$ is dense in $D_{A_q^\alpha}$, the imbedding $D_{A_q^\beta} \hookrightarrow D_{A_q^\alpha}$ being compact.

Let $(v_k^h), h = \frac{T}{K}, k = 0, ...K, K = 1, 2, ...,$ be the sequence of Rothe approximations $v_k^h \in D_{A_2}, ||v_k^h||_{H^2} \leq M$, from (3.3) which exist by Theorem 3.1 and which are of first order convergent to $u(kh)$ in L^2, uniformly with respect to k and h. Starting with v_k^h we will find a sequence (u_k^h) of approximations to u in $H^{2,q}$ from the linearized scheme

$$\frac{u_k^h - u_{k-1}^h}{h} + A_q u_k^h = -P v_{k-1}^h \cdot \nabla v_k^h = f_k, \tag{3.5}$$

$$k = 1, ..., K, \quad u_0^h = v_0^h.$$

Namely, after having found suitable bounds for f_k in X^q, inductively by interpolation methods we see the existence of $u_k^h \in D_{A_q} \hookrightarrow D_{A_2}$ and conclude $u_k^h = v_k^h$ from the uniqueness of the linear Stokes resolvent boundary value problem in D_{A_2}. Then using Ashyralyev and Sobolevskii's coercivity inequality [2 p. 93, (2.11)] we find uniform $H^{2,q}$ bounds for u_k^h. Following these lines, in [34] we have established

THEOREM 3.2: *For some $q \in [2, \infty)$ and $\zeta \in (0, 1/2q)$, let $u \in C^0([0, T], D_{A_q^{1+\zeta}})$ denote a solution of (3.4) with right hand side $Pf = 0$ and initial value $u_0 \in D_{A_q^{1+\zeta}}$. Then the approximations $v_k^h = u_k^h$ in Rothe's scheme (3.3), (3.5) with initial value $v_0^h = u_0$ converge in $H^{2,q}$ (and even in $D_{A_q^{1+\eta}} \hookrightarrow H^{2,q}$ for all $\eta \in [0, \zeta)$) to $u(k \cdot h)$ with $K \to \infty$ uniformly in $k = 1, ..., K$ and $h = T/K$.*

By interpolation between $L^2(\Omega)$ and $H^{2,q}$ with the help of a multiplicative inequality [10,p.27 Theorem 10.1], using a bootstrap argument, in [35] we have proved

COROLLARY 3.1: *Under the assumptions of Theorem 3.2, the approximations $u_k^h = v_k^h$ in Rothe's scheme (3.3), (3.5) with initial value $v_0 \in D_{A_q^{1+\zeta}}$ fulfil the error estimates*

$$\|v_k^h - u(kh)\|_{L^q} \leq b_0 h^{\kappa_0},$$

$$\|v_k^h - u(kh)\|_{H^{2,q}} \leq b_1 h^{\kappa_1},$$

$$\|v_k^h - u(kh)\|_{L^p} \leq b_2 h^{\kappa_2} \text{ for any } p \in (2, q],$$

where

$$\kappa_0 = \frac{4(1+\zeta)}{7+4\zeta-6/q},$$

$$\kappa_1 = \kappa_0 \frac{\zeta}{1+\zeta},$$

$$\kappa_2 = \frac{1}{1/2-1/q}\{1/p - 1/q + (1/2 - 1/q)\kappa_0\}.$$

The constants b_0, b_1, b_2 depend on $\sup_{t \in [0,T]} \|A_q^{1+\zeta} u(t)\|_{L^q}, T, q$, the constant b_2 additionally on p, too.

REFERENCES

[1] Alessandrini, G., Douglis, A., Fabes, E.: An approximate layering method for the Navier-Stokes equations in bounded cylinders, Ann. Math. Pura Appl. 135 (1983) 329-347.

[2] Ashyralyev, A., Sobolevskii, P.E.: Well-posedness of parabolic difference equations, Operator Theory Advances and Applications 69, Birkhäuser Basel, Boston, Berlin 1994.

[3] Beale, J.T.: The approximation of the Navier-Stokes equations by fractional time steps, Lecture at the conference: The Navier-Stokes equations, theory and numerical methods, Oberwolfach 18. - 24.8.91

[4] Beale, J.T., Greengard, C.: Convergence of Euler-Stokes splitting of the Navier-Stokes equations, IBM Research Report RC 18072 6/11/92, Commun. Pure Appl. Math. XL VII(1994) 1083-1115.

[5] Borchers, W.: A splitting algorithm for incompressible Navier-Stokes equations, in: H. Niki, M. Kawahara (Eds.): Int. conf. on computational methods in flow analysis, Okayama, Japan (1988) 454-461.

[6] Borchers, W.: On the characteristics method for the incompressible Navier-Stokes equations, in: E.H. Hirschel (Ed.): Finite Approximations in Fluid Mechanics II, Notes on Numerical Fluid Mechanics, Volume 25, Braunschweig 1989.

[7] Borchers, W.: Zur Stabilität und Faktorisierungsmethode für die Navier-Stokes-Gleichungen inkompressibler viskoser Flüssigkeiten, Habilitationsschrift für das Fach Mathematik im Fachbereich Mathematik-Informatik der Universität-GH Paderborn, November 1992.

[8] Chorin, A.J.: Numerical study of slightly viscous flow, J. Fluid Mechanics 57 (1973) 785-796.

[9] Chorin,A.J., Hughes, T.J.R., Mc Cracken, M.F., Marsden, J.E.: Product formulas and numerical algorithms, Comm. Pure Appl. Math. XXXI (1978) 205-256.

[10] Friedman, A.: Partial differential equations, Holt, Rinehart and Winston, New York 1964.

[11] Fujita, H.: On the semidiscrete finite element approximation for the evolution equation $u_t + A(t)u = 0$ of parabolic type, in Miller, J.J.(Ed.): Topics in Numerical Analysis III, Academic Press New York (1977) 143-157.

[12] Fujiwara, D., Morimoto, H.: An L_r -theorem of the Helmholtz decomposition of vector fields, J.Fac. Sci. Univ. Tokyo Sect. IA Math. 24 (1977) 685-700.

[13] Giga, Y.: The Stokes operator in L_r spaces, Proc. Japan Acad. 57 Ser. A (1981) 85-89.

[14] Giga,Y.: Analyticity of the semigroup generated by the Stokes operator in L_r spaces, Math. Z. 178 (1981) 297-329.

[15] Giga, Y.: Domains of fractional powers of the Stokes operator in L_p spaces, Arch. Rat. Mech. Anal. 89 (1985) 251-265.

[16] Hebeker, F.K.: Analysis of a characteristics method for some incompressible and compressible Navier-Stokes problems, Preprint 1126 des FB Mathematik, TH Darmstadt (1988)

[17] Heywood, J.G.,Rannacher, R.: Finite element approximation of the nonstationary Navier-Stokes problem I, Siam. J. Num. Anal. 19 (1982) 275-311.

[18] Lighthill, M.J.: Introduction, Boundary layer theory, in: Rosenhead, L.(Ed.), Laminar boundary layers, Oxford (1963) 46-113.
[19] Masuda, K.: Remarks on compatibility conditions for solutions of Navier-Stokes equations, J. Fac. Sci. Univ. Tokyo Sect. IA Math. 34 (1987) 155-164.
[20] Miyakawa, T.: On the initial value problem for the Navier-Stokes equations in L^p spaces, Hiroshima Math. J. 11 (1981) 9-20.
[21] Miyakawa, T.: On nonstationary solutions of the Navier-Stokes equations in an exterior domain, Hiroshima Math. J. 12 (1982) 115-140.
[22] Pironneau, O.: On the transport diffusion algorithm and its applications to the Navier-Stokes equations, Num. Math. 38 (1982) 309-332.
[23] Rautmann, R.: On the convergence rate of nonstationary Navier-Stokes approximations, in: Proc. IUTAM Symp. Paderborn 1979, Springer Lecture Notes in Math. 771 (1980) 435-449.
[24] Rautmann, R.: A semigroup approach to error estimates for nonstationary Navier-Stokes approximations, Proc. Conference Oberwolfach 1982, Methoden Verfahren Math. Physik 27 (1983) 63-77.
[25] Rautmann, R.: On optimum regularity of Navier-Stokes solutions at time $t = 0$, Math. Z. 184 (1983) 141-149.
[26] Rautmann, R.: Eine konvergente Produktformel für linearisierte Navier-Stokes-Probleme, Z. Angew. Math. Mech. 69 (1989) 181-183.
[27] Rautmann, R.: A convergent product formula approach to three dimensional flow computations, Finite Approximations in Fluid Mechanics II (1989) 322-325.
[28] Rautmann, R.: H^2-convergent linearizations to the Navier-Stokes initial value problem, in: Butazzo, G. Galdi, G.P. Zanghirati, L., (Eds.), Proc. Intern. Conf. on "New developments in partial differential equations and applications to mathematical physics", Ferrara 14. - 18. October 1991, Plenum Press New York (1992) 135-156.
[29] Rautmann, R.: H^2-convergence of Rothe's scheme to the Navier-Stokes equations, Journal of Nonlinear Analyis 24 (1995) 1081-1102.
[30] Rautmann, R.: Optimum regularity of Navier-Stokes solutions at time $t = 0$ and applications, Acta Mech. (1993) supp. 4, 1-11.
[31] Rautmann, R.: A remark on the convergence of Rothe's scheme to the Navier-Stokes equations, to appear in : Stability and Applied Analysis of Continuous Media 3 (1993).
[32] Rautmann, R., Masuda, K.: H^2-convergent approximation schemes to the Navier-Stokes equations, Comm. Math. Univ. Sancti Pauli 43 (1994) 55-108.
[33] Rautmann, R.: A direct construction of very smooth local Navier-Stokes solutions, Acta Appl. Math. 37 (1994) 153-168.
[34] Rautmann, R.: A regularizing property of Rothe's method to the Navier-Stokes equations, in: A. Sequeira (Ed.): Navier-Stokes Equations and Related Nonlinear Problems, Plenum Press New York (1995) 377-391.
[35] Rautmann, R.: Convergence rates in $H^{2,q}$ of Rothe's method to the Navier-Stokes equations, to appear.
[36] Rothe, E.: Zweidimensionale parabolische Randwertaufgaben als Grenzfall eindimensionaler Randwertaufgaben, Math. Ann. 102 (1930) 650-670.

[37] Rodenkirchen, J.: On optimum convergence rates of the Crank-Nicholson scheme to the Stokes initial value problem in higher order function spaces using realistic data. Thesis Paderborn 1995.

[38] Solonnikov, V.A.: On differential properties of the solutions of the first boundary-value problem for nonstationary of Navier-Stokes equations, Trudy Mat. Inst. Steklov 73 (1964) 221-291, Transl.: British Library Lending Div., RTS 5211.

[39] Solonnikov, V.A.: Estimates for the solutions of nonstationary Navier-Stokes equations, Zap. Nauchn. Sem. Leningrad Lemingrad Mat. Steklova 38 (1973) 153-231, J. Sov. Math. 8 (1977) 467-529.

[40] Süli, E.: Lagrange-Galerkin mixed finite element approximation of the Navier-Stokes equations, in: K.W. Morton and M.J. Baines (eds.): Numerical Methods for Fluid Dynamics, Oxford University Press (1985) 439-448.

[41] Süli, E.: Convergence and nonlinear stability of the Lagrange-Galerkin method for the Navier-Stokes equations, Numer. Math. 53 (1988) 459-483.

[42] Süli, E., Ware, A.: Analysis of spectral Lagrange-Galerkin method for the Navier-Stokes equations, in: Heywood, J.G., Masuda, K., Rautmann, R., Solonnikov, V.A., (Eds.), The Navier-Stokes Equations II, Springer Lecture Notes in Mathematics 1530 (1992) 184-195.

[43] Tanabe, H.: Equations of evolution, Pitman London 1979.

[44] Temam, R.: Navier-Stokes equations, North-Holland, Amsterdam 1977, rev. ed.1979, 3rd. rev.ed. 1984.

[45] Triebel, H.: Interpolation theory, function spaces, differential operators, North-Holland Amsterdam 1978.

[46] Varnhorn, W.: Time stepping procedures for the nonstationary Stokes equations, preprint 1353, Technische Hochschule Darmstadt 1991.

[47] Wahl, W. von: The equations of Navier-Stokes and abstract parabolic equations, Vieweg Braunschweig 1985.

Supported by Deutsche Forschungsgemeinschaft.

Parallelization Methods for a Characteristic's Pressure Correction Scheme

S. Blazy, W. Borchers, U. Dralle

Universität Paderborn, Fachbereich 17, Warburgerstr. 100, 33098 Paderborn

Summary

Pressure correction schemes and the method of characteristics are combined to obtain numerical approximation procedures for incompressible Navier-Stokes flows on massive parallel computers. The projection step is carried out with a new more efficient parallel preconditioned conjugate gradient method and for the computation of the characteristics we use a fast local Crank-Nicholson solver for linear finite elements over unstructered grids.

1. Introduction

In this paper we present a combination of the well-known method of characteristics [4, 5, 15, 18, 21, 24, 27] with the pressure correction method discussed in [10, 11, 12, 28]. The resulting approximation procedure is a special version of the general factorization or splitting method (see for example [1, 4, 5, 6, 10, 11, 12, 15, 25]).
By these methods the full Navier-Stokes equations are decomposed into simpler problems which allow the efficient construction of optimal parallelization methods with preconditioned conjugate gradient methods based on domain decomposition, local multigrid solvers (see Section 3) and a new method for the computation of the characteristics of the flow field (see Section 4).
It is well known that the method of characteristics reflects the hyperbolic nature of the equations for strongly convection dominated flows very well, i. e. this method remains stable at higher Reynolds numbers. It takes into account the upwind direction automatically and for such flows the discretization error in characteristic direction is usually smaller than in any other direction with a significant component perpendicular to a boundary layer (for example "t"-direction in instationary boundary layers). The dissipative nature is taken into account with a subsequent Poisson-resolvent solver for which efficient parallel solvers can be constructed. Moreover, it is shown (see Rautmann's contribution [24]) that these splitting schemes admit convergence in higher order norms.
After summarizing the known methods of domain decomposition in Section 3, we also present a new more efficient version, called CGBI-method. In a joint work with R. Friedrich we will apply the parallel Poisson solver to turbulent pipe flows (see also [16]). Certain implementation techniques are given in Section 4.

2. The characteristic's pressure correction formulation

An incompressible viscous flow in a smoothly bounded domain $\Omega \subset \mathbb{R}^n (n = 2, 3)$ is governed by the Navier-Stokes equation given in form (with normalized kinematic viscosity)

$$\left. \begin{array}{r} \frac{D}{Dt}v - \nu \Delta v + \nabla p = 0 \\ div\ v = 0 \end{array} \right\} \text{ in } \Omega, \tag{2.1}$$

completed by the boundary condition

$$v = 0 \text{ on } \partial\Omega \text{ (adhesion on the boundary } \partial\Omega)$$

and the initial condition $v(0) = v^0$. The symbol $\frac{D}{Dt}$ above denotes the Lagrangian (or material) derivative given by

$$\frac{D}{Dt}v = \left[\frac{d}{dt}v \circ X\right] \circ X^{-1}, \tag{2.2}$$

where X denotes the characteristic of v, i.e. $X = X(t, s, x)$ is the unique solution of the ordinary initial value problem

$$\frac{d}{dt}X = v \circ X := v(t, X), \qquad X(s, s, x) = x \qquad (x \in \Omega). \tag{2.3}$$

$X^{-1} = X(s, t, x)$ in (2.2) is the inverse mapping of X. We denote by v^n, p^n the approximation of v, p at time $t = t_n$. Then, assuming v^n, p^n to be known, the backward Euler discretization of the Lagrangian derivative leads to the following discrete equations for determining v^{n+1}

$$\left.\begin{array}{rcl}\frac{v^{n+1} - v^n \circ X_n^{-1}}{\Delta t} - \nu \Delta v^{n+1} + \nabla p^{n+1} &=& 0 \\ \operatorname{div} v^{n+1} &=& 0\end{array}\right\} \text{ in } \Omega \tag{2.4}$$
$$v^{n+1} = 0 \text{ on } \partial\Omega,$$

where Δt is the time step-size and X_n is the characteristic of v^n, i.e. the solution of the autonomous initial value problem associated with v^n. More accurately, we can also extrapolate v^n linear in time, but then the corresponding ordinary initial value problem is no longer autonomous. This will become important in Section 4.2.

The idea of the pressure correction method is to neglect first the pressure in (2.4). As a consequence, the resulting new velocity v^{n+1} violates the incompressibility constrain. Therefore, one has to project subsequently onto a subspace of divergence free functions. This projection is equivalent to a Poisson problem for the pressure. A more refined and also absolutely stable procedure results, if the actual pressure p^{n+1} in (2.4) is replaced by the pressure p^n already known from the old time step. In that case one gets a Poisson equation for the difference $\Pi = p^{n+1} - p^n$ (pressure correction).
This leads to the following splitting scheme:

Step 1: With v^n, p^n known, compute $v^n \circ X_n^{-1}$.

Step 2: Compute $v^{n+1/2}$ from the elliptic boundary value problem

$$\begin{array}{rcll}\frac{v^{n+1/2} - v^n \circ X_n^{-1}}{\Delta t} - \Delta v^{n+1/2} + \nabla p^n &=& 0 & \text{in } \Omega \\ v^{n+1/2} &=& 0 & \text{on } \partial\Omega.\end{array} \tag{2.5}$$

Step 3: Set $v^{n+1} = v^{n+1/2} - \Delta t \nabla \Pi$, $p^{n+1} = p^n + \nabla \Pi$,
where the pressure correction Π solves the Poisson equation ($\varepsilon = \Delta t$)

$$\begin{array}{rcll}-\varepsilon \Delta \Pi &=& -\operatorname{div} v^{n+1/2} & \text{in } \Omega \\ \frac{\partial \Pi}{\partial \nu} &=& 0 & \text{on } \partial\Omega \quad (\nu = \text{exterior normal on } \partial\Omega).\end{array} \tag{2.6}$$

The Poisson equation for Π in Step 3 is obtained by requiring the velocity v^{n+1} to be divergence free and tangential at $\partial\Omega$. Note also, that the boundary value problem for determining $v^{n+1/2}$ is a Poisson resolvent type problem and thus can be solved by the same code provided the different boundary conditions are taken into account. However, for sufficiently small time steps the system matrix corresponding to Step 2 is strongly diagonal dominant. So, the bottleneck in the above scheme is Step 3, for which an efficient parallelization method will be given in the next section.

3. Construction of the preconditioner

We consider the following problem resulting from Section 2. (pressure-correction-step).

$$\begin{aligned} -\Delta u &= f \text{ in } \Omega \\ u &= 0 \text{ on } \partial\Omega, \end{aligned} \qquad (3.1)$$

where we assumed homogeneous Dirichlet boundary conditions for simplicity [1]. Here, u denotes the unknown scalar valued solution of (3.1) and Ω is a bounded domain in $I\!R^2$ with a piecewise smooth boundary $\partial\Omega$. Multiplying (3.1) by $\phi \in H_0^1(\Omega)$ we find with familiar arguments $u \in H_0^1(\Omega)$ such that

$$a(u, \varphi) = (f, \varphi) \qquad (3.2)$$

for all $\varphi \in u \in H_0^1(\Omega)$. The standard Galerkin approximation leads to

$$\sum_{i=1}^{N} u_i a(\psi_i, \psi_j) = (f, \psi_j) \qquad (3.3)$$

for $j = 1, \ldots, N$ and for all $\psi_i \in Z_0^h$, where $Z_0^h := Z_0^h(\Omega)$ is a finite element subspace of $H_0^1(\Omega)$ consisting of continuous piecewise linear polynomials and ψ_i is a basis. It is well known that the condition number of the matrix $A = a(\psi_i, \psi_j)$ is growing of the order $O(h^{-2})$ as the step-size h tends to zero. Therefore, the basic iteration methods like Jacobi, Gauß-Seidel, SOR or CG are getting quickly inefficient. The problem is to to find a matrix B, which fulfills the following properties: First, the solution of $Bv = g$ should be easier to obtain and second B should be spectrally close to A. Therefore, we consider the equivalent system $B^{-1}Au = B^{-1}f$. In [8], B is represented by another bilinear form b satisfying $\gamma_0 b(V,V) \leq a(V,V) \leq \gamma_1 b(V,V)$ for all $V \in Z_0^h$. We shortly describe the main idea of this method. Let

$$a_i(u,v) = \int_{\Omega_i} \nabla u \cdot \nabla v \, dx \text{ then } a(u,v) = \sum_i a_i(u,v), \qquad (3.4)$$

where $\Omega = \bigcup_{i=1}^n \Omega_i$ (see fig. 1). In the situation of figure 1, we first decompose $W \in Z_0^h(\Omega)$ in the following way. $W = W_P + W_H$, where $W_P \in \bigoplus_i Z_0^h(\Omega_i)$ satisfies

$$a_i(W_P, \phi) = a_i(W, \phi), \text{ for all } \phi \in Z_0^h(\Omega_i) \qquad (3.5)$$

and $W_H \in Z^h(\Omega_i)$ is discrete harmonic

$$a_i(W_H, \phi) = 0, \text{ for all } \phi \in Z_0^h(\Omega_i). \qquad (3.6)$$

---
[1] These boundary condition for the pressure appears on the outflow boundary if natural boundary conditions are used.

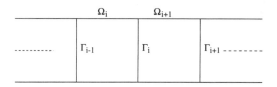

Figure 1: Partition of Ω

Thus, we have

$$a(W, \phi) = a(W_P, \phi_P) + a(W_H, \phi_H). \tag{3.7}$$

If $a(W_H, \phi_H)$ in (3.7) is replaced by $(l_i^{\frac{1}{2}} W_E, \phi_E)_{\Gamma_i}$ ($l_i = -\Delta_{\Gamma_i}$ is the Laplace-Beltrami operator on Γ_i with homogeneous Dirichlet boundary conditions) then the resulting bilinear form b is spectrally equivalent to a (see [8]). The term $(l_i^{\frac{1}{2}} W_E, \phi_E)_\Gamma$ may be additionally replaced by

$$J(W_E, \phi) = \int_{\tilde{\Gamma}} \int_{\tilde{\Gamma}} \frac{(W_E(x) - W_E(y))(\phi(x) - \phi(y))}{|x-y|^2} \, dx \, dy \tag{3.8}$$

without destroying the spectral equivalence. Here, $\Gamma \subsetneq \tilde{\Gamma}$ and the functions W_E, ϕ in (3.8) are assumed to be extended by zero. Note, that for $\phi \in Z_0^h(\Omega)$ we have $\phi_{|\Gamma} \in Z_0^h(\Gamma)$.

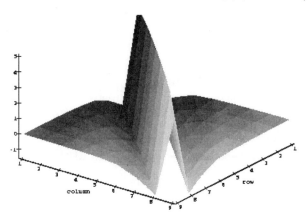

Figure 2: Matrixplot of C

For rectangular cuts we know the eigenvalues of the Laplace - Beltrami operator and the Fast Fourier Transform can be used for realizing the fractional power of the Laplace-Beltrami operator. However, even for 2D-problems this is inefficient if the elliptic operator has strongly varying coefficients. Denoting by $\psi_1, ..., \psi_M$ those basis functions whose supports intersect with Γ, we have for $W_E = \sum_{i=1}^M \alpha_i \psi_i$, $\phi = \sum_{i=1}^M \beta_i \psi_i$

$$J(W_E, \phi) = \sum_{i,j=1}^M C_{i,j} \alpha_i \beta_j \tag{3.9}$$

with the matrix C defined by $C_{ij} = J(\phi_i, \phi_j)$ $(i,j = 1, ..., M)$. The matrix C is dense (see Fig. 2) and the sum over all row entries tends to zero if $h \to 0$. If we replace C by the tridiagonal matrix $\tilde{C}$ defined by

$$\tilde{C}_{i,j} = \begin{cases} 2 + \alpha h & \text{for i=j} \\ -1 & \text{for } |i-j| = 1 \\ 0 & \text{for } |i-j| > 1, \end{cases} \quad (3.10)$$

then clearly the spectral equivalence is destroyed. However, one can show that the corresponding preconditioned CG-iteration has a contraction-number of order $1 - O(h^{\frac{1}{4}})$. Improvements adapted to the hierarchical discretization structure of the multigrid method as well as extensions to operators with variable coefficients or to unstructured grids are possible. On the other hand, our numerical tests (see Section 5) have shown that the use of $\tilde{C}$ is very efficient for space step-sizes up to 10^{-3}.

If artificial crosspoints are introduced by the partition (see Fig. 3), we use an additional decomposition. We decompose W_H as $W_H = W_E + W_V$, where $W_V \in Z_h(\Omega_i)$ is discrete a_i-harmonic whose values on $\partial \Omega_i$ are linear functions along each artificial boundary Γ_{ij} with same values W at the vertices. $W_E \in Z_h(\Omega_i)$ is discrete $\tilde{a}_i$-harmonic and $W_E|_{all\ vertices} = 0$. Let

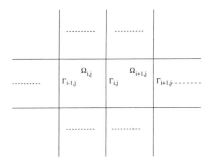

Figure 3: Partition with interior artificial crosspoints

$$\Theta(W_V, \phi_V) := \sum_{\Gamma_{i,j}} (W_V(v_i) - W_V(v_j))(\phi_V(v_i) - \phi_V(v_j)). \quad (3.11)$$

Then, B can be defined by the bilinear form

$$b(W, \phi) := a(W_P, \phi) + J(W_E, \phi_E) + \Theta(W_V, \phi_V). \quad (3.12)$$

The spectral equivalence is proved in [8] with condition number $O(ln(h^{-1}))$. With our approximation for C we get $O(h^{-\frac{1}{2}})$. For more details see [7, 8]. The algorithm to solve $Bv = g$, which is introduced in the Pre-CG algorithm (see Section 4), is then given by the following steps:

1. Find W_P on each Ω_i. This is can be done in parallel (with sequential multigrid methods for example) by solving

$$a(W_P, \phi) = (g, \phi) \text{ for all } \phi \in Z_0^h(\Omega_i)$$

on each subdomain Ω_i

2. Find W_E by solving
$$J(W_E, \phi) = (g, \phi) - a(W_P, \phi)$$
on each Γ_{ij}.

3. Find W_V by solving
$$\Theta(W_V, \phi) = (g, \phi) - a(W_P, \phi)$$
for all interior vertices.

4 With W_E and W_V known we solve
$$a(W_P, \phi) = 0 \text{ for all } \phi \in Z_0^h(\Omega_i).$$

Then the solution is given by $v = W_P + W_H$. Note that Step 2 above requires the solution of $CW_E = \hat{f}$ with the dense matrix C. In the next section we show how one can avoid this additional linear system.

3.1 The conjugate gradient boundary iteration

We will now present a new preconditioning technique [7], called "conjugate gradient boundary iteration" which is shown to be more efficient than the method described previously. The name "boundary iteration" coined from the fact, that the unknowns are distributed only on the artificial cuts Γ_i. For simplicity we will restrict us to the first case (see Fig.1) without interior crosspoints. The main idea is to represent the solution u as the sum

$$u = v_i + u_i \tag{3.13}$$

on each subdomain Ω_i, where v_i is the solution of the following pre-step

$$\begin{aligned}
-\Delta v_i &= f \text{ in } \Omega_i \\
v_i &= \text{ on } \partial\Omega_i - (\Gamma_{i-1} \cup \Gamma_i) \\
\frac{\partial v_i}{\partial \nu} &= 0 \text{ on } (\Gamma_{i-1} \cup \Gamma_i).
\end{aligned} \tag{3.14}$$

Then obviously u_i has to be harmonic on Ω_i. Denoting by φ_i the unknown normal derivative of u_i and requiring its continuity leads to the following problem for determining u_i

$$\begin{aligned}
-\Delta u_i &= 0 \text{ in } \Omega_i \\
u_i &= 0 \text{ on } \partial\Omega_i - (\Gamma_{i-1} \cup \Gamma_i) \\
\frac{\partial u_i}{\partial \nu} &= -\varphi_{i-1} \text{ on } \Gamma_{i-1} \\
\frac{\partial u_i}{\partial \nu} &= \varphi_i \text{ on } \Gamma_i,
\end{aligned} \tag{3.15}$$

with obvious modification of the boundary conditions on Ω_1 and Ω_n. The method is now to determine $\varphi = (\varphi_1, \ldots, \varphi_n)$ on the boundaries Γ_i (see Fig. 1). This is done by minimizing the quadratical functional

$$J(\varphi) = \sum_i \int_{\Gamma_i} (v_i - v_{i+1} + \frac{1}{2}(u_i(\varphi) - u_{i+1}(\varphi))\varphi_i \, do \to Min., \tag{3.16}$$

with u_i depending linearly on φ_{i-1} and φ_i. One can show that J can be extended to a continuous functional $J : H^{-\frac{1}{2}}(\Gamma) = (H_{0,0}^{\frac{1}{2}}(\Gamma))^* \to \mathbb{R}$, where Γ is the union of all Γ_i (see [20] for the definition of these spaces).

Since J (resp. J'') is coercive

$$\gamma_0 \|\varphi\|_{-\frac{1}{2}}^2 \leq J''(\varphi) \leq \gamma_1 \|\varphi\|_{-\frac{1}{2}}^2 \quad \text{for all } \varphi \in H^{-\frac{1}{2}}(\Gamma) \ (0 < \gamma_0, \gamma_1), \tag{3.17}$$

the above minimization problem has a unique solution φ_0. From

$$0 = J'(\varphi_0)\varphi = \sum_i \int_{\Gamma_i} \{(v_i + u_i(\varphi_0)) - (v_{i+1} + u_{i+1}(\varphi_0))\}\varphi_i \, do \text{ for all } \varphi \in H^{-1/2}$$

it follows that $v_i + u_i$ is continuous on Γ_i. On the other hand, the normal derivative is continuous as well by the above construction. Therefore the ansatz $u_{|\Omega_i} = v_i + u_i$ leads to the solution. We note that the constants γ_0, γ_1 depend only on the local norms of the corresponding trace and extension operators in $H^1(\Omega_i)$, $(i = 1, \ldots, n)$. To apply the conjugate gradient algorithm to the above minimization problem, we identify $H^{-1/2}$ with $H_{0,0}^{1/2}$ by the Riesz representation. Using this representation, the gradient of J is simply given by $C(u_i - u_{i+1})$ on Γ_i (discretized version). This leads to an pseudo algorithm similar to Step 1-4 but with C^{-1} replaced by C in Step 2. This is an important advantage compared to first method since the inversion problem with a dense matrix is avoided. The second improvement is that only one elliptic problem has to be solved on each subdomain per iteration step instead of two in the first method. Moreover, the matrix-vector multiplication is carried out only on the unknowns distributed on the artificial boundaries. This explains the much better results very well (see Section 5, Table 1). However, our numerical comparison displays that the above algorithm is only between two and three times faster. The reason for this is that our local multigrid method on Ω_i is less efficient for boundary value problems of mixed type. We note that this disadvantage vanishes if the original problem is of mixed type, since in that case the problems from Step 1 and Step 4 are also of this type. Improvements are under current investigation.

4. Implementation on massive parallel architectures

In this section we describe the implementations of the previous algorithms on massive parallel computers. The aim of this part is to design a flexible tool for fluid simulation combining the most advanced methods from mathematics and computer science to support highly efficient parallel numerical simulation including comfortable pre- and post-processing, meshgeneration and gridpartitioning for finite element codes on parallel supercomputers as well as on workstation clusters. These tools are implemented in a modular way with small and clearly defined interfaces, mainly on file-basis.

Tcl/Tk (Tool command language) is used to develop a portable prototype user-interface with a two-dimensional domain editor which allows the specification of arbitrary domains including boundary conditions, modification of meshes, modification of mesh-partitions and visualization of results [14].

In the next section's we describe the implementation of our preconditioning methods on structured grids and the implementation of the characteristic's method.

4.1 Implementation of the non-preconditioned CG-method and preconditioned CG-method

The simplest way for the partitioning of the domain Ω is to decompose it into stripes (see Fig. 4). This can be very useful to compute flows in long channels and other domains

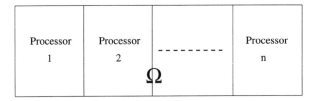

Figure 4: Decomposition of Ω (2D)

with a high ratio of the particular directions. For this type of decomposition a very simple communication structure follows. The link-structure is defined by pointers to the next and previous processor, which allows to use a short procedure to describe communication structure during the operations with high communication.

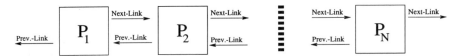

Figure 5: Link-Structure

Another decomposition of Ω (see Fig. 6) must be used if we want to test the preconditioning method with inner nodes described in Section 3.

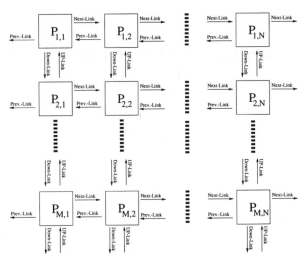

Figure 6: Link-Structure

To get also for this test problem a procedure which easily handles the communication structure, we decide to implement a grid-structure for the domain decomposition. The grid-structure can be implemented by using the same techniques as in the case of the stripe-decomposition. The bandwidth for the communication is only the half as in the case of the decomposition into stripes, because we have additionally to define the up and down direction for the communication.

In both cases we decompose the domain in subproblems with respect to the same number of gridpoints in the subdomains. For large numbers of unknown values we get good balanced processorload, because there is the same amount of mathematical operations for the subproblems. Even so the number of operations for the evaluation of the boundary condition gives another load, the load-difference between interior subdomains and boundary-subdomains is negligible small.

For the implementation of the parallel conjugate gradient method (similar the preconditioned method) for solving (3.1), we consider the following algorithm:

CG-Algorithm

CHOOSE $x_0, r^0 = Ax^0 - b \ ; \ p^1 = -r^0 \ ; \ \delta_0 := <r^0, r^0>$
IF $\delta_0 < \epsilon \Rightarrow$ STOP (Residual less enough)
WHILE $\delta_1 > \epsilon \{$
$$\begin{aligned}
h^k &:= Ap^{k-1} \\
\tau_k &:= \delta_0/<p^{k-1}, h^k> \\
x^k &:= x^{k-1} + \tau_k p^h \\
r^k &:= r^{k-1} + \tau_k h^k \\
\delta_1 &:= <r^k, r^k> \\
\alpha_{k-1} &:= \delta_1/\delta_0 \\
p^{(k)} &:= -r^k + \alpha_{k_1} p^{k-1}
\end{aligned}$$
$\}$
END.

The parallelization of this problem is actually done by one element wide overlapping at the artificial boundaries (see Fig. 7).

This storage scheme allows communication latency hiding, as the boundary exchange

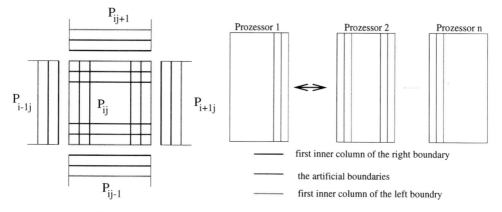

Figure 7: overlapping with and without interior crosspoints

and the computation of the interior matrix-vector product after the computation on the artificial boundaries can be done in parallel. For symmetric matrices the CG-method still requires two global scalar products during the iteration loop, realized as local scalar products on the subdomains followed by a global sum. These are the only global synchronization points of the CG-algorithm.

The preconditioned conjugate gradient method with preconditioning matrix B has the form:

PreCG-Algorithm

CHOOSE $x_0, r_0 = Ax_0 - b$; $s_0 = B^{-1}r_0$, $p_1 = -s_0$; $\delta_0 := <r_0, s_0>$
IF $\delta_0 < \epsilon \Rightarrow$ STOP (Residual less enough)
WHILE $\delta_1 > \epsilon$ {

$\quad h_k \quad := Ap_{k-1}$
$\quad \tau_k \quad := \delta_0 / <p_{k-1}, h_k>$
$\quad x_k \quad := x_{k-1} + \tau_k p_k$
$\quad r_k \quad := r_{k-1} + \tau_k h_k$
$\quad s_k \quad := B^{-1} r_k$
$\quad \delta_1 \quad := <s_k, r_k>$
$\quad \alpha_{k-1} := \delta_1 / \delta_0$
$\quad p_k \quad := -s_k + \alpha_{k_1} p^{k-1}$

}
END.

The construction of the implemented preconditioner B is described in detail in Section 3. The calculation of the preconditioning matrix is done by a multigrid algorithm with a Gaus-Seidel smoother. In the case of unstructured grids we combine the smoother with a broadsearch which gives a constant amount of iteration with respect to the step-size. Only the exchange of the information on the artificial boundaries is necessary. For the two global scalar products during the iteration loop, the same implementation as in the non-preconditioned CG-algorithm is used. In general, the use of preconditioning leads of course to a better ratio computation-time communication compared to conventional CG-methods, but it requires a more sophisticate load balancing. The chosen decomposition for the CG-method achieves the required good load balancing, because the computation load of each processor can be nearly estimated by the number of assigned data points in each subdomain. The amount of communication is proportional to the cut size. Optimal problem decomposition and mapping leads to highly efficient programs on large parallel systems like the GCel/1024 with 1024 processors (see [3]). The time needed for global communication is negligible small. Because of optimal domain partition, mapping and parallel execution of computation and communication during matrix-vector products only little overhead of the parallelization appears.

4.2 Implementation of a parallel algorithm for the characteristics

Besides the solver for the Poisson equation with mixed boundary conditions, which occurs from the elliptic part and the pressure correction step, we have to implement an algorithm for the computation of the characteristics [23]. Such an algorithm is equivalent to an algorithm for the transport equation proposed in [4, 5]. However, here we will present a direct characteristic's approach [15, 22] which is easy to parallelize. In order to have small damping in time, we will use a local Crank-Nicholson scheme for the computation of the characteristics. Here 'local' means, that the computation is carried out elementwise without inverting global matrices. Instead, only local matrices on each element have to be inverted. This leads to an efficient parallel algorithm even on unstructured grids.

Let v^n be an approximation of the Navier-Stokes solution v at time $t = t_n$. By the chain rule for the composition we have

$$v^n \circ X(t_n, t_{n+1}) = v^n \circ X(t_n, t_n + \tilde{k}) \circ \ldots \circ X(t_{n+1} - \tilde{k}, t_{n+1}))$$

with a subtime step $\tilde{k} = k/\tilde{m}$ ($\tilde{m} \in \mathbb{N}$). Thus, we are left with the computation of $X(t_n + m\tilde{k}, t_n + (m+1)\tilde{k})$ $m = 0, \ldots, \tilde{m} - 1$, where the subtime step $\tilde{k}$ is chosen such, that the curve lengths of the characteristics over the interval $[t_n + m\tilde{k}, t_n + (m+1)\tilde{k}]$ are smaller than the diameters of each element T of the triangulation. We remark, that $\tilde{k}$ (resp. $\tilde{m}$) can be estimated in terms of $h_{max} = \max_T diam\, T$ and the maximum-norm

$$\|v\|_{L^\infty([t_n, t_{n+1}] \times \Omega)} = \sup_{(t,x) \in [t_n, t_{n+1}] \times \Omega} |\hat{v}(t, x)|.$$

With each element T with nodalpoints x_T^1, x_T^2, x_T^3 (2D-case) we associate the 3×3-matrix

$$A = \begin{pmatrix} x_T^1 & x_T^2 & x_T^3 \\ 1 & 1 & 1 \end{pmatrix},$$

so that the barycentric coordinates $\lambda = (\lambda_1, \lambda_2, \lambda_3,)$ of $x \in T$ are given by the linear system $A\lambda = \begin{pmatrix} x \\ 1 \end{pmatrix}$. We assume that X is the above characteristic of the linear interpolation $\hat{v}$ of v^{n-1} and v^n in time, extrapolated up to $t = t_{n+1}$ (second order approximation). We also assume piecewise linear conforming finite elements. Now we consider a specific nodal point x_T^i of the element T. If the velocity has to be computed there, we first look whether we can use the linear function $\hat{v}$ on that element or not. To decide this, we solve the linear system

$$A\lambda = \begin{pmatrix} x_T^i - \tilde{k}\hat{v}_i \\ 1 \end{pmatrix},$$

where $\hat{v}_i$ is the value of $\hat{v}$ at the nodal point x_T^i of T at time $t_n + m\tilde{k}$. Thus the solution λ can be considered as a first order approximation of the characteristics and $\tilde{v} = \sum_{j=1}^3 \lambda_i v_i^n$ as an approximation of $v^n \circ X^{-1}$. Now if $\lambda_1, \lambda_2, \lambda_3 \geq 0$, then we know that $x_T^i - \tilde{k}\hat{v}_i \in T$, i.e. the particle starting backwards at x_T^i at time $t_n + (m+1)\tilde{k}$ stays in T at time $t_n + (m+1)\tilde{k}$ up to first order. In that case we recompute λ more accurately by the Crank-Nicholson scheme using the linear velocity $\hat{v}$ defined on T and allowing λ to 'leave' T (extrapolation of $\hat{v}$ outside T). Let $\lambda = A^{-1}\begin{pmatrix} X \\ 1 \end{pmatrix}$ denote the barycentric coordinates of the characteristics. From (2.3) and the fact that $X(t_n+(m+1)\tilde{k}, t_n+(m+1)\tilde{k}, x_T^i) = x_T^i$, we get the backwards linear initial value problem ($i = 1, 2, 3$)

$$\frac{d}{dt}\lambda = -A^{-1}\begin{pmatrix} \hat{v}^1(t) & \hat{v}^2(t) & \hat{v}^3(t) \\ 0 & 0 & 0 \end{pmatrix}\lambda, \tag{4.1}$$

$$\lambda(t_n + (m+1)\tilde{k}) = e_i, \tag{4.2}$$

where $\hat{v}^i(t)$ is the value of $\hat{v}(t)$ at the nodal point x_T^i and e_i is the ith unit vector in $I\!R^3$, (resp. $I\!R^4$ for 3D computation). The solution of (4.1),(4.2) gives the barycentric coordinates of that point x at which a particle has to start to reach x_T^i. Thus, the composition $v^n \circ X^{-1}$ in x_T^i at time t is simply given by the linear combination

$$v^n \circ X^{-1} = \sum_{\mu=1}^{3} \lambda_\mu v_\mu^n, \tag{4.3}$$

where $v_\mu^n (\mu = 1, 2, 3)$ are the nodal values of v^n, i.e. the values of v^n at x_T^μ. The Crank-Nicholson dicretization of (4.1),(4.2) leads to the following algorithm.

CN-Algorithm

Compute $\tilde{m} = [\frac{k}{h_{max}} \|\hat{v}\|_{L^\infty}], \tilde{k} = k/\tilde{m}$
{   for $(m = 1; m \leq \tilde{m}; m++)$
         for all T
            get T
            for $(i = 1; i \leq 3; i++)$
                Compute local CN-step
            }
         }
     }
     set $v^n = \sum_{\mu=1}^{3} \tilde{\lambda}_\mu v_\mu^n$
}
The local CN-Step is given by
Compute Matrix $A = \begin{pmatrix} x_T^1 & x_T^2 & x_T^3 \\ 1 & 1 & 1 \end{pmatrix}$,
Solve $A\lambda = x_T^i - \tilde{k}\hat{v}_i$)
if $(\lambda_1 \geq 0$ and $\lambda_2 \geq 0$ and $\lambda_3 \geq 0)$
     Compute new barycentric coordinates
     as the solution of the CN-system
$$[1 + \frac{\tilde{k}}{2}A^{-1}\begin{pmatrix} \hat{v}_m^1 & \hat{v}_m^2 & \hat{v}_m^3 \\ 0 & 0 & 0 \end{pmatrix}]\tilde{\lambda} = 2e_i$$
with $\hat{v}_m^i = \frac{1}{2}(\frac{2m-1}{\tilde{m}} + 2)v_i^n - \frac{1}{2}(\frac{2m-1}{\tilde{m}})v_i^{n-1}$
$\tilde{\lambda} = 2\tilde{\lambda} - e_i$.

Note that the new velocity is uniquely defined, because at each nodal point we have uniquely determined the element T by the if - command. Even for very small subtime steps the resulting parallel algorithm is very fast because of the local character. On the other hand, the Crank Nicholson scheme guarantees small damping and the optimal direction is taken into account automatically. Moreover the discretization error in streamwise direction is usually smaller (see [15]) . However, triangulations with high aspect ration lead to very small timesteps. The overall design of the Navier Stokes code then follows the guidelines of our FEM-Tool. Object oriented data structures are used to define the abstract data types needed in the various steps of the splitting scheme. A special new feature of the code is the adaption technique. The adaption of the mesh assimilates to the mesh generation strategy. For the mesh generation quad-tree based (for 2D-problems) and octree-based (for 3D-problems) methods [2, 13] are used, which create meshes with provable good quality. For the adaptive varying parts of the mesh given by an local error indicator, we now have to change the depth of the quad-tree at the indicated parts of the tree. With this strategy we reobtain the quality of the mesh independent from the mesh step-size. Concerning to the data structure only the redefined parts of the mesh lead to a reassembling of the parts of the matrices for the Poisson problems.

4.3 A finite element toolbox

A further task is the design of a flexible tool-box for FEM-problems on parallel computers including mesh generation, mesh decomposition, adaptive mesh modification and a comfortable pre- and postprcessing environment. This project is a common work together with the computer science department and the Paderborn Center for Parallel Computing. The parallel FEM-code is based on a abstract data-type describing the mesh and including all relevant information such as non-zero matrix entries, right hand sides and boundary values. Storing the matrix together with the mesh avoids the need for additional sparse storage schemes and allows an straight foreward extension to adaptive methods. The current implementation is able to solve the Poisson equation with mixed Dirichlet- and Neumann-boundary conditions in two dimensions using triangular finite elements. Conjugate Gradient, Preconditioned Conjugate Gradient and Multigrid iteration algorithms are implemented to solve the resulting system of linear equations. The basis for the parallel implementation is PVM running on workstation clusters, shared memory systems (Sparc-Center 1000, PowerChallenge) and MPPs (Parsytec GC/PP). The overall design of the toolbox follows the idea of *programming frames* which are skeletons with problem dependent parameters to be provided by the users. Programming frames focus on re-usability and portability as well as on small and easy-to-learn interfaces. Thus, non-expert users will be provided with tools to program and exploit parallel machines efficiently. The modular structure of our toolkit allows the easy exchange of certain parts of the library (the frame parameters) and abstract data types hide all details of the parallel systems from the user. The update of inner boundaries between subdomains is done automatically and the numerical code does not differ from sequential implementations using the same abstract data-type. All parts of the library can easily be exchanged, prototype developments can be done using a sequential version. More information about this FEM-tool can be found in [13, 14].

5. Numerical Results

We consider problem (3.1) with u(x,y) = $(x^2 - x) * (y^2 - y)$ as the solution on $\overline{\Omega}$ = $[0, Proc] \times [0, 1]$, where Proc is the number of processors (see Fig. 1). The break-off criteria for the conjugate gradient and the preconditioned conjugate gradient method ($\delta_1 < \varepsilon$, see Section 4) is given by $\varepsilon = 10^{-9}$. For the Dirichlet and Dirichlet-Neumann boundary-value problems we use a V-cycle multigrid method on each subdomain. The computations are carried out on a GC Power Plus System with 64 Processors.

Table 1: Method 1 is the usual DD-preconditioned CG-Method on a uniform grid using the sparse matix $\tilde{C}^{-1}$ for the transition condition. Method 2 is the conjugate gradient boundary iteration method (see Section 3) and for comparison we included computations with the unpreconditioned and parallelized CG-Method (see Section 4), here denoted by Method 3. The computation-time is given in seconds and Iter stands for the number of iterations.

Table 1

Proc.	Step-Size	Unknowns	Method 1		Method 2		Method 3	
			# Iter	Time	# Iter	Time	# Iter	Time
16	1/128	262144	11	6.73	5	5.64	465	19.28
	1/256	1048576	17	37.13	6	20.26	905	106.16
	1/512	4194304	22	190.76	6	76.13	1709	731.43
32	1/128	524288	17	9.23	6	6.18	476	22.28
	1/256	2097152	19	39.40	6	22.22	919	112.98
	1/512	8388608	24	194.20	7	91.10	1795	776.23
64	1/128	1048576	16	9.65	7	6.86	480	23.81
	1/256	4194304	20	39.87	8	26.22	927	116.74
	1/512	16777216	24	209.83	8	101.19	1805	786.96

If the step-size and thus the condition number is fixed one can observe that the number of iterations is nearly constant (see Fig. 8). This reflects the fact, that our replacement $\tilde{C}$ works well up to the given step-sizes. We also note the behavior mentioned already in Section 3 that Method 2 is only two times faster than Method 1.

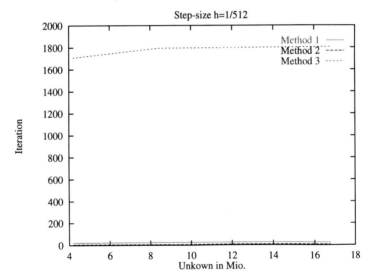

Figure 8: Iteration growth

Table 2 presents a comparison of Method 1 and Method 3 in case of interior crosspoints. The unit square $\Omega = [0,1] \times [0,1]$ is partitioned into 4, 16 resp. 64 congruent subsquares. The step-sizes are given by $h = \frac{1}{\sqrt{Unknowns}}$. Table 2 displays also that the number of iterations depends sensitive on the number of interior crosspoints. Moreover, for fixed number of interior crosspoints (see Fig. 9) the number of iterations increases (compare with Fig. 8). In a forthcoming paper this problem will be studied in detail.

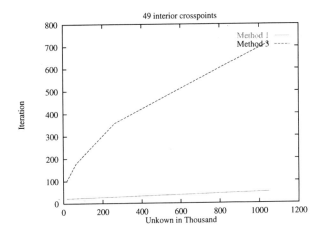

Figure 9: Iteration growth

Table 2

Proc.	Unknowns	Method 1		Method 3	
		Time	Iter	Time	Iter
(2 x 2)	16384	1.8068	14	1.8163	100
(2 x 2)	65536	6.3043	16	7.0674	179
(2 x 2)	262144	29.5822	18	44.5647	358
(2 x 2)	1048576	113.4382	20	328.4491	716
(4 x 4)	16384	0.9783	17	1.8348	100
(4 x 4)	65536	2.1935	19	4.4128	179
(4 x 4)	262144	8.9502	25	16.4439	358
(4 x 4)	1048576	41.8213	32	98.0449	716
(8 x 8)	16384	0.7968	23	2.3786	100
(8 x 8)	65536	1.1960	24	4.5216	179
(8 x 8)	262144	3.2221	29	11.0266	358
(8 x 8)	1048576	18.2573	51	38.0226	716

References

[1] Beale, J.T.: The approximation of the Navier-Stokes equations by fractional time steps, Lecture at the conference: The Navier-Stokes equations, theory and numerical methods, Oberwolfach 18. –24.8.91 [IBM Research Report RCI 18072 (79337) 6-1192].

[2] Bern, M., Eppstein, D., Gilbert, J.: Provably good mesh generation. Proc. 31st Symp. on Foundations of Computer Science (FOCS), (1990), 231-241

[3] Blazy, S., Dralle, U., Simon, J.: Parallel CG Poisson Solver for PowerPC 601, Power Explorer User Report. Application and Projects on Parsytec PowerXplorer Parallel Computers, 1st Edition, May (1995).

[4] Borchers, W.: A splitting algorithm for incompressible Navier-Stokes equations, in :H. Niki, M. Kawahara (eds.): Int. Conf. on computational methods in flow analysis, Okayama, Japan (1988), 454–461.

[5] Borchers, W.: On the characteristic's method for the incompressible Navier-Stokes-equations, In: Finite Approximations in Fluid Mechanics II; E. H. Hirschel (Ed.), Notes on Numerical Fluid Mechanics, Vol. 25, Vieweg, Braunschweig 1989, 43-50.

[6] Borchers, W.: Zur Stabilität und Faktorisierungsmethode für die Navier-Stokes Gleichungen inkompressibler viskoser Flüssigkeiten. Habilitationsschrift für das Fach Mathematik im Fachbereich Mathematik-Informatik der Universität-GH Paderborn (1992).

[7] Borchers, W.: A new conjugate gradient boundary iteration method for parallel numerical computation of elliptic boundary value problems. In preparation.

[8] Bramble, J. H., Pasciac, J.E., Schatz, A.H.: The construction of preconditioners for elliptic problems by substructering I, Math. Comp. 47, (1986).

[9] Bramble, J. H., Pasciac, J.E., Schatz, A.H.: The construction of preconditioners for elliptic problems by substructering II, Math. Comp. 49, (1987).

[10] Chorin, A. J., Hughes, T.J.R., Mc Cracken, M.F., Marsden, J.E.: Product formulas and numerical algorithms, Comm. Pure Appl. Math. XXXI (1978) 205–256.

[11] Chorin, A. J.: Numerical solution of the Navier-Stokes equation. Math. Comp. 22,(1968), 745-762.

[12] Chorin, A. J.: On the convergence of the discrete approximations of the Navier-Stokes equations. Math. Comp. 23, (1969), 341-353.

[13] Diekmann, R., Meyer, D., Monien, B.: Parallel Decomposition of Unstructured FEM-Meshes. Proc. of IRREGULAR '95, Springer LNCS 980, (1995), 199-215.

[14] Diekmann, R., Dralle, U., Neugebauer, F., Römke, T.: PadFem: A portable parallel FEM-Tool. Draft , Univ. of Paderborn, (1995) (submitted).

[15] Douglas, J., Russel, T.F.: Numerical methods for convection-dominated diffusion problems based on combining the method of characteristics with Finite Element or Finite Difference procedures. Siam J. Numer. Anal. Vol.19, No. 5, October (1982)

[16] Güntsch, E. Friedrich, R. : DNS of Turbulence Compressed in a Cylinder. In this publication.

[17] Hackbusch, W., Trottenberg, M.(eds.): Multigrid methods. Lecture Notes in Mathematics. Vol. 960. Springer Verlag, Berlin-Heidelberg-New York (1982).

[18] Hebeker, F.-K.: Analysis of a characteristics method for some incompressible and compressible Navier-Stokes problems. Preprint 1126 des Fb. Mathematik, TH Darmstadt (1988).

[19] Langer, U., Haase, G.: The approximate Dirichlet domain decomposition method. II. Application to 2nd-order elliptic BVPs. Comp. Arch. f. Inf. a. Numer. Comp. 47, (1991), 153-167.

[20] Lions, J. L., Magenes, E.: Non-Homogeneous Boundary Value Problems and Applications I, Springer Verlag, Berlin-Heidelberg-New York (1982).

[21] K. Masuda, Rautmann, R.: H^2-convergent approximation schemes to the Navier-Stokes equations. Commentarii Mathematici Universitatis Sancti Pauli 43, (1994), 55–108.

[22] Pironneau, O.: On the transport diffusion algorithm and its applications to the Navier-Stokes equations, Num. Math. 38, (1982), 309–332.

[23] Rautmann, R.: Ein Näherungsverfahren für spezielle parabolische Anfangsrandwertaufgaben mit Operatoren. Springer Lecture Notes in Math. 267, (1972), 187–231 (Habilitationsschrift).

[24] Rautmann, R.: Navier - Stokes approximations in high order norms. In this publication.

[25] Rannacher, R.: On Chorin's projection method for the incompressible Navier-Stokes equations, Lecture Notes in Math. Vol. 1530, (Heywood, J.G., Masuda, K., Rautmann, R., Solonnikov, S.A. Eds.), The Navier-Stokes Equations II - Theory and Numerical Methods, Proceedings, Oberwolfach (1991), Springer-Verlag.

[26] Shen, J.: On error estimates of projection methods for the Navier-Stokes equations: First order schemes, SIAM J. Numer. Anal. 29 (1992), no 1, 57–77.

[27] Süli, E., Convergence and nonlinear stability of the Lagrange-Galerkin method for the Navier-Stokes equations, Numer. Math. 53 (1988) 459–483.

[28] Temam, R.: Sur l'approximation de la solution des équations de Navier-Stokes par la méthode des pas fractionaires (I), Arch. Rational Mech. Anal., 32 (No 2), 1969, 135-153.

Supported by Deutsche Forschungsgemeinschaft

Weighted Particle Methods Solving Kinetic Equations for Dilute Ionized Gases

K. Steiner
Arbeitsgruppe Technomathematik
Fachbereich Mathematik
Erwin–Schrödinger–Straße
67653 Universität Kaiserslautern
Germany

Summary

The numerical treatment of kinetic equations for ionized gases has to handle two additional problems. The first one, the different time scales for the electrons and the heavy particle, is solved by asymptotic methods. The second one arises from small ionization rates and is treated with a weighted particle method. The derivation of a weighted particle method for general kinetic systems, the choice of the weights and the convergence properties are discussed in the first part of the paper. The use of variable weights in a particle scheme leads to nonconservative particle schemes which results in purely numerical instabilities. The second part of the paper presents possible modifications of the weighted particle method which conserve mass, charge, momentum and energy on the discrete level of description and lead to stable and convergent numerical methods. Numerical results for a reduced plasma system show the advantages in comparison to equiweighted particle method with respect to accuracy, computational time and computer memory.

1. Introduction

The first points on the re–entry trajectory of a space vehicle into the atmosphere of the earth or other planets are characterized as the rarefied gas regime. For a deeper understanding of the important physical and chemical effects in this first critical phase, it is necessary to develop more realistic models and powerful numerical methods.

The modelling has to include electric as well as magnetic effects by external and selfconsistent fields, the energy exchange between translational and internal degrees of freedom and chemical reactions, especially ionization and recombination reactions. Extenstions of the kinetic theory of gases to the case of a plasma — a mixture of ionized gases including all these phenomena — can be found in various books (i.e. [7]). But, for example, a concrete formulation of the ionization and recombination collison terms are rather rare and we use here the notations of [24].

Two typical properties of plasmas lead to stiff numerical problems:
First, the characteristic time scale for the electrons is much smaller than the time scale of the ions or neutrals. This is due to the small mass of the electrons compared to the heavy particles — the ions and neutrals. Other different time scales arise due to different reaction rates, i.e., for ionization and recombination reactions. The problem of the different time scales can be treated by an appropriate scaling of the underlying equations [10]. The use of asymptotic methods yields mathematical models, which are valid on different physical scales. The

interesting model for the re–entry problem is the heavy particle scale (see also [24]). This leads to a coupled system of macroscopic equations for the electron density and temperature and kinetic equations for the ions and neutrals.

The second difficulty arises from the small ionization rates of plasmas in the re–entry phase. But, due to the resulting electro-magnetic fields and chemical reactions, the small concentrations of charged particles cannot be neglected. If we use particle methods — and we have to use them by other reasons — the numerical difficulty arises from an appropriate representation of these small concentrations. A well–known modification is the use of variable weights, which we apply here to the different species only. Other applications lead to the same problem, e.g., dissociation of molecules in a rarefied flow where the atom concentration is very small, or evaporation and sputtering processes.

The paper concentrates on these weighted particle methods for systems of coupled kinetic and macroscopic equations with strongly different concentrations. The treated systems can result from the extended kinetic theory as well as from simplification by asymptotic methods.

In the next chapter, we explain the idea of a weighted particle method and apply this method to the general form of a kinetic equation. Here, we derive the measure–valued formulation of the kinetic system, which cannot be obtained directly. Therefore, we first split the equation into three fractional steps; a Vlasov equation, describing the drift of the particles by the flow, a Boltzmann equation, representing the collisions and reactions of the particles, and the macroscopic equations. Then, well–known results for both types of kinetic equations can be extended and result in a convergent weighted particle method coupled with a finite element method for the Poisson equation.

Conservation of mass, charge, momentum and energy on the discrete level of description is necessary to avoid numerical instabilities. Charge and energy conservation for the drift term is obtained by a consistent choice of the mollifier and the ansatz function for the finite element method. Due to the different weights, momentum and energy conservation in the collision step can be no longer satisfied using a symmetric collision process. Hence, we develop a new algorithm to choose collision parameters which guarantees momentum and energy conservation. The mass conservation is violated during reactive collisions when using the standard weighted particle approach. A suitable transformation of the equations makes it possible to derive a mass conserving particle method such that weights are at each time proportional to the concentrations. Finally, we present some numerical results and give some concluding remarks.

2 Weighted particle methods for kinetic equations

2.1 Kinetic equations for dilute ionized gases

In this paper we consider kinetic equations of the general form

$$\partial_t f + div_{x,v}(D[f]f) = J[f]f. \qquad (2.1)$$

Here, f is the probability density of the gas ensemble at time $t \in I\!R_+$ at position $x \in \Lambda$ in the state $z \in \Gamma$; therefore, $f : I\!R_+ \times \Pi \to I\!R_+$ with the phase space $\Pi = \Lambda \times \Gamma$. In classical kinetic theory the state z coincides with the velocity $v \in I\!R^3$ of the particle. But in realistic applications one has to add internal energy variables e which can be discrete and/or continuous [21] and the sorting index $s \in I\!I$ of the particle ($I\!I$ is index set of the species). Here, we assume $z = (v, e, s)$ with one internal energy variable $e \in I\!R_+$ representing rotational degrees of freedom, $s \in I\!I$ and $\Gamma = I\!R^3 \times I\!R_+ \times I\!I$.

Macroscopic quantities like the mass, charge, momentum and energy densities of species s and the mixture are given by moments of the density function f. A moment of f with respect to $\varphi : \Gamma \to I\!R$ is defined by

$$\mathcal{M}[\varphi](t,x) = \int_\Gamma \varphi(z) f(t,x,z) dz = \sum_s \int_{R^3} \int_{R_+} \varphi(v,e,s) f(t,x,v,e,s) de dv. \qquad (2.2)$$

For example, the number density ν_s of the species s is the moment of f with respect to $\varphi = (0,...0,1,0,...0)$ and the number density ν of the mixture is given by $\nu = \mathcal{M}[(1,...,1)]$.
The evolution equation (2.1) consists of the drift of particles given by the flux $D[f] = (v,a)$ (a is the acceleration of a particle) and the instantaneous interactions between two particles like collisions and reactions described by the collision operator J.
For an ionized gas, the acceleration a of a particle is determined by the Lorenz force

$$m_s a = q_s (E[f] + v \times B[f])$$

where m_s (q_s) is the mass (charge) of a particle of species s. The electric field E and the magnetic induction B are the sum of external forces and selfconsistent fields given by Maxwell equations.
In the following, we assume that magnetic effects can be neglected. Furthermore, we introduce the potential $U = U[f]$ by $E = -\nabla U$ which satisfies Poisson's equation

$$-\Delta_x U = \frac{1}{\epsilon_0} \tau, \qquad (2.3)$$

where $\tau = \mathcal{M}[(q_s)_{s \in I}]$ denotes the charge density of the mixture. In the noncollisional case ($J[f] = 0$) the system (2.1), (2.3) reduces to the Vlasov–Poisson system.
The component J_s of the collision operator J is the sum of reactive and nonreactive Boltzmann like collision operators of the form

$$J_s[f] f(z) = \int_\Gamma \int_{\Lambda'} \Theta^{s',s'*}_{s,s*}(E,\omega;\omega') \left\{ \left(\frac{m_s m_{s*}}{m_{s'} m_{s'*}} \right)^3 f(z') f(z'^*) - f(z) f(z^*) \right\} d\varpi' dz^*, \qquad (2.4)$$

where $\Theta^{s',s'*}_{s,s*}$ is the scattering kernel for the reaction

$$s + s^* \longrightarrow s' + s'^*$$

and z' and z'^* are the postcollisional states of the pair (z, z^*) due to the collision parameter ω'. In the case $s = s'$ and $s'^* = s^*$ a nonreactive collision occurs. In the following, it is not necessary to know the detailed form of J_s. The only essential property is the principle of detailed balance connecting the forward and backward collision processes via

$$\left(\frac{m_s m_{s*}}{m_{s'} m_{s'*}} \right)^3 \Theta^{s',s'*}_{s,s*}(E,\omega;\omega') = \mathcal{J}(\mathcal{C}^{s',s'*}_{s,s*}) \Theta^{s,s*}_{s',s'*}(E',\omega';\omega). \qquad (2.5)$$

Here, E and E' are the total collision energies before and after the reaction and $\mathcal{J}(\mathcal{C}^{s',s'*}_{s,s*})$ denotes the Jacobian of the collision transformation $\mathcal{C}^{s',s'*}_{s,s*}(z,z^*,\omega') = (z',z'^*,\omega)$ for the forward process. The detailed balance relation (2.5) is the fundamental equation to prove the so–called H-theorem (see [6]), which implies the irreversibility of the system. A detailed description and the extension to ionization and recombination reactions is given in [24].
The complete system under consideration can be written as a system of kinetic equations for each phase space density f_s ($f_s(t,x,v,e) = f(t,x,v,e,s)$)

$$\partial_t f_s + v \cdot \nabla_x f_s - \frac{q_s}{m_s} \nabla U[f] \cdot \nabla_v f_s = J_s[f] f_s, \quad s \in I\!I, \qquad (2.6)$$

coupled with the Poisson equation (2.3) and therefore, is called Boltzmann–Vlasov–Poisson system.

The system has to be completed by initial as well as boundary conditions. By $f_s^{(0)}$ we denote the initial densities for species s. Realistic boundary conditions in the case of rarefied ionized flows are only modeled in few special cases [11] and totally unknown for re-entry problems. Here, we use specular reflecting boundary conditions for each f_s and fixed Dirichlet conditions for the potential U, which simplifies the theory. Although other boundary conditions like diffuse reflection of molecules or varying potentials are possible.

In the case of a plasma the index set $I\!I$ consists of the electrons e and different kinds of ions and neutrals (denoted by i respectively o). We only remark that a scaling of the Boltzmann–Vlasov–Poisson system with the mass ratio of electrons and heavy particles yields several plasma models valid on different physical time scales. This scaling was suggested by the work of P. Degond and B. Lucquin-Desreux [9, 10] and is treated in detail in [24]. The resulting systems belong to the class of kinetic systems coupled with macroscopic equations and can be solved numerically using the methods developed in this paper.

2.2 Weighted particle methods

The common idea of all particle methods is the interpretation of the density f as a continuous measure $f(t,p)dp$ varying in time. Therefore, every nonnegative integrable function f can be approximated in the sense of measures by a sum of weighted Dirac-measures [20]

$$f(t,p)dp \approx \sum_{i=1}^{N} \alpha_i(t)\delta(p - p_i(t)) =: \delta_{\alpha,p}^N, \qquad (2.7)$$

where p_i is the position in the phase space $I\!I$ and $\alpha_i > 0$ the weight of particle i. A particle approximation of $f(t,\cdot)$ is determined if the evolution equations for the positions and the weights of all particles are known. The aim of the next sections is to derive a particle method for the Boltzmann–Vlasov–Poisson system.

First, we explain in which way we want to determine the weights in our particle method. Nearly all particle methods use identically weighted particles, since the evolution equation of these identically weighted particles roughly coincide with the physical trajectories. Sometimes there is need for a weighting in the position space, i.e., in axisymmetric geometries [26] or in high density regions [22].

In the case of a weakly ionized plasma there are strong differences in the concentrations of the charged particles and the neutrals. An identical weighting of all particles causes a large number of particles, if we assume that a minimal number of particles per cell is necessary to represent the distribution adequately. For example, in a single plasma with 1% degree of ionization and a minimal local particle number of 10 leads to a total particle number of about 1000 per cell.

A quite natural way to overcome these large particle numbers is the use of weights proportional to the local concentrations. Then, the total particle number per cell is of the order of the minimal local particle number multiplied by the number of different species. Therefore, we want to ensure the relation

$$\alpha_i^N(t) = \frac{\gamma_s(t)}{N_s} \qquad \text{for} \quad s_i = s, \qquad (2.8)$$

where γ_s is the concentration and N_s the number of particles of species s. For the rest of the paper we assume that the weights of particles of one species are equal, but may differ between the species. Therefore, the weights may be different locally in space which leads to the serious problem of discrete conservation discussed in chapter 3.

2.3 Time splitting

To obtain the evolution equations for the positions and weights we need a measure–valued formulation of our system (2.1). In general, a measure form of a kinetic system cannot be obtained directly, especially when the system involves nonlinear terms. A typical example is the inhomogeneous Boltzmann equation. The quadratic collision term can only be viewed as a measure defined on Γ with parametrical dependence on either x or t [1, 3]. For the Vlasov–Poisson system, the singularity of the kernel has to be smeared out to obtain a corresponding measure–valued equation [17].

The first step in our approach is a splitting of the complete kinetic equation (2.1) into three partial systems – a system of Vlasov equations, a homogeneous Boltzmann system and the Poisson equation. For a fixed time interval $[0, T]$ and a partition into intervals $T_k = \frac{k}{n}T$, $k = 1, ..., n$, $n \in \mathbb{N}$ of size $\Delta T = \frac{1}{n}T$, we define the splitting scheme of (2.6) by

$$\partial_t g_s^k = -v \cdot \nabla_x g_s^k + \frac{q_s}{m_s} \nabla_x U^{k-1} \cdot \nabla_v g_s^k, \qquad g_s^k(T_k) = f_s^{k-1}(T_k), \qquad (2.9)$$

$$\partial_t f_s^k = J_s[f^k] f_s^k, \qquad f_s^k(T_k) = g_s^k(T_{k+1}), \qquad (2.10)$$

$$-\Delta_x U^k = \frac{1}{\epsilon_0} \sum_s \int_{\Gamma_s} q_s f_s^k dz, \qquad T = T_k, \qquad (2.11)$$

where U^0 is the solution of (2.11) with $f^0(0, p) = f^{(0)}(p)$. The ordering of the splitting scheme is arbitrary. The charge conservation of the collision operator implies

$$\sum_s \int_{\Gamma_s} q_s g_s^k dz = \sum_s \int_{\Gamma_s} q_s f_s^k dz, \qquad (2.12)$$

which allows the computation of the potential by (2.11) before the Boltzmann step (2.10). The convergence of the splitting scheme can be shown using the results of [12] and [13].

Now, we can separately derive measure–valued forms of the Vlasov and Boltzmann systems leading to a particle method which iterates the drift of the particles with fixed forces and the collisions between the particles. The forces have to be updated after each time step ΔT by solving the Poisson equation.

2.4 Vlasov equation

We may study the Vlasov equation separately for each species and neglect the internal energy variables without loss of generality. The force F is defined by $F(x) = qE(x)$ with the electric field E at a fixed time.

We can solve the Vlasov equation along the characteristics

$$\begin{aligned} \frac{d}{dt}x(t) &= v(t), & x(0) &= x_0, \\ \frac{d}{dt}v(t) &= a(x(t)) = \frac{1}{m}F(x(t)), & v(0) &= v_0, \end{aligned} \qquad (2.13)$$

using, e.g., specular reflection at the boundary of the domain Λ, i.e.,

$$v(t_+) = v(t_-) - 2\left(v(t_-) \cdot n_{x(t_-)}\right) n_{x(t_-)} \qquad (2.14)$$

for all $x(t_-) \in \partial \Lambda$, $v(t_-) \cdot n_{x(t_-)} > 0$.

If we assume that E is Lipschitz continuous then (2.13) has a unique global solution denoted by $\Phi_{t,0}(x_0, v_0) = (x, v)(t)$. $\Phi_{t,\tau} : \Pi \to \Pi$ is a measure preserving group homomorphism [17] ($\Phi_{t,t} = Id$ and $\Phi_{t,\tau}^{-1} = \Phi_{\tau,t}$) and, with continuous initial conditions $g^{(0)} \in \mathcal{C}(\Pi)$, we obtain the solution of the Vlasov equation by

$$g(t, x, v) = (g^{(0)} \circ \Phi_{0,t})(x, v). \tag{2.15}$$

Now, we interpret $g(t, \cdot)$ as a density of a continuous measure μ_t (see (2.7)). Then, for every Borel set $M \subset \Gamma$,

$$\mu_t(M) = \int_M (g^{(0)} \circ \Phi_{0,t})(p) dp = \int_{\Phi_{0,t}(M)} g^{(0)}(q) dq = \mu(\Phi_{t,0}^{-1}(M)).$$

With the definition of the image of a measure μ under the mapping Φ by $\Phi(\mu)(M) = \mu(\Phi^{-1}(M))$ we get the measure form of the Vlasov equation

$$\mu_t = \Phi_{t,0}(\mu). \tag{2.16}$$

In [17] it is shown that (2.16) has a unique solution for every probability measure which coincides for a continuous measure with the weak solution.
In the case of a discrete measure $\mu^N = \delta_{\alpha,p}^N$, $p = (x, v)$, it is easy to see that $\Phi_{t,0}(\mu^N)$ is also a discrete measure with the same weights and the evolution of the points given by the characteristic equations (2.13)

$$(x_i^N, v_i^N)(t) = \Phi_{t,0}(x_i^N, v_i^N). \tag{2.17}$$

Moreover, the solution is unique if E is at least continuous [8].
Due to relation (2.17) it is only necessary to solve the ordinary differential equations (2.13) over a time interval $[0, \Delta T]$. Therefore, we approximate $\Phi_{t,0}$ by some difference operator $\Phi_t^{\Delta t}$ and obtain the time discretized measure form of the Vlasov equation

$$\mu_{\Delta t} = \Phi_0^{\Delta t}(\mu). \tag{2.18}$$

Convergence is proved under the assumption of a uniform convergent difference scheme $\Phi_t^{\Delta t}$ and a Lipschitz–continuous force F [17, 24].
A simple and convergent method is the so–called leap–frog scheme [4], which is explicit, symmetric and second order in time, and belongs to the class of symplectic methods for hamiltonian systems [14].
The simulation scheme for the Vlasov equation using the leap–frog scheme for a given initial approximation $\mu^N = (\alpha^N; (x, v)^N)$ reads

(V) SIMULATION SCHEME FOR THE VLASOV EQUATION:

For every time step $t_k = k\Delta t$, $k = 0, 1, ...$.

$$v_i^N(t_{k+1}) = v_i^N(t_k) + \Delta t \frac{q_{s_i^N}}{m_{s_i^N}} E(x_i^N(t_k))$$

$$x_i^N(t_{k+1}) = x_i^N(t_k) + \Delta t \, v_i^N(t_{k+1})$$

$$\alpha_i^N(t_{k+1}) = \alpha_i^N(t_k) \, .$$

2.5 Boltzmann equation

Equation (2.10) is a homogeneous Boltzmann system with, in general, inhomogeneous initial condition. In the next section we show how to discretize (2.10) in position space Λ and assume here locally homogeneous initial conditions. Therefore, we can restrict our considerations to the phase space $\Pi = \Gamma$.

Different from the approach of the last section, we first discretize the Boltzmann system in time by a simple explicit Euler step

$$f_{k+1}(z) = f_k(z) + \Delta t J[f_k]f_k(z) \qquad (2.19)$$

where $f_k(z) = f(t_k, z)$ and $t_k = k\Delta t$.

In recent publications [5, 28] it is also shown how to derive implicit and second order particle methods for the Boltzmann equation.

In the following, we explain the principle of the derivation of the measure form of the time discretized homogeneous Boltzmann system (2.19) using the detailed balance relation (2.5) for each collision process. A detailed description can be found in [24].

For simplicity we only consider one collision process (2.1) — reactive or nonreactive — with the corresponding inverse process

$$s' + s'^* \longrightarrow s + s^*$$

satisfying relation (2.5).

For any test function $\phi : \Gamma \to I\!R$ we can transform the collision operator into the following form:

$$\int_\Gamma \phi(z) J[f] f(z) dz = \int_\Gamma \int_\Gamma \int_{\Omega'} \Theta_{s,s^*}^{s',s'^*}(E,\omega;\omega') \phi(z) \{f(z'^*)f(z') - f(z^*)f(z)\} d\varpi' dz^* dz$$

$$= \int_\Gamma \int_\Gamma \int_{\Omega'} \{\phi(z') - \phi(z)\} \Theta_{s,s^*}^{s',s'^*}(E,\omega;\omega') d\varpi' f(z^*) dz^* f(z) dz.$$

The last step is obtained by a coordinate transformation of the postcollisional variables in the gain term with the collision transformation $(z', z'^*, \omega) = C_{s,s^*}^{s',s'^*}(z, z^*, \omega')$ and the use of the detailed balance relation (2.5).

Then the weak form of the time discretized Boltzmann equation reads

$$\int_\Gamma \phi(z) f_{k+1}(z) dz = \int_\Gamma \int_\Gamma \int_{\Omega'} \left(\phi(z') \Delta t \Theta_{s,s^*}^{s',s'^*}(E,\omega;\omega') \right. \qquad (2.20)$$

$$\left. + \phi(z) \left(1 - \Delta t \Theta_{s,s^*}^{s',s'^*}(E,\omega;\omega')\right) \right) d\varpi' f_k(z^*) dz^* f_k(z) dz,$$

where we have used the property that f_k is the density of a normalized measure for all $k \in I\!N$. Under the restriction that

$$\int_{\Omega'} \Delta t \Theta_{s,s^*}^{s',s'^*}(E,\omega;\omega') d\varpi' < 1, \qquad (2.21)$$

which is a consequence of the explicit time discretization, (2.20) leads to the measure–valued form of the time discretized Boltzmann equation replacing $f_k(z)dz$ by $d\mu_k$. Introducing an additional variable $\lambda \in [0;1]$, first proposed in [25], the measure equation 2.19) can be written in a closed form

$$\int_\Gamma \phi(z) d\mu_{k+1} = \int_\Gamma \int_\Gamma \int_{\Omega'} \int_0^1 \left(\mathcal{X}_{[0,\Delta t\Theta(E,\omega;\omega')]}(\lambda) \phi(z') + \mathcal{X}_{[\Delta t\Theta(E,\omega;\omega'),1]}(\lambda) \phi(z) \right) d\lambda d\varpi' d\mu_k^* d\mu_k$$

$$= \int_\Gamma \int_\Gamma \int_{\Omega'} \int_0^1 \phi \circ \Psi_{\Delta t}(\lambda, \omega', z^*, z) d\lambda d\varpi' d\mu_k^* d\mu_k.$$

The mapping $\Psi_{\Delta t}: [0,1] \times \Omega' \times \Gamma \times \Gamma \to \Gamma$ is defined as

$$\Psi_{\Delta t}(\lambda, \omega', z^*, z) = \begin{cases} z', & \lambda \in [0, \Delta t \Theta_{s,s^*}^{s', s'^*}(E, \omega; \omega')] \\ z, & \lambda \in [\Delta t \Theta_{s,s^*}^{s', s'^*}(E, \omega; \omega'), 1] \,. \end{cases}$$

Applying the transformation theorem for measures gives the time discretized measure form of the Boltzmann equation

$$\mu_{k+1} = \Psi_{\Delta t}(\lambda \otimes \varpi' \otimes \mu_k^* \otimes \mu_k). \tag{2.22}$$

The question of the construction of a discrete product measure $\mu_k^N \otimes \mu_k^N$ with N particles is a serious problem first solved in [3] for identically weighted particle approximations and extensively studied in [25] for arbitrary weights. Here we use this result which leads to the following

(**B**) SIMULATION SCHEME FOR THE BOLTZMANN SYSTEM:

For every time step $t_k = k\Delta t$, $k = 0, 1, \ldots$.

- choose for every particle with index i a collision partner with index $c(i)$ due to the target weight distribution.

- choose an identically weighted particle approximation $(\lambda, \omega')^N \in [0,1] \times \Omega'$ of $\lambda \otimes \varpi$.

- define the postcollisional state $(z_i^N)(t_{k+1})$ by

$$(z_i^N)(t_{k+1}) = \Psi_{\Delta t}(\lambda_i^N, \omega_i'^N, z_{c(i)}^N, z_i^N,)(t_k) \tag{2.23}$$

$$\alpha_i^N(t_{k+1}) = \alpha_i^N(t_k).$$

The weights are kept constant in both simulation schemes but the index of a particle may change during a reactive collision. This destroys the identical weights in the species as far as not all weights are identical. The problem is discussed in section 3.4 in connection with the question of mass conservation.

The convergence of the simulation scheme (**B**) is proved in the case of a general reactive system in [24] assuming that the scattering kernel is bounded.

2.6 Space–discretization: Mollifiers

To complete our numerical method we have to answer several questions arising from a necessary discretization in position space Λ. First, the solution of the homogeneous Boltzmann equation by scheme (**B**) assumes the homogeneity of the initial distribution. Second, the Poisson solver is defined on a grid on Λ to avoid the N^2-effort for the direct solution of the N-body problem. Therefore, one has to interpolate the forces from the grid points to the particle positions and the charge of particles to the grid. Another question related to the latter one is the definition of macroscopic quantities like the charge density. All these problems may be handled by the concept of a mollifier in position space only.

For a discrete measure $\mu^N = \delta_{\alpha,p}^N$ on the phase space $\Pi = \Lambda \times \Gamma$ we define the mollified measure $\mu^{\Delta x, N}$ by the convolution with a kernel $\beta^{\Delta x}(\cdot, \cdot)$

$$\mu^{\Delta x, N} = \int_\Lambda \beta^{\Delta x}(x, x_*) d\mu_*^N = \sum_{i=1}^N \alpha_i^N \beta^{\Delta x}(x, x_i^N) \delta_{z_i^N} \lambda_x. \tag{2.24}$$

Therefore, $\mu^{\Delta x,N}$ is a continuous measure with respect to x and discrete in z. The kernel $\beta^{\Delta x} : \Lambda \times \Lambda \to I\!R_+$, also called mollifier, is assumed to be continuous, bounded, symmetric, normalized and decaying in the following sense: There exists a constant C_β, such that

$$\int_\Lambda \|x - x_*\| \beta^{\Delta x}(x, x_*) dx_* \leq C_\beta \Delta x \qquad \text{for all } x \in \Lambda. \tag{2.25}$$

Then it is easy to prove that $\mu^{\Delta x,N}$ converge to μ^N in the weak sense of measures if Δx goes to zero [24].

Typical examples are constructed by symmetric, continuous, bounded, normalized generating functions $\mathcal{G} : I\!R^3 \to I\!R_+$ with compact support:

$$\beta_\mathcal{G}^{\Delta x}(x, x_*) = \frac{1}{\Delta x} \mathcal{G}\left(\frac{x - x_*}{\Delta x}\right). \tag{2.26}$$

The connection to standard discretization techniques on a mesh can be seen by the following mollifiers. Let $\bigcup_{k \in K} Z_k^{\Delta x} = \Lambda$ be a triangulation with tetrahedrons $Z_k^{\Delta x}$, $diam(Z_k^{\Delta x}) \leq \Delta x$, and corresponding nodes $y_j, j = 1, ..., M$. Then the charactristic function $\mathcal{X}_{Z_k^{\Delta x}}$ on a cell as well as the piecewise linear ansatz functions $\mathcal{Y}_j^{\Delta x}(x)$ defined on the nodes by $\mathcal{Y}_j^{\Delta x}(y_k) = \delta_{kj}$ generate mollifiers by

$$\beta_\mathcal{X}^{\Delta x}(x, x_*) = \sum_{k \in K} \frac{\mathcal{X}_{Z_k^{\Delta x}}(x) \mathcal{X}_{Z_k^{\Delta x}}(x_*)}{\int \mathcal{X}_{Z_k^{\Delta x}}(y) dy} \quad \text{and} \quad \beta_\mathcal{Y}^{\Delta x}(x, x_*) = \sum_{j=1}^M \frac{\mathcal{Y}_j^{\Delta x}(x) \mathcal{Y}_j^{\Delta x}(x_*)}{\int \mathcal{Y}_j^{\Delta x}(y) dy}. \tag{2.27}$$

Meaningful discrete moments can be defined by the mollified measure as

$$\mathcal{M}_\beta^{\Delta x}[\varphi](t, x) = \int_\Gamma \varphi_s(z) d\mu_t^{\Delta x,N} = \sum_{i=1}^N \alpha_i^N \beta^{\Delta x}(x, x_i^N(t)) \varphi_s(z_i^N(t)). \tag{2.28}$$

Since the kernel is normalized the discrete moments are globally exact, i.e.,

$$\int_\Lambda \mathcal{M}_\beta^{\Delta x}[\varphi](t, x) dx = \int_\Lambda \mathcal{M}[\varphi](t, x) dx. \tag{2.29}$$

Moreover the mollifiers $\beta_\mathcal{X}^{\Delta x}$ and $\beta_\mathcal{Y}^{\Delta x}$ produce locally exact moments on each cell respectively each node, which can be seen by the decomposition of the moments in the basis of the ansatz functions. For example,

$$\mathcal{M}_\mathcal{Y}^{\Delta x}[\varphi](t, x) = \sum_{j=1}^M M_j[\varphi] \mathcal{Y}_j^{\Delta x}(x); \quad M_j[\varphi] = \sum_{i=1}^N \alpha_i^N \varphi(z_i^N(t)) \frac{\mathcal{Y}_j^{\Delta x}(x_i^N(t))}{\int \mathcal{Y}_j^{\Delta x} dy}, \tag{2.30}$$

where $M_j[\varphi]$ is the moment defined on node j.

Due to the convergence of $\mu^{\Delta x,N}$ to μ^N we can replace discrete measures in the simulation schemes (**V**) and (**B**) by mollified measures without changing the convergence properties [24]. The substitution of the target measure μ_*^N in (2.22) by a mollified measure $\mu_*^{\Delta x,N}$ answers the first question. The use of mollifier $\beta_\mathcal{X}^{\Delta x}$ results in local homogeneous particle approximations and simulation scheme (**B**) can be directly applied. With an additional step distributing the particles to the nodes y_j the mollifier $\beta_\mathcal{Y}^{\Delta x}$ leads to a generalized Boltzmann scheme. Mollifiers $\beta_\mathcal{G}^{\Delta x}$ produce grid free simulation schemes. For details see [24].

The second problem depends on the Poisson solver. For a finite element method with linear ansatz functions we show in section 3.2 that a mollification with the kernel $\beta_\mathcal{Y}^{\Delta x}$ is the right way to obtain a consistent method.

3 Discrete conservation laws

3.1 Introduction

Kinetic systems of the general form (2.1) as well as the reduced asymptotic systems [24] are conservative physical systems with constant total mass, charge, momentum and energy at least in the interior of the domain Λ. The Boltzmann–Vlasov–Poisson system together with the boundary conditions above specified satisfies mass and charge conservation, in other words,

$$\frac{d}{dt}\int_\Lambda \mathcal{M}[\varphi]dx = 0 \qquad (3.1)$$

with $\varphi = (m_s)_{s\in I}$ respectively $\varphi = (q_s)_{s\in I}$. Moreover, the total energy is conserved, i.e.,

$$\frac{d}{dt}\left\{\int_\Lambda\int_\Gamma \left(\frac{m}{2}|v|^2 + e\right)fdzdx + \frac{1}{2}\int_\Lambda \epsilon_0 |\nabla_x U(x)|^2 dx\right\} = 0. \qquad (3.2)$$

It is a well–known principle in doing numerics that properties of the original system have to take over to the discrete algorithm. Discrete N–particle systems arising from a particle approximation can show totally different behavior if one pays no attention to the conserved quantities. In the case of the classical Boltzmann equation, Greengard and Reyna [15] have proved that the nonconservative Nanbu scheme always runs into the zero temperature state, although it conserves momentum and energy in the mean. In collisionless plasma simulation schemes it is well–known [4] that a consistent interpolation between the particle positions and the grid of the electric field is necessary to neglect purely numerical instabilities.

In the next sections modifications and extensions of the algorithms (**V**) and (**B**) are discussed which fulfill exactly the conservation equations for fixed particle number.

3.2 Conservation for Vlasov–Poisson systems

During the drift step, kinetic and electric energy is exchanged which makes a common analysis of the particle scheme for the Vlasov equation and the Poisson solver necessary. In this section we also answer the question how to interpolate between quantities defined on the grid and the particle positions. The resulting method is an extension of well–known schemes to unstructured meshes.

Let $\bigcup_{i\in I} Z_i^{\Delta x}$ be a triangulation of Λ with nodes $y_j, j = 1,...,M$ and linear ansatz functions $\mathcal{Y}_j^{\Delta x}$ as in (2.27). We use a Galerkin method, e.g the test functions coincide with the ansatz functions, and obtain the well-known finite element form (for details see [16, §4])

$$\sum_{k=1}^M \int_\Lambda u_k \nabla \mathcal{Y}_k^{\Delta x}(x) \cdot \nabla \mathcal{Y}_j^{\Delta x}(x)dx = \epsilon_0^{-1} \int_\Lambda \mathcal{Y}_j^{\Delta x}(x)\tau(x)dx \qquad \forall\, j=1,...,M, \qquad (3.3)$$

where the potential is defined by $U^{\Delta x}(x) = \sum_k u_k \mathcal{Y}_k^{\Delta x}(x)$.
In the discrete case we substitue the right hand side of (3.3) by the discrete charge density $\tau^N = \sum_{i=1}^N \alpha_i^N q_{s_i} \delta_{x_i^N}$ and obtain

$$\int_\Lambda \mathcal{Y}_j^{\Delta x} d\tau^N = \sum_{i=1}^N \alpha_i^N q_{s_i} \mathcal{Y}_j^{\Delta x}(x_i^N) = \int_\Lambda \mathcal{Y}_j^{\Delta x}(x)dx\,\tau_j \qquad (3.4)$$

where τ_j denotes the total charge evaluated at the node y_j (compare with (2.30)). Then, the charge density on the grid is

$$\tau^{\Delta x}(x) = \sum_{j=1}^M \tau_j \mathcal{Y}_j^{\Delta x}(x), \qquad (3.5)$$

which is consistent with the charge density given by the particles, i.e.,

$$\int_\Lambda \tau^{\Delta x}(x)dx = \int_\Lambda \int_\Gamma q\, d\mu^N. \tag{3.6}$$

Moreover, the total charge in the discrete system is conserved.
The electric field E is obtained for all $x \in \Lambda$ by the negative gradient of the potential

$$E(x) = -\sum_{j=1}^{M} u_j \nabla \mathcal{Y}_j^{\Delta x}(x), \tag{3.7}$$

and therefore constant on each tetrahedron. A separate interpolation to the coordinates of the particles is not necessary.
Moreover, the energy conservation is valid in discrete form, if we neglect the error of the time discretization of the simulation scheme for the Vlasov equation. We have

$$\frac{d}{dt}\left\{\int_\Lambda \int_\Gamma \left(\frac{m}{2}|v|^2 + e\right) d\mu^N + \frac{1}{2}\int_\Lambda \epsilon_0 \left|\nabla_x U^{\Delta x}(x)\right|^2 dx\right\} = 0, \tag{3.8}$$

which is shown using the properties of the simulation scheme, of the mollifier and the finite element method.
For a rectangular grid the prescribed method reduces to the area–weighted method by Lewis (see [4, §X-3]).

3.3 Conservation for Boltzmann equations

The simulation scheme (B) is of Nanbu type, which means that in every collision only one particle state is changed. Therefore, it conserves momentum and energy only in the mean and produces the wrong stationary state for finite particle numbers [15]. In [2], Babovsky derived a symmetric scheme conserving exactly momentum and energy in every pairwise collision and therefore, during the hole collision step. These pairwise collisions change the state of both collision partners due to the collision transformation and reduce the numerical effort by a factor two. But, the weights are assumed to be identical. An extension of the Babovsky method to particles with different weights seems to be the simplest way to achieve conservation. Why this is impossible can be seen by the following arguments.
The change of momentum ΔI_2 and energy ΔE_2 during a collision of two particles with velocity v and v^*, with different weights α and α^*, is

$$\begin{aligned}\Delta I_2 &:= \alpha\, mv' + \alpha^* m^* v'^* - \alpha\, m\, v - \alpha^* m^* v^*, \\ \Delta E_2 &:= \alpha\, \mathcal{E}(z') + \alpha^* \mathcal{E}(z'^*) - \alpha\, \mathcal{E}(z) - \alpha^* \mathcal{E}(z^*)\end{aligned} \tag{3.9}$$

with particle energy $\mathcal{E}(z) = \frac{m}{2}|v|^2 + e$. For an elastic collision the internal energies e are constant and the collision transformation $\mathcal{C}: \mathbb{R}^3 \times \mathbb{R}^3 \times S^2 \to \mathbb{R}^3$ reads $v' = \mathcal{C}(v, v^*, \omega')$, where

$$\mathcal{C}(v, v^*, \omega') = G(v, v^*) + \frac{\mu}{m}|v - v^*|\omega', \tag{3.10}$$

with reduced mass $\mu = \dfrac{m\, m^*}{m + m^*}$ and center of mass velocity $G = \dfrac{mv + m^* v^*}{m + m^*}$. From this we find the relations $v'^* = \mathcal{C}(v, v^*, \omega'^*)$, $v = \mathcal{C}(v, v^*, \omega)$ and $v^* = \mathcal{C}(v, v^*, -\omega)$ with the direction of the relative velocity $\omega = \dfrac{v - v^*}{|v - v^*|}$. Then the change of momentum and energy (3.9) can be written as

$$\Delta I_2 = \alpha m(\mathcal{C}(v,v^*,\omega') - \mathcal{C}(v,v^*,\omega)) + \alpha^* m^*(\mathcal{C}(v,v^*,\omega'^*) - \mathcal{C}(v,v^*,-\omega))$$
$$= \mu|v - v^*|\left(\alpha\omega' + \alpha^*\omega'^* + (\alpha^* - \alpha)\omega\right) \quad (3.11)$$

and

$$\Delta E_2 = \alpha \frac{m}{2}(\mathcal{C}(v,v^*,\omega')^2 - \mathcal{C}(v,v^*,\omega)^2) + \alpha^* \frac{m^*}{2}(\mathcal{C}(v,v^*,\omega'^*)^2 - \mathcal{C}(v,v^*,-\omega)^2)$$
$$= \mu|v - v^*|\, G(v,v^*) \cdot \left(\alpha\omega' + \alpha^*\omega'^* + (\alpha^* - \alpha)\omega\right). \quad (3.12)$$

Therefore, $\Delta I_2 = 0$ and $\Delta E_2 = 0$, if and only if

$$\alpha\omega' + \alpha^*\omega'^* = (\alpha - \alpha^*)\omega, \quad (3.13)$$

which implies $\omega' = -\omega'^*$ and therefore, by (3.13), $\omega' = \omega$, if $\alpha \neq \alpha^*$. Hence, no collision occurs and extending Babovsky's method is impossible.

To guarantee conservation of momentum and energy it is not necessary to impose the restrictions $\Delta I_2 = 0$ and $\Delta E_2 = 0$. It is rather sufficient to fulfill

$$\Delta I_N := \sum_{i=1}^{N} \alpha_i m_i v'_i - \sum_{i=1}^{N} \alpha_i m_i v_i = 0, \quad (3.14)$$

$$\Delta E_N := \sum_{i=1}^{N} \alpha_i m_i |v'_i|^2 - \sum_{i=1}^{N} \alpha_i m_i |v_i|^2 = 0, \quad (3.15)$$

with $v'_i = \mathcal{C}(v_i, v^*_{c(i)}, \omega'_i)$ for $i = 1, ..., N$.

Now, the problem can be formulated as follows: Is it possible to choose collision parameters $\omega'_i \in S^2$, which, e.g., are uniformly distributed (for an isotropic collision law), such that (3.14) and (3.15) holds?

The answer is yes and we sketch here the derivation of the algorithm for the energy equation only. For extensions to the momentum equation, to the energy equations with internal energy exchange and the convergence results we refer to [19, 24].

The energy equation (3.15) can be written in the form

$$\sum_{i=1}^{N} \alpha_i \mu_i g_i\, G_i \cdot \omega'_i = E_N := \sum_{i=1}^{N} \alpha_i \mu_i g_i\, G_i \cdot \omega_i \quad (3.16)$$

where $g_i = |v_i - v_{c(i)}|$ is the magnitude of the relative velocity. Equation (3.16) is linear in $G_i \cdot \omega'_i$. If we choose a local orthonormal system (e_i^1, e_i^2, e_i^3) with polar axis $e_i^3 = G_i/|G_i|$, we can represent ω'_i by

$$\omega'_i = \sin\theta'_i(\cos\varphi'_i\, e_i^1 + \sin\varphi'_i\, e_i^2) + \cos\theta'_i\, e_i^3 \quad (3.17)$$

with polar angle $\theta'_i \in [0, \pi]$ and azimuth angle $\varphi'_i \in [0, 2\pi]$. Hence, the energy equation (3.16) reduces to a linear equation in the cosine of the polar angle $c'_i = \cos\theta'_i$,

$$\sum_{i=1}^{N} b_i c'_i = E_N \quad (3.18)$$

with strict positive coefficients $b_i = \alpha_i \mu_i g_i |G_i|$. Especially, E_N can be written in the form $E_N = \sum_{i=1}^{N} b_i c_i$ with $c_i = \eta_i \cdot e_i^3$.

Therefore, the problem has been reduced to finding uniformly distributed points $c'_i, i = 1,...,N$ in $[-1;1]$ which fulfill equation (3.18). Equation (3.18) has at least one solution: the trivial one with $c'_i = c_i$, such that no collision occurs.

Moreover, it is obvious that it is not possible to find uniformly distributed points on a hyperplane given by (3.18) in $[-1;1]^N$ for arbitrary, but fixed coefficients b_i and right hand side E_N. E.g., one may choose $b_i = 1$, $i = 1,...,N$, und $E_N = 1$. Then, all N-dimensional unit vectors e^i, $i = 1,...,M$ are solutions and consequently all c'_i must be nonnegative and cannot be uniformly distributed in $[-1,1]$. But, varying the right hand side E_N, we can impose the correct distribution by the following recursive algorithm:

(E) ALGORITHM CONSERVING ENERGY:

Recursively for $k = 1,...,N$:
choose c'_k in $[L_k; R_k] \subset [-1;1]$ uniformly distributed where $[L_k; R_k]$ is the maximal interval such that the reduced constraint equation for the remaining variables $c'_{k+1},....c'_N$

$$\sum_{i=k+1}^{M} b_i c'_i = E_{M-k} \quad \text{with} \quad E_{M-k} := E_M - \sum_{i=1}^{k} b_i c'_i \qquad (3.19)$$

is solvable.

Due to the linearity of the constraint equation (3.18) the interval $[L_k; R_k]$ is uniquely determined. In the N-th step algorithm (**E**) stops with

$$c'_N = \frac{1}{b_N} \left(E_M - \sum_{i=1}^{N-1} b_i c'_i \right), \qquad (3.20)$$

which implies the constraint energy equation $\Delta E_N = 0$ (3.15). The proof of the uniform distribution of c'_k in $[-1;1]$ and hence the convergence of the simulation scheme (**B**) with algorithm (**E**) is shown in [24]. There are some modifications in algorithm (**E**) necessary to prove the convergence, which also improve the algorithm from a numerical point of view.

As mentioned above the momentum and energy equations in the inelastic case can also be satisfied by transforming the system of constraint equations to a linear one of form (3.18). For details and numerical tests of algorithm (**E**) compare [19, 24]. In the last section we present some applications of algorithm (**E**) for the reduced plasma system as discussed in the next section.

3.4 Conservation in reactive systems

In this section we answer the remaining problems questioned in this paper. We explain how to derive weighted particle methods for reactive systems, which change the weights of the particles in such a way that they are always proportional to the concentrations and also conserve mass and charge. The idea is shown by means of the following reduced plasma system

$$\partial_t C = -C \left(k_R(T)C^2 - k_I(T)(1-C) \right), \qquad (3.21)$$

$$\partial_t T = \left(T + \frac{2}{3}I \right) \left(k_R(T)C^2 - k_I(T)(1-C) \right), \qquad (3.22)$$

$$\partial_t f_i = J_{i,i}[f_i]f_i + J_{i,o}[f_o]f_i + f_o k_I(T)C - f_i k_R(T)C^2, \qquad (3.23)$$

$$\partial_t f_o = J_{o,o}[f_o]f_o + J_{o,i}[f_i]f_o + f_i k_R(T)C^2 - f_o k_I(T)C. \qquad (3.24)$$

This space homogeneous kinetic system for the ions i (3.23) and neutrals o (3.24) is coupled with macroscopic equations for the electron density C (3.21) and temperature T (3.22). $J_{\cdot,\cdot}$ denote the nonreactive Boltzmann collision operators, k_I and k_R the rate constants for the ionization and recombination reaction

$$A + e \longleftrightarrow A^+ + e + e. \qquad (3.25)$$

The derivation of this reduced plasma system as an asymptotic limit for the mass ratio of electrons and heavy particles going to zero is described in [24].

For the numerical solution of the plasma system we point out that the equations for the electron density and temperature can be solved first — independent of the kinetic equations. Then it is possible to discretize the kinetic system by an explicit Euler step in time and to write down the measure–valued form analogeous to (2.22) (see [24]). It is not surprising that the corresponding particle method causes the same disadvantage as mentioned in section 2.5. The weights are always constant and therefore mixed through the different species if a reaction occurs and the proportionality of the weights and the concentrations is distroyed, which is the essential point of our weighted particle method.

To derive evolution equations for the particle positions in the phase space, such that at every time step the weights are proportional to the concentration of the species, we write the phase density f_s of species s as a product of the concentration γ_s and a remaining density g_s:

$$g_s = \frac{1}{2\gamma_s} f_s, \qquad s = i, o. \qquad (3.26)$$

The factor 2 is chosen such that normalization of f implies normalization of g.

Integration of the kinetic equations (3.23), (3.24) with respect to the state variable z results in equations for the concentrations of the form

$$\partial_t \gamma_i = k_I(T) C(T) \gamma_o - k_R(T) C(T)^2 \gamma_i,$$
$$\partial_t \gamma_o = k_R(T) C(T)^2 \gamma_i - k_I(T) C(T) \gamma_o. \qquad (3.27)$$

Using (3.27) yields the kinetic equations for g

$$\partial_t g_i = 2\gamma_i J_{i,i}[g_i] g_i + 2\gamma_o J_{i,o}[g_o] g_i + k_I(T) C(T) \frac{\gamma_o}{\gamma_i}(g_o - g_i),$$
$$\partial_t g_o = 2\gamma_o J_{o,o}[g_o] g_o + 2\gamma_i J_{o,i}[g_i] g_o + k_R(T) C(T)^2 \frac{\gamma_i}{\gamma_o}(g_i - g_o). \qquad (3.28)$$

Next, we write the kinetic system (3.28) for g in measure form and derive a corresponding particle algorithm, again in the way described above. If we now choose all weights identical then we obtain a particle method solving system (3.28) with constant weights. This equiweighted particle approximation of g may be transformed to a particle approximation for f multiplying the weights by the concentrations which are known from system (3.27). Therefore, the weights evolve due to equations (3.27).

The simulation scheme in detail is of no interest here and can be found in [24]; nevertheless several nice properties of the resulting particle algorithm for the kinetic system (3.23) and (3.24) should be mentioned in the following. First, the weights are by construction at every timestep proportional to the concentrations. Second, the particle number of each species and consequently the total particle number is constant in time. Hence the total mass and charge are conserved; i.e., the discrete mass $\rho_s^N(t)$ of species s at time t is

$$\rho_s^N(t) := \sum_{\substack{i=1 \\ s_i^N = s}}^{N} m_s \alpha_i^N(t) = m_s \gamma_s(t) =: \rho_s(t), \qquad (3.29)$$

where we have used the proportionality relation (2.8) between weights and concentrations.

These properties are the main reasons why the factorization of the phase density f into the concentrations and a remainder is the right way to find conservative numerical methods for arbitrary reactive systems.

3.5 Numerical examples

In this section we solve the reduced plasma system for one species of ions and neutrals with the weighted particle method derived in the last section. The conservation of momentum and energy is guarenteed by the generalization of algorithm (**E**) (see section 3.4). As initial condition we choose two different Maxwellians with parameters

$$\gamma_i = 0.2 \qquad\qquad \gamma_o = 0.8$$
$$u_i = (2.2, 0.0, 0.0) \qquad\qquad u_o = (-0.55, 0.0, 0.0)$$
$$T_i = 1.0 \qquad\qquad T_o = 0.5 \ .$$

The input data are chosen such that the stationary state is at temperature $T_s^\infty = 1$ and mean velocity $u_s^\infty = 0$ for $s = i, o$. The concentrations and the electron temperature T are obtained by Saha's equation with values $C^\infty = \gamma_i^\infty = 1 - \gamma_o^\infty = 0.249$ and $T = 0.403$.

The calculations denoted by C1 – C3 are done for different total particle numbers as shown in table 1. The calculation C0 is a simulation using a standard particle method with identically weighted particles and pairwise collisions. Hence, the partition of the particle numbers per species is due to the concentrations.

Table 1: Comparison of particle numbers and CPU-times

	N	N_i	N_o	CPU
C0	1000	200	800	1.00
C1	200	100	100	0.45
C2	500	250	250	1.04
C3	1000	500	500	2.15

The computatonal effort for C0 is only half compared to the weighted particle method with the same total number of particles because of the pairwise collisions.

In figure 1 the evolution of the ion concentrations for C0 and C1 are compared with the electron concentration C. Ion and electron concentrations should be the same due to the neutrality of the plasma. The results obtained with the weighted particle method agree exactly. The standard method runs into troubles if the reaction rate per time step is less than $1/N$.

Other macroscopic quantities like the mean velocity or the temperature are not sensitive to the error in the concentrations for C0. On the other hand, the weighted schemes show no instabilities since momentum and energy are conserved. Figure 2 shows the good agreement of the temperatures for C0 and C1 on a logarithmic scale.

Differences can be seen in the fluctuations of the macroscopic quantities. The variances of the temperature calculated using 1000 independent runs are shown in figure 3. The dotted lines correspond to the computations with the weighted particle method (C1 – C3) and the straight line to C0. With the same computational effort and half number of particles (see table 1) the variance for C2 is slightly lower than for C0.

The standard particle method for realistic small concentrations of ions (about 1%) leads to extremely high particle numbers. Then, the advantage of the weighted particle method is obvious.

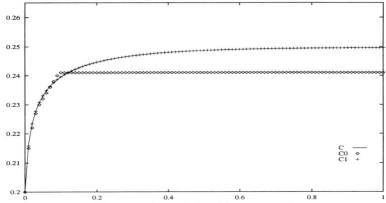

Figure 1 Time evolution of ion concentrations

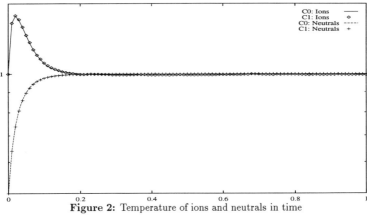

Figure 2: Temperature of ions and neutrals in time

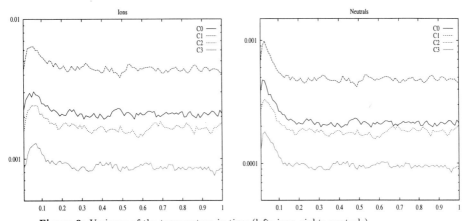

Figure 3: Variance of the temperature in time (left: ions, right: neutrals)

337

4 Conclusion

During the four year research project founded by the DFG we have developed two main approaches to solve kinetic systems arising in the simulation of plasma effects during the re-entry of a space vehicle.
The first tool is the use of asymptotic methods to handle the problem of different time scales for electrons and heavy particles. Different reduced kinetic systems are found, which are valid on different characteristic scales. These results are presented in [24].
The second tool is the weighted particle method explained in this paper. The weights are chosen proportional to the concentrations of the species, which allow the use of less particles. The problems arising from the nonconservative form of the collision scheme are solved, and the convergence of the method, including the modifications in the collision step, is proved. Numerical tests show the advantages of the weighted particle method in comparison to equi-weighted methods, if strong differences in the concentrations occur.

References

[1] Arkeryd, L.; Cercignani, C.; Illner, R.: MEASURE SOLUTIONS OF THE STEADY BOLTZMANN EQUATION IN A SLAB, Commun. Math. Phys. 142, 285–296 (1991).

[2] Babovsky, H.: ON A SIMULATION SCHEME FOR THE BOLTZMANN EQUATION, M^2AS 8, 223 – 233 (1986).

[3] Babovsky, H.: A CONVERGENCE PROOF FOR NANBU'S BOLTZMANN SIMULATION SCHEME, European J. Mech., B/Fluids 8, No. 1, 41 – 55 (1989).

[4] Birdsall, C.K.; Langdon, A.B.: PLASMA PHYSICS VIA COMPUTER SIMULATION, McGraw-Hill, New York (1981).

[5] Bobylev, A.V.; Struckmeier, J.: IMPLICIT AND ITERATIVE METHODS FOR THE BOLTZMANN EQUATION, AGTM-Report Nr. 123 (1994), printed in TTSP.

[6] Cercignani, C.; Illner, R.; Pulvirenti, M.: THE MATHEMATICAL THEORY OF DILUTE GASES, Springer, New York (1994).

[7] Chapman, S.; Cowling, T.G.: THE MATHEMATICAL THEORY OF NON-UNIFORM GASES, University Press, Cambridge (1970).

[8] Cottet, G.-H.; Raviart, P.-A.: PARTICLE METHODS FOR THE ONE-DIMENSIONAL VLASOV-POISSON EQUATIONS, SIAM J. Numer. Anal. Vol. 21, No.1, 52–76 (1984).

[9] Degond, P.; Lucquin-Desreux, B.: THE ASYMPTOTICS OF COLLISION OPERATORS FOR TWO SPECIES OF PARTICLES OF DISPARATE MASSES, submitted to TTSP.

[10] Degond, P.; Lucquin-Desreux, B.: TRANSPORT COEFFICIENTS OF PLASMAS AND DISPARATE MASS BINARY GASES, submitted to M^3AS.

[11] Degond, P.; Raviart, P.A.: AN ASYMPTOTIC ANALYSIS OF THE ONE-DIMENSIONAL VLASOV-POISSON SYSTEM: THE CHILD-LANGMUIR LAW, Asymp. Anal. 4, 187–214 (1991).

[12] Desvillettes, L.: ON THE CONVERGENCE OF SPLITTING ALGORITHMS FOR SOME KINETIC EQUATIONS, Asymp. Anal. 6, 315–333 (1993).

[13] Desvillettes, L.; Mischler, S.: ABOUT THE SPLITTING ALGORITHM FOR BOLTZMANN AND B.G.K. EQUATIONS, submitted to Asymp. Anal.

[14] Feng Kang; Qin Meng–shao: THE SYMPLECTIC METHODS FOR THE COMPUTATION OF HAMILTONIAN EQUATIONS, 1–37, in: Zhu you–lan; Gu Ben–yu (Eds.): Proc. of the 1st Chinese Conf. on Numerical Methods of PDE's (1986), Springer, Berlin (1987).

[15] Greengard, C.; Reyna, L.G.: CONSERVATION OF EXPECTED MOMENTUM AND ENERGY IN MONTE CARLO PARTICLE SIMULATION, Phys. Fluids A 4 (4) 849–852 (1992).

[16] Großmann, C.; Roos, H.-G.: NUMERIK PARTIELLER DIFFERENTIALGLEICHUNGEN, Teubner, Stuttgart (1992).

[17] Neunzert, H.: MATHEMATICAL INVESTIGATIONS ON PARTICLE–IN–CELL METHODS, Proc. of XIII Symp. advanc. Probl. Meth. Fluid Mech., Fluid Dyn. 9 (1978).

[18] Neunzert, H.: AN INTRODUCTION TO THE NONLINEAR BOLTZMANN–VLASOV EQUATION, 60–110, in: Cercignani, C. (Ed.): KINETIC THEORIES AND THE BOLTZMANN EQUATION, Springer, Berlin (1984).

[19] Neunzert, H.; Steiner, K.; Wick, J.: ENTWICKLUNG UND VALIDIERUNG EINES PARTIKELVERFAHRENS ZUR BERECHNUNG VON STRÖMUNGEN UM RAUMFAHRZEUGE IM BEREICH VERDÜNNTER IONISIERTER GASE, DFG–Bericht Ne 269/8-1 (1993).

[20] Neunzert, H.; Struckmeier, J.: PARTICLE METHODS FOR THE BOLTZMANN EQUATION, Acta Numerica (1995).

[21] Sack, W.: MODELLIERUNG UND NUMERISCHE BERECHNUNG VON REAKTIVEN STRÖMUNGEN IN VERDÜNNTEN GASEN, PhD Thesis, University of Kaiserslautern (1995).

[22] Schreiner, M.: WEIGHTED PARTICLES IN THE FINITE POINTSET METHOD, TTSP 22 (6), 793–817 (1993).

[23] Spatschek, K.H.: THEORETISCHE PLASMAPHYSIK, B.G. Teubner, Stuttgart (1990).

[24] Steiner, K.: KINETISCHE GLEICHUNGEN ZUR BESCHREIBUNG VERDÜNNTER IONISIERTER GASE UND IHRE NUMERISCHE BEHANDLUNG MITTELS GEWICHTETER PARTIKELVERFAHREN, PhD Thesis, University of Kaiserslautern (1995).

[25] Struckmeier, J.: DIE METHODE DER FINITEN PUNKTMENGEN — NEUE IDEEN UND ANREGUNGEN — PhD Thesis, University of Kaiserslautern (1994).

[26] Struckmeier, J.; Steiner, K.: BOLTZMANN SIMULATIONS WITH AXISYMMETRIC GEOMETRY, AGTM–Report Nr. 83 (1992).

[27] Struckmeier, J.; Steiner, K.: A COMPARISON OF SIMULATION METHODS FOR RAREFIED GAS FLOWS, Physics of Fluids A (1995).

[28] Struckmeier, J.; Steiner, K.: SECOND ORDER SCHEME FOR THE SPATIALLY HOMOGENEOUS BOLTZMANN EQUATION WITH MAXWELLIAN MOLECULES, AGTM–Report Nr. 127 (1995), printed in M^3AS.

Multiple Pressure Variable (MPV) Approach for Low Mach Number Flows Based on Asymptotic Analysis

K.J. Geratz, R. Klein
RWTH Aachen
Institut für Technische Mechanik
Templergraben 64, 52056 Aachen, Germany

C.D. Munz, S. Roller
Forschungszentrum Karlsruhe, Technik und Umwelt
Institut für Neutronenphysik und Reaktortechnik
PO-Box 3640
76021 Karlsruhe, Germany

SUMMARY

An asymptotic analysis of the compressible Euler equations in the limit of vanishing Mach numbers is used as a guideline for the development of a low Mach number extension of an explicit higher order shock capturing scheme. For moderate and large Mach numbers the underlying explicit compressible flow solver is active without modification. For low Mach numbers, the scheme employs an operator splitting technique motivated by the asymptotic analysis. Advection of mass and momentum as well as long wave acoustics are discretized explicitly, while in solving the sonic terms, the scheme uses an implicit pressure correction formulation to guarantee both divergence-free flow in the zero Mach number limit and appropriate representation of weakly nonlinear acoustic effects for small but finite Mach numbers. This asymptotics based approach is also used to show how to modify incompressible flow solvers to capture weakly compressible flows.

1. INTRODUCTION

1.1 Motivation

One motivation of the work to be presented stems from combustion applications. Reliable simulation of slow flames, flame acceleration and the deflagration-to-detonation transition (DDT) require an "all Mach number capability" of the simulation code: Laminar premixed combustion generates flows with Mach numbers $M_{lam} = 10^{-4} \ldots 10^{-3}$, while a detonation wave attains lead shock Mach numbers $M_{deto} = 5\ldots 10$. A considerable variety of numerical techniques for zero and/or low Mach number and weakly compressible flows have been proposed in recent years, e.g., [1],[3],[4],[5],[6],[7]. Nevertheless, each of these schemes fails to satisfy at least one of the following necessary requirements, which are motivated by the deflagration-to-detonation transition phenomenon. The scheme should

1. yield an acceptable numerical approximation for the equations of zero Mach number flow with large amplitude density variations [12], as $M = 0$,
2. provide an accurate representation of the long wave acoustics responsible for flow acceleration to higher Mach numbers, and it should
3. reduce to a reasonable explicit full compressible Euler solver for characteristic flow Mach numbers of order $M \approx 1$ and larger.

The principal idea behind the new class of schemes is to first perform an asymptotic analysis for small Mach numbers and to use the insight gained for designing a suitable numerical discretisation [8],[9],[10],[11]. It needs to be emphasized, that this procedure yields a scheme for the full Euler equations, not just one for solving some reduced asymptotic system. In this sense, the asymptotic analysis is used here in a quite unusual fashion: Rather than providing simplified limit equations, which are then solved by highly specialized methods with restricted applicability, the asymptotics serve as a guideline in designing numerical methods for the full equations, which operate efficiently even in a certain singular limit regime.

1.2 Reactive Euler equations

The compressible reactive Euler equations in nondimensional form read as

$$
\begin{aligned}
\rho_t &+ \nabla \cdot \vec{m} &= 0 \\
\vec{m}_t &+ \nabla \cdot (\vec{m} \circ \vec{v} + \tfrac{1}{M^2} p \overline{\overline{I}}) &= 0 \\
e_t &+ \nabla \cdot (\vec{v}[e + p]) &= 0 \\
(\rho Y)_t &+ \nabla \cdot (\vec{m} Y) &= -\rho \omega
\end{aligned}
\quad (1)
$$

where ρ, $\vec{m}$, e and Y are the conserved quantities mass, momentum, total energy per unit volume and the mass fraction of the unburned species, $\vec{v} \equiv \vec{m}/\rho$ is the flow velocity, p the pressure, ω the local reaction rate, $\overline{\overline{I}}$ denotes the unit tensor and the 'o' symbol indicates the tensorial product. A premixed fuel is assumed with only two species present, unburned and burned, whose thermodynamics are governed by the same γ-gas law and have the same molecular weights. Furthermore, the ratio of specific heats is given by $\gamma = c_p/c_v = $ constant, and the heat release per unit mass of unburned gas by q_0. The pressure is then related to the conserved quantities by the equation of state for mixtures of perfect gases,

$$ p = (\gamma - 1) \left[e - M^2 \frac{\vec{m}^2}{2\rho} - \rho Y q_0 \right] \quad \text{with} \quad \gamma = \text{const.} \quad (2) $$

The reference values used, are density ρ_{ref}, velocity v_{ref}, length l_{ref}, species Y_{ref} and an independent reference for the pressure p_{ref}. The reference velocity v_{ref} is then independent of $c_{ref} = (p_{ref}/\rho_{ref})^{1/2}$ and the nondimensional velocities $\vec{v}$ and $\vec{c}$ remain well-defined and finite in the limit of vanishing Mach number

$$M = \frac{v_{ref}}{\sqrt{p_{ref}/\rho_{ref}}} \to 0 \quad. \tag{3}$$

The parameter M is called the global Mach number and characterizes the nondimensionalization but not the local flow Mach number. The incompressible flow is then clearly distinguished from the compressible flow by the vanishing global Mach number (M=0). The wave speed vectors of the nondimensional Euler equations (1) are

$$\vec{a}_1(\vec{n}) = \vec{v} + \frac{1}{M}c\vec{n}, \qquad \vec{a}_2 = \vec{v} \quad. \tag{4}$$

The $\vec{a}_1(\vec{n})$ wave speed vectors become arbitrarily large as M converges to zero. As a consequence, the hyperbolic Euler equations converge towards a hyperbolic/elliptic system with infinite propagation rates of perturbations. Theoretical insight into this singular limit is obtained by a singular perturbation analysis for $M \to 0$.

2. SUMMARY OF ASYMPTOTIC ANALYSIS

2.1 Expansion ansatz

In a deflagration-to-detonation process the flow acceleration from low to high Mach numbers is caused by the action of acoustic pulses, that are generated by the unsteady premixed turbulent flame. All flow phenomena are then dominated by the time scale of the propagation of the unsteady flame front. The relevant asymptotic regime involves a single time scale (t), but multiple length scales represented by differently scaled spacial coordinates $(\vec{x}, \vec{\xi})$. The small scale variable $\vec{x}$ resolves entropy fluctuations and vortex structures, while the large scale variable $\vec{\xi}$ resolves acoustic pressure changes. The acoustic waves as well as the small scale incompressible flow both have a leading order influence on the velocity field.

An appropriate asymptotic expansion for the terms in the Euler equations, (e.g. density, velocity, pressure, mass fraction total energy per volume and reaction rate), $\underline{U} = (\rho, \vec{v}, p, Y, e, \omega)$ reads as

$$\underline{U}(\vec{x}, \vec{\xi}, t) = \underline{U}^{(0)}(\vec{x}, \vec{\xi}, t) + M\underline{U}^{(1)}(\vec{x}, \vec{\xi}, t) + M^2\underline{U}^{(2)}(\vec{x}, \vec{\xi}, t) + \cdots; \qquad M \ll 1, \tag{5}$$

with $\vec{\xi} = M\vec{x}$. This expansion is introduced into the compressible reactive Euler equations (1) and the equation of state (2) by using the relation $\nabla = \nabla_{\vec{x}} + M\nabla_{\vec{\xi}}$ for spatial derivatives. Using large scale differencing and spatial averaging, the following results are obtained (see [8] for details).

2.2 Pressure expansion

The pressure expansion exhibits three distinct physical roles of "the pressure" in a single time - multiple space scale regime:

1. The leading order pressure $p^{(0)}$ is spatially homogeneous and plays the role of a thermodynamic variable, i.e. $p^{(0)} = P_0(t)$. It is also called mean or background pressure.
2. The first order pressure $p^{(1)}$ varies spatially only on the acoustic length scale, i.e., $p^{(1)} = P^{(1)}(\vec{\xi}, t)$, and reflects the influence of long wave acoustics [8]. Flows that undergo considerable changes of the characteristic flow Mach number involve acoustic pressure amplitudes, which produce leading order effects on the velocity field. Therefore, the term $Mp^{(1)}$ in the pressure expansion must be considered and may not be excluded, as usually done in asymptotic derivations of the incompressible limit [12].
3. The second order pressure $p^{(2)}$ has small scale and acoustic scale spatial structure, i.e., $p^{(2)} = p^{(2)}(\vec{x}, \vec{\xi}, t)$. It acts as a local ballance of forces agent and guarantees the divergence constraint of incompressible flows as $M \to 0$.

Hence, the pressure expansion scheme reads as

$$p = P_0(t) + MP^{(1)}(\vec{\xi}, t) + M^2 p^{(2)}(\vec{x}, \vec{\xi}, t) + O(M^3). \tag{6}$$

2.3 Short wavelength dynamics

On the small convective length scale, which is identical to the reference length l_{ref}, the leading order continuity, the second order momentum and the leading order energy conservation equations yield the following system

$$\begin{aligned}
\rho_t^{(0)} + \nabla_{\vec{x}} \cdot (\rho \vec{v})^{(0)} &= 0 \\
(\rho \vec{v})_t^{(0)} + \nabla_{\vec{x}} \cdot (\rho \vec{v} \circ \vec{v})^{(0)} + \nabla_{\vec{x}} p^{(2)} &= -\nabla_{\vec{\xi}} P^{(1)} \\
\nabla_{\vec{x}} \cdot \vec{v}^{(0)} &= -\frac{1}{\gamma P_0} \frac{dP_0}{dt} + \frac{\gamma-1}{\gamma P_0} q_0 \omega^{(0)}
\end{aligned} \tag{7}$$

where $\nabla_{\vec{x}}$ and $\nabla_{\vec{\xi}}$ refer to the gradient concerning the $\vec{x}$-scale and $\vec{\xi}$-scale, respectively. These are the inviscid variable density flow equations for zero Mach number, supplemented by a long wave acoustic pressure gradient source term in the momentum equation $(7)_2$ and global compression as well as thermal expansion source terms in $(7)_3$. The small scale average (i.e., with respect to $\vec{x}$) of the divergence constraint $(7)_3$ over a domain Ω yields an evolution equation for the background pressure P_0

$$\frac{dP_0}{dt} = \frac{(\gamma-1)q_0}{Vol(\Omega)} \int_\Omega \omega^{(0)} \partial\Omega - \frac{\gamma P_0}{Vol(\Omega)} \int_{\partial\Omega} \vec{v}^{(0)} \cdot \vec{n} ds \quad . \tag{8}$$

The Gauss-Green theorem has been applied to the $\vec{v}^{(0)}$-term, where $\vec{n}$ denotes the outward directed unit normal vector on the boundary $\partial\Omega$ of the domain Ω with volume $Vol(\Omega)$. Two cases have to be distinguished now: In the limit, when $Vol(\Omega)$ becomes large and $\vec{v}^{(0)}$ is bounded, the second term on the right hand side of (8) vanishes. The mean pressure P_0 is then determined by the thermal source term. In the case that there is no thermal source term present, the incompressible divergence constraint $\nabla_{\vec{x}} \cdot \vec{v}^{(0)} = 0$ is obtained and P_0 is constant in space and time. The flow is then divergence-free, except for termal expansion due to chemical reactions. Another situation occurs when the domain of the fluid flow is bounded. The second term on the right hand side of (8) then does not vanish, but rather describes an overall pressure rise due to compression from the boundary. The velocity $\vec{v}^{(0)}$ in a non-reeactive flow is then not necessarily divergence-free and the right hand side of $(7)_3$ can be viewed as a distributed volume source, where a positive contribution results in an expansion of the gas and P_0 changes with time.

It is interesting to note, that the divergence constraint $(7)_3$ results from the *energy equation* and not from the continuity equation as usually in the derivation of the incompressible Euler equations. Depending on the overall system dimensions, only one of the supplementary pressure source terms can be present: Systems with dimensions L of order $O(l_{ref})$ are in the *single length scale regime* and no long wave acoustics can be accomodated and hence $\nabla_{\vec{\xi}} P^{(1)} \equiv 0$. The global compression term in $(7)_3$ is then readily determined by the boundary conditions of the system and the thermal source term. Systems with dimensions $L = O(l_{ref}/M)$ are large enough to accomodate long wave acoustics and therefore lie in the *multiple length scale regime*. With no chemical source term present, the source term of the divergence constraint vanishes, $dP_0/dt = \nabla_{\vec{x}} \cdot \vec{v} \equiv 0$, while $\nabla_{\vec{\xi}} P^{(1)} \neq 0$. Leading order pressure changes then occur only due to accumulation of first order pressure wave effects on time scales $O(l_{ref}/Mv_{ref})$ and the local flow divergence is $O(M)$ only. The second order pressure gradient, $\nabla_{\vec{x}} p^{(2)}$, adjusts to guarantee satisfaction of the small scale divergence constraint, $(7)_3$. Thus, the second order pressure in a low Mach number expansion attains the role of "the pressure" of the incompressible flow equations as $M \to 0$, [12], [8].

2.4 Long wave length dynamics

The $\vec{x}$-scale-averages of the second order momentum equation $(7)_2$ and the first order energy equation yield a system, that describes the long wave acoustic momentum exchange:

$$\begin{aligned} \overline{\vec{m}}_t + \nabla_{\vec{\xi}} P^{(1)} &= 0 \\ P_t^{(1)} + \nabla_{\vec{\xi}} \cdot (\overline{c^2} \, \overline{\vec{m}}) &= \nabla_{\vec{\xi}} \cdot (\overline{c^2} \, \overline{\tilde{\rho}\tilde{v}}) + \tfrac{(\gamma-1)q_0}{Vol(\Omega)} \int_{\Omega} \omega^{(1)} \partial \Omega \end{aligned} \qquad (9)$$

with $\tilde{\rho} = \rho^{(0)} - \overline{\rho}$, $\tilde{v} = v^{(0)} - \overline{v}$. Here $\overline{\rho}, \overline{\vec{m}}, \overline{v}$ denote the $\vec{x}$-averaged leading order density, momentum and velocity. The square of the averaged sound speed is denoted by $\overline{c}^2 = \gamma P_0 / \overline{\rho}$. This is the system of linearized acoustics except for the inhomogeneous source terms in $(9)_2$. The first source term of $(9)_2$ is generated by the $\vec{x}$-scale fluctuation correlation of density and velocity of multidimensional flows and introduces nontrivial coupling between the small-scale, quasi-incompressible flow and the superimposed long wave acoustics. The second source term describes the influence of acoustics on the chemical reaction rate. Acoustic waves, emitted by a turbulent flame on the $\vec{x}$-scale, can lead to autoignition of premixed reactive gases by accumulation of acoustic pressure wave effects and their influence on the reaction rate. For one-dimensional flows, the acoustic subsystem decouples from the small scale quasi-incompressible flow and there is no feedback from the small scale flow to the long wave acoustics.

3. NUMERICAL METHODS

3.1 Operator Splitting

The compressible hyperbolic Euler equations degenerate as $M \to 0$ and the incompressible limit is approached. The convection terms for mass and momentum remain nonsingular but the divergence constraint $(7)_3$ leads to an elliptic Poisson-type equation for $p^{(2)}$, while the acoustic system (9) predicts infinitely fast signal propagation in the limit $M = 0$. This justifies the following numerical operator splitting of the Euler equations (1):
System I describes the advection of mass, momentum and species and their changes due to flow divergence,

$$\left. \begin{array}{rcl} \rho_t + \nabla_{\vec{x}} \cdot (\vec{m}) &=& 0 \\ \vec{m}_t + \nabla_{\vec{x}} \cdot (\vec{v} \circ \vec{m}) &=& 0 \\ (\rho Y)_t + \nabla_{\vec{x}} \cdot (\vec{m} Y) &=& -\rho \omega \end{array} \right\} \text{System I.} \qquad (10)$$

System II describes the flow acceleration due to pressure forces and energy conservation,

$$\left. \begin{array}{rcl} \vec{m}_t + \frac{1}{M^2} \nabla_{\vec{x}} p &=& 0 \\ e_t + \nabla_{\vec{x}} \cdot (H \vec{m}) &=& 0 \end{array} \right\} \text{System II,} \qquad (11)$$

where $H = (e+p)/\rho$ and $p = (\gamma - 1)(e - M^2 \vec{m}^2 / 2\rho - \rho Y q_0)$.

3.2 Discretization of System I

System I (in one space dimension) is nonstrictly hyperbolic and has a double eigenvalue and only one eigenvector. The term $\nabla_{\vec{x}} p$ of order $O(M^2)$ is therefore introduced into the momentum equation $(10)_2$ in order to regularize System I.

$$\left. \begin{array}{rcl} \rho_t + \nabla_{\vec{x}} \cdot (\vec{m}) &=& 0 \\ \vec{m}_t + \nabla_{\vec{x}} \cdot (\vec{v} \circ \vec{m} + p) &=& 0 \\ (\rho Y)_t + \nabla_{\vec{x}} \cdot (\vec{m} Y) &=& -\rho \omega \end{array} \right\} \text{System I*} \qquad (12)$$

The new System I* (12) is strictly hyperbolic and is solved by use of a higher order compressible flow solver with Strang-type splitting [14]. The error, which is introduced by the regularization, is compensated for in System II* $(13)_1$ up to second order by subtracting $\nabla_{\vec{x}} p$ from $(11)_1$. The pressure and density of System I* are coupled by assuming a locally isentropic relationship for density and pressure.
Embedding a Godunov-type compressible flow solver into the numerical scheme for System I* guarantees, that the scheme operates reliably for low and high Mach number flows including strong shocks.

3.3 Discretization of System II

System II is of elliptic/hyperbolic character and contains three distinct physical mechanisms. By introducing the asymptotic expansions for the pressure, the equation of state and the species the following System II* is obtained:

$$\left. \begin{array}{rcl} \vec{m}_t + \nabla_{\vec{x}}\left(p^{(2)} - p\right) &=& -\nabla_{\vec{\xi}} P^{(1)} \\ M^2 e_t^{(2)} + \nabla_{\vec{x}} \cdot (\vec{v}[e+p]) &=& -\frac{P_{0_t}}{\gamma-1} - M\frac{P_t^{(1)}}{\gamma-1} - (\rho Y q_0)_t \end{array} \right\} \text{System II*}, \qquad (13)$$

with $\nabla_{\vec{x}}(\frac{1}{M^2}p - p) = \nabla_{\vec{x}}(p^{(2)} - p) + \nabla_{\vec{\xi}} P^{(1)}$. The time derivative of the total energy was transformed by the relation

$$e = \frac{P_0}{\gamma-1} + M\frac{P^{(1)}}{\gamma-1} + M^2 e^{(2)} + \rho Y q_0 \ . \qquad (14)$$

In (13), the distinct physical meanings of "the pressure" show up clearly:

1. The pressure $p^{(2)}$ accounts for the local ballance of forces and satisfies the divergence constraint of incompressible flows.
2. In the single length scale regimes a global background compression due to changes in the system volume or chemical reactions is introduced by the P_0-term.
3. In the multiple length scale regime the propagation of acoustic pressure changes with correct signal speeds is included. The $P^{(1)}$-terms of the right hand side are obtained from the wave equations (9).

System II* is solved in a predictor-corrector fashion. Therefore, the changes in momentum and energy are split up into explicit hyperbolic and implicit elliptic contributions. The explicit predictor step then yields either the leading order pressure changes of P_0 due to a global background compression, or the dominant first order O(M) pressure changes of $P^{(1)}$ through a non-dissipative discrete solution of the long wave acoustic system (9). The implicit corrector step then computes the small scale $O(M^2)$ pressure fluctuations and thereby guarantees compliance with the small scale divergence constraint as $M \to 0$. Furthermore, the corrector step produces the weakly nonlinear acoustic effects which, for small but nonzero Mach numbers, are neglected in the linear acoustic predictor.

3.4 Predictor step

The predictor step determines preliminary updates for momentum and energy in acoustic and chemical predictors. The prediction involves

(i) the determination of small scale averaged momenta $\overline{\vec{m}}$ and of approximations to the asymptotic first order pressure $P^{(1)}$ by suitable averaging and summation procedures,
(ii) the solution of the linearized acoustic equations (9), and
(iii) for more than one space dimension, the inclusion of the coupling term on the right hand side of $(9)_2$.
(iv) In reactive flows, the influence of the heat release is evaluated.

The results of the acoustic predictor are preliminary acoustic updates $\delta^t \overline{\vec{m}}$ and $\delta^t_{ac} e = M \delta^t P^{(1)}/(\gamma-1)$ of momentum and energy. They are then employed subsequently in the final correction step. The chemical source term in $(13)_2$ is treated explicitly as well and contributes to the preliminary energy update by $\delta^t e = \delta^t_{ac} e + \delta^t_{ch}(\rho Y q_0)$.

3.5 Acoustics on a coarse grid

In solving the linearized acoustic system (9), two strategies can be followed:
Either, a., the large scale derivative ∇_ξ can be replaced by $\nabla_{\vec{x}}/M$. It is then necessary to apply an implicit scheme in order to comply with the CFL condition or to use an explicit large time step method, [23].

$$\frac{\Delta t}{\Delta x}\max\left(\frac{c}{M}\right) \leq 1. \qquad (15)$$

Or, b., the acoustics are solved on a coarse grid with space increments $\Delta\xi = \Delta x/M$. The CFL condition then reads as

$$\frac{\Delta t}{\Delta \xi}\max\left(\frac{c}{M}\right) = \frac{\Delta t \cdot M}{\Delta x}\max\left(\frac{c}{M}\right) \leq 1 \quad, \qquad (16)$$

where the Mach number cancels out. System (9) can then be solved explicitly in an efficient way by the following algorithm:

1. Restrict $\overline{\vec{m}}, \overline{c^2}, P^{(1)}$ to the coarse grid.

2. Solve (9) by an explicit approximation.

3. Prolongate the obtained values for $\overline{\vec{m}}, P^{(1)}$ back to the fine grid by using an appropriate interpolation.

In cases with periodic boundary conditions, trigonometric interpolation proved to be superior to linear and cubic interpolation, but all were stable. Small scale disturbances may be introduced by the averaging procedure or by interpolation to the fine grid, and may cause small oscillations in the numerical solution of the acoustic equations. But the solution only serves as prediction of the long wave effects and will be compensated for in the corrector step.

3.6 Corrector step

Given the linear acoustic and chemical predictions $\delta^t\overline{\vec{m}}$ and $\delta^t e$, the corrector solves for the second order $O(M^2)$ updates of the total energy and pressure, and for the final momentum changes. The here presented version is first order in time for simplicity of exposition:

$$\begin{aligned}\vec{m}^{n+1} - (\vec{m}^* + \delta^t\overline{\vec{m}}) &= -\Delta t \; \tilde{\nabla}_1\left[p^{(2),n+1} - p^*\right] \\ M^2\left[\left(\frac{p^{(2)}}{\gamma-1} + \frac{\vec{m}^2}{2\rho}\right)^{n+1} - e^{(2),*}\right] + \delta^t e &= -\Delta t \; \tilde{\nabla}_2 \cdot [H^{**}\vec{m}^{n+1}],\end{aligned} \qquad (17)$$

where

$$H^{**} = \frac{1}{\rho^*}\left[(e+p)^* + \frac{\gamma}{2}\delta^t e\right]. \qquad (18)$$

Here a single *-superscript denotes data after the first split step, i.e., after the solution of System I*, while a double star ** indicates data additionaly modified by the predictor step. The symbols $\tilde{\nabla}_1, \tilde{\nabla}_2\cdot$, denote suitable discrete approximations of gradient and divergence. The term $\Delta t \tilde{\nabla}_1 p^*$ makes up for the $O(M^2)$ difference between System I and System I*, as discussed above. Upon elimination of $\vec{m}^{n+1}$ from (17) a discrete Poisson-type equation is obtained for $p^{(2)}$ with a weak $O(M^2)$ nonlinearity due to the time dependence of the kinetic energy. This can be accounted for, e.g., by an outer iteration as described in [8]. The Poisson-type equation itself is solved using a classical CG-method from the LINSOL-package [17].

3.7 Compressible projection method

The above presented approach can be understood as a compressible projection method or fractional step scheme. Incompressible projection methods have first been introduced by Chorin [2],[3], while the fractional step scheme approach was introduced by Temam [18],[19]. In the first step, density and velocity are transported by solving two advection equations without strictly enforcing the divergence constraint of incompressible flows

$$\begin{aligned} \vec{v}^* &= \vec{v}^n - \Delta t\,[(\vec{v} \cdot \nabla)\vec{v}] \\ \rho^* &= \rho^n - \Delta t\,[\vec{v} \cdot \nabla \rho]. \end{aligned} \quad (19)$$

The superscript ^{n} denotes values at the old time level. In the second step, the intermediate velocity field $\vec{v}^*$ is then projected onto the space of divergence-free vector fields. The Hodge-Helmholtz decomposition [15] views an arbitrary vector field as being composed of two orthogonal components, one divergence-free and the other the curl-free gradient of a scalar field. The vector field $\vec{v}^*$, defined on a spatial domain Ω, can thus be written as

$$\vec{v}^* = \vec{v}_d + \nabla \phi \quad (20)$$

where $\nabla \cdot \vec{v}_d = 0$, and $\vec{v}_d$ satisfies the boundary conditions. ϕ is given as the solution to the elliptic equation

$$L_\rho \phi = \nabla \cdot \left(\frac{1}{\rho^*}\nabla \phi\right) = \nabla \cdot \vec{v}^*,$$

$$\left.\frac{\partial \phi}{\partial n}\right|_{\partial \Omega} = \vec{v}^* \cdot n\big|_{\partial \Omega}. \quad (21)$$

The velocity at the new time level t^{n+1} is then obtained from $\vec{v}^{d,n+1} = \vec{v}^* - \nabla \phi$. In the above framework the scalar ϕ can be identified with the second order pressure term $p^{(2)}$. In the zero Mach number limit and under the assumption, that there is neither a global compression nor a chemical reaction, the predictions vanish ($\delta^t \vec{m} = \delta^t e = 0$) and (17) yields (see also [10])

$$\tilde{\nabla}_2 \cdot \left(\frac{\Delta t}{\rho^*}\tilde{\nabla}_1 p^{(2),n+1}\right) = \tilde{\nabla}_2 \cdot \vec{v}^{**} \quad \text{with} \quad \vec{v}^{**} = \vec{v}^* + \frac{\Delta t}{\rho^*}\tilde{\nabla}_1 p^*. \quad (22)$$

This is the pressure equation of a projection method for incompressible flows [6].

Consider now a projection method that is extended to the compressible equations without the introduction of multiple pressure variables. In the limit $M \to 0$ the Poisson-type equation will take the form [10]

$$\tilde{\nabla}_2 \cdot \left(\frac{1}{\rho^*}\tilde{\nabla}_1 p^{n+1}\right) = 0. \quad (23)$$

Hence the $O(M^2)$ pressure fluctuations that are known to occur have to be represented numerically by subtle cancellations of large numbers and the performance of such a scheme will depend in an undesired way on the machine accuracy for small Mach numbers. In order to guarantee the proper limit behavior $M \to 0$ of the incompressible equations, it is necessary to decompose and thereby separate the pressure p into a second order pressure $p^{(2)}$ and the actual thermodynamic pressure P_0 within the numerical framework.

3.8 Extension of incompressible methods

Besides compressible flow solvers also incompressible solvers can be extended to the regime of low Mach number flows. Numerical schemes for incompressible flows are usually based either on projection methods (e.g., Chorin [2],[3]) or on segregated methods (e.g., the SIMPLE algorithm of Patankar and Spalding [20]). The principle of both methods is a decoupling of pressure and velocity at the new time level t^{n+1}.

Since the above presented scheme can be understood as a *compressible* projection method, it is possible to apply techniques originally developed for incompressible flows to the low Mach number compressible regime. In the following, an outline of a compressible projection method in terms of primitive variables is given. The compressible Euler equations in primitive variables read as

$$\begin{aligned} \rho_t + \nabla \cdot (\rho v) &= 0 \\ v_t + (v \cdot \nabla) v + \tfrac{1}{M^2 \rho} \nabla p &= 0 \\ p_t + v \cdot \nabla p + \gamma p \nabla \cdot v &= 0 \end{aligned} \quad . \tag{24}$$

The fractional step scheme for (24) separates the convective terms, which are explicitly discretized,

$$\begin{aligned} \rho_t + \nabla \cdot (\rho v) &= 0 \\ v_t + (v \cdot \nabla) v &= 0 \\ p_t + v \cdot \nabla p &= 0 \end{aligned} \quad , \tag{25}$$

from the sonic terms

$$\begin{aligned} v_t + \tfrac{1}{M^2 \rho} \nabla p &= 0 \\ p_t + \gamma p \nabla \cdot v &= 0 \end{aligned} \quad . \tag{26}$$

The following considerations are focused on the second step (26) and first order accuracy in time. A fully implicit approximation to (26) with pressure decomposition has the form

$$\begin{aligned} \tfrac{v^{n+1}-v^n}{\Delta t} &+ \tfrac{1}{M^2 \rho^{n+1}} \tilde{\nabla}_1 \left(P_0 + M P^{(1)} + M^2 p^{(2)} \right)^{n+1} = 0 \\ \tfrac{P_0^{n+1}-P_0^n}{\Delta t} &+ M \tfrac{P^{(1),n+1}-P^{(1),n}}{\Delta t} + M^2 \tfrac{p^{(2),n+1}-p^{(2),n}}{\Delta t} + \gamma p^{n+1} \tilde{\nabla}_2 \cdot v^{n+1} = 0 \end{aligned} \quad . \tag{27}$$

The variables after the convection step are represented by v^n and p^n, as long as no mistaking for the values at the old time level can occur. Because the density remains unchanged in this step, $\rho^{n+1} \equiv \rho^n$.

A common procedure to solve system (27) is to separate v^{n+1} in equation $(27)_1$ and insert the result into $(27)_2$. This gives a nonlinear elliptic equation for the pressure, (due to the coefficient γp of $\tilde{\nabla}_2 \cdot v^{n+1}$) which needs to be linearized. This pressure-correction algorithm can be formulated as:

1. Decompose the pressure p^n into $p^n = P_0^n + MP^{(1),n} + M^2 p^{(2),n}$ by suitable averaging procedures.
2. Estimate global wave length effects during the time step to obtain $\delta P_0, \delta P^{(1)}, \delta \overline{v}$.
3. Estimate the pressures p^* and $p^{(2),*}$, at the next time level, e.g., $p^* = p^n + \Delta t \delta P_0 + \Delta t M \delta P^{(1)}$ and $p^{(2),*} = p^{(2),n}$.
4. Calculate the corresponding velocity v^* from $(27)_1$ by use of $\nabla_x P^{(1)}/M = \nabla_\xi P^{(1)} = \overline{\rho}\,\overline{v}_t$ as

$$v^* = v^n + \Delta t \frac{\overline{\rho}^{n+1}}{\rho^{n+1}} \overline{v}_t - \frac{\Delta t}{\rho^{n+1}} \tilde{\nabla}_1 p^{(2),*}. \tag{28}$$

5. Introduce the corrections $p^{(2),n+1} = p^{(2),*} + \delta p^{(2)}$ and $v^{n+1} = v^* + \delta v$ into (27), eliminate δv, and solve the obtained elliptic pressure-correction equation for $\delta p^{(2)}$:

$$\gamma p^* \tilde{\nabla}_2 \cdot \left(\tfrac{1}{\rho^{n+1}} \tilde{\nabla}_1 \delta p^{(2)} \right) - \left(\tfrac{M}{\Delta t} \right)^2 \delta p^{(2)} = \tfrac{1}{\Delta t^2} \left(\delta P_0 + M \delta P^{(1)} \right) + \tfrac{\gamma p^*}{\Delta t} \tilde{\nabla}_2 \cdot v^*. \tag{29}$$

6. Add the pressure-corrections

$$p^{(2),n+1} = p^{(2),*} + \delta p^{(2)} \quad \text{and} \quad p^{n+1} = p^* + M^2 \left(p^{(2),n+1} - p^{(2),n} \right), \tag{30}$$

and calculate the corresponding velocity v^{n+1}.

The factor γp^{n+1} in $(27)_2$ has been replaced by γp^*. Therefore, steps 4. to 6. should be applied iteratively until convergence is reached. A relaxation parameter $\alpha \in [0,1]$ can be introduced in step 6. to set $p^{(2),n+1} = p^{(2),*} + \alpha \delta p^{(2)}$.

4. NUMERICAL SIMULATIONS

4.1 Inviscid vortex transport at M=0

For zero Mach number no acoustic pressure waves can be accomodated in the domain and the $\vec{\xi}$-scale variable becomes void. The numerical method then reduces to an incompressible projection method. Gresho and Chan [22] proposed the inviscid transport of a vortex as test case for incompressible constant density flows. A vortex is transported in x-direction through a rectangular domain. The exact solution to such a problem is known, i.e., a pure translation at unit speed of the initial conditions. The boundary conditions for the test case are inlet to the left, outlet to the right, and solid frictionless walls at the top and bottom. The initial conditions are that of a triangle vortex. The tangential velocity component describes a solid body rotation in the core of the vortex. At $r = R$ the velocity then switches to a decreasing linear function of r until $r = 2R$, where the tangential velocity returns to zero. The vortex initial conditions, before adding the translational velocity, read as

$$
\begin{aligned}
u_\phi(r) &= u_0 r/R & & & 0 &\le r \le R, \\
u_\phi(r) &= u_0(2 - r/R) & &\text{for} & R &< r \le 2R, \\
u_\phi &= 0 & & & r &> 2R,
\end{aligned}
\qquad (31)
$$

where $u_0 = 1$. A uniform mesh of 80x20 elements is used to cover the domain that is one unit high and four units long. The radius of the triangle vortex is resolved by $2R = 8\Delta x$. In compliance with [22], a CFL number of 0.1 was chosen. Figure 1 shows the relative streamlines ψ of the vortex as it moves through the mesh. The vortex remains centered along the x-axis, and no phase lag can be observed. The influence of the outlet region deforms the vortex, which is due to reflections from the outlet. The results presented by Gresho and Chan, [22], show a falling vortex, while their SCM-projection method slightly better preserves the maximum of the velocity.

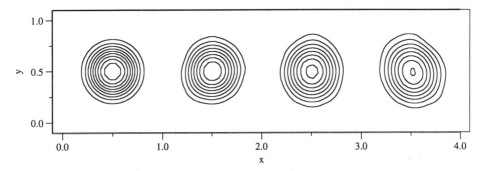

Fig.1: Relative streamlines at t = 0, 1, 2, 3; $\psi = 0.02$ (0.02) 0.18; Mach number M=0 with MPV-Approach.

The test case clearly demonstrates the zero Mach number capability of the MPV-Approach. Improved outlet boundry conditions for the pressure $p^{(2)}$ are being developed to obtain low reflection boundaries.

4.2 Generation of baroclinic vorticity at M=1/20

For nonzero Mach numbers, the acoustic scale $\vec{\xi}$ is coupled with the small scale $\vec{x}$ by the linearized acoustic source term in the momentum equation $(7)_2$. An acoustic pulse with amplitude $P^{(1)}$ will lead to the creation of baroclinic vorticity due to compressibility effects.

In a double-periodic domain of size 40x8 units, the following initial data are given:

$$\left.\begin{array}{rl} \rho(x,y,0) &= \bar{\rho}_0 + M\tilde{\rho}_0^{(1)}\frac{1}{2}(1.0+\cos(\pi x/L)) + \Phi(y) \\ p(x,y,0) &= \bar{p}_0 + M\tilde{p}_0^{(1)}\frac{1}{2}(1.0+\cos(\pi x/L)) \\ u(x,y,0) &= \tilde{u}_0 \frac{1}{2}(1.0+\cos(\pi x/L)) \\ v(x,y,0) &= 0.0 \end{array}\right\} \qquad (32)$$

$$\text{for} \quad \begin{array}{c} -L \leq x \leq L = \frac{1}{M} \\ -L_y \leq y \leq L_y = L/5 \end{array} ; \quad M = \frac{1}{20},$$

and the coefficients

$$\begin{array}{ll} \bar{\rho}_0 = 1.0 & \bar{p}_0 = 1.0 \quad \tilde{u}_0 = 2\sqrt{\gamma} \\ \tilde{\rho}_0^{(1)} = 0.4 & \tilde{p}_0^{(1)} = 2\gamma \end{array} \qquad (33)$$

The function $\Phi(y)$ in (32) is defined by

$$\Phi(y) = \begin{cases} \rho_y y & , \quad 0 \leq y \leq \frac{1}{2}L_y \\ \rho_y(y-\frac{1}{2}L_y) - \tilde{\rho}_0^{(0)} & , \quad \frac{1}{2}L_y < y \leq L_y \end{cases}, \text{ with } \rho_y = 2\tilde{\rho}_0^{(0)}/L_y \quad . \quad (34)$$

The data from (32) represent a periodic train of long-wave right-running acoustic pulses, that set a saw-tooth like density layering into motion. The numerical solution uses 401 x 80 grid points within the interval of $(32)_2$, which gives a constant background pressure P_0. The acoustic equations (9) are solved using a Crank-Nicholson type approximation, while the coupling term has been neglected. The different density of two neighboring particles leads to a different speed under the influence of the same acoustic pressure wave. This results in the occurence of a Kelvin-Helmholtz instability [16] where one fluid is moving at a different rate relative to another. Figure 2 shows, that the initially horizontal interface starts rolling up into large vortical structures, creating vorticity. The large scale acoustics thereby feed energy into the small scale structures. The computations were performed on a CRAY J90.

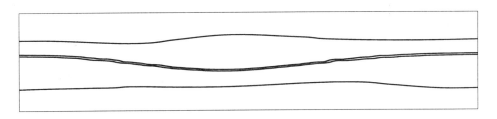
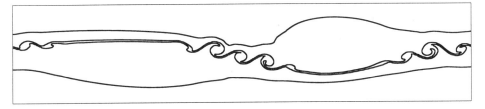

Fig.2: Density isocontours at values of 0.8 and 1.2 on a 401x80 grid. Mach number M=1/20 and times t = 7.0, 15.0 with MPV-Approach.

REFERENCES

[1] Patnaik G., Guirguis R.H., Boris J.P., Oran E.S., "A Barely Implicit Correction for Flux Corrected Transport", JCP, **71** (1987), pp. 1-20.

[2] Chorin A.J., "Numerical solution of the Navier-Stokes equations", Math. Comp., **22** (1968), pp. 745-762.

[3] Chorin A.J., "On the convergence of Discrete Approximations to the Navier-Stokes Equations", Math. Comput., **23** (1969), pp. 341-353.

[4] Gustafsson B., Stoor H., "Navier-Stokes Equations for almost Incompressible Flow", SIAM J. Num. Anal., **28** (1991), pp. 1523-1547.

[5] Yee H.C., Shinn J.L., "Semi-Implicit and Fully Implicit Shock Capturing Methods for Hyperbolic Conservation Laws with Stiff Source Terms", AIAA Jnl., **27** (1989), pp. 299.

[6] Bell J.B., Colella P., Glaz H.M., "A Second-Order Projection Method for the Incompressible Navier-Stokes Equations", JCP, **85** (1989), pp. 257-283.

[7] Weimer M., Meinke M., Krause E., "Numerical Simulation of Incompressible Flows with the Method of Artifical Compressibility", In this publication.

[8] Klein R., "Semi-Implicit Extension of a Godunov-Type Scheme Based on Low Mach Number Asymptotics I: One-dimensional flow", JCP, **121** (1995), pp. 213-237.

[9] Klein R., Munz C.D., "The Multiple Pressure Variable (MPV) Approach for the Numerical Approximation of Weakly Compressible Fluid Flow", Proc. of the Int. Conf. on Num. Meth. in Cont. Mech., Prague, June 1994.

[10] Klein R., Lange K., Willems W., Munz C.D., "Semi-Implicit High Resolution Schemes for Low Mach Number Flows", Proc. of Fifth Int. Conf. on Hyperbolic Problems, Theory, Numerics, and Applications, Stony Brook, 1994.

[11] Klein R., Munz C.D., "The Extension of Incompressible Flow Solvers to the Weakly Compressible Regime", submitted to Theoretical and Numerical Fluid Mechanics, Sept. 1995.

[12] Majda A.J., Sethian J., "The derivation and Numerical Solution of the Equations of Zero Mach Number Combustion", Comb. Sci. Technol., **42** (1985), pp. 185-205.

[13] Klainerman S., Majda A.J., "Compressible and Incompressible Fluids", Comm. Pure Applied Math., **35** (1982), pp. 629.

[14] Strang, G., "On the Construction and Comparison of Difference Schemes", SIAM, J. Num. Anal., **5** (1967), pp. 506-517.

[15] Chorin A.J., Marsden J.E., "A Mathematical Introduction to Fluid Mechanics", Springer, New York, 1990.

[16] Mulder W., Osher S., "Computing Interface Motion in Compressible Gas Dynamics", JCP, **100** (1992), pp. 209-228.

[17] LINSOL, Solver for Large and Sparce Linear Systems,
Rechenzentrum der Universität Karlsruhe,
http://www.rz.uni-karlsruhe.de/Uni/RZ/Forschung/Numerik/linsol/index.htmp .

[18] Temam R., "Sur l'approximation de la solution des equations de Navier-Stokes par la methode des pas fractionaires", Arch. Rat. Mech. Anal., **32** (1969), pp. 135-153.

[19] Temam R., "Navier-Stokes Equations", North-Holland, Amsterdam, New York, Oxford, 1977.

[20] Patankar S., "Numerical Heat Transfer and Fluid Flow", McGraw-Hill, Washington D.C., 1980.

[21] Patankar S., Spalding D., "A Calculation Procedure for Heat, Mass and Momentum Transfer in Three-Dimensional Parabolic Flow", Int. J. Heat Mass Transfer, **15** (1972), 1787-1806.

[22] Gresho P.M., Chan S.T., "On the Theory of Semi-Implicit Projection Methods for Viscous Incompressible Flow and its Implementation via a Finite Element Method That Also Introduces a Nearly Consistent Mass Matrix. Part2: Implementation", Int. J. for Num. Meth. in Fluids, **11** (1990), pp. 621-659.

[23] LeVeque R.J., "A Large Time Step Generalization of Godunov's Method for Systems of Conservation Laws", SIAM J. Num. Anal., **22** (1985), pp. 1051-1073.

Numerical Simulation of Incompressible Flows with the Method of Artificial Compressibility

M. Weimer, M. Meinke, E. Krause

Aerodynamisches Institut, RWTH Aachen,
Wüllnerstraße zw. 5 u. 7,
52062 Aachen, Germany

Abstract

The Navier-Stokes equations for three-dimensional, unsteady, and incompressible flows are solved numerically. The applied method is based on the concept of artificial compressibility, combined with a dual-time stepping scheme. The solution within each physical time step is carried out with both an implicit and explicit scheme with possible multi-grid acceleration. The algorithm is formulated for a non-staggered, node-centered, curvilinear, block-structured grid, enabling the simulation of flows in complex geometries. Results are presented and compared with each other for the flow around a circular cylinder mounted in a square duct. For the flow in a 90°-T-junction results are compared to experimental flow visualization.

Introduction

A large number of flow problems, which are of technical or scientific interest are characterized by small Mach numbers so that they can be assumed to be incompressible. Simulating such flows with methods for compressible flows, setting the Mach number to small values, leads to poor convergence rates due to the small characteristic time-steps, resulting from the increased stiffness of the set of equations. Therefore, algorithms for the numerical solution of the Navier-Stokes equations for incompressible flow are required. In contrast to the equations for compressible flows, pressure and velocity components are decoupled, so that solution techniques for compressible flows cannot be utilized directly. In [7] and [10] several approaches for simulating incompressible flows are described and compared with each other. Here, the method of artificial compressibility is applied, for which solution schemes known from compressible problems can be adapted. This method was applied successfully e. g. in [1]. The artificial compressibility concept can also be used as a preconditioning method for the simulation of small, but non vanishing Mach number flows, [18], for which the results presented here may also be representative.

The simulation of unsteady flow becomes an increasing subject of investigations. For that purpose it is desirable to choose large time-steps with a still sufficient time accuracy of the solution scheme. This can be achieved with a dual-time stepping scheme, in which the accuracy is solely determined by the discretization of the partial derivative with respect to the physical time, which can be chosen independently from the other partial derivatives. Additionally, unsteady flows often appear in complex geometries, so that for structured grids it is necessary to include a multi-block capability in order to avoid grid singularities. In this paper two different solution techniques for both steady and unsteady incompressible flows based on the concept of artificial compressibility and a dual-time stepping scheme are presented. The conservation equations for mass and momentum are integrated with an implicit line-relaxation method as well as a five-step Runge-Kutta scheme. A FAS multi-grid method was formulated and applied in both algorithms. The equations are discretized for three-dimensional curvilinear coordinates by using primitive variables in node-centered block-structured grids. The efficiencies of both schemes in terms of convergence and memory requirements are compared for the steady and unsteady flow around a circular cylinder mounted in a square duct. For a 90° T-junction results of a flow simulation are presented and compared to the flow visualization of an in-house experiment.

Governing Equations

The motion of viscous Newtonian fluids is governed by the Navier-Stokes equations. They describe the conservation of mass, momentum and energy. For incompressible flows and constant viscosity the conservation equations for mass and momentum can be solved decoupled from the energy equation. In dimensionless form these equations read for three-dimensional, unsteady, and incompressible flows in curvilinear coordinates:

$$I_{4,t} Q_t + F_\xi + G_\eta + H_\zeta - \frac{1}{Re}\left(R_\xi + S_\eta + T_\zeta\right) = 0 \tag{1}$$

$$Q = J^{-1}\left(p, u, v, w\right)^\top \qquad I_{4,t} = diag\left(0,1,1,1\right) \tag{2}$$

$$F = J^{-1}\begin{pmatrix} U \\ uU + \xi_x p \\ vU + \xi_y p \\ wU + \xi_z p \end{pmatrix} \qquad R = J^{-1}\begin{pmatrix} 0 \\ g_1 u_\xi + g_2 u_\eta + g_3 u_\zeta \\ g_1 v_\xi + g_2 v_\eta + g_3 v_\zeta \\ g_1 w_\xi + g_2 w_\eta + g_3 w_\zeta \end{pmatrix}$$

$$G = J^{-1}\begin{pmatrix} V \\ uV + \eta_x p \\ vV + \eta_y p \\ wV + \eta_z p \end{pmatrix} \qquad S = J^{-1}\begin{pmatrix} 0 \\ g_2 u_\xi + g_4 u_\eta + g_5 u_\zeta \\ g_2 v_\xi + g_4 v_\eta + g_5 v_\zeta \\ g_2 w_\xi + g_4 w_\eta + g_5 w_\zeta \end{pmatrix}$$

$$H = J^{-1}\begin{pmatrix} W \\ uW + \zeta_x p \\ vW + \zeta_y p \\ wW + \zeta_z p \end{pmatrix} \qquad T = J^{-1}\begin{pmatrix} 0 \\ g_3 u_\xi + g_5 u_\eta + g_6 u_\zeta \\ g_3 v_\xi + g_5 v_\eta + g_6 v_\zeta \\ g_3 w_\xi + g_5 w_\eta + g_6 w_\zeta \end{pmatrix}$$

$$U = \xi_x u + \xi_y v + \xi_z w \qquad V = \eta_x u + \eta_y v + \eta_z w \qquad W = \zeta_x u + \zeta_y v + \zeta_z w$$

$$g_1 = \xi_x^2 + \xi_y^2 + \xi_z^2 \qquad g_2 = \xi_x \eta_x + \xi_y \eta_y + \xi_z \eta_z \qquad g_3 = \xi_x \zeta_x + \xi_y \zeta_y + \xi_z \zeta_z$$
$$g_4 = \eta_x^2 + \eta_y^2 + \eta_z^2 \qquad g_5 = \eta_x \zeta_x + \eta_y \zeta_y + \eta_z \zeta_z \qquad g_6 = \zeta_x^2 + \zeta_y^2 + \zeta_z^2$$

in which Q is the vector of the primitive variables, while F, G, H are the convective, and R, S, T the viscous flux vectors, respectively. U, V, W represent the contravariant velocities, J refers to the metric Jacobian, and the metric coefficients g_i are introduced for simplification. The Reynolds number Re is based on appropriate reference values.

Numerical Method

The numerical method is based upon the concept of artificial compressibility combined with a dual-time-stepping technique [8], [15]. For source free incompressible flow fields the coefficient matrix $I_{4,t}$ becomes singular. Therefore an artificial equation of state

$$p = \beta \varrho \qquad (3)$$

is introduced, in which β is a parameter controlling the (artificial) compressibility. To achieve an equation for the conservation of mass similar to that for compressible flows, a derivative of the pressure with respect to an artificial time τ is added to the continuity equation. With this modification the Navier-Stokes equations read:

$$I_{4,\tau} Q_\tau + I_{4,t} Q_t + F_\xi + \ldots - \frac{1}{Re}(R_\xi + \ldots) = 0 \quad, \quad I_{4,\tau} = diag\left(\frac{1}{\beta}, 1, 1, 1\right). \qquad (4)$$

Thus, all dependent variables become coupled and solution algorithms similar to those for compressible flows can be applied to solve Eq. (4). Here, two different approaches are used to carry out the solution in each physical time step, which are described subsequently. In both cases the system of equations is discretized with a node-centered finite difference scheme, so the discretized flux vector, e. g. F_ξ, reads:

$$F_\xi|_{i,j,k} = \frac{1}{\Delta \xi}\left(F_{i+1/2,j,k} - F_{i-1/2,j,k}\right). \qquad (5)$$

Implicit Relaxation Method

First, an implicit relaxation method is presented. The spatial derivatives of Eq. (4) are formulated for an unknown artificial time level $\nu + 1$. The resulting non-linear equations are linearized using a Taylor series expansion of first order, e. g.:

$$F^{\nu+1} = F^\nu + A \cdot \Delta Q \quad ; \quad A = \frac{\partial F}{\partial Q} \quad, \quad \Delta Q = Q^{\nu+1} - Q^\nu. \qquad (6)$$

The flux Jacobians are split into matrices containing only non-negative and negative eigenvalues by a similarity transformation, e. g. for A:

$$A = M_A \Lambda M_A^{-1} = M_A \Lambda^+ M_A^{-1} + M_A \Lambda^- M_A^{-1} = A^+ + A^- \qquad (7)$$

$$\Lambda = diag(\lambda_1, \lambda_2, \lambda_3, \lambda_4) \quad \lambda_{1,2} = U \mp S \quad \lambda = \lambda_{3,4} = U \quad S = \sqrt{U^2 + g_1 \beta}.$$

Herein M represents the matrix of the right eigenvectors and Λ the diagonal matrix of the eigenvalues. The latter is split into Λ^+ and Λ^-, containing either the non-negative or negative values. For the case of curvilinear coordinates the Jacobian matrices, e. g. A^+ and A^-, are given by:

$$A^\pm = \frac{1}{2S^2}\begin{pmatrix} \pm g_1 \beta S & \pm \xi_1 \beta \lambda_2 S & \pm \xi_2 \beta \lambda_2 S & \pm \xi_3 \beta \lambda_2 S \\ \pm g_1 a_1 - b_1 & \pm \xi_1 \lambda_2 a_1 + c_{23} & \pm \xi_2 \lambda_2 a_1 - d_{12} & \pm \xi_3 \lambda_2 a_1 - d_{13} \\ \pm g_1 a_2 - b_2 & \pm \xi_1 \lambda_2 a_2 - d_{21} & \pm \xi_2 \lambda_2 a_2 + c_{13} & \pm \xi_3 \lambda_2 a_2 - d_{23} \\ \pm g_1 a_3 - b_3 & \pm \xi_1 \lambda_2 a_3 - d_{31} & \pm \xi_2 \lambda_2 a_3 - d_{32} & \pm \xi_3 \lambda_2 a_3 + c_{12} \end{pmatrix} \qquad (8)$$

$$H_i = v_i U + \xi_i \beta \qquad v_i = (u,v,w) \qquad \xi_i = (\xi_x, \xi_y, \xi_z)$$
$$a_i = v_1 S \pm H_i \qquad b_i = 2\lambda\left(g_1 v_i - \xi_i U\right) \qquad c_{ij} = 2\lambda(\xi_i H_i + \xi_j H_j) \qquad d_{ij} = 2\lambda \xi_j H_i \,.$$

Rearranging terms for the known artificial time level ν on the right and unknown terms on the left hand side results in a delta-formulation:

$$\left(\frac{I_{4,\tau}}{\Delta \tau} + \frac{I_{4,t}}{\Delta t} + \left(\delta_\xi(A^+ + A^-) + \ldots + Re^{-1}(\ldots)\right)\right)\Delta Q^\nu = -\frac{\partial Q}{\partial t} + RHS$$

where the implicit operator can be seen on the left and the discretized Navier-Stokes equations for incompressible flows on the right hand side. In order to achieve good convergence rates the left hand side is approximated by using first order upwind differences according to the sign of the eigenvalues (see [7] for details). The discretization of the implicit operator only influences the rate of convergence, while the discretization of the right hand side determines the accuracy of the solution. Therefore, second order accurate discretizations were applied to the temporal and spatial differences. The time derivative is carried out with a backward difference:

$$\frac{\partial Q}{\partial t} = \frac{3Q^n - 4Q^{n-1} + Q^{n-2}}{2\Delta t} + O(\Delta t^2) \,.$$

The convective terms are approximated with the QUICK interpolation [14],[16], while the viscous terms are discretized with central differences.

The resulting system of equations is solved by an implicit Line-Jacobi relaxation method with alternating directions, which involves the solution of block-tridiagonal equations. A vectorization of this algorithm was achieved by solving all equations on the planes perpendicular to the direction of relaxation simultaneously. The line relaxation was programmed for one coordinate direction only. The alternating directions were realized by a cyclic change of the storage indices of the corresponding data arrays in order to obtain a vector stride of 1 for all arithmetic operations. This involves a certain overhead, because data has to be shifted in the computer memory. Numerical tests proved however, that the additional effort of restoring the data is more than compensated by a higher vector performance.

The solution method described above is very robust and possesses favourable smoothing properties and can easily be rewritten for the application to compressible flows. The work of developing and programming such an algorithm, however, is considerable. Furthermore, this algorithm requires a large amount of computer memory for the intermediate storage of e. g. the split flux Jacobians.

Explicit Time Stepping Method

An alternative approach for the simulation of unsteady flows following the concept in [12], [2] is now presented. Starting from Eq. (4) a semi-discrete formulation reads:

$$\frac{\Delta Q}{\Delta \tau} = -\frac{\partial Q}{\partial t} + RHS \,. \tag{9}$$

Eq. (9) can be solved numerically within each physical time-step with the help of a five-step Runge-Kutta scheme:

$$Q^{(0)} = Q^{\nu}$$
$$Q^{(l)} = Q^0 - \alpha_l \Delta\tau \; J \left(\left.\frac{\partial Q}{\partial t}\right)\right|^{(l-1)} + RHS(Q^{(l-1)}) \right) \quad \Big\} \quad l = 1, 2, .., 5 \,. \quad (10)$$
$$Q^{(\nu+1)} = Q^N$$

Since the discretization in physical time is fully implicit, the time-step can be chosen arbitrarily. For the integration in the artificial time τ a Full Approximation Storage direct multi-grid concept [4], [5], [11], can be applied. The formulation of the spatial and temporal discretization of the right hand side are equivalent to that of the implicit relaxation method. In both algorithms explicit boundary conditions with respect to the artificial time are imposed.

For steady state problems a large physical time step is chosen in both methods. For this case the term $\frac{\partial Q}{\partial t}$ vanishes and the explicit time stepping method becomes identical to that described in [6].

Multi-Block Method

For the computation of flows in complex geometries the physical domain is divided into sub-domains, in which the solution algorithms described above are applied locally. Here, composite grids are used, for which grid points coincide at block boundaries. After each update of the variables within the iterative solution of the equations, information is exchanged between neighbouring blocks, also on coarse grid levels. For that purpose a unique number is assigned to each block, which can possess an arbitrary number of windows. A window represents a rectangular surface on a bounding surface of a block, to which a physical boundary condition is specified or which is connected to another block. The information, which has to be provided for each window, consists of the extent of the window, the type of boundary condition, and, in the case of a multi-block boundary, the number, the face, and the orientation of the adjacent block.

For a conservation of mass it is necessary to preserve the physical fluxes across block boundaries. This can be achieved in the following manner: Additional grid points are added at each boundary containing a block connection, so that a flux balance can be computed locally also for grid points located on a block boundary. In order to ensure flux conservation also for corner points, which can possess control volumes of arbitrary shape, see Fig. 1, only the local block contribution of the flux balance is retained. In an exchange algorithm the local flux contributions of all neighbouring blocks are added, so that the overall flux balance is obtained.

To avoid special treatments for the grid points on block boundaries and to achieve good vectorization rates, the local flux contributions are computed with the same algorithm as for the inner grid points. The only difference being, that metric terms, which can be interpreted as surfaces of the control volume cell, are set to zero, if they are located outside of the considered block, as illustrated in Fig. 2. This requires, that the control volumes are computed appropriately, as shown by the shaded areas in Fig. 1.

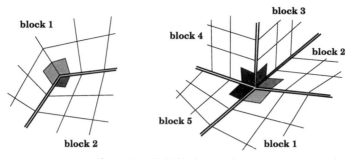

Figure 1: Multi-block examples

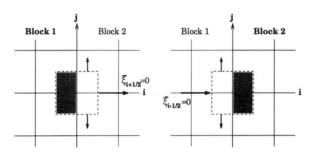

Figure 2: Flux conservation at block boundary

Results

Circular cylinder in a square duct

The above described algorithms are applied to the simulation of the flow around a circular cylinder mounted in a square duct, a test case defined in [9]. Fig. 3 shows a sketch of the physical domain and one plane of the numerical grid, which consists of five blocks with a total of 165000 grid points. The Reynolds number is based on the diameter of the cylinder D_C and the mean inflow velocity. The duct has a height and width of 4.1 D_C, the length amounts to 25 D_C.

As a first test-case a steady flow was simulated at a Reynolds number of 20. As initial conditions the velocity profile of a fully developed channel-flow is prescribed at the inflow cross-section, zero velocities are prescribed in the rest of the flow field. On all rigid walls the no-slip condition and a zero pressure gradient normal to the wall are imposed. At the outflow plane the pressure is held fixed and all velocity components are extrapolated. For both algorithms and all results presented, the same boundary conditions are used. Lift- and drag-coefficients of $c_L = 0.0095$ and $c_D = 6.123$, respectively, were obtained. In Fig. 4 a the histories of the maximum residuals are plotted versus work-units (WU). Herein one work-unit corresponds to the work for one iteration (Runge-Kutta time step or line-relaxation) on the finest grid-level. For simulations with the multi-grid method the additional operations for the computation of the defect correction, the injection and interpolation are not taken into account. Different multi-grid strategies within the implicit scheme did not improve the convergence, therefore only single grid results are presented for this scheme. It should be noted, that 1 WU of the implicit line-relaxation scheme is more expensive in terms of CPU time than 1 WU in the Runge-Kutta algorithm, due to the time-consuming solution of the block-tridiagonal systems of equation. For the implementation here, 1 WU of the implicit scheme used approximately the same CPU time as 2 WU of the explicit scheme. A direct comparison of the two methods is possible with Fig. 4 b, in which the residual is plotted versus the CPU-time. Both algorithms were programmed in a similar way, they achieved the same MFlop rates on the VPP500 on the same grid. It can clearly be seen that for this test case, the best convergence rates are obtained with the explicit, multi-grid accelerated scheme, although the MFlop-rates decrease by about 25% due to the shorter vector lengths on the coarser grid levels. It is almost four times faster than the explicit single-grid algorithm and about two times faster than the implicit scheme.

A time-dependent flow in the above described domain is obtained, when the Reynolds number is increased to 100. A solution of the steady-state computation was used as an initial condition. For both algorithms the solution in each physical time was considered to be sufficiently approximated after the residual dropped four orders of magnitude within the artificial time. In Fig. 9 a plot of the absolute value of the velocity vector in the plane of symmetry at a dimensionless time of $t = 171$ is shown. A von Kármán vortex street is clearly visible behind a large separated flow region behind the cylinder. The vortex street generates oscillating lift- and drag-coefficients with a small amplitude as shown in Fig. 8. A Strouhal number of 0.339 is obtained for this Reynolds number. In Fig. 6 the time history of the maximum residual for one representative physical time-step is shown. The dotted horizontal line marks the convergence criterion in artificial time. The comparison of the different schemes now shows a different tendency. The implicit scheme converges faster than both explicit methods, additionally the difference between the multi- and single grid methods becomes smaller. For the chosen number of grid points, the single grid method consumes even less computing time than the multi-grid scheme, because of its higher MFlop performance. The multi-grid method will be again of advantage, if a larger number of grid points is used.

90°-T-junction

The flow in branched tubes is of major interest, because it can be found in nearly all technical devices as well as in the blood circulation of human beings. Therefore these flows, which are characterized by strong three dimensional structures and unsteadiness, are subject of current investigations, see e. g. [17].

In this paper the result of a numerical simulation of a flow in a 90° T-junction of two pipes with different diameter is presented. The two-block grid contains approximately 380.000 grid points, with a diameter-ratio of the pipes $D_0/D_1 = 2.35$. This grid was taken from [13].

In Fig. 11 the surface grid is shown, the flow enters the pipe from the left (0) and from below (1), and leaves the junction to the right. The initial conditions are approximated by using fully developed pipe-flow relations for both pipes and a global conservation of mass. The ratio of volume flux is chosen to be $\dot{Q}_0/\dot{Q}_1 = 5/6$. The Reynolds number based on D_0 and the mean inflow velocity of the larger pipe is $Re = 309$. The boundary conditions are the same as for the test case described above. The computation was carried out on a SNI S600/20 by using the line relaxation algorithm, which required 210 MByte of memory.

Fig. 12 displays above an experimental flow visualization [3], based on the laser-light-sheet technique. That experiment was conducted with water with the same flow parameters, as in the numerical solution. The figure shows streak-lines in the plane of symmetry, which were injected in tube 1. The lower picture shows the result of the numerical simulation at a dimensionless time of $t = 4.19$ and depicts gray-scaled the amount of vorticity in the same plane. It is seen that the numerically simulated vortex structures agree well with the experimental data, though the inclination of the jet is slightly larger in the simulation. This effect might result from a slightly different volume-flux ratio.

In Fig. 13 the temporal development of the flow from $t = 3.86$ to $t = 5.85$ is presented. All of the pictures show the amount of vorticity. In the left column the above mentioned plane is depicted, the right column shows a cut along the axis of tube 0 perpendicular to the axis of tube 1. The position of tube 1 is sketched by a small circle in the right column. The sequence in the left row shows a small vortex at the wall opposite to tube 1 and upstream of the incoming jet ($t = 3.86$), which was identified to be a horse-shoe vortex in the experiment. That vortex grows in time ($t = 4.19$), in the course of which several smaller vortices separate ($t = 4.86$) traveling along the incoming jet ($t = 5.20$). The vorticity of these separating vortices becomes smaller ($t = 5.53$) until they vanish completely and a state similar to the first one is reached again ($t = 5.85$). After that the cycle starts again. Resulting from the blockage of tube 0 by the incoming jet of tube 1, a vortex-street-like structure evolves in the plane perpendicular to that jet, as can be seen in the right column. The black boundary of the half-moon-like structure marks the shear-layer of the joining streams. On both sides of that region vortices separate alternatingly and travel downstream. The characteristic time of this vortex structure is much smaller than the period of the horse-shoe vortex.

Conclusion

For the numerical solution of the Navier-Stokes equations for three-dimensional, unsteady, and incompressible flows both an implicit line-relaxation procedure and a five-step Runge-Kutta scheme were developed. Both algorithms are based on the concept of artificial compressibility combined with a dual-time stepping and utilize a fully implicit formulation for the discretization in the physical time. The conservation equations were approximated with second order accuracy in curvilinear coordinates by using the QUICK-interpolation for the convective and central differences for the viscous terms. A multi-block data-structure in combination with flux-conservation across adjacent blocks permits the use of block-structured grids, as it is necessary for the simulation of flows in complex geometries. The time-dependent flow around a circular cylinder in a square duct is simulated as a test-case for a comparison of the described algorithms. The converged solution is identical in both cases. With a FAS multi-grid method the convergence could be accelerated within each time step for the explicit scheme. For the implicit scheme two different indirect multi-grid formulations were tested, but no improvement in convergence could be achieved for the line-relaxation scheme. Both solution methods, the implicit and multi-grid accelerated explicit schemes, showed comparable efficiency for the considered flow problem. The implicit scheme becomes more advantageous for a small number of large blocks, while the explicit scheme will be more efficient for a large number of small blocks.

As an application the flow in a 90° T-junction was simulated. A comparison showed good qualitative agreement with an experimental flow visualization.

References

[1] W. Althaus, Ch. Brücker, and M. Weimer. Breakdown of slender vortices. In S. I. Green, editor, *Fluid Vortices*. Kluwer Academic Publishing, 1995.

[2] A. Arnone, M. S. Liou, and L. A. Povinelli. Multigrid time-accurate integration of Navier-Stokes equations. Technical Memorandum 106373, NASA, November 1993.

[3] B. Bartmann and R. Neikes. Experimentelle Untersuchung verzweigender Innenströmung II. FKM/AIF-Vorhaben Nr.113, Abschlußbericht, 1991.

[4] A. Brandt. Multi-level adaptive solutions to boundary-value problems. *Mathematics of Computation*, 31(138):333–390, 1977.

[5] A. Brandt. Guide to multigrid development. In *Lecture Notes in Mathematics*, pages 220–312. Springer Verlag Berlin, 1981.

[6] M. Breuer and D. Hänel. A dual time-stepping method for 3-d, viscous, incompressible vortex flows. *Computers Fluids*, 22(4/5):467–484, 1993.

[7] Michael Breuer. *Numerische Lösung der Navier-Stokes Gleichungen für dreidimensionale inkompressible instationäre Strömungen zur Simulation des Wirbelplatzens*. Dissertation, Aerodynamisches Institut der RWTH-Aachen, Juni 1991.

[8] A.J. Chorin. A Numerical Method for Solving Incompressible Viscous Flow. *J. Comput. Phys.*, 2:12–26, 1967.

[9] F. Durst, M. Schäfer, K. Wechsler, R. Becker, R. Rannacher, and S. Turek. Definition of benchmark problems (incompressible laminar flow. In *Flow Simulation with High Performance Computers*. DFG, 1995.

[10] Dieter Hänel. Computational techniques for solving the Navier-Stokes equations. In *CFD Techniques for Propulsion Applications*, pages 1.1–1.26. AGARD, May 1991. AGARD CP-510.

[11] A. Jameson. Solution of the Euler equations for two dimensional transonic flow by a multigrid method. *Applied Math. and Comp.*, 13:327–355, 1983.

[12] A. Jameson. Time dependent calculations using multigrid, with applications to unsteady flows past airfoils and wings. Technical Report AIAA paper No. 91-1596, American Institute of Aeronautics and Astronautics, 1991.

[13] E. Krause, W. Limberg, Ch. Brücker, M. Meinke, and A. Wunderlich. Direkte Numerische Simulation verzweigender Innenströmungen. AIF-Abschlußbericht, Forschungsvorhaben AIF-Nr. 9163, FKM-Nr. 690176, 1996. to be published.

[14] B. P. Leonard. A stable and accurate convective modelling procedure based on quadratic upstream interpolation. *Computer Methods in Applied Mechanics and Engineering*, 19:59, 1979.

[15] C. L. Merkle and M. Athavale. Time-accurate unsteady incompressible flow algorithms based on artificial compressibility. *AIAA-Paper, AIAA 87-1137*, 1987.

[16] Y. Nakamura and Y. Takemoto. Solutions of Incompressible Flows Using a Generalized QUICK Method. In *Numerical Methods in Fluid Mechanics II, Proc. of the Int. Symp. on Comput. Fluid Dynamics, Tokyo*, Sept. 9-12 1985.

[17] J. Ong, G. Enden, and A.S. Popel. Converging three-dimensional Stokes flow of two fluids in a t-type bifurcation. *J. Fluid Mech.*, 270:51–71, 1994.

[18] E. Turkel. Review of preconditioning methods for fluid dynamics. *ICASE Report*, No. 92–47, 1992.

Figures

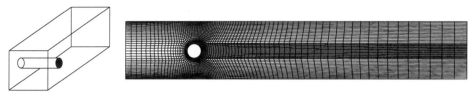

Figure 3: Sketch of the domain of integration for a circular cylinder in a square duct (left) and a plane of the numerical grid normal to the cylinder axis, 5 blocks, 165000 grid points (right)

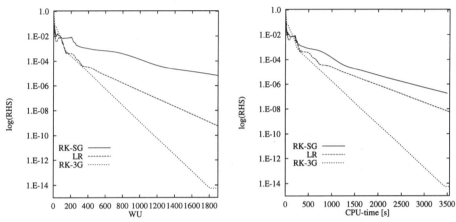

Figure 4: Maximum residual vs. WU, circular cylinder in a square duct, Re=20

Figure 5: Maximum residual vs. CPU-time, circular cylinder in a square duct, Re=20

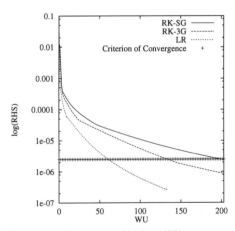 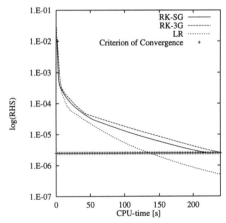

Figure 6: Maximum residual vs. WU, circular cylinder in a square duct, Re=100

Figure 7: Maximum residual vs. CPU-time, circular cylinder in a square duct, Re=100

365

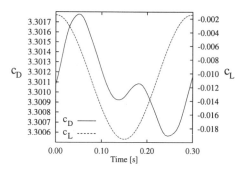

Figure 8: Time history of the lift- and drag-coefficients

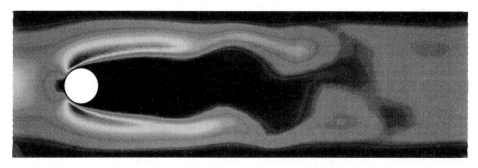

Figure 9: Absolute value of velocity in the plane of symmetry, Re=100

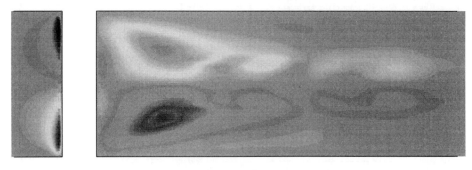

Figure 10: velocity component parallel to the axis of the cylinder, Re=100

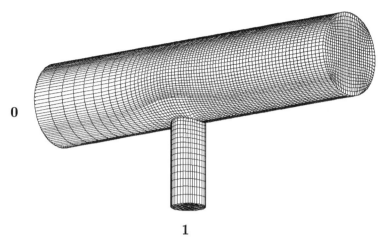

Figure 11: Domain of integration for the flow in a T-junction
380.000 grid points, ratio of tube-diameters $D_0/D_1 = 2.35$

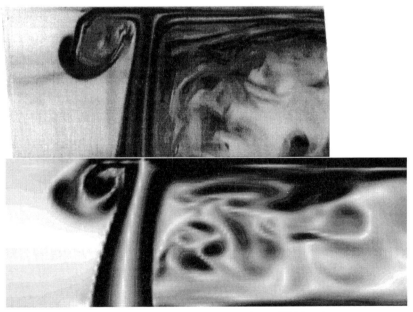

Figure 12: Comparison of flow visualization from the experiment [3] (above) and numerical solution (below), Re=309
above : streaklines, visualized by laser-light-sheet technique
below : absolute value of vorticity in the plane of symmetry ($t = 4.19$)

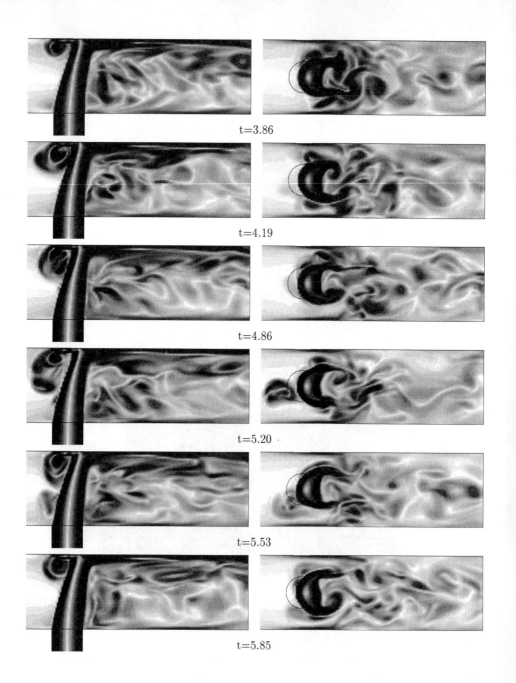

Figure 13: Absolute value of vorticity for different time levels
left column : in the plane of symmetry
right column : in the plane perpendicular to the axis of tube 1

Higher Order Upwind Schemes on Unstructured Grids for the Nonstationary Compressible Navier–Stokes Equations in Complex Timedependent Geometries in 3D

Monika Wierse, Dietmar Kröner

Institut für Angewandte Mathematik, Universität Freiburg,
Hermann–Herder–Str. 10, 79104 Freiburg, Germany
email: mwierse@mathematik.uni-freiburg.de

Summary

The aim of this project was to develop a higher order upwind finite volume scheme on unstructured grids of tetrahedra to solve the nonstationary compressible Navier–Stokes equations with high Reynolds numbers. The new aspects in this numerical scheme are a new definition of a limiter function for a special class of linear reconstruction function to get an upwind finite volume scheme of higher order on an unstructured grid of simplices, a new criterion to adapt the unstructured grid locally and a new kind of discretization of second derivatives. With several numerical tests we will show that we can obtain better results with these new approaches than with existing ones, which we have also used in numerical tests. So it turns out that the new upwind scheme of higher order for conservation laws is really of higher order in regions where the solution is smooth and has no oscillations at discontinuities. The quality of the new discretization of second derivatives will be shown also on grids with large aspect ratios. We will apply the solver for the compressible Euler equations to the flow in a simplified two–stroke engine with a moving piston in 3D. For the modelling of the moving boundary we developed a new technique which guarantees the conservation of mass during the calculations.

Introduction

The flexibility to discretize complex geometries and to adapt the grid locally according to special features (like shocks, vortices, and singularities) of the fluid flow motivates the use of unstructured grids. With local adaption of the grid we mean the time dependent increase and decrease of the grid fineness by refinement and coarsening in certain regions. For the considered nonstationary flow problems it is essential that the fineness of the grid can not only be increased but also decreased locally. With appropriate adaption methods and adaption criteria the fineness of the grid can be varied without changes in the numerical scheme.

But on these unstructured grids it is more difficult to define (theoretically motivated) an upwind scheme of higher order in space and to discretize second derivatives. By 'theoretically motivated' we mean that for the developed scheme applied to a scalar equation a proof of convergence to the correct physical solution can be shown or at least a maximum principle. Therefore the main emphasis of this project was to develop a new upwind finite volume method of higher order on an unstructured grid of simplices. We concentrated our considerations on cell–centered upwind finite volume schemes working on tetrahedra as finite volumes. Some advantages for us to use the tetrahedra as finite volumes and not dual cells on a grid of tetrahedra are: most theoretical considerations are done first for those schemes and have been available first; the bandwidth of a matrix for an implicit scheme is much smaller and the new technique to model the moving boundary is easier to implement. A numerical scheme on dual cells is discussed in [19], in this publication. Most of the content of this paper is part of the author's Ph. D. thesis [21]. Another publication related to this project will be [22].

The System of Equations to Solve

The system of the three–dimensional nonstationary compressible Navier–Stokes equations in dimensionless form can be written as

$$\partial_t u + \nabla \cdot f(u) - \frac{1}{Re} \nabla \cdot h(u) = 0 \text{ in } \Omega \times [0, T], \text{ where}$$

$$u = \begin{pmatrix} \rho \\ \rho v_1 \\ \rho v_2 \\ \rho v_3 \\ e \end{pmatrix}, \quad f_i(u) = \begin{pmatrix} \rho v_i \\ \rho v_i v_1 + \delta_{i1} p \\ \rho v_i v_2 + \delta_{i2} p \\ \rho v_i v_3 + \delta_{i3} p \\ (e+p) v_i \end{pmatrix}, \quad h_i(u) = \begin{pmatrix} 0 \\ \tau_{i1} \\ \tau_{i2} \\ \tau_{i3} \\ f \cdot \partial_k T + \sum_k v_k \tau_{ki} \end{pmatrix}.$$

In addition the following equations of state hold:

$$p = (\gamma - 1)\left(e - \frac{\rho}{2}(v_1^2 + v_2^2 + v_3^2)\right), \quad T = \frac{(v^d)^2 \cdot T^d}{(\gamma - 1) \cdot c_v} \frac{p}{\rho}.$$

The quantity ρ describes the density, $\mathbf{v} = (v_1, v_2, v_3)^T$ the velocity, e the total energy, p the pressure and T the temperature in the fluid. These sizes are made dimensionless with $\rho^d, v^d, \rho^d(v^d)^2, \rho^d(v^d)^2$ and T^d. We set γ equal to 1.4, the value of the specific gas constant corresponding to air. With $\tau_{ij} = \mu\{(\partial_i v_j + \partial_j v_i) - \frac{2}{3}(\nabla \cdot \mathbf{v})\delta_{ij}\}$ the stress tensor is defined. The appearing constants are $f = \frac{c_p \cdot T^d \cdot \mu}{(v^d)^2 \cdot Pr}$, Pr the Prandtl number, c_p (c_v) the specific heat capacity under constant pressure (constant volume), $Re = \frac{\rho^d v^d L^d}{\mu^d}$ the Reynolds number with L^d the reference length $\mu = \mu(T) = T^{1.5}\frac{1+S}{T+S}$ with $S = \frac{110°K}{T^d}$.

For an arbitrary finite volume $V(t)$ with a boundary moving with velocity $w(x,t) = (w_1, w_2, w_3)^T$ the system of equations to be discretized are

$$\frac{d}{dt} \int_{V(t)} u(x,t)dx = -\int_{\partial V(t)} \tilde{f}(u) \cdot n + \frac{1}{Re} \int_{\partial V(t)} h(u) \cdot n \quad \text{with} \tag{1}$$

$$\tilde{f}_i(u) = \begin{pmatrix} \rho(v_i - w_i) \\ \rho v_1(v_i - w_i) + \delta_{i1} p \\ \rho v_2(v_i - w_i) + \delta_{i2} p \\ \rho v_3(v_i - w_i) + \delta_{i3} p \\ (e+p)(v_i - w_i) + p w_i \end{pmatrix}.$$

Development of the Numerical Scheme Close to the Convergence Theory

Let *dim* be the dimension of space, which can be here 2 or 3 and I be an index set of simplices. In the literature a lot of approaches can be found to define higher order upwind finite volume methods in space for conservation laws and convection diffusion problems. Since for most of the existing approaches no theoretical basis in form of a convergence result or at least a maximum principle exists, we want to define approaches which are motivated by this theoretical constraints. To explain the ideas in 2D we use the following notations:

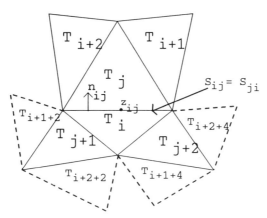

We want to use the following explicit cell–centered upwind finite volume scheme to get piecewise constant data on simplices:

The First Order Scheme:

$$u_i^{n+1} := u_i^n - \frac{\Delta t^n}{|T_i|} \sum_{j=0}^{dim+1} g_{ij}(u_i^n, u_j^n) + \frac{1}{Re} \frac{\Delta t^n}{|T_i|} \sum_{j=0}^{dim+1} h_{ij}(u_h). \tag{2}$$

The Higher Order Scheme in Space:

$$u_i^{n+1} := u_i^n - \frac{\Delta t^n}{|T_i|} \sum_{j=0}^{dim+1} g_{ij}(L_i^n(z_{ij}), L_j^n(z_{ij})) + \frac{1}{Re} \frac{\Delta t^n}{|T_i|} \sum_{j=0}^{dim+1} h_{ij}(u_h), \tag{3}$$

where

- g_{ij} is the numerical flux, which must fulfill Assumption 2.1.

- L_i^n and L_j^n are linear reconstruction functions on the simplices T_i and T_j respectively depending on $\{u_i^n, i \in I\}$ (see Definition 2.3).

- $h_{ij}(u_h)$ is a function which is on a simplex T_i a linear combination of the piecewise constant data $u_h := \{u_i^n, i \in I\}$. With $h_{ij}(u_h)$ we want to discretize the first derivatives at each edge (face) S_{ij} multiplied by components of n_{ij} and $|S_{ij}|$. $h_{ij}(u_h)$ must be conservative. $h_{ij}(u_h)$ will be used with the gradients $G_{ij}(u_h^n)$ defined in Definition 2.4.

ASSUMPTION 2.1 *(on the numerical flux $g_{ij}(u,v)$)*

The numerical flux $g_{ij}(u,v)$ should be consistent with $f(u) \cdot \nu_{ij}$, conservative, Lipschitz-continuous and monotone.

In the case of the system of compressible Euler or Navier–Stokes equations we use the three numerical fluxes of Steger and Warming [17], Dick [5] and van Leer [20] respectively. In the case of the scalar test examples we use the numerical flux of Engquist and Osher. The first step in this project was to develop a higher order scheme in space for conservation laws motivated by the onedimensional MUSCL idea. Following the result in [10] we have to define linear reconstruction functions L_i to a simplex T_i such that the scheme in (3) with $h_{ij}(u_h) = 0$ fulfills a maximum principle. With the next Lemma we could proof that this is possible for a special class of reconstruction functions.

LEMMA 2.2 *(Maximum principle for a special class of reconstruction functions)* Denote by z_{ij} the edge midpoint (face gravity center) on S_{ij} of a simplex T_i. We assume that for each simplex a linear reconstruction function L_i^n is given for a fixed time t^n with

$$u_i^n = \frac{1}{dim+1} \sum_j^{dim+1} L_i^n(z_{ij}), \qquad |L_i^n(z_{ij})| \leq M, \quad \forall z_{ij} \in \partial T_i$$

where $M = \sup_i |u_i^0|$, the numerical flux fulfills Assumption 2.1 and Δt^n used in (3) (with $h_{ij}(u_h) = 0$) satisfies

$$\sup_i \frac{\Delta t^n}{|T_i|} C_g(M) \sum_j^{dim+1} |S_{ij}| \leq \frac{1}{dim+1}$$

with $C_g(M)$ the Lipschitz constant of the numerical flux.
Then $\sup_{i \in I} |u_i^{n+1}| \leq M$ holds for u_i^{n+1} calculated with (3) with $h_{ij}(u_h) = 0$.

Motivated by this result and the other conditions in [10] to prove convergence of the approximated solution to the physical solution we define the following limiter function for the twodimensional case.

DEFINITION 2.3 *(the limiter function C in 2D, dim = 2)*
Construct for each simplex $T_i, i \in I$ a linear reconstruction function of the form

$$F_i^n(x) = u_i^n + s_i^n \cdot (x - x_i)$$

with s_i^n describing the slope of a plane constructed by the piecewise constant data $\{u_i^n, i \in I\}$, see for example the approach in [2], [6], [8]. Let $i \in I, n \in \mathbb{N}$ be fixed and $j = 1, .., dim + 1$. Let

$$u_i^{max} := max\{u_i^n, u_{ik}^n, \ldots, u_{idim+1}^n\} \text{ and } u_i^{min} := min\{u_i^n, u_{ik}^n, \ldots, u_{idim+1}^n\}$$

the maximum and minimum of the values of the values on simplex T_i and its neighbours.

We then define the values $L_i^n(z_{ij})$ by $L_i^n(z_{ij}) := u_i^n + \beta \cdot term(j)$ where β and $term(j)$ are defined with the following algorithm:

$$term(j) := \begin{cases} s_i^n \cdot (z_{ij} - x_i), & \text{if } s_i^n \cdot (z_{ij} - x_i) \cdot (u_i^n - u_{ij}^n) \leq 0 \\ 0, & \text{otherwise}. \end{cases}$$

If $term(j) = 0$ only for one j_1 and $term(j_1 + 1) \cdot term(j_1 + 2) < 0$ then

$$term(j_1 + 1) := sign(term(j_1 + 1)) \cdot minvalue$$

and

$$term(j_1 + 2) := sign(term(j_2 + 1)) \cdot minvalue$$

where $minvalue = min(|term(j_1 + 1)|, |term(j_2 + 1)|)$.
If $term(j) = 0$ for two j's then set $\beta = 0$ and for the other cases
determine largest $\beta \in [0, 1]$ such that $\forall j$ and a fixed $\alpha \in (\frac{1}{2}, 1]$
a) $\beta \cdot |term(j)| \leq C_1 h^\alpha$ and b) $u_i^{min} \leq u_i^n + \beta \cdot term(j) \leq u_i^{max}$.

The 3D version of this limiter can be found in [21]. To solve the compressible Navier-Stokes equations we have to discretize first derivatives at the the edge midpoint (face gravity center) z_{ij}. In the following definition we will give the discretization of the gradient at the edge midpoint z_{ij} we want to use.

DEFINITION 2.4 (discretization of the gradient at the edge midpoint z_{ij} in 2D) Let L_{ij}^n a linear function with $L_{ij}^n(\mathbf{x}_i) = u_i^n$, $L_{ij}^n(\mathbf{x}_{j+k}) = u_{j+k}^n$, $L_{ij}^n(\mathbf{x}_j) = u_j^n$ with $k \in \{1, 2\}$ and $\mathbf{x}_i = \binom{x_i}{y_i}$ the gravity center of T_i. We choose $k \in \{1, 2\}$ such that

$$\frac{v_{ji}^+ \cdot n_{ij}}{det A_{j,j+k,i}} \leq 0, \text{ where } v_{ji}^+ := \begin{pmatrix} -(y_j - y_i) \\ x_j - x_i \end{pmatrix}, A_{j,j+k,i} := \begin{pmatrix} x_j - x_i & y_j - y_i \\ x_{j+k} - x_i & y_{j+k} - y_i \end{pmatrix}.$$

Then we define $G_{ij}(u_h^n) := \nabla L_{ij}^n$. The gradient $G_{ji}(u_h^n)$ is defined similar as the gradient of the linear function $L_{ji}^n(x)$ with $L_{ji}^n(\mathbf{x}_j) = u_j^n$, $L_{ji}^n(\mathbf{x}_{i+k'}) = u_{i+k'}^n$, $L_{ji}^n(\mathbf{x}_i) = u_i^n$ with $k' \in \{1, 2\}$, such that $(G_{ij}(u_h^n) + G_{ji}(u_h^n)) \cdot n_{ij}$ is conservative.

The 3D version can be found in [22] and some further comments to this definition for simplices with high aspect ratios.

NOTE 2.5 For this kind of discretization of gradients in 2D we can show a maximum principle for certain kinds of triangulations [22] for the first order scheme in (2) applied to

$$\frac{\partial}{\partial t} u(x, t) + \nabla \cdot f(u(x, t)) - \epsilon \Delta u(x, t) = 0.$$

The only convergence result, which could be found in the literature [9], with an error estimate for scheme (2) uses $h_{ij}(u_h) := \frac{|S_{ij}|}{|d_{ij}|}(u_j^n - u_i^n)$ with $|d_{ij}| := |x_j - x_i|$ and x_i the circum center of T_i. This result holds only for

$$div(u(x)v(x)) - div(k(x)\nabla u(x)) + b(x)u(x) - f(x) = 0$$

(k, v, b given functions) and is restricted to triangulations with angles less than 90°. The same expression for $h_{ij}(u_h)$ is used in a recent paper [11], in which we could proof an error estimate for the scheme (3) for certain L_i^n's and a special triangulation.

Numerical Order of Convergence of the New Approaches

To evaluate the new approaches we performed some calculations in 2D for the scalar Burgers equations, the compressible Euler equations, the convection diffusion equation and the compressible Navier–Stokes equations. We chose examples where the solutions are known such that the "experimental order of convergence" (EOC) can be determined. Denote by u_h the approximated solution of the first order scheme, by u_h^C, u_h^S, u_h^F the approximated solutions of the second order scheme using limiter function Ansatz C (Scheme C), the superbee approach in [6] and the unlimited interpolation approach in [8]. Some more numerical results can be found in [21],[22].

The results are obtained on the following grids:

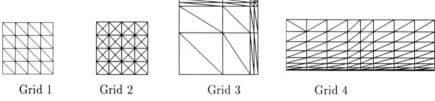

Grid 1   Grid 2   Grid 3   Grid 4

The aspect ratio in grid 3 is 1:11.

Numerical Tests for the Scalar Burgers Equation in 2D with a Smooth Solution:
We consider the initial value problem $\partial_t u + \partial_x u^2 + \partial_y u^2 = f$ in $[0,1]^2$, $u(x,0) = u_0$, with a chosen smooth $u(x,t)$. Here the higher order scheme in space is used with the TVD-Runge–Kutta method of second order in time presented in [15].

Results on Grid 1

h	$\|u - u_h\|_{L^1}$	EOC	$\|u - u_h^C\|_{L^1}$	EOC	$\|u - u_h^S\|_{L^1}$	EOC	$\|u - u_h^F\|_{L^1}$	EOC
0.35	0.0635		0.0284		0.0261		0.0281	
0.18	0.0299	1.09	0.0082	1.79	0.0118	1.14	0.0070	2.02
0.09	0.0148	1.01	0.0020	2.07	0.0046	1.35	0.0017	2.01
0.04	0.0075	0.98	0.0005	2.04	0.0012	1.92	0.0005	1.94
0.02	0.0038	0.99	0.0001	2.01	0.0003	2.03	0.0001	1.94

Numerical Tests for the System of the Compressible Euler Equations in 2D:
We considered the system of the Euler equations in two space dimensions where we used different numerical fluxes. We considered a quasi-twodimensional shock tube problem, where the solution can be calculated to any order of accuracy with a Newton solver [4]. In the following tables the numerical errors in the density for the different schemes are presented. The calculations are done on a triangulation like that of grid 2.

Results with the Numerical Flux of Steger and Warming

h	$\|\rho - \rho_h\|_{L^1}$	EOC	$\|\rho - \rho_h^C\|_{L^1}$	EOC	$\|\rho - \rho_h^S\|_{L^1}$	EOC
0.25	0.3910		0.2790		0.2757	
0.13	0.2975	0.39	0.1345	1.05	0.1407	0.97
0.06	0.2132	0.48	0.0677	0.99	0.0774	0.86
0.03	0.1438	0.57	0.0346	0.97	0.0446	0.79
0.02	0.0938	0.62	0.0172	1.01	0.0327	0.45

Results with the Numerical Flux of Dick

h	$\|\rho - \rho_h\|_{L^1}$	EOC	$\|\rho - \rho_h^C\|_{L^1}$	EOC	$\|\rho - \rho_h^S\|_{L^1}$	EOC
0.25	0.3252		0.1983		0.2086	
0.13	0.2302	0.50	0.1061	0.90	0.1068	0.97
0.06	0.1584	0.54	0.0534	0.99	0.0639	0.74
0.03	0.1051	0.59	0.0286	0.90	0.0390	0.71
0.02	0.0669	0.65	0.0150	0.93	0.0279	0.48

For the numerical flux of van Leer we obtained errors very similar to those with the numerical flux of Dick. Note that the improvement we got with the flux of Dick against that of Steger and Warming in combination with the higher order scheme Scheme C is not as large as the results of the first order schemes.

Numerical Tests for the Convection Diffusion Equation: Now we want to consider the initial value problem $\partial_t u - \epsilon \Delta u + \binom{1}{1} \cdot \nabla u = f$ in $\Omega; u(x,0) = u_0$ on $\Omega; u(x,t) = 0$ on $\partial\Omega$ with $u(x,t) = t \cdot x \cdot y \cdot (1 - e^{\frac{x-1}{\epsilon}}) \cdot (1 - e^{\frac{y-1}{\epsilon}})$.

We will compare the new approach in Definition 2.4 to discretize $\int_{S_{ij}} \nabla u \cdot n_{ij}$ by

$$h_{ij}(u_h) := \frac{1}{2} \cdot (G_{ij}(u_h^n) + G_{ji}(u_h^n)) \cdot n_{ij}|S_{ij}| \quad \text{with} \quad h_{ij}(u_h) := \frac{|S_{ij}|}{|d_{ij}|}(u_j^n - u_i^n)$$

(we will denote this by the **simple approach**) where $|d_{ij}| := |x_j - x_i|$ and x_i the gravity center of T_i. For the case $\epsilon = 0.1$ we used the TVD–Runge–Kutta method we mentioned above for the higher order upwind scheme in space.

Results with the simple approach

Grid 2: $\epsilon = 10^{-4}$				Grid 3: $\epsilon = 10^{-4}$			
$\|u - u_h\|_{L^1}$	EOC	$\|u - u_h^C\|_{L^1}$	EOC	$\|u - u_h\|_{L^1}$	EOC	$\|u - u_h^C\|_{L^1}$	EOC
3.143		2.765		8.065		6.259	
1.795	0.80	1.206	1.20	4.892	0.72	3.347	0.92
0.961	0.90	0.530	1.19	2.648	0.89	1.563	1.10
0.510	0.91	0.242	1.13	1.393	0.93	0.680	1.20
0.275	0.89	0.117	1.05	0.723	0.94	0.299	1.19

The errors in the L^1-norm are multiplied by 10^3.

Grid 2: $\epsilon = 10^{-1}$				Grid 3: $\epsilon = 10^{-1}$			
$\|u - u_h\|_{L^1}$	EOC	$\|u - u_h^C\|_{L^1}$	EOC	$\|u - u_h\|_{L^1}$	EOC	$\|u - u_h^C\|_{L^1}$	EOC
3.204		1.484		8.919		1.109	
1.862	0.78	0.498	1.57	8.685	0.04	1.016	0.13
1.062	0.81	0.179	1.47	9.333	-0.10	0.993	0.03
0.577	0.88	0.061	1.56	9.532	-0.03	0.987	0.01
0.302	0.93	0.021	1.52	9.657	-0.02	0.983	0.01

Results with the new approach

Grid 3: $\epsilon = 10^{-4}$				Grid 3: $\epsilon = 10^{-1}$			
$\|u - u_h\|_{L^1}$	EOC	$\|u - u_h^C\|_{L^1}$	EOC	$\|u - u_h\|_{L^1}$	EOC	$\|u - u_h^C\|_{L^1}$	EOC
8.089		8.047		6.285		6.396	
4.919	0.72	4.872	0.72	3.877	0.70	2.397	1.42
2.676	0.88	2.627	0.89	2.116	0.87	0.909	1.40
1.423	0.91	1.371	0.94	1.109	0.93	0.290	1.65
0.760	0.90	0.703	0.97	0.564	0.98	0.081	1.85

The results on grid 2 for the new approach are the same as for the simple approach.

Numerical Tests for the System of the Compressible Navier–Stokes Equations in 2D:

To test the solver for the compressible Navier-Stokes equations we determined a right hand side f to an exact solution corresponding to a parabolic velocity field and a linear pressure. The calculations are performed on grid 2.

$\|v_1 - v_{1h}\|_{L^1}$	EOC	$\|\rho - \rho_h\|_{L^1}$	EOC	$\|p - p_h\|_{L^1}$	EOC
0.004750		0.000358		0.000753	
0.003208	0.57	0.000240	0.58	0.000508	0.57
0.001842	0.80	0.000146	0.72	0.000310	0.71
0.000982	0.91	0.000084	0.80	0.000179	0.79

Another wellknown test example is the flow over a flat plate for low mach numbers. Following the ideas of H. BLASIUS in [14] we can write the solutions for the velocity components v_1 und v_2 and the temperature T in the boundary layer with the space dependent variable $\eta := y\sqrt{\rho\, U\, Re/(\mu\, x)}$. In the following figure we compare the results calculated with the numerical flux of Steger and Warming and that of Dick, where U is the absolute value of the velocity at the inflow boundary and T_∞ the temperature at the inflow boundary. The position of comparision is 0.4 unit length from the beginning of the flat plate and the Reynold number is 3000. The calculations are done on a triangulation like that in grid 4, with an aspect ratio 1:150.

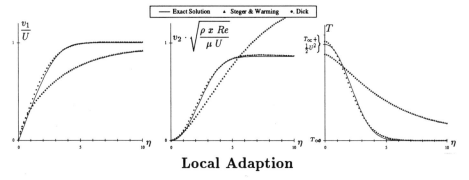

Local Adaption

We are using the local adaption algorithm proposed in [1] on a grid of tetrahedra, with which it is possible to refine and coarse locally. With this technique of bisection of tetrahedra it is very easy to project the data between two grids under the constraint of conservation. A criterion is necessary to control the mesh size according to the error of the numerical scheme. In contrast to elliptic or parabolic problems no a-posteriori error estimator is available for the system of the Euler equations. Therefore a lot of heuristic criteria, for instance oriented on large gradients, differences or second derivatives, are used. An approach based on the residual is well motivated for scalar conservation laws in 1D with a result of Tadmor [18]. Therefore we also used residuals in the case of the system of Euler equations.

DEFINITION *(of the adaption criterion used) Denote by L_i^n the linear reconstruction function to each conservative variable on the tetrahedron T_i determined with the unlimited reconstruction given in [8]. Set*

$$S_i^n := \text{largest face of } T_i, \qquad r_i^n := \frac{u_i^{n+1} - u_i^n}{\Delta t^n} + \sum_{j=1}^{3} f_j'(u_i^n) \partial_j L_i^n(x_i)$$

with f_j the flux vector in the system of the Euler equations. Denote by $(r_i^n)_l$ the lth component of the local residual r_i^n. We then mark the tetrahedron T_i for refinement or coarsening depending on the value of $\sqrt{|S_i^n|}|T_i|\sum_{l=1}^{5}|(r_i^n)_l|$.

The additional power of h ($\sqrt{|S_i^n|}$) is motivated in [21] to get the same scale for $||r_i^n||_{L^1(T_i)}$ and $||u(t^n) - u_h^n||_{L^1(T_i)}$.

Numerical Algorithm for the Engine Flow Problem

To model the exchange process in a simplified two-stroke engine we considered the geometry depicted in the following figure.

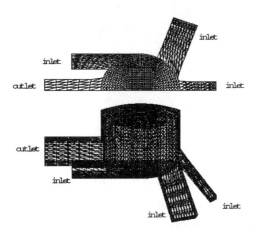

Figure 1: Geometry of a simplified two–stroke engine

We can restrict the calculation to the half geometry by a symmetry assumption. In this geometry a piston is moving down and up in a prescribed manner. Depending on the piston position parts of the discretization domain can be disconnected.

We discussed in [21] existing approaches to model the piston motion: moving discretization points (see [7], [12]) and overlapping grids (see [16]). A lot of effort would be necessary to administrate the cell connectivities at the cell faces between cylinder and ports while the points of an unstructured grid are moving. Furthermore the conservation property of the scheme is complicated to handle because of the interpolation between different meshes and because the degeneration of the tetrahedra must be avoided by a grid smoother. Therefore we propose a new approach on a grid of tetrahedra: the piston motion is modelled by cutting the piston out of a fixed grid at each time step. Let $\mathcal{T}$ be a set of tetrahedra describing the geometry of the simplified two–stroke engine, where the discretisation of the ports is connected to the cylinder in a suitable fashion. Denote by $\Pi(t)$ the closed part of $\Omega = \bigcup_{T_i \in \mathcal{T}} T_i$ occupied by the piston at time t and $\Omega(t) := \Omega \backslash \Pi(t)$. Depending on the piston position at time t, the discretization domain can be disconnected. The process of cutting the piston out of the grid at a certain time is nothing else than to determine the partition of tetrahedra by a plane (= piston crown). After dividing a tetrahedron by a plane we can get again a tetrahedron or a prism.

We denote the resulting discretization of $\Omega(t^n)$ by $\mathcal{T}^n = \bigcup_{T_i \in \mathcal{T}, T_i \cap \Omega(t^n) \neq \{\}} T_i^n$, which does not only consist of complete tetrahedra ($T_i^n := \{T_i \cap \Omega(t^n) : T_i \in \mathcal{T}, T_i \cap \Omega(t^n) \neq \{\}\}$). It can be seen that $\mathcal{T}^n$ may contain very small cells which would cause very small time steps if we applied an explicit finite volume scheme to the elements of $\mathcal{T}^n$. Therefore we introduced the approach of **collected cells**, which means that we put some finite volumes T_l^n together to define a new finite volume $C_i^n := \bigcup_{l \in m(i,n)} T_l^n, m(i,n)$: a set of tetrahedron indices, where we get values on C_i^n by $u_i^n := \frac{\sum_{j \in m(i,n)} |T_j^n| u_j^n}{\sum_{j \in m(i,n)} |T_j^n|}$. Some examples of collected cells are:

Denote the resulting discretization by $\bar{T}^n$. The collection process is steered such that the cut tetrahedra do not influence Δt^n and the collected cells are collections of cut and uncut tetrahedra with at least one common vertex not in the piston.

Figure 2: Applying the approach of collected cells on T^n to get $\bar{T}^n$

To define the finite volume scheme we need $\bar{T}^n$ at time t^{n+1} and set therefore $\bar{S}^{n+1}$ equal to the set of finite volumes of $\bar{T}^n$ at time t^{n+1}. With the approach of **collected cells** we also avoid $T_i^{n+1} = \{\}$ and $C_i^{n+1} = \{\}$ with T_i^n and C_i^n finite volumes of $\bar{T}^n$. In the case of upward piston motion $\bar{S}^{n+1}$ is a discretization of $\Omega(t^{n+1})$, this need not be the case if the piston is moving down, see Figure 3. This is the second problem of this approach: cells of the fixed grid in the previous timestep completely lying inside the piston can be exposed in the next timestep. This means that we have no new values in the new timestep for those exposed cells. To get values in the exposed cells at time t^{n+1} the approach of collected cells is also used to put those cells together with neighbouring cells containing values at time t^{n+1}.

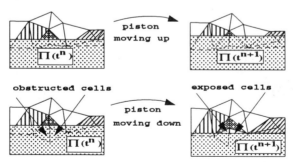

Figure 3: $\bar{T}^n$ to $\bar{S}^{n+1}$

DEFINITION: *(Basic scheme to calculate values u_i^{n+1} on $V_i^{n+1} \in \bar{S}^{n+1}$ with $V_i^n \in \bar{T}^n$ to approximate (1) in the case of the compressible Euler equations) Let us assume that at time t^n the piston is at a certain position and we have calculated values u_i^n for $V_i^n \in \bar{T}^n$.*

$$u_i^{n+1} := \frac{|V_i^n|}{|V_i^{n+1}|} u_i^n - \frac{\Delta t^n}{|V_i^{n+1}|} \sum_{j, S_{ij}^n \in \partial V_i^n} g_{ij}(u_i^n, u_{ij}^n)$$

The size of the time step Δt^n is determined such that

$$\sup_{V_i^n \in T^n} \frac{\Delta t^n}{|V_i^n|} \max_{j, S_{ij}^n \in \partial V_i^n} \lambda_{ij}(u_i^n) \leq CFL \qquad (CFL < 1 \text{ is fixed})$$

with $\lambda_{ij}(u_i^n) = (|w_i^n \cdot n_{ij}| + c_i^n)|S_{ij}^n|$, where $c_i^n = \sqrt{\frac{\gamma \cdot p_i^n}{\rho_i^n}}$, ρ_i^n, p_i^n and w_i^n are the approximated sound speed, density, pressure and velocity in the finite volume V_i^n.

It can easily be proved that this scheme is conservative in mass. In [3] the same idea of merging cells is used on structured grids. How to include the local adaption process is also described in [21].

Numerical Results
Examples with the Nonstationary Compressible Euler equations

a) Forward Facing Step Problem in 3D

Since the forward facing step problem [23] is often used to analyse the quality of schemes we calculated this problem in 3D as an quasi–twodimensional problem. The maximal number of tetrahedra used for this calculation with local adaption of the grid is 200000.

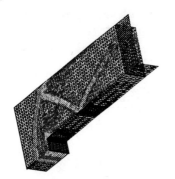
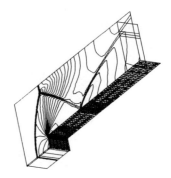

b) Flow in the Complex Geometry with a fixed piston in 3D

Now we want to compare the velocity on a cutting plane through the complex geometry: We see from left to the right the result of the first order scheme on a grid of 50000, on a grid of 300000 tetrahedra and the results of the second order scheme with Ansatz C on a grid of 50000 tetrahedra.

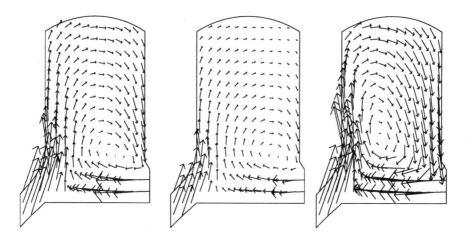

The velocity vectors are projected onto this plane.

c) Flow in the Complex Geometry with a Moving Piston in 3D

Since we are not modelling the combustion process we want to start with the simulation after the combustion took place, that means the piston is on a position such that the outlet and the inlets are closed. The piston moves down and back up to this position. The number of tetrahedra had been restricted to 80000 during the calculation with local adaption.

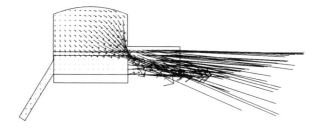

Figure 4: Piston motion down, just after the start of the piston motion

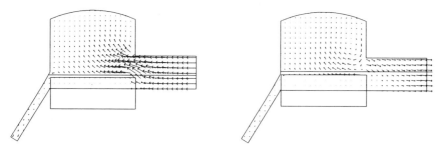

Figure 5: Piston motion up again   Figure 6: Piston motion up again

381

To analyse the exchange process of "old" and "new" gas we can look on the development of isosurfaces of the entropy or particle traces in time.

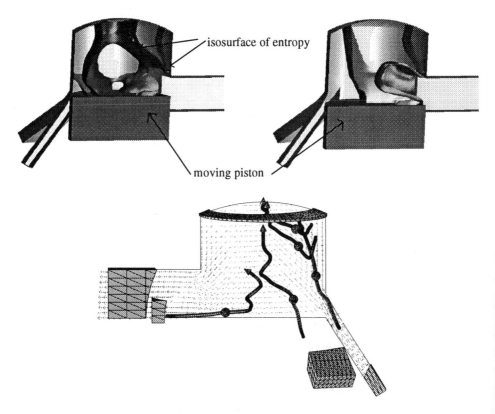

Figure 7: The little balls lying on the particle traces correspond to the position of the particles at the moment where the piston is completely down. The same holds for the depicted velocity field. The particles starting in the upper right corner of the cylinder indicate that this geometry is not optimal with respect to the gas exchange. Only some parts of the model geometry are depicted.

Examples with the Compressible Navier–Stokes equations

In the complex geometry we solved the compressible Navier–Stokes equations on a grid of 300.000 and 700.000 tetrahedra. In the following figure we see the velocity field through the corresponding grids on the same cutting plane as above.

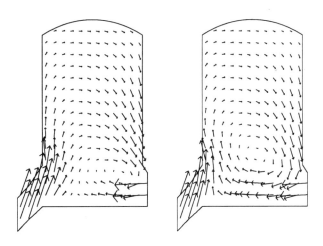

Acknowledgement For the newer part in this paper concerning the solution of the compressible Navier–Stokes equations I want to thank Ralph Schwörer who implemented uncountable ideas and calculations in 2D which can be found in his diplom thesis. The results on grids with more than 200.000 tetrahedra could only be obtained in a reasonable time by using the parallel concept of Bernhard Schupp, where I had more or less to put in my serial code.

References

[1] Bänsch, E.: *An Adaptive Finite-Element Strategy for the Three-dimensional Time-dependent Navier-Stokes equations.* Journal of Computational and Applied Mathematics 36 (1991), 3-28.

[2] Barth, T. J.; Jesperson, D. C.: *The Design and Application of Upwind Schemes on Unstructured Meshes.* AIAA-89-0366.

[3] Bayyuk, S. A.; Powell, K. G.; v. Leer, B.: *A Simulation Technique for 2-D Unsteady Inviscid Flows Around Arbitrarily Moving and Deforming Bodies of Arbitrary Geometry.* AIAA 93-3391

[4] Chorin, A. J.: *Random Choice Solution of Hyperbolic Systems.* Journal of Computational Physics 22 (1976), 517–533.

[5] E. Dick, *Multigrid Methods for Steady Euler- and Navier–Stokes equations based on polynomial Flux–Difference Splitting,* in: International Series of Numerical Mathematics, Vol. 98, Birkhäuser, 1991.

[6] Durlofsky, L. J.; Engquist, B.; Osher, S.: *Triangle Based Adaptive Stencils for the Solution of Hyperbolic Conservation Laws.* Journal of Computational Physics 98 (1992), 64–73.

[7] Epstein, P. H.; Reitz, R. D.; Foster, D. E.: *Computations of a Two-Stroke Engine Cylinder and Port Scavenging Flows.* SAE 910672.

[8] Frink, N. T.; Parikh, P.; Pirzadeh, S.: *A Fast Upwind Solver for the Euler Equations on Three-Dimensional Unstructured Meshes.* AIAA-91-0102.

[9] Herbin, R.: *An Error Estimate for a Finite Volume Scheme for a Diffusion-Convection Problem on a Triangular Mesh*. Numerical Methods for Partial Differential Equations 11, 165-173, 1995.

[10] Kröner, D.; Noelle, S.; Rokyta, M.: *Convergence of Higher Order Finite Volume Schemes on Unstructured Grids for Scalar Conservation Laws in Two Space Dimensions*. Numerische Mathematik 71, 1995.

[11] Kröner, D.; Rokyta, M.; Wierse, M.: *Apriori Error Estimates for Upwind Finite Volume Schemes in Several Space Dimensions*.(in preparation)

[12] Lai, Y. G.; Przekwas, A. J.; Sun, R. L. T.: *CFD Simulation of Automative I. C. Engines with Advanced Moving Grid and Multi-Domain Methods*. AIAA-93-2953.

[13] Pan, D.; Cheng, J.-C.; *Upwind Finite-Volume Navier-Stokes Computations on Unstructured Triangular Meshes*. AIAA Journal Vol. 31, No. 9, 1993.

[14] Schlichting, H.: *Grenzschicht-Theorie*. Verlag G. Braun, Karlsruhe (1965)

[15] Shu, C. W.; Osher, S.: *Efficient Implementation of Essentially Non-Oscillatory Shockcapturing Schemes*. Journal of Computational Physics 77 (1988), 439-471.

[16] Steger, J.: *On the Use of Composite Grid Schemes in Computational Aerodynamics*. Computer Methods in Applied Mechanics and Engineering 64 (1987), 301-320.

[17] Steger, J. L.; Warming R. F.: *Flux Vector Splitting of the Inviscid Gasdynamic Equations with Application to Finite-Difference Methods*. Journal of Computational Physics 40 (1981), 263-293. Pitman Press, London (1979).

[18] Nessyahu, H.; Tassa, T.; Tadmor, E.: *The Convergence Rate of Godunov Type Schemes*. SIAM J. Numer. Anal., Vol. 31, No. 1, February 1994.

[19] Vilsmeier, R., Hänel D.: *Computational Aspects of Flow Simulation on 3-D, Unstructured, Adaptive Grids*. In this publication.

[20] van Leer, B: *Flux Vector Splitting for the Euler Equations*, Proc. 8th International Conference on Numerical Methods in Fluid Dynamics, Berlin. Springer Verlag, 1982.

[21] Wierse, M.: *Higher Order Upwind Schemes on Unstructured Grids for the Compressible Euler Equations in Timedependent Geometries in 3D*. Dissertation, Freiburg, 1994. Preprint 393, SFB 256, Bonn.

[22] Wierse, M.: *Cell-Centered Upwind Finite Volume Scheme on Triangles for Scalar Convection Diffusion Equations*. in preparation.

[23] Woodward, P. R.; Colella, P.: *The Numerical Simulation of Two-Dimensional Flow with Strong Shocks*. Journal of Computational Physics 54 (1984), 115-173.

Improvement and Application of a Two Stream–Function Formulation Navier–Stokes Procedure

Leiping Xue, Thomas Rung, Frank Thiele

Hermann–Föttinger–Institut für Strömungsmechanik
Technische Universität, Berlin
Straße des 17. Juni 135, D–10623 Berlin, Germany

Summary

An improved structured–grid multiblock Navier–Stokes procedure, based on the two stream–function Euler–Potential formulation, is presented for the simulation of three–dimensional incompressible flows in complex geometries. The algorithm is based on general non–orthogonal coordinates and employs a fully co–located storrage arrangement for all transport properties. In this, diffusion terms are approximated using second–order central differences, whereas advective fluxes are approximated using upwind biased schemes. The solution is iterated to convergence employing an ILU type (SIP) procedure. The numerical procedure has been parallelized by means of a domain decomposition strategy. Results are reported in comparison to measurements for a laminar curved duct flow and the simulation of three–dimensional wing flow at low angle of attack. The principal aim of the paper is to convey the capabilities of the pure stream–function formulation for the simulation of internal and external aerodynamic flows.

Introduction

Most numerical methods which compute viscous flows are based on the primitive variable formulation due to a straightforward extension to the simulation of complex three–dimensional flow fields. However, the classical stream–function formulation of the momentum equations has a number of advantages over the primitive variable formulation, at least when attention is drawn to two–dimensional incompressible flows. In contrast to this, the general three–dimensional extension of the stream–function approach is associated with a number of difficulties arising from the Euler–Potential formulation itself and the formulation of appropriate boundary conditions. The nature and rational of the three–dimensional stream–function approach, including a discussion of specific problems arising from the formulation, have previously been outlined in great detail by Thiele et al.[1,2]. The objective of the present paper is to present improvements obtained with the Euler–Potential formulation. Furthermore, attention is drawn to the implementation of a parallelized procedure on the massively parallel processing (MPP) system CRAY T3D as well as on work station clusters, by means of a domain decomposition technique. Examples included refer to 90^0 bended square duct and three–dimensional wing flow at low angle of attack (2^0).

Within this Priority Research Program "Flow Simulation on Supercomputers" the two stream–function formulation Navier–Stokes procedure has been developed in connection with the finite–volume approximation of Lilek et al. [3], the solution techniques for linear equation systems of Durst et al. [4] and the parallelization strategies of Hofhaus et al. [5] and Durst et al. [6]. Parts of investigations performed in this project have been published by Xue et al. [7] and Bärwolff et al. [8]. In addition the work contributed to the stability analysis of wake flows [9] as well as to the benchmark computations within this project.

Mathematical Model

The motion of an incompressible viscous fluid is governed by the solution of the continuity and momentum equations. Introducing the two scalar stream functions Ψ and Φ of a solenoidal three–dimensional velocity field $\underline{U}$ by

$$\underline{U} = \nabla \Psi \times \nabla \Phi , \qquad (1)$$

the continuity constraint is satisfied identically. Substituting expression (1) into the momentum equations yields the so called Euler–Potential formulation [2]

$$\nabla \cdot \left\{ \left[\frac{\partial}{\partial t} + (\nabla \Psi \times \nabla \Phi) \cdot \nabla - \frac{1}{Re} \Delta \right] (\nabla \Phi \nabla \Psi - \nabla \Psi \nabla \Phi) \right\} = 0 , \qquad (2)$$

where Re denotes the Reynolds number, and pressure has been eliminated by taking the curl. Specific aspects of the numerical integration of this fourth–order differential equation are outlined in [1]. Severe difficulties arising from its strict non–linearity are avoided by introducing a new vector potential $\underline{\psi}$

$$\underline{\psi} = \nabla \cdot (\nabla \Phi \nabla \Psi - \nabla \Psi \nabla \Phi) . \qquad (3)$$

Substituting equation (3) into the Euler–Potential formulation (2) results in a system of transport equations for $\underline{\psi}$

$$\frac{\partial \underline{\psi}}{\partial t} + \nabla \cdot [(\nabla \Psi \times \nabla \Phi)\underline{\psi} - \underline{\psi}(\nabla \Psi \times \nabla \Phi)] - \frac{1}{Re} \Delta \underline{\psi} = \underline{0} . \qquad (4)$$

Using the definition of the vector potential (3), the two scalars Ψ and Φ are determined from the following equations

$$\underline{\psi} \cdot \nabla \Phi = \nabla \cdot (\nabla \Phi \nabla \Psi \cdot \nabla \Phi - \nabla \Psi \nabla \Phi \cdot \nabla \Phi) , \qquad (5)$$

$$\underline{\psi} \cdot \nabla \Psi = \nabla \cdot (\nabla \Phi \nabla \Psi \cdot \nabla \Psi - \nabla \Psi \nabla \Phi \cdot \nabla \Phi) . \qquad (6)$$

The two independent components of $\underline{\psi}$ follow from the solution of (4), whereas the third component is evaluated from the compatibility condition (3). Assuming initial values for

Ψ and Φ the differential equations (4) to (6) are consecutively iterated to convergence. The advantage of the present approach over the original Euler–Potential formulation is the linearity of the occuring differential equations. It also benefits from the fact that the resulting system consists of second–order instead of fourth–order differential equations.

The boundary conditions are specified by reference to physical considerations. Along rigid walls no–slip boundary conditions $\underline{U} = 0$ are imposed, which, in terms of the two stream functions, read

$$\begin{array}{l} \Phi_{,n} = 0.; \ \Psi_{,nn} = 0 \quad \text{or,} \\ \Psi_{,n} = 0.; \ \Phi_{,nn} = 0 \ . \quad n = \text{ wall} - \text{normal coordinate} \ . \end{array} \tag{7}$$

The boundary condition for ψ follows directly from the compatibility condition (3). At the inlet plane, the values for Ψ and Φ are determined in conformity with a prescribed inlet velocity vector

$$\underline{U}_\infty = \nabla\Psi \times \nabla\Phi \ . \tag{8}$$

The outflow conditions applied within the present study are non–reflective, hence all transport properties are convected through the outlet plane via

$$\frac{\partial \Psi}{\partial t} + \nabla \cdot (\underline{U}_\infty \Psi) = 0 \ , \tag{9}$$

$$\frac{\partial \Phi}{\partial t} + \nabla \cdot (\underline{U}_\infty \Phi) = 0 \ , \tag{10}$$

$$\frac{\partial \psi}{\partial t} + \nabla \cdot [(\nabla\Psi \times \nabla\Phi)]\psi] = \underline{0} \ . \tag{11}$$

The latter has succesfully been applied to the simulation of unsteady flows around streamlined bodies [10].

Numerical Solution Procedure

The two stream–function formulation of the Navier–Stokes equations derived requires the numerical solution of the differential equations (4) to (6). In order to solve the unsteady transport equation (4) the transient term is discretized by a flexible three ($\alpha = 1$)/two point ($\alpha = 0$) second–order implicit scheme

$$\frac{\partial F(t)}{\partial t} + G(t) = 0 \ \Rightarrow$$
$$\frac{1}{\Delta t}[(1+0.5\alpha)F(t) - (1+\alpha)F(t-\Delta t) + 0.5\alpha F(t-2\Delta t)] + G(t) = 0 \ , \tag{12}$$

where Δt denotes the time increment.

The resulting system of equations forms a linear boundary–value problem. In the following, attention is drawn to the weak formulation obtained from the volume integration of equations (4) to (6). Applying Green's theorem, the system of second–order partial differential equations is reduced to a system of first-order integro-differential equations

$$\psi: \quad \int_V \frac{\partial \underline{\psi}}{\partial t} dV + \oint_{dA} d\underline{A} \cdot \left\{ [(\nabla \Psi \times \nabla \Phi)\underline{\psi} - \underline{\psi}(\nabla \Psi \times \nabla \Phi)] - \frac{1}{Re} \nabla \underline{\psi} \right\} = \underline{0} , \quad (13)$$

$$\Phi: \quad \oint_{dA} d\underline{A} \cdot (\nabla \Phi \nabla \Psi \cdot \nabla \Psi - \nabla \Psi \nabla \Phi \cdot \nabla \Psi) = \int_V \underline{\psi} \cdot \nabla \Psi dV , \quad (14)$$

$$\Psi: \quad \oint_{dA} d\underline{A} \cdot (\nabla \Phi \nabla \Psi \cdot \nabla \Phi - \nabla \Psi \nabla \Phi \cdot \nabla \Phi) = \int_V \underline{\psi} \cdot \nabla \Phi dV . \quad (15)$$

In order to evaluate the volume integrals, the physical domain is subdivided into finite volumes as illustrated in Fig. (1). To solve the resultant set of equations all transport properties are stored in the center of each volume cell. The finite–volume approximation applied is based on the techniques proposed by Lilek et al. [3]. Assuming the values of the dependent variables to vary linearly between cell centers, the cell–surface expressions f and ∇f can be approximated by algebraic finite–difference expressions based upon the surface–orientated coordinate system ξ_i:

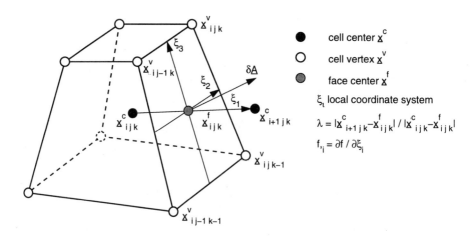

Figure 1: Finite volume and local coordinate system at an east face.

$$f = f^c_{ijk}(1 - \lambda) + f^c_{i+1jk}\lambda; \quad \frac{\partial f}{\partial x_i} = \frac{\partial \xi_1}{\partial x_i} f_{,1} + \frac{\partial \xi_2}{\partial x_i} f_{,2} + \frac{\partial \xi_3}{\partial x_i} f_{,3} . \quad (16)$$

Here, face–normal derivatives are treated implicitly

$$f_{,1} \approx f^c_{i+1jk} - f^c_{ijk} , \quad (17)$$

whereas cross derivatives are explicitly evaluated from the interpolated cell–vertex values. At odds with this, advective terms are approximated by high–order upwind biased formulae, e.g.

$$f = \underbrace{f^c_{ijk}}_{UDS} + 0.25 \underbrace{\left[(1 + \kappa)(f^c_{i+1jk} - f^c_{ijk}) + (1 - \kappa)(f^c_{ijk} - f^c_{i-1jk}) \right]}_{antidiffusive\ gradient} . \quad (18)$$

The spatial scheme (18) involves a number of alternatives which differ in the choice of the numerical parameter κ. Unbounded approximation schemes corresponding to central differencing (CDS), second–order linear upwind differencing (LUDS) and Leonard's quadratic upstream–weighted differencing (QUICK) arise by setting $\kappa = 1, -1, 0.5$. In order to preserve the bandwidth of the resulting algebraic equation system and to keep the numerical procedure flexible, the above mentioned high–order spatial scheme consists of an implicit upwind part (e.g. f_{ijk}^c) and an explicitly treated antidiffusive gradient term ("deferrd correction procedure"). Hence, the boundary value problem results in a septa-diagonal system of algebraic equations

$$A_{ijk}^W f_{i-1jk} + A_{ijk}^E f_{i+1jk} + A_{ijk}^S f_{ij-1k} + A_{ijk}^N f_{ij+1k} + A_{ijk}^B f_{ijk-1} + A_{ijk}^T f_{ijk+1} + A_{ijk}^P f_{ijk} = S_{ijk} \ . \tag{19}$$

Structured–Grid Multiblock Matrix Algorithm

With regards to the geometrical flexibility of an unstructured–grid approach and the numerical efficiency of a globally structured–grid method, the structured–grid multiblock approach, involving globally unstructured blocks with locally structured grids, can be considered as a suitable compromise between numerical efficency and computational effort to generate the grid. Futhermore, the structured–grid multiblock approach is predestinated for an efficient parallelization of the overall solution procedure.

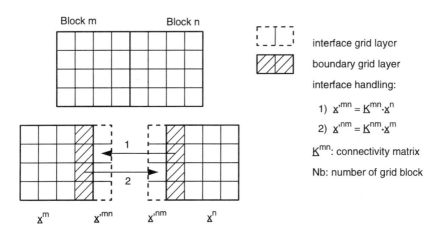

Figure 2: structured–grid multiblock arrangement and interface

The matrix algorithm used in the present study originates from the "Strongly Implicit" ILU decomposition procedure (SIP) of Stone [11]. At the interface layer of two adjacent blocks the values of the neighbouring block will be stored additionally for each block as shown in Fig. (2). The modified SIP for the solution of linear equation systems applied

389

herein reads

$$\underline{\underline{A}}^m \cdot \underline{x}^m + \sum_{n=1}^{Nb} \underline{\underline{A}}'^{mn} \cdot \underline{x}'^{mn} = \underline{S}^m , \ m=1,...,Nb . \qquad (20)$$

The algorithm of the modified multiblock SIP has been developed in connection to the techniques proposed by Durst et al. [4, 6] and can be summarized as follows:

(1) Guess initial values for $\underline{x}^{m(0)}, m=1,...,Nb$; l=1.

(2) Calculate residuals of (20):

$$\underline{r}^m = \underline{S}^m - \underline{\underline{A}}^m \cdot \underline{x}^{m(l-1)} - \sum_{n=1}^{Nb} \underline{\underline{A}}'^{mn} \underline{x}'^{mn(l-1)}, \ m=1,...,Nb ; \qquad (21)$$

$$\text{convergence check}: \ \text{if} \ \sum_{m=1}^{Nb} \|\underline{r}^m\|_\infty < \varepsilon \ \text{then stop.} \qquad (22)$$

(3) Iterate SIP for each block:

$$\underline{x}^{m(l)} = \underline{x}^{m(l-1)} - (\underline{\underline{R}}^m)^{-1} \cdot (\underline{\underline{L}}^m)^{-1} \cdot \underline{r}^m , m=1,...,Nb ; \qquad (23)$$

where $\underline{\underline{L}}^m$, $\underline{\underline{R}}^m$ are derived from Stone's incomplete LU decomposition for block m.

(4) Recalculate all variables at the interface grid layer:

$$\underline{x}'^{mn(l)} = \sum_{n=1}^{Nb} \underline{\underline{K}}^{mn} \cdot \underline{x}^{n(l)} , \ m=1,...,Nb . \qquad (24)$$

(5) Return to step (2).

In the solution procedure applied the variables at the interface layer are updated after each SIP iteration for each block. Convergence is achieved if the residuals for all finite volumes are below a given criterion.

Parallelization

The parallelization of the above described algorithm has been performed on the massively parallel processing system CRAY–T3D at the Konrad–Zuse–Zentrum für Informationstechnik Berlin (ZIB). The distributed memory system consists of 256 node DEC alpha 21064 processors, each with 64 MByte core memory and a peak performance of 150 MFlops. The communication through message passing between the node processors is based on the CRAY–T3D specific software SHAREDMEMORY GET & PUT. In this way, by changing the communication routines, the parallelized algorithm can also be transfered to any other message–passing system, e.g. a workstation cluster. An efficient parallelization requires a distribution of tasks, each of which are of common size to the processor nodes. The decomposition of the domain into partitions with about the same amount of grid points is well suited to implement a parallel solution method. It has to be emphasized that the efficiency depends on the decision of how to distribute the data and workload among the processors and their local memory. Data dependencies occur at block interfaces where information from neighbouring blocks is needed to advance the iteration.

Each processor is assigned to one or several blocks of the domain such that the SIP iteration can be performed. The load balancing of the processors is achieved by the decomposition of the domain into blocks of the same grid size. The parallelization process follows the spatial distribution of the grid blocks.

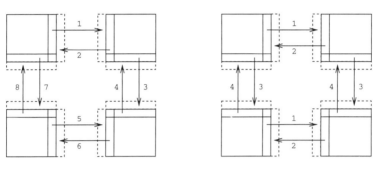

synchronized communicationstep: 1: receiving from left (1,5) 2: receiving from right (2,6)
3: receiving from up (3,7) 4: receiving from down (4,8)

Figure 3: Communication steps for 2x2 processor array.

The communication between the processors is determined by the exchange of the values at the interface grid layer which are required by the SIP iteration from the neighbouring processes (22). For a processor array arranged in two dimensions (see Fig. (3)) the communication described is independent of the number of processes and can be carried out in four steps with a simple technique which follows the ideas of Hofhaus et al. [5]:

Step 1: asynchronous sending to right, synchronous receiving from left
Step 2: asynchronous sending to left, synchronous receiving from right
Step 3: asynchronous sending to bottom, synchronous receiving from top
Step 4: asynchronous sending to top, synchronous receiving from bottom

Fig. (3) shows the eight communication steps which are necessary to complete the communication of a 2 × 2 processor array. Such a distribution of processes can be easily synchronized and needs only four steps.

This simple technique can not be applied to an unstructured distribution of blocks which arises from complex flow problems. As an example shown in Fig. (4) the flow around a circular cylinder in a channel is considered (see in this publication, benchmark test 3D–Z). Performing the 44 communication steps sequentially would result in a wait status for several of the processors. The other possibility is a prescribed structure in synchronized steps as demonstrated in Fig. (5) for 16 blocks. The communication between the blocks is automatically generated by a special procedure described as follows. For each process a communication table is created which assigns the synchronous steps from

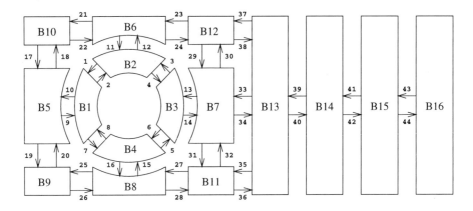

Figure 4: Communication steps for 16 randomly ordered processes.

or to which neighbouring processor the data will be received or sent.

For example the process B 13 consist of the following communication steps:
- Step 1: Sending to process B_{12}
- Step 2: Receiving from process B_{12}
- Step 3: Receiving from process B_{14}
- Step 4: Sending to process B_{14}
- Step 5: Sending to process B_{11}
- Step 6: Receiving from process B_{11}
- Step 7: Receiving from process B_7
- Step 8: Sending to process B_7

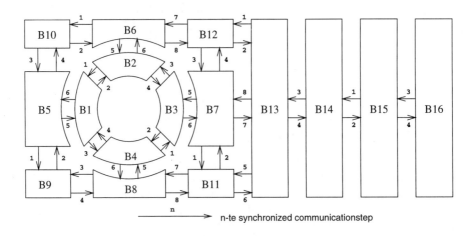

Figure 5: Synchronized communication steps.

Altogether, the 44 communication steps can be reduced to 8 synchronized steps for 16 blocks. For the three–dimensional wing flow the communication between the 16 grid blocks is presented in Fig. (12).

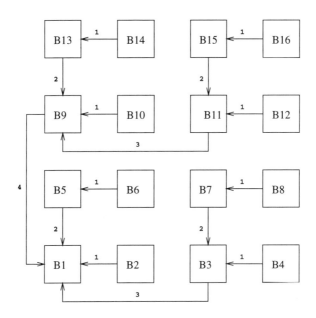

Figure 6:   Global communication, snowball system.

Besides the communication between the grid blocks, an overall step has to be considered where the data exchange includes all processors. This task does not depend on the distribution of the processes and is evaluated by reference to a snowball system as indicated in Fig. (6). In this way n communication steps can achieve 2^n processes.

Results

The validation of the numerical procedure has been carried out with reference to the 90° curved square bend, where experiments are reported by Taylor et al. [12]. Calculations have been performed on a 161 × 31 × 31 (axial, radial, z) grid arrangement, subdivided into 16 blocks of 11 × 31 × 31 nodes. Using 16 processors, the computing time required on CRAY-T3D was 505 CPU minutes.

In accord with the measurements a constant bulk mean velocity $V_c = 0.0198[m/s]$ is imposed at the inlet cross section. The testcase is particularly meaningful for the validation of a numerical procedure, as it involves complex secondary flow phenomena due to both, corner as well as streamline curvature effects.

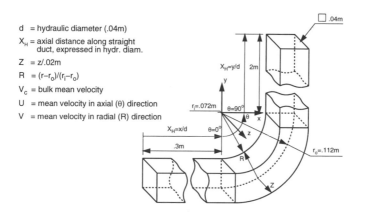

Figure 7: Curved square bend; geometry and coordinate system

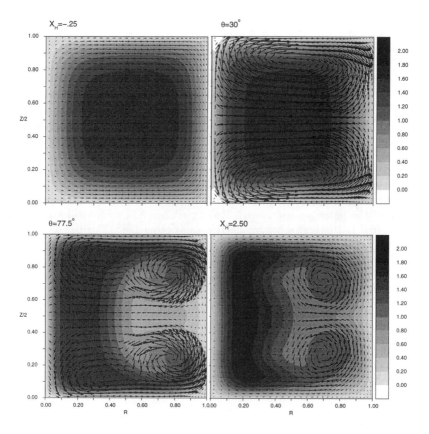

Figure 8: Curved bend, $Re = 790$; contour plot of the normalized axial velocity U/V_c and secondary motion vectors.

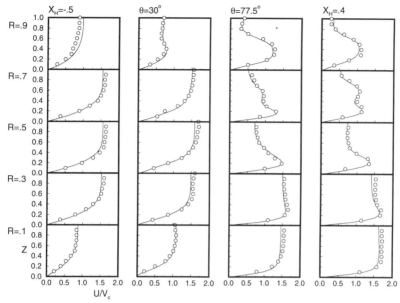

Figure 9: Curved bend, $Re = 790$; axial velocity U/V_c profiles at different positions, o measurement, − present calculation.

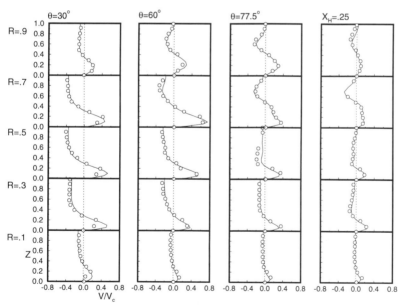

Figure 10: Curved bend, $Re = 790$; radial velocity V/V_c profiles at different positions, o measurement, − present calculation.

Results given in Fig. (8) show the secondary motion and the isolines of the axial velocity component U/V_c at four different cross sections. A strong secondary motion originates from the bending section and is convected in the far downstream. Between $\theta = 0°$ and $\theta = 45°$ negative values of the axial velocity component occur at the outer corner as indicated by the white spots for $\theta = 30°$. The reverse flow area should not appear according to the measurements. However, an earlier experimental investigation [13] indicates that flow reversal occures provided that a fully developed pipe flow is present at the entrance of the bending section, thus indicating a strong sensitivity of the flow with respect to seemingly subtle details of the inlet conditions. Figs. (9) and (10) give an overview of the predictive accuracy returned by the present numerical method. Results are generally in good agreement with the experimental data, which proves the capability of the numerical procedure. Inaccuracies of the computational model are confined to the core regime of the profiles caused by an inappropriate coarse mesh.

The final application to be presented herein, is the laminar flow around the DLR–F5 wing mounted on a plate [14]. As shown in Fig. (11), this testcase is more challenging with respect to the geometry involved. Calculations have been performed for $U_\infty = [10m/s]$ at low angle of attack $\alpha = 2°$. Using 16 processors, the computational domain was again subdivided into 16 blocks with a total of $145 \times 32 \times 25$ grid nodes. In order to achieve load balancing of the processors each of the processor allocated blocks did consist of $10 \times 33 \times 25$ grid nodes.

Fig. (12) gives an overview of the domain decomposition and the sychronized communication steps. In Figs. (13) and (14) results for the DLR–F5 wing are reported in terms of pressure and skin–friction lines. Here, due to the very low incidence of the airfoil, only a small area of flow reversal was detected on the pressure side of the airfoil close to the wing root.

Concluding remarks

A Navier–Stokes procedure based on the two stream–function formulation has been applied to three-dimensional flows in complex geometries. The prior application of a two stage (algebraic) manipulation of the governing Euler–Potential equation facillitates improved numerical properties. The stuctured–grid multiblock finte–volume algorithm presented herein, is based on general non-orthogonal coordinates and employs a fully co-located storage arrangement. Attention was drawn to a variety of parallelization aspects. The parallel procedure has succesfully been validated against a curved square duct experiment involving severe secondary flow phenomena. Here, the predictive accuracy obtained by the present approach is at least as good as results obtained by other researchers using a primitive variable approach. Moreover, the simulated wing flow conveys the capabillities of present method with respect to complex geometries. Present investigations are concerned with the simulation of engineering wing flows at high angle of attack.

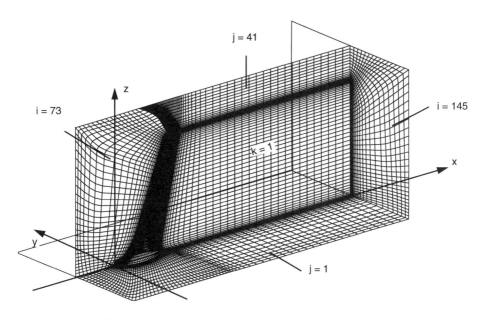

Figure 11: DLR–F5 wing with surface grid.

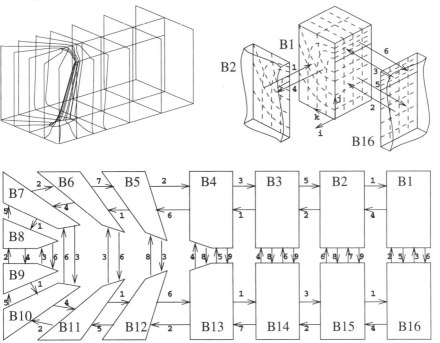

Figure 12: DLR–F5 wing; domain decomposition and synchronized communication steps.

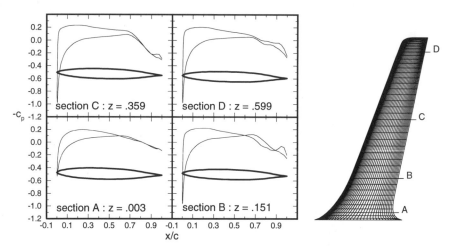

Figure 13: DLR–F5 wing, $\alpha = 2^0, U_\infty = 10m/s$; chordwise C_p-distribution

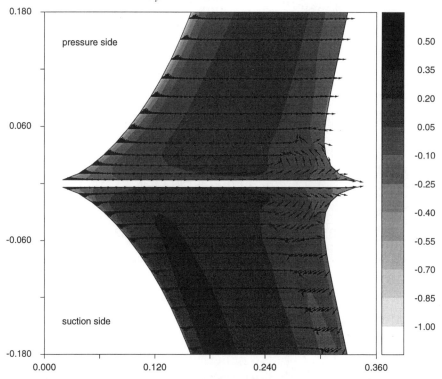

Figure 14: DLR–F5 wing, $\alpha = 2^0, U_\infty = 10m/s$; skin–friction lines and pressure contours

References

[1] Xue, L., Rung, T., Thiele, F.: "Simulation of Unsteady Viscous Three–Dimensional Flow Fields Using a Two Stream Function Formulation", Notes on Numerical Fluid Mechanics, 38 (1993), pp. 393–406.

[2] Schütz, H., Thiele, F.: "Zur Stromfunktionsformulierung der Navier–Stokes Gleichungen für dreidimensionale Strömungen", ZAMM, 69 (1989), pp. 573–575.

[3] Lilek, Ž., Perić, M., Seidl, V.: "Development and Application of a Finite Volume Method for the Prediction of Complex Flows", in this publication.

[4] Durst, F., Perić, M., Schäfer, M., Schreck, E.: "Parallelization of Efficient Numerical Methods for Flows in Complex Geometries", Notes on Numerical Fluid Mechanics, 38 (1993), pp. 79–92.

[5] Hofhaus, J., Meinke, M., Krause, E.: "Parallelization of Solution Schemes for the Navier–Stokes Equations", in this publication.

[6] Durst, F., Schäfer, M., Wechsler, K.: "Efficient Simulation of Incompressible Viscous Flows on Parallel Computers", in this publication.

[7] Xue, L., Rung, T., Thiele, F.: "Entwicklung eines finiten Volumenverfahrens zur Lösung der 3-D, inkompressiblen Navier–Stokes–Gleichung in der Euler–Potentialformulierung", ZAMM 73 (1993), pp. T589–T592.

[8] Bärwolff, G., Ketelsen, K., Thiele, F.: "Parallelization of a Finite–Volume Navier–Stokes Solver on a T3D Massively Parallel System", Proc. 6th International Symposium on Computational Fluid Dynamics (1995), pp. 63–68, Lake Tahoe, Nevada, USA.

[9] Morzynski, M., Thiele, F.: "Numerical Investigation of Wake Instabilities", Bluff–Body Wakes, Dynamics and Instabilities. Springer–Verlag, Berlin 1993, pp. 135–142.

[10] Lugt, H. J., Haussling, H. J.: "Laminar Flow Past an Abruptly Accelerated Elliptic Cylinder at 45^0 Incidence", J. Fluid Mech., 65 (1974), pp. 711–734.

[11] Stone, H. L.: "Iterative Solution of Implicit Approximations of Multidimensional Partial Differential Equations", SIAM J. Numer. Anal., 5 (1968), pp. 530–558.

[12] Taylor, A. M. K. P., Whitelaw, J. H., Yianneskis, M.: "Curved Ducts With Strong Secondary Motion: Velocity Measurements of Developing Laminar and Turbulent Flow", ASME J. Fluids Engineering, 104 (1982), pp. 350–359.

[13] Humphrey, J. A. C., Taylor, A. M. K. P., Whitelaw, J. H.: "Laminar Flow in a Squared Duct of Strong Curvature", J. Fluid Mech., 83 (1977), pp. 509–527.

[14] Schwamborn, D.: "Simulation of the DLR–F5 Wing Experiment Using a Block Structured Explicit Navier–Stokes Method", Notes on Numerical Fluid Mechanics, 22 (1988), pp. 244–268.

AMRFLEX3D — Flow Simulation Using a Three-Dimensional Self-Adaptive, Structured Multi-Block Grid System

R. Hentschel and E.H. Hirschel
Universität Stuttgart, Institut für Aerodynamik und Gasdynamik
Pfaffenwaldring 21, D-70550 Stuttgart, Germany

Summary

The first section gives a description of the numerical procedure which is used for the integration of structured grid blocks. Such blocks are the basic elements of the self-adaptive grid structure. A hierarchically ordered level system, which reflects different refinement stages, serves as a scaffold. After a description of sensors used for refining interesting flow phenomena, an overview about the integration steps is given. The application of AMRFLEX3D to the cases of a three-dimensional, inviscid, supersonic corner flow and of a delta wing at transonic flow conditions is presented. For the latter simulation, both Euler and Navier-Stokes equations are used. Conclusions that can be drawn from experiences with grid refinement investigations are given.

Introduction

Computing powers are rising more and more. To enable a reasonable use of resources, self-adaptive flow simulation has shown advantages in memory and CPU time aspects. Among several strategies as shown in [9] and [17], there is an approach, which seems to be an easy modification for existing multi-block codes. It enables the reuse of verified present-day flow solvers and in this way saves time and money. Furthermore, the grid-generation process is simplified and accelerated, because the user only has to adapt the grid to the geometric configuration.

The procedure described in this paper, represents an extension of Fischer's algorithm [6, 8], which was developed within this project, to three spatial dimensions. In addition to the results in [11], viscous flow simulation has been shown to be possible in the same way. Experiences from several flow cases confirm the considerable simplification of the grid-generation process for complex flow-fields. Adaptation to both shear layers and compression shocks has strong influences on flow fields and surface pressure distributions. The high resolution which is possible, allows an investigation of the grid influence on the solution and therefore supports the reliability of the results. In addition, small cell sizes allow the detection of weak phenomena and a comparison with flow theory leads to a better understanding of flow kinematics. Further publications of this project are [5, 7, 10, 11].

Solution Method

The Euler or Navier-Stokes equations are solved on single mesh blocks. Using the well-known set of conservative variables $\rho, \rho u, \rho v, \rho w, e$ and Cartesian coordinates x, y, z, the Navier-Stokes equations may be written in the following form:

$$\tilde{U}_{,t} + \tilde{E}_{,x} + \tilde{F}_{,y} + \tilde{G}_{,z} = 0,$$

where the comma indicates a partial differentiation by the given variable; u, v, w are the Cartesian velocity components and the flux-vector $\tilde{E}$ reads with its inviscid and viscous part ($\tilde{F}, \tilde{G}$ analogously):

$$\tilde{E} = \begin{pmatrix} \rho u \\ \rho u^2 + p \\ \rho u v \\ \rho u w \\ (e+p)u \end{pmatrix} + \begin{pmatrix} 0 \\ -\tau_{xx} \\ -\tau_{xy} \\ -\tau_{xz} \\ -\tau_{xx}u - \tau_{xy}v - \tau_{xz}w + q_x \end{pmatrix}.$$

After a transformation to general curvilinear coordinates ξ, η, ζ, the system of equations reads as follows:

$$U_{,t} + E_{,\xi} + F_{,\eta} + G_{,\zeta} = 0,$$

with

$$\begin{pmatrix} E \\ F \\ G \end{pmatrix} = \frac{1}{J} \frac{\partial(\xi, \eta, \zeta)}{\partial(x, y, z)} \begin{pmatrix} \tilde{E} \\ \tilde{F} \\ \tilde{G} \end{pmatrix},$$

$$J = \det\left[\left(\frac{\partial(\xi, \eta, \zeta)}{\partial(x, y, z)}\right)^{-1}\right] \text{ and } U = J \cdot \tilde{U}.$$

The discretization of the time-dependent solution vector U is done by a 'backward Euler' scheme of first order (the upper index indicates the time level):

$$\frac{U^{n+1} - U^n}{\Delta t} + E_{,\xi}^{n+1} + F_{,\eta}^{n+1} + G_{,\zeta}^{n+1} = 0.$$

Linearization of the flux vectors E, F, G around time level n yields (F, G analogously)

$$E^{n+1} = E^n + A^n \cdot \Delta U,$$

with

$$A^n = \frac{\partial E^n}{\partial U^n} \text{ and } \Delta U = U^{n+1} - U^n,$$

resulting in the following system of equations:

$$\frac{U^{n+1} - U^n}{\Delta t} + (A^n \Delta U)_{,\xi} + (B^n \Delta U)_{,\eta} + (C^n \Delta U)_{,\zeta} = -\left(E^n_{,\xi} + F^n_{,\eta} + G^n_{,\zeta}\right).$$

The inviscid part of the fluxes E, F, G is calculated by a homogenous Riemann-solver according to Eberle [2], involving an interpolation scheme up to third-order accuracy for calculating the cell-face fluxes. This cell-centred scheme is gradually switched down to first order in regions of strongly varying flow variables. Furthermore, to avoid oscillations at strong shocks, the calculated Riemann flux is mixed with a Steger and Warming split flux.

For calculating the viscous fluxes, a central discretisation is used [15]. The Jacobi matrices A, B, C on the left side are discretized in their inviscid part by an approximate upwind formulation [15]. The equation system resulting from the discretisation is solved by a Red-Black Gauss-Seidel relaxation technique.

Sensors

Shock sensor: Starting with the information given in [6], several modifications led to a reliable three-dimensional sensor. It is based on a density criterion which enables a pre-sorting of the spatial directions around a cell-centre by one-sided differences. From the six possibilities, the direction showing the largest value of the quotient

$$\alpha = \frac{|\rho_2 - \rho_1|}{\min(\rho_1, \rho_2)}$$

is chosen and taken as a first guess on the presence of a shock. Indices 1 and 2 stand for the flow values in front of and behind the shock. Thereby, a fixed minimum value of $\alpha > \alpha_{min}$ has to be exceeded. This limit was lowered for the three-dimensional case to incease the sensitivity.

However, strong density gradients are also a characteristic of isentropic expansion and compression phenomena. In consequence, one has to apply a second criterion to decide whether the detected feature is isentropic or not. Across a shock the following equation holds, with p being the thermodynamic pressure, ρ the density, γ the ratio of the specific heats and M_{n1} the shock-normal Mach number:

$$\left(\frac{p_2 - p_1}{\rho_2 - \rho_1}\right) \frac{\rho_1}{p_1} = \frac{\gamma \left(2 + (\gamma - 1) M_{n1}^2\right)}{\gamma + 1}.$$

Because the right-hand side of this equation increases monotonically with the Mach number, a threshold for the detection of shocks above a certain strength can be set. User interaction was simplified in such a way that only the shock-normal Mach number is required.

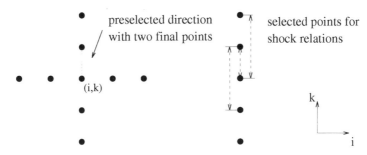

Figure 1: Example of selected shock-sensor points.

In general, the shock may be smeared over several cells and care has to be taken calculating the pressure and density jumps, which are determined at three stations, namely across the pre-selected cell-centres and across the cell face between them (for clarity see Figure 1). Note that no gradients are calculated in the above criterion, but possible pressure and density jumps across a shock, which is supposed to lie in a preselected direction. These jumps are defined by the status of the flow before and behind the shock, and therefore the reliability of the sensor depends on the resolution capabilities of the integration procedure. For example, if the shock is smeared over three cells, the jumps will have to be calculated over a distance of three cells or more to obtain the desired flow values. Up to now, the shock sensor has proven its accuracy and reliability in two-dimensional as well as in three-dimensional calculations from the transonic to the hypersonic flow regime.

Entropy sensor: The entropy sensor is based on a simple three-dimensional gradient calculation with centred differences. Several attempts to find better functions failed. It seems that this gradient is an optimum regarding reliability and expense in the area of phenomenological based sensors. As a measure of the entropy the ratio p/ρ^γ was chosen. This sensor allows to recognize entropy or vorticity layers in both Euler and Navier-Stokes calculations. Detected shock cells are excepted during the calculation of the entropy criterion. The experiences with three-dimensional calculations are very satisfactory.

Density sensor: As a sensor for the detection of temperature gradients in viscous layers, a simple density sensor is used. It is based on the maximum of six possible one-sided differences in the directions of the calculation space. The calculation of the temperature gradient is avoided, because in viscous layers with small pressure gradients, temperature changes have their main effect on the density. To simplify user interaction, the sensor values in all cells are referred to the maximum value on the basic grid. This procedure is also used in case of the entropy sensor.

During the development period, many other sensors were investigated. A list of them, including a brief description of their functionality, is given in [10].

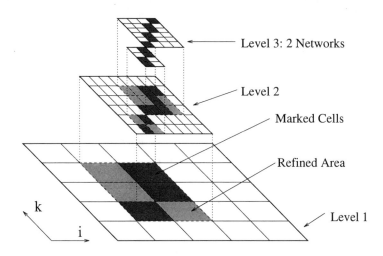

Figure 2: Hierarchical ordering of networks and levels.

Grid Structure

The grid system is based on a usual three-dimensional and structured multi-block grid. On this grid, a hierarchically ordered level system serves as scaffold (see Figure 2). Each level is afflicted with a constant refinement factor and represents a defined refinement stage. To enable an adaptation to flow-field features, many blocks can be distributed automatically on the levels. Their position is only limited by the condition, that blocks on finer levels have to lie within the blocks of the underlying level. There is no further condition. The position of a block is fixed by the coordinates of its lower left and upper right corners in the calculation space. Using these points, neighbours of a block can be detected on the same, respectively on other levels. The geometry of the cells is generated during the mesh adaptation process by subseqent linear or higher-order interpolation from the basic grid.

Boundary information is supplied by two layers of so-called dummy cells around each block. Those cells enable any possible grid topology and are a crucial element of the adaptive strategy. Four types of boundary conditions are possible: fine – fine, coarse – fine, very coarse – fine and external boundary conditions. In case of a fine – fine boundary, data from a neighbouring block of the same level is copied into the boundary cells of a network. The coarse – fine condition requires data-interpolation from the next lower level. This linear interpolation is MINMOD-limited to avoid the induction of oscillations. In rare cases, data from more than one level below is required (very coarse - fine grid case). Because an interpolation is supposed not to provide better information, the value from a corresponding cell is directly copied to the dummy cell. External boundary conditions are set in the usual manner. For example, to achieve a supersonic outflow condition, values from the border of the computational domain are extrapolated into the dummy layers.

As experience shows, the number of all dummies is nearly as large as the number of the so-called real cells. This is a characteristic of three-dimensional cases. Therefore, a special storage technique is applied at the expense of a small amount of CPU time. In this way, only memory for dummies of the biggest block possible has to be reserved.

A grid structure with several levels is established by calculating an initial solution on the basic grid. Further levels are then constructed by marking cells in interesting regions of the flow domain with the help of sensors. Dark-grey cells in Figure 2 correspond to these. Subsequently, signed cells are grouped to networks by a segmentation procedure, which allows a certain amount of non-marked cells in the network areas. The latter are represented by grey shaded cells in Figure 2. Data for a special level is found during the adaptation process by a transfer of

1. coarse-grid information of the level below to the fine grid by linear interpolation as it is done in case of coarse-fine boundaries. This interpolation is only necessary to gain an initial guess in regions that were not refined during previous adaptations.

2. Interpolated data is then overwritten by information which had been present on the same refinement level of the old grid structure before the adaptation process started.

The second step ensures that data of a special refinement level is conserved in areas, which are marked by the sensors during every grid adaptation. In case of an increase in levels, the second point is omitted for obvious reasons.

At the beginning of an integration step (global iteration), the boundary values for the meshes have to be set as described above. On the entire grid system, the integration is carried out with a global timestep, consisting of several sub-timesteps on the higher levels according to the refinement factors connected to the levels.

These sub-timesteps are performed in a recursive manner. For example let us imagine a grid system with two levels and a refinement factor of three for the second level. Before the basic grid is integrated for one timestep, three timesteps are performed on the second level to ensure a consistent timelevel of the solution on both levels. Nevertheless local timestepping is used.

After three sub-timesteps on level 2, the solution is averaged cell by cell according to the refinement factor and copied down on level number 1.

Now, continuing the integration of level 1, the solution of the overlapped region is assumed to fulfill exactly the model-equations due to its finer discretisation on the level above and is blanked out to avoid disturbances emanating from the overlapped region.

The global timestep is finished after one timestep on level 1, because the refinement factor for the lowest level is always 1:1.

During the approach to the final steady solution, the entire grid structure is dynamically adapted every few iterations, and flow features are followed as they evolve and move in the solution process.

The above description is kept very short and therefore highlights only the main aspects of the integration scheme. For further information the reader is referred to [8, 10].

Example 1: Supersonic, Inviscid Corner Flow

As a first three-dimensional test case, a supersonic inviscid corner flow was chosen, which is generated by two unswept compression ramps with a 9.5° wedge angle as can be seen from Figure 3. The flow is characterized by the free stream values $Ma_\infty = 3.0$ and $T_\infty = 300K$. In addition to that, the fluid is taken as perfect gas.

The three-dimensional shock pattern is formed by two-dimensional wedge shocks, which are connected across a so-called corner shock. Emanating from the intersection lines, the surfaces of secondary shocks and contact discontinuities dominate the flow structure close to the corner region.

The flow field phenomena are of conical structure with their origin at the leading edges as proposed by several authors (see,for example, Marsilio [13]).

Using sensors, described in the previous sections, a grid system was created by the adaptive scheme, consisting of three levels, one basic and two additional ones. For the basic discretisation, 23 grid lines in every space direction were chosen in a nearly cubic computational domain to simplify the grid generation process and to exploit the refining capabilities of the procedure. Only half a day was needed to generate the initial grid.

The shock sensor was tuned to mark shocks with a normal Mach number of 1.06 and larger, and the threshold of the entropy sensor was set to mark cells showing a minimum entropy gradient of 8 percent of the maximum gradient in the whole field. In the calculation of the entropy function, regions with previously detected compression shocks were excepted.

Table 1 shows basic data of the grid system at the end of the simulation. Because of two additional refinement levels, the effective resolution is equivalent to a grid with 198 cells in each coordinate direction. However, a complete refinement of the computational domain would require about 7.8 million cells. This shows that a memory saving factor of 5.65 in comparison to a complete refinement is achieved. Because of the hierarchical structure causing a multiple overlap of some areas, 3.7 per cent of the real cells are overlapped by subsequent levels. Obviously, the resulting memory overhead is negligible. The number of dummy cells needed would be about as large as the number of real cells. Therefore applying the special data-structure, mentioned in a preceeding section, halves the memory requirements.

As can be seen from Figure 4, the shock system is captured very well by the adaptive grid structure. The large refined regions near the corner stem from strongly smeared

Table 1: Data of grid structure in corner-flow case.

	basic grid	1. additional refinement level	2. additional refinement level
Refinement factor (in each direction)	1:1	1:3	1:3
Percentage of entire computational domain (cells)	100 per cent	41.0 per cent	16.2 per cent
Number of real cells	10 648	117 855	1 259 982
Number of networks	1	24	501
Average number of cells in one network	10 648	4 910.6	2 514.9
Resultant refinement factor			1:9
Sum of real cells			1 388 485

contact discontinuities in that area. The improved discretisation on the second level leads to a sharper detection of the contact discontinuities with help of the entropy sensor. Therefore, only 17 per cent of the computational domain are refined by the third level (refinement factor of 1:3 in every space direction). The grid is slightly asymmetric because of internal program limits, which are regarded by the segmentation procedure.

Figure 5 shows Mach-number isolines in the cross section at $x = 0.95$ in conical coordinates y/x and z/x. The direct comparison to the initial solution is very impressive. Furthermore, it should be kept in mind that Figure 5 shows a two-dimensional cut through a three-dimensional flowfield. A slight curvature of the secondary shock and entropy surfaces can be clearly seen, whereas the wedge shocks and the corner shock seem to be plane surfaces. The resolution is comparable to that of Marsilio [13], who utilized the hyperbolic character of the flow and performed quasi two-dimensional calculations, but one should keep in mind that the present method allows to compute also three-dimensional subsonic and transonic flow situations. There are a few isolines in the smooth regions between the contact-discontinuities and the secondary shock surfaces. As a thorough investigation of the flowfield shows, these slight variations in the Mach-number are caused by continuous expansion and compression phenomena. This is confirmed by surface pressure distributions and a plot of pressure isolines in the bisector plane, which are shown in [10]. To the authors knowledge, it is the first three-dimensional simulation, which clearly resolves such phenomena. Slight variations of the Mach-number near the solid body wall stem from an entropy-layer. The reason for this is to be sought in the numerical boundary condition, as investigations in [1] confirm. The entropy-layer is refined because of a very sensitive adjustment of the sensors.

The adaptation process consumed less than 0.5 percent of the total computational time in case of three active levels and a complete recalculation of the grid structure every 5 global iterations as described in the previous sections. The resolution achieved is so high that the development of an interaction model for this type of flow was possible. This model gives an explanation for the weak phenomena in the interaction region of the corner. It is an example for a situation, where an adaptive grid structure and high resolution of flow-phenomena enable a deeper insight into flow kinematics.

Example 2: Delta Wing with Sharp Leading Edges

A second flow situation, which was used as a test case during the period of development is the delta wing configuration shown in Figure 6. Free stream conditions were set to $Ma_\infty = 0.85$ and angle of attack $\alpha = 10°$. Free stream temperature was set to $T_\infty = 300K$. Both, inviscid and viscous simulations were performed. Differences between laminar flow and simulations using the algebraic turbulence model from Baldwin and Lomax on the leeside of the wing were investigated. This test case is interesting because of great difficulties in generating a grid, which resolves the vortical layers feeding the leeside vortices. Up to now, prediction of nonlinear lift with sufficient accuracy is an important problem in aerospace industry.

The initial grid was generated using Poisson-equations [16]. The sharp leading edge was regarded by a special grid topology described in [10]. Cross sections of the wing show an O-topology. Cuts in chord direction show a C-topology. An idea of the three-dimensional grid is given by Figure 7. During the simulation two refinement levels were added. These improve the initial grid-resolution by a factor of 12 in every coordinate direction.

Table 2 shows characteristic data of the grid structure, which was established at the end of the Euler-simulation. Only 8 per cent, respectively 2.3 per cent of the cells are refined by subsequent levels. This is due to the limited extension of the vortices near the wing. As a consequence, memory requirements are reduced by a factor of 38 in comparison to a complete refinement of the computational domain. This again illustrates the great potential inherent in the method.

Figure 8 shows a cross section of the wing. The grid structure was adapted to the vortical layers leaving the sharp edge. As can be seen, the entropy sensor is able to follow the feeding sheet. In addition, the vortex core is refined because of high losses in total pressure ΔP_t. The cross flow shock is captured as well, which is due to the implemented shock sensor. Because of the reduction of numerical diffusion, Mach numbers over $Ma = 2$ are reached in the vortex core. Wiggles in the Mach distribution stem from velocity gradients in the discretised vortical layer and are characteristic of the low level of diffusion. Looking at Figure 8, one should keep in mind that this is a three-dimensional

Table 2: Grid data of delta wing case (Euler).

	basic grid	1.additional refinement level	2.additional refinement level
Refinement factor	1:1	1:4	1:3
Percentage of entire computational domain (cells)	100 per cent	8.0 per cent	2.3 per cent
Number of real cells	77 760	397 248	3 026 295
Number of networks	2	52	1 392
Average number of cells in one network	38 880	7 639	2 174
Resultant refinement factor			1:12
Dummy cells / real cells			0.851
Sum of real cells			3 501 303
Average shape of grid blocks			$i \times j \times k =$ $10.7 \times 8.9 \times 8.0$

simulation. The sharp resolution of flow-field phenomena is impressive. They allow to investigate the correlation between the vorticity-content in the feeding sheet and the amount of circulation in the flow. More details about this topic are again given in [10].

A comparison of the surface C_p distribution at 60 per cent root chord is shown in Figure 9. In the left part, one can see great differences between experimental values and the result of an Euler-simulation. The suction peak is much too high and too far outboard. The reason for that is a secondary separation, arising in the experiment. Grid convergence seems to be achieved for the surface pressure distribution, because there are only minor changes between one additional and two additional refinement levels. The agreement between experiment and Euler-model on the lower side of the wing is satisfactory. The right part of Figure 9 shows the result of a Navier-Stokes simulation. On the basic level, both a laminar and a turbulent simulation agree with the experiment in the maximum value of the suction peak. Major differences in the spanwise position of the maximum maybe due to a wrong simulation of the secondary separation. A subsequent refinement level changes the picture. The maximum value is now much too high, but closer to the experiment as in the Euler simulation. Despite this, the pressure gradient on the inboard side of the peak is now the same as in the experiment. Obviously, the influence of the turbulence model is negligible in comparison to effects of grid resolution which is confirmed by a laminar simulation with one additional refinement level (not shown here). Because grid convergence is not achieved, one further refinement level should be added. This was not possible due to memory requirements for 35 million points. Several investigations in [10] show, that an excellent simulation of the separation at the leading edge is a basic requirement for predicting the lift coefficient, respectively the vortical structures on the leeside of delta wings to a sufficient degree. The solid body boundary condition can have a great influence on the flowfield in Euler simulations, too. Especially, this is the case, if fluid leaves the solid body at separation lines and subsequently interacts with fluid in the vicinity of the body again. This nonlinear interaction is the reason for the generation of nonlinear lift in case of a delta wing geometry. Quite similar kinematics are present at the tip of a transport aircraft wing. As the development of winglets show, losses by tip vortices are no longer negligible. In this context, the method presented may lead to practical improvements. The importance of kinematic investigations by high resolution of phenomena is also realized in the area of rotorcraft development [12].

Table 3 shows a comparison of the predicted lift coefficient C_a, drag coefficient C_w and moment coefficient C_{m_y}. One can see that a Navier-Stokes simulation yields improved val-

Table 3: Integrated wing coefficients

	C_a	C_w	C_{m_y}	C_a/C_w
experiment (NLR,1985) from [16]:				
	0.4573	0.0846	-0.271	5.405
Euler simulation:				
2 additional levels	0.5034	0.0822	-0.3101	6.125
Navier-Stokes:				
1 additional level	0.4765	0.0890	-0.2846	5.354

ues for C_a, C_{m_y} and C_a/C_w, but there are still differences which are not tolerable. Regarding the requirements in computing time and problems in grid generation, improvements gained by a Navier-Stokes calculation are too expensive in a practical design process. Nevertheless, the results are encouraging to continue developments in this direction.

Conclusions

The method presented has shown to produce very good results in both supersonic and transonic viscous flow cases. It is a tool for investigating effects of grid refinement and grid topology. Therefore results of numerical flow simulation can be verified efficiently with regard to grid influence. The method has shown that turbulence models may have a minor effect in comparison to the former. The self-adaptive grid structure simplifies three-dimensional grid-generation enormously. Because of high resolution of flowfield phenomena, new insights can be gained into the kinematics of flows. However, an important problem still remains. Highly sophisticated post-processing of large data sets is not established. However, this is an inevitable task if a method for numerical flow simulation has to produce results with sufficient accuracy and reliability. Therefore, further efforts are necessary to exploit the full potential of this self-adaptive flow solver. Convergence acceleration procedures and portation to massively parallel computer architectures (see ,e.g., [14]) will have to be regarded then. Nevertheless, the present method has proved to be suitable for industrial applications.

Acknowledgement

The authors thank the Deutsche Forschungsgemeinschaft (DFG) for supporting these studies in its Priority Research Programme Strömungssimulation mit Hochleistungsrechnern (Hi 342/2).

References

[1] DADONE, A., AND GROSSMAN, B. Surface Boundary Conditions for the Numerical Solution of the Euler Equations. *AIAA Journal 32*, 2 (1994), 285 – 293.

[2] EBERLE, A. Characteristic Flux Averaging Approach to the Solution of Eulers Equations. VKI-Lecture Series 1987-04, 1987.

[3] ELSENAAR, A. Summary of NLR Wind Tunnel Tests on the International Vortex Flow Model. Tech. Rep. Memorandum AC-87-023 L, NLR, 1987.

[4] ELSENAAR, A., AND HOEIJMAKERS, H. An Experimental Study of the Flow over a Sharp-Edged Delta Wing at Subsonic and Transonic Speeds. In *Vortex Flow Aerodynamics* (1991), pp. 15–1 – 15–19. AGARD-CP-494.

[5] FISCHER, J. *Sensors for Self-Adapting Grid Generation in Viscous Flow Computations*, Vol. 35 of *NNFM*. Vieweg, Braunschweig/Wiesbaden, 1992, pp. 365 – 375.

[6] FISCHER, J. Selbstadaptive, lokale Netzverfeinerung für die numerische Simulation kompressibler, reibungsbehafteter Strömungen. Doctoral Thesis, 1993. Universität Stuttgart.

[7] FISCHER, J. Self-adaptive mesh refinement for the computation of steady, compressible, viscous flows. *Z. Flugwiss. Weltraumforsch. 18* (1994), 241 – 252.

[8] FISCHER, J., AND HIRSCHEL, E. H. *Adaptive Navier-Stokes Calculations Using a Combination of an Implicit Finite-Volume Method with a Hierarchically Ordered Grid Structure*, Vol. 38 of *NNFM*. Vieweg, Braunschweig/Wiesbaden, 1993, pp. 279 – 294. Flow Simulation with High Performance Computers I.

[9] GREZA, H., BIKKER, S., AND KOSCHEL, W. Efficient FEM Flow Simulation on Unstructured Adaptive Meshes. In this publication.

[10] HENTSCHEL, R. Entwicklung und Anwendung eines dreidimensionalen selbstadaptiven Verfahrens auf der Basis strukturierter Gitter. Doctoral Thesis, 1996. Universität Stuttgart.

[11] HENTSCHEL, R., AND HIRSCHEL, E. H. Self Adaptive Flow Computations on Structured Grids. In *Computational Fluid Dynamics '94* (September 1994), S. Wagner, E. H. Hirschel, J. Periaux, and R. Piva, Eds., John Wiley & Sons, pp. 242 – 249.

[12] LANDGREBE, A. J. New Directions in Rotorcraft Computational Aerodynamics Research in the U.S. In *Aerodynamics and Aeroacoustics of Rotorcraft* (1995), pp. 1-1 – 1-12. AGARD-CP-552.

[13] MARSILIO, R. Vortical Solutions in Supersonic Corner Flows. *AIAA Journal 31* (1993), 1651 – 1658.

[14] MICHL, T. Effiziente Euler- und Navier-Stokes Löser für den Einsatz auf Vektor-Hochleistungsrechnern und massiv-parallelen Systemen. Doctoral Thesis, 1995. Universität Stuttgart.

[15] SCHMATZ, M. A. *Three-Dimensional Viscous Flow Simulations Using An Implicit Relaxation Scheme*, Vol. 22 of *NNFM*. Vieweg, Braunschweig/Wiesbaden, 1987, pp. 226 – 243. Numerical Simulation of the Transonic DFVLR-F5 Wing Experiment.

[16] SCHWARZ, W. IEPG-TA15 Aerodynamische Berechnungsverfahren, Teil 2. Tech. Rep. DASA/LME211/S/R/1619, DASA-LM, 1993.

[17] VILSMEIER, R., AND HÄNEL, D. Computational Aspects of Flow Simulation on 3-D, Unstructured, Adaptive Grids. In this publication.

Figures

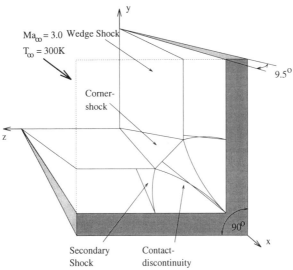

Figure 3: Sketch of geometry and position of several flow-field phenomena.

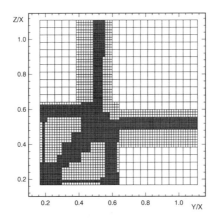

Figure 4: Cut with x=constant through adaptive grid structure.

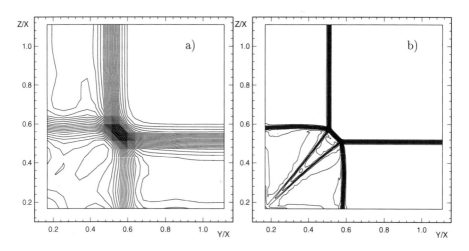

Figure 5: Isolines of local Mach-number in a cutting plane x=constant:

a) basic grid; 40 lines	b) adapted grid; 50 lines
min./max. 2.10396 / 2.97807	min./max. 2.03136 / 2.98105

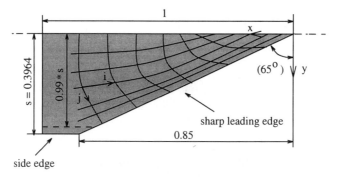

Figure 6: Projection of right half of symmetric wing.

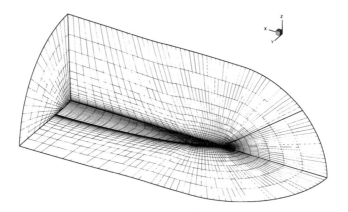

Figure 7: Computational grid of delta wing case (Euler).

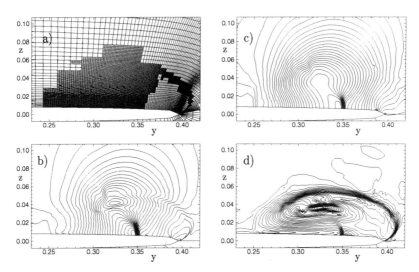

Figure 8: Cross section through flow field at wing leading edge. Euler simulation, in every picture 30 isolines, $x \approx 0.86$:

a)	Grid	c)	C_p-distribution: $-1.6, \ldots, 0.02$
b)	Mach-number: $2.02, \ldots, 0.7$	d)	$1 - P_t/P_{t,\infty}$: $0.0, \ldots, 0.35$

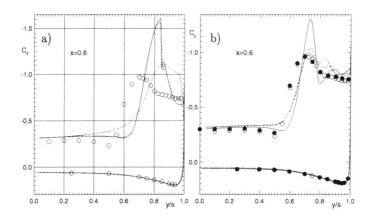

Figure 9: C_p-distribution at 60 per cent root chord:

a) Euler	b) Navier-Stokes
○ exp. [3]	●,○ exp. [3, 4]
— 2nd add. level	— — basic grid, lam.
– – 1st add. level	– – – basic grid, turb.
- ·· - basic grid	—— 1st add. level, turb.

415

DEVELOPMENT AND APPLICATION OF A FINITE VOLUME METHOD FOR THE PREDICTION OF COMPLEX FLOWS

Ž. Lilek, M. Perić and V. Seidl

Institut für Schiffbau der Universität Hamburg,
Lämmersieth 90, D-22305 Hamburg, Germany

SUMMARY

In this paper the development and application of a finite volume method for the prediction of compressible and incompressible, laminar and turbulent, steady and unsteady flows in complex geometries is presented. The method uses quadrilateral (2D) or hexaedral (3D) control volumes, block-structured grids (which may not fit at block interfaces) and a colocated (cell-centered) arrangement of variables on the grid. The conservation equations for mass, momentum, energy, turbulent kinetic energy and its dissipation rate are solved iteratively in a sequential manner. Discretization methods up to fourth order were tested, but second order centered approximations, together with local grid refinement, were found to be the best compromise between accuracy, efficiency and ease of implementation. The efficiency is increased by using multigrid methods and parallel computing. Results of several example calculations are presented to demonstrate the efficiency and accuracy of the method.

INTRODUCTION

Computation of fluid flow is becoming to play an important role in all kinds of industrial applications: mechanical, civil, bio-mechanical, environmental, chemical and other engineering branches. The computational methods are therefore required to be applicable to a wide range of problems, and first of all, to be applicable to flows in complex geometries. The accuracy and efficiency of the method are also important, as well as its robustness. In addition, flows in engineering applications often involve complicated boundary conditions which require special attention. This is e.g. the case with solution domains which change in time, like in the case of flows with free surface.

In the last ten years a significant improvement has been achieved in the development of solution methods for complex fluid flows. The most important developments were aimed at increasing the efficiency (new linear equation solvers were developed, like CGSTAB [22], and multigrid methods were adapted to solution algorithms for fluid flow problems, see [8]), applicability to large problems (through the adaptation to parallel computing, see [24]), and to the treatment of complex boundary conditions (moving grids [5], flows with free surface [15, 12], etc). An increased use of more complex turbulence models has also been seen, although the improvements in this area were not so spectacular.

In this paper a finite volume solution method is presented, which was developed with the aim of being applicable to a wide range of flow problems. The starting point was a method for incompressible, steady flows and structured grids with a colocated arrangement of variables [17]. It has been extended to compressible and unsteady flows, allowing moving grids and determination of the form and position of a free surface. Also, block-structured grids, higher-order discretization, multigrid method and parallel computing were introduced. These extensions are based on the experience of the present authors and co-workers [2, 3, 9, 12, 13, 20, 21] and on the experience shared with other projects within the research programme, especially [7, 1, 23]. The method is briefly described in the next section, while the following section presents results of some representative applications.

SOLUTION METHOD

The starting point are the conservation equations for space, mass, momentum and scalar quantities in integral form:

$$\frac{d}{dt}\int_\Omega d\Omega - \int_S \mathbf{v}_b \cdot \mathbf{n}\, dS = 0, \quad (1)$$

$$\frac{d}{dt}\int_\Omega \rho\, d\Omega + \int_S \rho(\mathbf{v}-\mathbf{v}_b)\cdot \mathbf{n}\, dS = 0, \quad (2)$$

$$\frac{d}{dt}\int_\Omega \rho u_i\, d\Omega + \int_S \rho u_i(\mathbf{v}-\mathbf{v}_b)\cdot \mathbf{n}\, dS = \int_S (\tau_{ij}\mathbf{i}_j - p\mathbf{i}_i)\cdot \mathbf{n}\, dS + \int_\Omega b_i\, d\Omega, \quad (3)$$

$$\frac{d}{dt}\int_\Omega \rho\phi\, d\Omega + \int_S \rho\phi(\mathbf{v}-\mathbf{v}_b)\cdot \mathbf{n}\, dS = \int_S \Gamma_\phi \operatorname{grad}\phi \cdot \mathbf{n}\, dS + \int_\Omega q_\phi\, d\Omega, \quad (4)$$

where Ω represents the volume bounded by a closed surface S, $\mathbf{v}_b$ is the velocity of the volume surface, ρ is the fluid density, $\mathbf{v}$ is the fluid velocity with its Cartesian velocity components u_i, ϕ stands for any scalar quantity and Γ_ϕ for its diffusion coefficient. $\mathbf{b}$ and q_ϕ are the volumetric sources for momentum and scalar quantity, respectively.

The above equations are applied to each control volume (CV); the surface and volume integrals are calculated using either midpoint rule (second order; used in both 2D and 3D) or Simpson rule (fourth order; used only in 2D) [13]. Both methods require the values of variables and their gradients at cell face centers, while for the Simpson rule also the corner values are needed. These have to be obtained from assumed profile shapes (shape functions); linear and cubic profiles are used for the midpoint rule and Simpson rule integration, respectively. The velocity of the cell faces is not calculated explicitly; instead, the volume flux due to the movement of the cell face is calculated using the volume swept by the face during one time step, $\delta\Omega_c$, see Fig. 1. The swept volume can be calculated when the position of CV vertices at successive time steps are known [4].

The integration in time is performed using a fully implicit second order scheme with three time levels; all surface and volume integrals are evaluated at the new time level only, while a quadratic profile in time is used to approximate the time derivative at the new time level. The integration in time is then performed using midpoint rule, i.e. the integration interval is centered around the new time level. This scheme has proven to be less prone to oscillations than the Crank-Nicolson scheme. The grid velocity at the new time level needed to compute the mass fluxes is obtained using the space conservation law and the swept volumes defined above:

$$\dot{\Omega}_c = \int_{S_c} \mathbf{v}_b \cdot \mathbf{n}\, dS \approx (\mathbf{v}_b \cdot \mathbf{n})_c S_c = \frac{3\,\delta\Omega_c^n}{2\,\Delta t} - \frac{\delta\Omega_c^{n-1}}{2\,\Delta t}. \quad (5)$$

The mass flux through a cell face 'c' can therefore be calculated as:

$$\dot{m}_c = \int_{S_c} \rho(\mathbf{v}-\mathbf{v}_b)\cdot \mathbf{n}\, dS \approx \rho_c(\mathbf{v}\cdot \mathbf{n})_c S_c - \rho_c \dot{\Omega}_c. \quad (6)$$

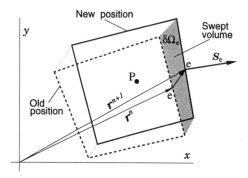

Fig. 1: A typical 2D CV at two time steps and the volume swept by a cell face

The solution method involves two iteration levels: *outer* and *inner* iterations. One outer iteration comprises assembly of (linearized) algebraic equation system for each dependent variable and its solution by a suitable solver to a certain tolerance; while solving for one variable, others are treated as known. Inner iterations are those performed in the linear equation solver. The efficiency and stability of the method is increased by using deferred correction approach [19]; only the nearest neighbor nodes, as resulting from the use of simple low-order approximations, are included in the implicit algebraic equation, while the difference between the higher-order approximation and the lower-order one is calculated explicitly and added to the source term; see [5, 13] for details of implementation for convective and diffusive terms.

The mass conservation equation is transformed into an equation for pressure correction using the SIMPLE methodology [16]. In the case of incompressible flows, the mass fluxes through CV faces are corrected by correcting the velocity alone; the velocity correction is proportional to the gradient of pressure correction and, by requiring that the corrected velocities satisfy the continuity equation, a Poisson equation for pressure correction is obtained (see [13] for a detailed derivation of the pressure-correction equation for both second and fourth order methods).

When the flow is compressible, the mass fluxes depend on both velocity and density; the mass flux correction involves then correcting both of these variables. The density correction is proportional to the pressure correction, the proportionality coefficient resulting from the equation of state. The velocity correction is the same is in the case of incompressible flows. The pressure-correction equation in this case has the form of a convection/diffusion equation; the diffusion-like part stems from the velocity correction, while the convection-like part stems from the density correction. The relative importance of the two terms depends on the local Mach number; in high Mach number regions, the convection-like term dominates, while in low Mach number regions, the Poisson equation as in the case of an incompressible flow is recovered. Details about the extension of the method to compressible flows are described in [3].

The method was extended to predict flows with free surfaces where two additional boundary conditions must be considered: the kinematic and dynamic condition. The kinematic boundary condition (no mass flow through the free surface) and the space conservation law are used to determine the surface position while the dynamic condition (balance of forces) serves to calculate pressure and velocity at the free surface; see [12] for more details.

To accelerate the convergence of the iterative solution method, a full approximation, full multigrid scheme has been implemented. The coupled set of equations is discretized consistently on a hierarchy of systematically refined grids, with coarse grid solutions serving to eliminate low frequency errors on fine grids. For more details on implementation and performance analysis see [9].

In order to be applicable to large problems on parallel computers, the algorithm has been adapted exploiting data parallelism in both space and time [20, 10, 21]. For space parallelism the solution domain is decomposed into non-overlapping subdomains with each subdomain being assigned to one processor. A memory overlap one CV layer wide along interfaces has to be updated after each inner iteration by a local communication between neighbor subdomains. Global communication due to convergence checks and global reduction of inner products is also required. The iterative nature of implicit time schemes allows parallelism in time. As soon as one processor produces an estimate for the new solution after the first outer iteration, another processor can start computation for the next time step. The preliminary solution from the preceding time step has to be updated once per outer iteration in a local one-way communication. As long as the number of concurrently processed time steps is smaller than the number of outer iterations per time step, high numerical efficiency is achieved. Especially when too many subdomains in space would lead to a significant performance degradation, a combined space- and time-parallel solution strategy has proven more efficient [21]. Communication both in space and time can be overlapped with computation, leading to an increased efficiency on parallel computers which support concurrency of computation and communication [18].

EXAMPLES OF APPLICATION

In this section several application examples are presented to demonstrate the efficiency, accuracy and robustness of the developed method.

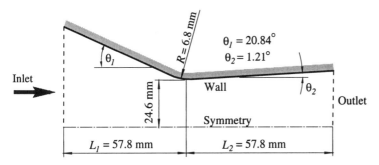

Fig. 2: Geometry and boundary conditions for the compressible channel flow

The first example shows results of prediction of a compressible inviscid flow through a plane, symmetric converging/diverging channel. The geometry and boundary conditions are shown in Fig. 2. At the inlet, the total pressure $p_t = 1.99$ bar and temperature $T_t = 300$ K were specified; at the outlet, all quantities were extrapolated. Five grids were used: the coarsest had 42×5 CVs, the finest 672×80 CVs.

The lines of constant Mach number are shown in Fig. 3. A shock wave is produced behind the throat, since the flow cannot accelerate due to the change in geometry. The shock wave is reflected from the walls twice before it exits through the outlet cross-section.

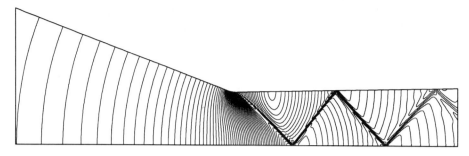

Fig. 3: Mach number contours in the compressible channel flow (from minimum Ma = 0.22 at inlet to maximum Ma = 1.46, step 0.02)

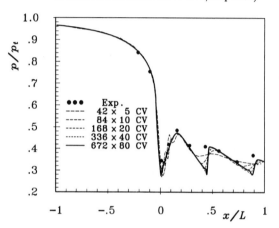

Fig. 4: Comparison of predicted and measured ([14]) distribution of pressure along channel wall

In Fig. 4, the computed pressure distribution along the channel wall is compared with experimental data of from [14] . Results on all grids are shown. On the coarsest grid, the solution oscillates; it is fairly smooth on all other grids. As in all other computations of compressible flows with the present method (see [3]), the locations of the shocks do not change with grid refinement – only the steepness is improved as the grid is refined. The numerical error is low everywhere except near the exit, where the grid is relatively coarse; the results on the two finest grids can hardly be distinguished. Agreement with the experimental data is also quite good. Several other test cases were studied and published in [3].

Due to the use of a second order centered scheme, strong oscillations appear at the shocks. While it is possible to obtain solutions even for inviscid flows using the centered scheme (equivalent to central differencing and resulting in very sharp resolution of shocks), a small amount of the first-order upwind scheme is usually added (typically 5%). For ultimate accuracy this should be done locally; however, we applied the blending globally and found little degradation in accuracy while suppressing most of the oscillations and improving the convergence rate.

The solution method described above tends to converge faster as the Mach number is

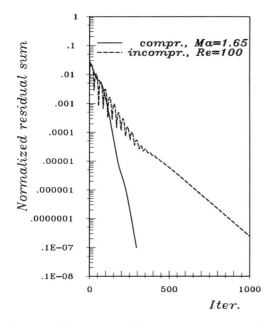

Fig. 5: Convergence of the pressure-correction method for laminar flow at Re = 100 and for supersonic flow at Ma_{in} = 1.65 over a bump in a channel (160 × 80 CV grid)

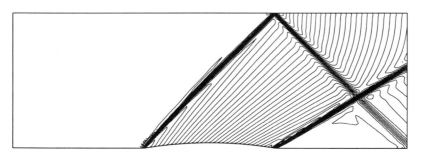

Fig. 6: Predicted Mach number contours for supersonic flow at Ma_{in} = 1.65 over a bump in channel (160 × 80 CVs grid)

increased. In Fig. 5 the convergence of the method for the solution of laminar incompressible flow at Re = 100 and for the supersonic flow at Ma = 1.65 over a bump in a channel (symmetric channel constriction) is shown. The result of calculation for the supersonic case is shown in Fig. 6. The same grid and under-relaxation parameters are used for both flows. While in the compressible case the rate of convergence is nearly constant, in the incompressible case at low Reynolds number it gets lower as the tolerance is tightened. At very high Mach numbers, the computing time increases almost linearly with the number of grid points as the grid is refined (the exponent is about 1.1, compared to about 1.8 in the case of incompressible flows). However, as shall be demonstrated below, the convergence of the method for elliptic problems can be substantially improved by using the multigrid approach, making the method efficient for all flow speeds. The compressible version of the method is also suitable for both steady and unsteady flow problems.

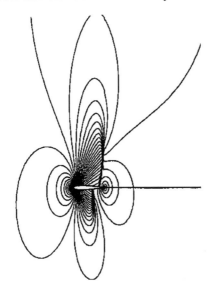

Fig. 7: Predicted isobars around NACA 0012 airfoil, $\alpha = 1°$, $M_\infty = 0.85$

The method was also applied to the transonic flow around NACA 0012 airfoil at 1° angle of attack and an inlet Mach number of 0.85, and the performance of the method was compared with methods designed specifically for compressible flows. Figure 7 shows predicted isobars around the airfoil, calculated using a C-type grid.

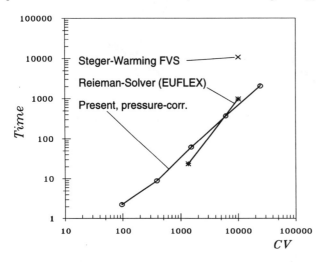

Fig. 8: Comparison of CPU-time for different methods applied to flow around NACA 0012 airfoil

Drikakis [6] performed calculations for the same problem using a code employing Steger-Warming flux vector splitting, and a Riemann solver EUFLEX. The computing times on various grids for the present method and two comparison methods are shown in Fig. 8. In the present method, the steady conservation equations were solved until the residual levels for all variables were reduced five orders of magnitude. The convergence criteria were different in comparison methods, which use time marching; however, the criterion used in the present method is likely to be more stringent. The computing times with the present method and the EUFLEX code are comparable, although with grid refinement the present method leads to a less pronounced increase in computing time.

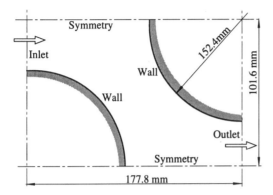

Fig. 9: Symmetry segment in a staggered tube bank showing geometry of the solution domain and boundary conditions

The above examples demonstrated that the present method, when applied to compressible flows, is performing equally well as the methods designed specially for compressible flows. It is also efficient when applied to laminar and turbulent incompressible flows. As an example of application to such flows, the computation of flow in a symmetry segment of a staggered tube bank is considered. The geometry of the problem is given in Fig. 9. The calculation was made at $Re_D = 140000$ using the $k - \epsilon$ turbulence model with wall functions in their standard form. Figure 10 shows the predicted streamlines and the contours of turbulent kinetic energy.

A multigrid method was applied to outer iterations, and five grid levels were used; the coarsest grid had 11×3 CV, while the finest had 176×48 CV. The highly non-linear problem of solving the conservation equations for mass, momentum, and turbulence quantities k and ϵ is an especially hard test case for the multigrid method, since the effective viscosity varies in the solution domain by three orders of magnitude. It was found that the best efficiency is achieved when the turbulent viscosity is updated only on the finest grid and is frozen during one multigrid V-cycle.

Figure 11 shows variation of residuals for u_x and kinetic energy of turbulence k for the single grid and multigrid method. The rate of convergence of the multigrid method is nearly constant, while in the case of the single-grid method the residuals are reduced fast initially and attain a much lower reduction rate at a later stage. For a typical convergence tolerance (reduction of residual levels four orders of magnitude), the computing time is reduced one to two orders of magnitude, depending on the problem and grid size. The multigrid approach was used to obtain very accurate solutions for several test problems, as documented in [2].

The extension of the method to allow moving grid and free surfaces was tested first by studying small amplitude sloshing in a tank. In a 2D simulation the free surface of a

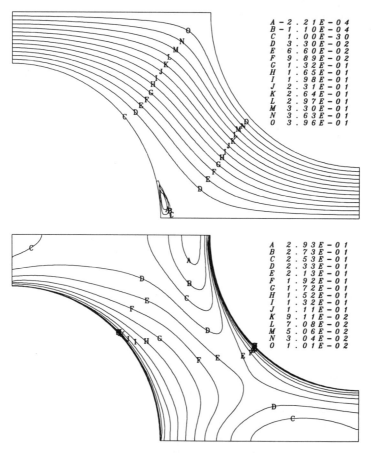

Fig. 10: Details of predicted turbulent flow in a staggered tube bank at $Re_D = 140000$ using standard $k - \epsilon$ model of turbulence: streamlines (above) and contours of turbulence intensity $\sqrt{k}/u_{\text{in}}$ (below)

1 m water column in a 1 m wide tank was initialized with a sinusoidal elevation of 0.01 m amplitude. Figure 12 shows the calculated free surface elevation along the side walls for both an inviscid and a viscous test case. The predicted period P for the inviscid case is in very good agreement with linear theory which gives $P = 2\pi/\sqrt{\pi \tanh \pi} = 3.55$. The rate at which the oscillations are dumped in the viscous case is also correctly predicted.

In order to validate the method for three-dimensional turbulent flows with free surface it has been applied to the flow around Wigley ship hull at Re= 4.5×10^3 and Fr= 0.267. Figure 13 shows the predicted wave contour map around the hull. The wave patterns, originating from both bow and stern, are in good agreement with the patterns predicted by the linear Kelvin theory. Comparison with experimental wave profile along ship surface (cf. Fig. 14) also shows good agreement, except near the wake where the behavior of the turbulence model is very important. More details on this and other free-surface flow calculations are given in [12].

The last example demonstrates the suitability of the method for direct numerical

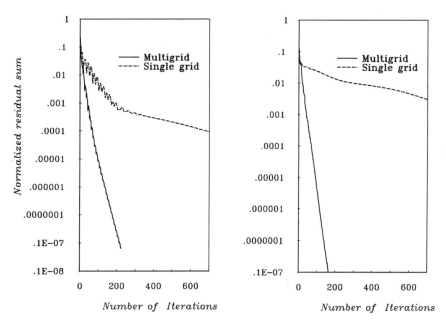

Fig. 11: Convergence history for the turbulent flow in a staggered tube bank: u_x (left), k (right)

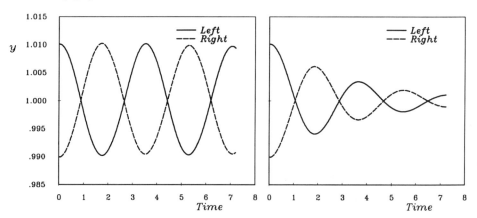

Fig. 12: Predicted free surface elevation along side walls for a small amplitude sloshing in a tank, using a grid with 40 × 40 CVs and $\Delta t = 0.002$: inviscid flow (left), viscous flow (right)

simulations (DNS) of turbulent flows, requiring large computational capacities which are only available on parallel computers. The turbulent flow over a square rib placed on the bottom wall of a plane channel has been simulated at a Reynolds number Re= $3.7 \cdot 10^3$, based on the mean velocity U_m above the rib and the step height h. Both in spanwise and streamwise direction periodic boundary conditions were applied with a fixed pressure drop in streamwise direction. The same conditions (except for the pressure drop) were

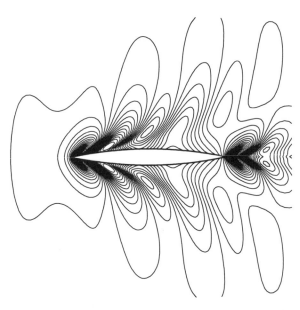

Fig. 13: Wave contour map around Wigley ship hull (Fr= 0.267) calculated with standard $k - \epsilon$ model of turbulence using a grid with $208 \times 48 \times 48$ CVs at Re= 4.5×10^6

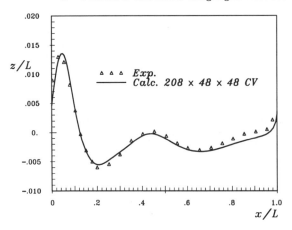

Fig. 14: Comparison of predicted wave profile on the surface of a Wigley ship hull (Fr= 0.267) at Re= 4.5×10^6 with experiment [11]

used in [25].

Figure 15 shows a detail of the non-uniform Cartesian grid with blockwise refinement around the rib. The non-matching block interfaces are treated according to the domain decomposition approach underlying parallelisation in space [21]. Thus, the regular structure within each block can be preserved while the grid is globally unstructured. The simulation has been performed with $1.13 \cdot 10^6$ CVs on a Cray T3D parallel computer at the Konrad-Zuse-Zentrum in Berlin with a total efficiency of about 90 % using 64 processors.

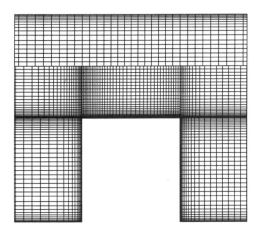

Fig. 15: Detail of the grid around the rib (the simulated section of the channel was $2h$ wide and $31h$ long; the distance between channel walls was $2h$)

Fig. 16: Contours of the instantaneous streamwise velocity in channel center-plane

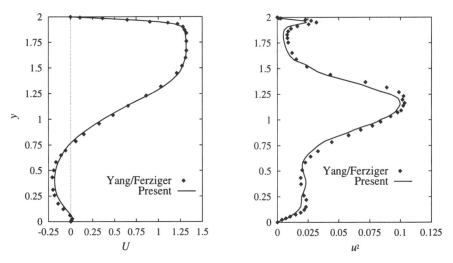

Fig. 17: Profiles of the normalized mean and fluctuating streamwise velocity component one step height behind the rib

The instantaneous contours of the streamwise velocity component in Fig. 16 illustrate the turbulent flow regime. The principal re-attachment length, averaged in the spanwise direction and in time over 37 characteristic time units h/U_m, is $6.4\,h$. Taking into account the difference in Reynolds numbers (3700 vs. 3200), the results compare well with those of Yang and Ferziger [25]. A comparison of profiles of the mean and fluctuating streamwise velocity component is shown in Fig. 17.

CONCLUSIONS

The main results of the project can be summarized as follows:

- A fully-conservative FV discretization up to the fourth order in space was implemented and tested on Cartesian grids. It was found that the second-order method based on linear interpolation and central differences, combined with local grid refinement, is the best compromize between accuracy, efficiency, and ease of implementation in complex geometries.

- The multigrid method for outer iterations was implemented, and it was found to be essential for efficient computation of steady state flows; computing times compared to single-grid method were reduced by one to two orders of magnitude using four to six grid levels. In the case of unsteady flows which require small time steps for accuracy reasons, the multigrid method can not increase the efficiency substantially.

- The pressure-correction method has been extended to treat flows at all speeds. It was found to be efficient for both incompressible and highly compressible flows (flows with Mach numbers up to 10 were calculated). The method is especially attractive for flows in which regions of both high and low Mach numbers exist (e.g. flows around bodies with turbulent wake), since the method needs no adjustments regarding the type of flow.

- The method has been extended to treat flows with moving boundaries and free surfaces. The space conservation law is obeyed, thus guaranteeing full conservativenes of the scheme while using a stable second-order fully implicit time integration scheme.

- The method has been parallelized in both space and time and applied to direct numerical simulation of turbulent flows. It can use block-structured grids which do not match at interfaces, thus allowing a block-wise local grid refinement.

Future extensions of the method will include the use of unstructured grids with arbitrary control volume shape and cell-wise local refinement, coupling of fluid flow and solid stress analysis and fluid-structure interaction.

ACKNOWLEDGEMENTS

This project was initially carried out at the Lehrstuhl für Strömungsmechanik of the University of Erlangen-Nürnberg. The authors acknowledge the contribution to the success of the project by Prof. F. Durst and co-workers. DFG provided access to parallel computers in Aachen and Heidelberg. The authors also thank Prof. F. Thiele and the personnel of Konrad-Zuse-Zentrum in Berlin for their support and making computations on Cray T3D computer possible.

REFERENCES

[1] BURMEISTER, J., AND HACKBUSCH, W. On a time and space parallel multi-grid method including remarks on filtering techniques. In *Notes on Numerical Fluid Mechanics* (1996), E. H. Hirschel, Ed., Vieweg, Braunschweig. In print.

[2] DEMIRDŽIĆ, I., LILEK, Ž., AND PERIĆ, M. Fluid flow and heat transfer test problems for non-orthogonal grids: Bench-mark solutions. *International Journal for Numerical Methods in Fluids 15* (1992), 329–354.

[3] DEMIRDŽIĆ, I., LILEK, Ž., AND PERIĆ, M. A collocated finite volume method for predicting flows at all speeds. *International Journal for Numerical Methods in Fluids 16* (1993), 1029–1050.

[4] DEMIRDŽIĆ, I., AND PERIĆ, M. Space conservation law in finite volume calculations of fluid flow. *International Journal for Numerical Methods in Fluids 8* (1988), 1037–1050.

[5] DEMIRDŽIĆ, I., AND PERIĆ, M. Finite volume method for prediction of fluid flow in arbitrary shaped domains with moving boundary. *International Journal for Numerical Methods in Fluids 10* (1990), 771–790.

[6] DRIKAKIS, D. Private communication, Lehrstuhl für Strömungsmechanik, Universität Erlangen, 1992.

[7] DURST, F., SCHÄFER, M., AND WECHSLER, K. Efficient simulation of incompressible viscous flows on parallel computers. In *Notes on Numerical Fluid Mechanics* (1996), E. H. Hirschel, Ed., Vieweg, Braunschweig. In print.

[8] HACKBUSCH, W. *Multi-Grid Methods and Applications*. Springer Series in Computational Mathemathics. Springer-Verlag, 1985.

[9] HORTMAN, M., PERIĆ, M., AND SCHEUERER, G. Finite volume multigrid prediction of laminar natural convection: Bench-mark solutions. *International Journal for Numerical Methods in Fluids 11* (1990), 189–207.

[10] HORTON, G. TIPSI - a time-parallel SIMPLE-based method for the incompressible Navier-Stokes equations. In *Proceedings of the Parallel CFD 1991, Stuttgart 1991* (1991), K. R. et al., Ed., Elsevier Science Publisher B.V.

[11] ITTC. Cooperative Experiments on Wigley Parabolic Models in Japan, 17th ITTC resistance Comittee Report, 1993.

[12] LILEK, Ž. Ein Berechnungsverfahren für dreidimensionale, viskose Strömungen mit freien Oberflächen. Institut für Schiffbau der Universität Hamburg, Bericht Nr. 553, ISBN 3-89220-553-1.

[13] LILEK, Ž., AND PERIĆ, M. A fourth-order finite volume method with colocated variable arrangement. *Computers & Fluids 24*, 3 (1995), 239–252.

[14] MASON, M. L., PUTNAM, L. E., AND RE, R. The effect of throat contouring on two-dimensional converging-diverging nozzle at static conditions. *NASA Techn. Paper No. 1704* (1980).

[15] MIYATA, H., ZHU, M., AND WATANABE, O. Numerical study on a viscous flow with free-surface waves about a ship in steady straight course by a finite-volume method. *Journal of Ship Research 36*, 4 (1992), 332–345.

[16] PATANKAR, S. V., AND SPALDING, D. B. A calculation procedure for heat, mass and momentum transfer in three-dimensional parabolic flows. *International J. Heat Mass Transfer 15* (1972), 1787–1806.

[17] PERIĆ, M. *A finite volume method for the prediction of three-dimensional flow in complex ducts.* PhD thesis, University of London, 1985.

[18] PERIĆ, M., AND SCHRECK, E. Analysis of efficiency of implicit CFD methods on MIMD computers. In *Parallel CFD '95, Pasadena* (june 1995).

[19] RUBIN, S. G., AND KHOSLA, P. K. A diagonally dominant second-order accurate implicit scheme. *Computers & Fluids 2* (1974), 207.

[20] SCHRECK, E., AND PERIĆ, M. Computation of fluid flow with a parallel multigrid solver. *International Journal for Numerical Methods in Fluids 16* (1993), 303–327.

[21] SEIDL, V., PERIĆ, M., AND SCHMIDT, S. Space- and time-parallel Navier-Stokes solver for 3D block-adaptive cartesian grids. In *Parallel Computational Fluid Dynamics – Implementation and Results Using Parallel Computers* (1996), P. Fox, Ed., Elsevier Science B.V.

[22] VAN DEN VORST, H. A. BI-CGSTAB: A fast and smoothly converging variant of BI-CG for the solution of non-symmetric linear systems. *SIAM J. Sci. Stat. Comput. 13* (1992), 631–644.

[23] VILSMEIER, R., AND HÄNEL, D. Computational aspects of flow simulation on 3-D, unstructured, adaptive grids. In *Notes on Numerical Fluid Mechanics* (1996), E. H. Hirschel, Ed., Vieweg, Braunschweig. In print.

[24] WAGNER, S., Ed. *Computational Fluid Dynamics on Parallel Systems* (1995), Vieweg, Braunschweig.

[25] YANG, K., AND FERZIGER, J. H. Large-eddy-simulation of turbulent flow using a dynamic subgrid-scale model. *AIAA Journal 31*, 8 (1993), 1406–1413.

COMPUTATIONAL ASPECTS OF FLOW SIMULATION ON 3-D, UNSTRUCTURED, ADAPTIVE GRIDS

R. Vilsmeier, D. Hänel
Institut für Verbrennung und Gasdynamik, Universität Duisburg
D-47048 Duisburg, Germany

SUMMARY

Solution concepts on unstructured, adaptive grids for the 3-D Navier-Stokes equations are considered. The solution scheme is a finite-volume method with upwind or central discretization of inviscid fluxes and explicit integration in time. Generation of tetrahedral meshes and their adaption are based on a common algorithm. Several numerical and hardware aspects for different discretization and data structures are analysed and discussed. A perspective is given for future extension towards hybrid grid concepts.

INTRODUCTION

Solution methods on unstructured grids for Euler equations and Navier-Stokes equations using Finite Element or Finite Volume methods have been applied to a wide range of flow problems A review about the state of computational techniques is given in e.g. [1]. Several attempts and applications, related to computations on unstructured grids are presented in this publication, [2, 3, 4, 5]. Essential advantages of this grid concept are the geometrical flexibility and the ease of adapting meshes to local requirements of accuracy. By adaption, grid cells can be added, removed or deformed during the solution according to criteria derived from the solution. Therefore such adaptive methods are in principle ideal methods to deal with flow problems of different characteristic scales, where a high resolution is required in parts of the integration domain, whereas in other parts moderate resolution is sufficient. Thus the generally higher effort per grid point for unstructured grids can be compensated by the sparse, effective use of grid cells.

The difficulties in formulations and applications of unstructured, adaptive methods depend strongly on the character of the solution. Regarding inviscid flows, where besides geometrical features only distinct discontinuities, e.g. shocks, appear, formulations on unstructured grids are well suited and widely used.

Solutions of the Navier-Stokes equations at high Reynolds numbers are generally more complex. A crucial problem is the presence of very different viscous scale lengths in different directions, as they are e.g. body length, boundary layer thickness or vortex extensions. An appropriate resolution results in anisotropic meshes to meet the corresponding solution, e.g. in the thin shear layers. The resulting deformed, very flat cells introduce numerical inaccuracies and additional stiffness which impairs the performance of the solution algorithms. Actually, high efforts are made to reduce this drawback, for example by developing hybrid grid concepts.

An additional problem arises, especially for compressible, viscous flows, that often more than one adaption criterion, different in value and direction, has to be taken into account. To satisfy this requirement, an adaption concept based on local transformation (virtual stretching) was developed by the authors and applied to solutions for 2-D, steady, viscous flow in [6, 7, 8, 9], for unsteady vortex separation in [10], for chemical flow and detonation structures in [11] and extended to three dimension in [12, 13].

The favourable properties of unstructured, adaptive methods, i.e. flexibility and adaptivity, are of particular interest in computations of flow problems in three dimensions. However, compared to 2-D, the efficiency of the algorithm plays a much more important role. Improvements require the analyses of the constructs of the algorithm with respect to their influence on the performance. The present paper deals with an investigation of different algorithmic elements and their influence. After a brief description of the Finite-Volume method for the Navier-Stokes equations, numerical and hardware aspects are discussed for different arrangements of control volumes and data structures. The principle of grid generation and adaption is outlined and a few computational examples are presented finally.

METHOD OF SOLUTION

Governing Equations

The time-dependent equations for the conservation of mass, momentum and energy in integral form read:

$$\frac{\partial}{\partial t} \int_V \vec{Q} d\hat{V} + \oint_{\delta V} \vec{H} \vec{n} dA = 0. \tag{1}$$

$\vec{Q}$ is the vector of conservative variables, $\vec{H}$ is the generalized flux vector. V is any volume within the flow field and δV is its boundary, whose local normal vector is $\vec{n}$.

$$\vec{Q} = \begin{bmatrix} \rho \\ \rho \vec{v} \\ \rho E \end{bmatrix} \qquad \vec{H} = \begin{bmatrix} \rho \vec{v} \\ \rho \vec{v} \vec{v} + \sigma \\ \rho \vec{v} E + \sigma \vec{v} + \vec{q} \end{bmatrix}.$$

ρ, $\vec{v}$ and E are the density, speed vector and specific total Energy. $\vec{q}$ is the heat-flux vector and σ the stress tensor. More detailed explanations may be found elsewhere. The flux vector $\vec{H}$ can be subdivided in an inviscid and a viscous part, $\vec{H} = \vec{H}_{inv} + \vec{H}_{visc}$, with:

$$\vec{H}_{inv} = \begin{bmatrix} \rho \vec{v} \\ \rho \vec{v} \vec{v} + pI^d \\ \rho \vec{v}(E+p) \end{bmatrix} \qquad \vec{H}_{visc} = \begin{bmatrix} 0 \\ \sigma - pI^d \\ (\sigma - pI^d)\vec{v} + \vec{q} \end{bmatrix}$$

where p is the pressure and I^d is the unit matrix in the dimension d of the problem.

Finite Volume Approach

A discrete formulation of the integral conservation equation for any discrete Volume Vd reads:

$$\left. \frac{\Delta \vec{Q}}{\Delta t} \right|_{Vd} + \vec{Res}_{\Delta,Vd} = 0. \tag{2}$$

Since unstructured grids do not offer a natural ordering of nodes and elements, an artificial ordering, a data structure, is required. An element of the data structure will be called a molecule. To perform a time step for the equation given above, the discrete residual $\vec{Res}_{\Delta,Vd}$ has to be constructed upon such molecules:

$$\vec{Res}_{\Delta,Vd} = \frac{1}{V_{Vd}} \sum_{i=1}^{nr(Vd)} \vec{H}_{\tilde{i}(i)} \vec{n}_{\tilde{i}(i)} \Delta A_{\tilde{i}(i)} . \tag{3}$$

$nr(Vd)$ is the number of molecules contributing to the residual, V_{Vd} is the corresponding volume. $\vec{H}_{\tilde{i}(i)}$ is the flux vector, $\vec{n}_{\tilde{i}(i)}$ the normal vector and $\Delta A_{\tilde{i}(i)}$ the area of the control interface supported by a molecule $\tilde{i}(i)$. The control interfaces have to enclose the volume completely.

Control Volumes and Data Structures

The very general description of the finite volume approach offers a large variety of possibilities how to define volumes and corresponding data structures. On unstructured grids, the data structure is the base of the whole flow solver. The following criteria may be regarded for its choice:

- Memory requirements to store the data. Compared to structured methods, high additional storage is required.

- Flexibility concerning grid type and the possibility of discretizing all terms is required.

- Computational efficiency. This item includes the number of operations to perform the discretizations and the efficiency of the memory access, being a typical bottleneck on RISC-computers. On vector and parallel machines it is useful to minimize data dependencies.

For the present paper Cell-Vertex (CV) and Node-Centred (NC) approaches are used. In both cases the flow variables are stored at the nodes.

Cell-Vertex (CV) The control volume for a node P consist of the set of elements, having this node as a vertex. The control volumes of neighbouring nodes overlap. Corresponding control interfaces consist of natural element faces. In the case of a tetrahedral mesh, these are the triangular faces of the tetrahedra.

A useful data structure for a CV computation is based on the triangular faces of the tetrahedra. For a corresponding molecule the addresses of the three nodes $K1, K2, K3$ forming the triangular face and the opposite nodes $K4$ and $K5$ of the adjacent tetrahedra are stored. Fluxes are computed at the triangular faces and contribute to the residuals of the opposite nodes, Fig 1, left.

The face-based data structure requires large amounts of memory. It may therefore be sometimes useful to employ the much smaller tetrahedron-node structure. For each tetrahedron, being a molecule of the structure, only the four node addresses are required. However, beside the restriction to central computations, all fluxes have to be computed twice, Fig 1, right.

Node-Centred (NC) Meshes are decomposed in a set of non overlapping control volumes around the nodes. This requires the definition of polyhedral regions and corresponding cell interfaces between neighbouring nodes.

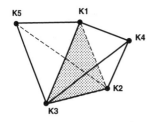
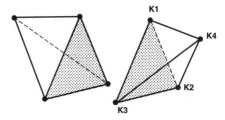

Figure 1: Flux computation in Cell-Vertex arrangements for molecule(s) of the face-based data structure (left) and the tetrahedron-based data structure (right).

NC computations on 3-D tetrahedral meshes are not very common in use. The reason is probably the difficulty of finding a suitable discrete data structure to support the flux computations. Again the tetrahedron-node structure may be used. However, inside a single tetrahedron six cell interfaces according to the six edges are located. Computational efforts are therefore very high, Table 1.

For the present paper a mixed discrete-analogue edge-based data structure is employed. For each edge of the mesh the addresses of the two ending nodes $K1, K2$ and the control interface area-normal vectors $\vec{n}\Delta A$ between the control volumes of both the nodes are stored. The segments of this control interface consist of the set of triangles surrounding the edge through the centres of faces and volumes of the adjacent elements. Flux computations are performed on the cell interface and contribute to the residuals of the two nodes $K1, K2$. In summary, four basic concepts have been discussed:

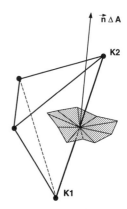

Figure 2: Molecule of the edge-based Node-Centred concept.

- Cell-Vertex, face-based data structure (faces → nodes)
- Cell-Vertex, tetrahedron-based data structure (tetrahedra → nodes)
- Node-Centred, tetrahedron-based data structure (tetrahedra → nodes)
- Node-Centred, edge-based data structure (edges → nodes and $\vec{n}\Delta A$)

Requirements for the Flux Discretizations

In this sub-chapter the evaluation of the required quantities for central, upwind and viscous computations, employing the different data structures are discussed.

Scalar averages at the cell interfaces Scalar averages of the flow variables at the cell interfaces are required for the discretizations of central inviscid and viscous fluxes. Since the flow variables are not available at the cell interfaces but only at the nodes, interpolations are required.

In the Cell-Vertex case, the interpolations have to be performed in lateral direction, in the Node-Centred case the direction along the connecting edges between neighbouring nodes are used. Interpolations can be performed on all four previously discussed basic concepts.

Gradient vectors at the nodes Gradient vectors at the nodes are required in the projection step for higher order upwind discretizations. The gradients can be computed via discretization on the control volumes themselves. For a central node in a volume Vd these read:

$$\vec{\nabla} K_{\Delta,Vd} = \frac{1}{V_{Vd}} \sum_{i=1}^{nr(Vd)} K_{\tilde{i}(i)} \vec{n}_{\tilde{i}(i)} \Delta A_{\tilde{i}(i)} \qquad (4)$$

where $K_{\tilde{i}(i)}$ is the scalar average of the variables K at the interface supported by the molecule $\tilde{i}(i)$. Since beside the metrics only these scalar averages are required, a computation of the gradient vectors at the nodes is again possible for all four basic concepts. However Cell-Vertex arrangements are less useful for upwind computations. One reason is the large region of influence ranging up to the second neighbours for a first order and third neighbours for a second order discretization. Employing the tetrahedron-based data structure it is even impossible to discretize the upwind formulations, since only one-sided informations are available.

In contrast, upwind formulations on Node-Centred arrangements have a smaller region of influence, one neighbourship lower, and are therefore prefered for upwind discretizations of the inviscid Fluxes.

Gradient vectors at the cell interfaces Gradient vectors at the cell interfaces are required for the formulation of the viscous terms. They can be obtained by various ways, depending on the basic concept. For the tetrahedron-based Cell-Vertex concept and the edge-based Node-Centred concept the gradients at the cell interfaces are obtained by averaging the node wise computed gradient vectors in the same way as done for the scalar averages. For Node-Centred edge-based computations the average gradients can be corrected as in the direction of the edge a central difference is possible.

For the Cell-Vertex discretization with face-based data structure an average gradient may be computed upon the gradients within the two tetrahedra adjacent to the common face, thus avoiding the intermediate storage of the gradient vectors at the nodes. The tetrahedron-based Node-Centred concept allows the computation of the gradient vectors directly at the tetrahedron, also avoiding the storage of nodes-wise gradient vectors.

Comparison of the Basic Concepts

Table 1 below shows the average storage and CPU-time requirements for the previously discussed control volume types and data structures on tetrahedral meshes. The number of elements per node is supposed to be six in average. Accordingly, seven edges per node are assumed. Both, integer I and real numbers R are stored as 4-Byte numbers. A precision correction for double precision computations is possible. The meaning of the abbreviations used is:
C: suitable for a central computation of the inviscid fluxes
U: suitable for an upwind computation of the inviscid fluxes
N: suitable for the computation of viscous fluxes
H: suitable for hybrid grids with a uniform formulation

The edge-based Node-Centred concept shows the best compromise for our purposes. Major advantages are the flexibility concerning the grid type and the low amount of flux

Table 1: Average storage and CPU-time requirements comparing the four basic concepts. ([1]) Restrictions due to large region of influence)

Basic Concept	Molecules per node	Storage for data-structure per node	Flux comp. per node	Flexibility
CV, face-nodes	12	12*5I = 240 Bytes	12*1 = 12	C,U[1]),N
CV, tet.-nodes	6	6*4I = 96 Bytes	6*4 = 24	C,N
NC, tet.-nodes	6	6*4I = 96 Bytes	6*6 = 36	C,U,N
NC, edge-$\vec{n}\Delta A$	7	7*(2I+3R) = 140 Bytes	7*1 = 7	C,U,N,H

computations. The memory requirements are medium. However, only the tetrahedron-based concepts require less memory but these algorithms are inflexible (CV) or extremely CPU-time consuming (NC). Nevertheless The latter is of interest, since it enables a viscous upwind computation with remarkable low memory.

Discretization of Inviscid Terms

Central discretizations A central discretization is the easiest way to compute the inviscid fluxes, since only the scalar averages are required. The computational costs are low. The disadvantage however is that artificial damping terms are required.

The computation and control of damping terms is the essential difficulty for central schemes. For the present computations quasi-second and quasi-fourth derivatives of the variables are used. Formulations correspond in essential to that originally proposed by Mavriplis for 2-D computations, [14].

Upwind discretizations Due to the difficulties in handling the damping terms and the neglect of characteristic directions when using central discretizations, it is sometimes more useful to employ upwind formulations for the inviscid fluxes.

On 2-D triangular meshes several upwind schemes, Van-Leer/Hänel [15], HLL [16], Roe [17] and AUSM [18] were studied. Best results concerning the numerical viscosity and stability were obtained with a slightly modified version of the AUSM method. This result is remarkable since the AUSM discretization is the simplest of the above mentioned methods.

The idea of the AUSM scheme is to split the inviscid flux vector in a convective part and a part containing the pressure terms for the momentum equations: $\vec{H}_{inv} = \vec{H}_{inv,c} + \vec{H}_{inv,p}$ with:

$$\vec{H}_{inv,c} = \vec{v}_1 \begin{bmatrix} \rho \\ \rho\vec{v}_2 \\ \rho(E+p) \end{bmatrix} \qquad \vec{H}_{inv,p} = \begin{bmatrix} 0 \\ pI^d \\ 0 \end{bmatrix}.$$

The convective part is written as the product of a convection speed with the corresponding convected quantities, where $\vec{v}_1$ is the convection speed while $\vec{v}_2$ is a part of the convected quantities. For the convected quantities a simple one sided upwind splitting is used, depending only on the sign of the scalar product $\vec{n} \cdot \vec{v}_1$ at the control interface. The formulation of the convection speed $\vec{v}_1$ allows several possibilities; lowest viscosity was observed for a central formulation. The pressure part of the inviscid flux vector also allows a variety of possible formulations. Among these, a linear weighting, according to the Mach number in the direction of the cell interface normal vector is used.

For our application however the scheme showed some disadvantages on strong shocks. The reason is probably the approximation of the pressure wave characteristics. To overcome these difficulties, the scheme is locally blended by the HLL discretization.

A formal second order is obtained with Van-Leers MUSCL approach, [19]. Average gradients at the nodes are used for the linear projections of variables. An example of an inviscid solution, the subsonic flow past a sphere, Fig 3, reveals the low numerical viscosity of the second order AUSM scheme on a rather coarse mesh. This is indicated by a by vanishing drag or qualitatively by the symmetry of isolines.

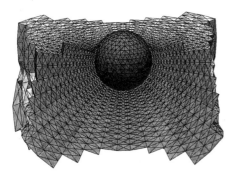 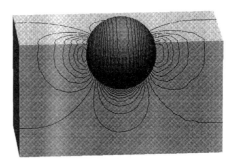

Figure 3: Coarse, prismatic mesh for a sphere, broken partition, left. Corresponding lines of const. pressure, right ($Ma_\infty = 0.2$, inviscid, 2. ord. AUSM. Node centred)

Flows with embedded discontinuities require a local reduction of the discretization order. The formulation of corresponding limiters showed to be a severe problem. Conventional projection limiters as the van Albada limiter were, in our case, unsuitable for strong shocks and produced unnecessary additional viscosity in smooth regions. Experiments (in 2-D) with the FCT concept showed, apart the high computational expenses, other remarkable problems concerning shock position and smoothness. At present, a two-sided projection limitation is in development, which is based on the comparison of the discretized gradients at both sides of a control interface with the ones obtained by a parabolic approach along the edge. Good results were obtained within the field but some stability problems remain near the boundaries.

Discretization of Viscous Terms

Due to the elliptic nature of the viscous terms a central formulation is employed. Since the orientation to the walls is difficult on unstructured grids, only full viscous terms may be used. However, the results on tetrahedral meshes were poor. Using isotropic clustering of nodes in the boundary layer results in immense computational efforts. The computation of the boundary layer flow past a square flat plate at a Reynolds number of 200 for example, requires an amount of grid cells in the same order as the inviscid adaptive computation in Fig 5/6.

By way of contrast, the discretization of the viscous terms on anisotropically clustered cells showed severe stability problems. The reason is the bad behaviour of these terms on elements with obtuse angles, [20]. On 2-D triangular meshes these difficulties were affordable. For the 3-D counterpart it will probably be impossible to efficiently compute viscous flows at high Reynolds numbers employing tetrahedra in the boundary layer.

Time Integration

For the present computations the explicit Runge-Kutta method is employed. To integrate flows with very different stability conditions for different nodes, a multi-sequence version of the algorithms [10] can be employed. However, this integration method is not yet fully available in the version for vector computers.

HARDWARE ASPECTS

Present computations and developments were performed on RISC-workstations and vector computers. These two types of computers show very different behaviours. The optimal use of their performances requires individual coding and, still more important, different data sorting.

Optimizations for RISC-based systems

RISC based systems have already quite powerful processors. The bottleneck, however, is the data throughput between memory and CPU. It is thus useful to minimize the data transfer. This may be done in two ways: First, the algorithms may be optimized, coding as few loops as possible but with a large amount of operations per iteration. Second, the access sequence of array elements has to be optimized. This is a difficult but indispensable task, when running large problems with indirect addressing.

Optimized memory access for RISC computers The previously mentioned bottleneck of memory access on RISC-computers is severely aggravated, if jumps in the indirect addressing occur. The problem arises due to the "on request" architecture of the cache concept. If an access to an array element is required, a whole memory region is copied from the main memory to the cache. Another array element, subsequently required, may or may not be already resident in the cache memory. In the latter case, again a request for the main memory is required and this takes some time.

It is therefore useful to load data for many subsequent loop iterations at once, providing a so called data locality or data continuity. A simple method, to provide data continuity is the "nearest neighbour sort". Starting at a boundary node of the mesh, all elements sharing this node are inserted at the top of the element list. The recurrent process is continued at another vertex of these previously found elements up to the whole elements and nodes are sorted. The disadvantage however is, that since a one-dimensional data continuity is impossible, randomly distributed address jumps occur in the later flow computations. An attempt to solve the problem is to introduce a multiple data locality.

For 2-D problems the meshes may be sorted in a spiral way, winding up molecules and nodes. The addresses of the molecules and addressed nodes rise according to their position in the one-dimensional stripe. Computations to be performed for the molecules within a loop may then address nodes that may be located or fully inside the current layer of the stripe, or in two neighbouring layers. In the latter case multiple data continuity is obtained.

The 3-D version of this sorting type is difficult, since several two-dimensional layers of the coil exist. To obtain multiple data continuity it is essential, that the stripes in the layers on top of each other show a parallel alignment. A corresponding sorting algorithm is in development, but the success is not yet sure.

For multiple continuity of data and thus more organized jumps in the addressing, the previously mentioned nearest neighbour sort may be used introducing a grouping concept. Inside geometrical groups, a reasonable data locality is obtained. Discretizations over the group boundaries require data from both sides, corresponding to the idea of multiple continuity.

Optimizations for vector computers

The strength of vector computers is the high data throughput due to their memory concept. Instead of a cache, these machines offer vector registers to feed the processors. These high data throughputs enable the use of very powerful processors. Since vector computers are well balanced machines, very high sustained performances are possible. Applications range between 1/5 to more than 1/2 of the theoretical performances. Prerequisite is a good vectorization. In addition, maximum performances require optimized memory access.

Data independence for vector computers Equations (3) and (4) require the summation of molecule contributions. Since the list of molecules may address any control volume, these summations bear potential recursions. To allow vectorization, the whole list of molecules is sorted in non recursive groups. The amount of molecules in each group are restricted to address each control volume at most once and vector compilation is forced via directive.

In the vector version of the present algorithm all floating-point operations on array elements are vectorized without any exception. Vector lengths are if ever possible chosen as multiples of the machine register length.

Optimized memory access for vector computers The memory of vector-machines is interleaved, that is split in a number n_{bank} of banks. Consecutive array elements are placed in different banks. An access to an array element blocks the corresponding bank for a non negligible latency time.

On unstructured grids the indirectly addressed array elements may show any order (considered as random). It is therefore possible, that an access occurs to a memory bank, that is still in state of latency, thus blocking the load/store units. To avoid this problem, it is useful to resort the molecules and to obtain an access sequence respecting the latencies of all banks. Each memory bank once hit via indirect addressing is saved at least for a number $n_{c,lat}$ of access cycles.

A disadvantage however is, that the access sequences must be optimized individually for each machine. Not only the number of banks and latency cycles is variable, also the interleaving concepts vary. As an exotic example, the memory of the Fujitsu VPP 500 may be considered, consisting of two interleave levels: a number of memory groups exist, consisting each of several banks. The sorted molecule sequence should therefore respect the interleaving of the memory groups and their banks individually.

Computational performances Test computations were performed on several computers. As an example the results for the Fujitsu S600 (vector), VPP 500 (vector) and the HP 9000/735/99 (RISC) are presented. Tests are based on a node-centred, inviscid second order upwind computation for a mesh with 33'956 nodes, Table 1 . Larger Applications show slightly higher performances on the vector machines.

Table 2: Computational performances depending on sorting strategies in M-Flops. (**: no influence)

Computer	Data Dependence	Data Access	Performance*
HP-735/99	recursive**	random	10.2
HP-735/99	recursive**	nearest neighbour	38.3
HP-735/99	recursive**	group, nearest neighb.	42.2
S600	recursive	random	97.5
S600	independent	random	869.4
S600	independent	interleaved	1119.1
VPP 500	independent	random	420.3
VPP 500	independent	interleaved (banks only)	509.7
VPP 500	independent	twin interleaved	556.6

MESH GENERATION AND ADAPTIVITY

A field method for the automatic generation of meshes is employed. After a first triangulation of the computational domain between the boundaries, meshes are optimized by local transformations.

Tetrahedral grids

Since the methods, concerning the tetrahedral mesh generation have previously been presented, [12], the descriptions here are restricted to a short overview.
As inputs for the generation system triangulated surfaces, bounding the computational domain are required. These surfaces are obtained from a modified version of a 2-D unstructured mesh generator. Coupling of this surface generation method with CAD programs is subject of ongoing investigations.

First triangulation of the computational domain Starting at given triangulated surfaces, tetrahedra are build using a front Delaunay, rising bubble type mesh generation algorithm. At this stage of generation only the given nodes at the surfaces are used. The rising bubble algorithm may be summarized as follows:
I)   The boundary triangles of the domain are the initial front triangles.
II)   For each front triangle (3 nodes) an additional fourth node is sought, in order to obtain a tetrahedron, whose circumsphere does not contain other nodes in the triangulation direction. The chosen node is the one, which yields the tetrahedron with the rear-most centre of its circumsphere. Some checks for the new tetrahedron produced are performed.

III) The sides of the new tetrahedra are inserted in the list of front triangles. Front triangles touching each other are deleted from the list. The process continues at II) until the front list is empty.

Due to the existence of a polyhedron with triangulated surfaces, that cannot be decomposed in tetrahedra unless using an additional central node, Fig 4 the above described algorithmic sequence may be blocked. This polyhedron has the form of a distorted prism with triangulated quadrilateral sides. If such a polyhedron or agglomerated groups of them remains while generating the mesh, no further tetrahedron can be produced with the above described rising bubble method. Another blocking reason is a possible accuracy hazard. In any cases of blocking an additional field node is required, being inserted automatically at an optimized position and the triangulation process continues.

Figure 4: Polyhedron with triangulated surfaces, that may not be decomposed in tetrahedra, unless using additional, internal nodes. Imagination: Light source from the top.

Mesh optimization After the initial triangulation the mesh is optimized employing a set of local tools. These are employed in a recurrent structure up to medium convergence of the mixed discrete analogue optimization problem:

I) The local mesh density is represented by the locally preferred length G_k of the edges. This local quantity provides the information required for a smooth change of the element sizes between the boundaries of the domain. The quantity is computed solving a boundary value problem during the development of the mesh.

II) Additional nodes are inserted in the centre of tetrahedra or surface triangles according to the local edge length compared to the mesh density function G_k.

III) Mesh reconnection by a swapping algorithm. Local groups of tetrahedra with at all five nodes are analyzed. These are two neighbouring tetrahedra, three ones around a common edge or four ones around a common node. Possible swaps are able to transform the two tetrahedra version in the one with three tetrahedra and vice-versa. Decisions are made upon the Delaunay criterion.

IV) Smoothing by moving nodes to optimized positions. The smoothing criterion is formulated as the minimization of a sensitive quantity, based on the the circumsphere volumes and own volumes of the tetrahedra.

Mesh adaptation Adaptivity is introduced via virtual stretching. According to some criteria, the physical space is transformed in a wider virtual space. The transformations are performed employing local symmetric 3x3 matrices. Mesh optimization proceeds at the wider virtual space employing the previously introduced generation algorithm, and in physical space the adaptive triangulation is obtained. Since the transformation matrices are very flexible, isotropically or anisotropically adaptive regions can be produced.

Extensions towards hybrid grids

Due to the bad properties of stretched tetrahedral elements when computing viscous flows, the use of bilinear elements, namely prisms and hexahedra, will be preferable. Hybrid grids may than consist mainly of these two element types. Pyramids are required as intermediate elements between quadrilateral and triangular faces of other elements.

Tetrahedra are still required, to provide maximum geometric flexibility. However the internal angles of the pyramids and tetrahedra have to be controlled due to the mentioned stability problems. Another reason for the use of prisms and hexahedra is the higher efficiency of these elements. Compared to the number of nodes, tetrahedral grids require far most elements and also most edges. One possibility to generate geometrically complex grids with bilinear elements is the construction of structured or semi structured blocks. Depending on the formulations it may or may not be required to generate unstructured intermediate regions between these blocks. Such an example is shown in Fig 7.

Subject of the ongoing investigation is the trial to generate such hybrid grids starting at fully unstructured, tetrahedral ones, generated by automatic algorithms. Upon local transformations, the desired elements can be created within the tetrahedral mesh. Once such cells are created, they may act as seeds for a crystallization of whole mesh regions. For the orientation of the bilinear cells, the stretching matrices (mesh adaption) can be used. Preferable directions are the corresponding eigenvectors of the matrices. However, the optimal crystallization process is not yet clear, since many promising possibilities exist.

COMPUTATIONAL EXAMPLES

3-D, inviscid flow on adaptive, tetrahedral grids

As an example for the flow solver and the tetrahedral mesh generation concept, the flow past a simplified model of a supersonic aircraft is presented. Fig 5 shows the adaptive grid (obtained after the first mesh adaptation), Fig 6 corresponding isolines of the density. The Node-Centred, edge-based concept has been employed in conjunction with a central flux formulation and artificial damping.

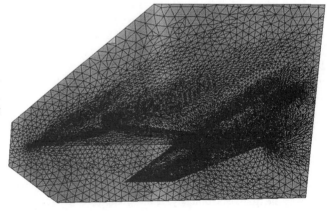

Figure 5: Anisotropically adaptive mesh on surface and symmetry plane for a model of an aircraft.

Viscous unsteady laminar flow on a hybrid grid

As a first example for a flow computation on a hybrid grid, the flow past a sphere is presented. Fig 7 shows a broken partition of the grid. The surface of the sphere is triangulated and consists of 6368 surface triangles. For the boundary layer around the sphere a concentric prismatic layer is employed, structured in normal direction with

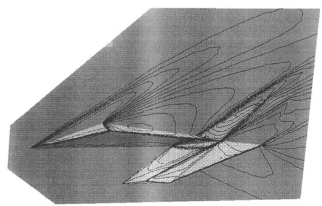

Figure 6: Lines of constant density for inviscid flow, $Ma_\infty = 3.0$, Node centred, central. Obtained on the mesh from Fig 5.

$24 \times 6368 = 152832$ prisms. Behind the sphere a Cartesian, structured block is located, consisting of $60 \times 40 \times 40 = 96000$ hexahedra. The quadrilateral faces of this block are covered by pyramids, thus 12800 of them are required. These mesh partitions are placed in a box with triangulated faces. Finally, the empty spaces are filled with tetrahedra employing the tetrahedral mesh generation techniques. It results in 740703 tetrahedra, which is a large number, considering that the length of their edges is in the same order as the edges of the other elements nearby. The remarkable high amount of tetrahedra confirms the low efficiency of these elements.

A steady solution for $Re = 100$ and $Ma = 0.2$ is shown in Fig 8. The separation length matches well with the ones observed experimentally by Taneda [21]. A solution for a time dependent flow at a higher Reynolds number of $Re = 1000$ is presented in Fig 9. Lines of constant pressure on the symmetry plane and the surface are shown.

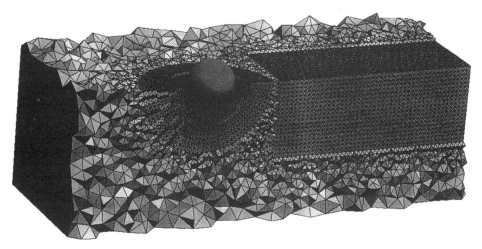

Figure 7: Broken partition of a hybrid grid for the computation of flows past a sphere. Mesh consisting of prisms, Cartesian hexahedra, pyramids and tetrahedra.

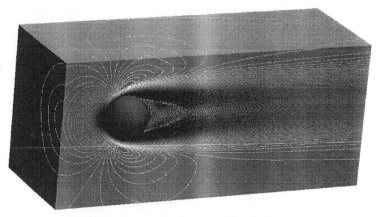

Figure 8: Lines of constant speed for a laminar compressible flow at $Re = 100$ and $Ma_\infty = 0.2$, computed on the mesh above. (AUSM, NC)

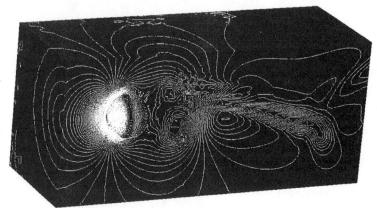

Figure 9: Lines of constant pressure for a laminar, time dependent compressible flow at $Re = 1000$ and $Ma_\infty = 0.2$, computed on the mesh above. (AUSM, NC)

References

[1] V. Venkatakrishnan, A Perspective on Unstructured Grid Flow Solvers. NASA CR-195025 ICASE Report No. 95-3 ,(1995).

[2] F. Lohmeyer, O. Vornberger: CFD with Adaptive FEM on Massively Parallel Systems. In this publication (1996).

[3] M. Wierse: Higher Order Upwind Schemes on Unstructured Grids for the Nonstationary Compressible Navier-Stokes Equations in Complex Timedependent Geometries in 3-D. In this publication (1996).

[4] C. Roehl, H. Simon: Flow Simulations in Aerodynamically Highly Loaded Turbomaschines Using Unstructured Adaptive Grids. In this publication (1996).

[5] H. Greza, S. Bikker, W. Koschel: Efficient FEM Flow Simulation on Unstructured Adaptive Meshes. In this publication (1996).

[6] R. Vilsmeier, D. Hänel: Adaptive Solutions of the Conservation Equations on Unstructured Grids. Ninth GAMM Conf. on Numerical Methods in Fluid Mechanics, Lausanne (Switzerland), Sept. 1991. In: Notes on Numerical Fluid Mechanics, vol. 35, pp. 321-330, Vieweg Verlag, (1992).

[7] D. Hänel, R. Vilsmeier: Computations of Flow on Adaptive Unstructured Grids. In: E.H. Hirschel (Ed.): Flow Simulation with High Performance Computers I, Notes on Numerical Fluid Mechanics, vol. 38, pp 295-307, Vieweg Verlag, Wiesbaden, (1993).

[8] R. Vilsmeier, D. Hänel: Adaptive Solutions for Compressible Flows on Unstructured, Strongly Anisotropic Grids. In: C. Hirsch, J. Periaux, W. Kordulla (Ed.): Computational Fluid Dynamics '92, vol. II, pp 945-952. Elsevier Science Publisher, Amsterdam, (1992).

[9] R. Vilsmeier, D. Hänel: Adaptive Methods on Unstructured Grids for Euler and Navier-Stokes Equations. Computer & Fluids, vol. 22, pp. 485-499, (1993).

[10] R. Vilsmeier, D. Hänel: Adaptive Solutions for Unsteady Laminar Flows on Unstructured Grids. Int. Journal for Numerical Methods in Fluids, vol. 21, (1995).

[11] D. Hänel, I. Keilhauer, U. Uphoff, R. Vilsmeier: Shock Capturing and Dynamic Grid Adaption for Reactive Flow. Proc. of 6th Int. Symp. on Comp. Fluid Dynamics, Lake Tahoe, USA, (1995).

[12] Vilsmeier R. and Hänel D.: A Field Method for 3-D Tetrahedral Mesh Generation and Adaption. Proc. of 14th Int. Conf. on Num. Meth. in Fluid Dynamics, Bangalore, India (1994).

[13] R. Vilsmeier, D. Hänel: Solutions of the Conservation Equations and Adaptivity on 3-D Unstructured Meshes. Proc. of 9th International Conference on Numerical Methods in Laminar and Turbulent Flow, Atlanta, USA, (1995).

[14] Mavriplis D. J.: Solution of the Two-Dimensional Euler Equations on Unstructured Triangular Meshes, Thesis, Princeton University, (1987).

[15] Schwane, R., Hänel, D.: An Implicit Flux-Vector Splitting Scheme for the Computation of Viscous Hypersonic Flow. AIAA-paper No. 89-0274, (1989).

[16] Harten, A., Lax, D., van Leer, B.: On upstream differencing and Godunov schemes for hyperbolic conservation laws. SIAM Review, vol 25, (1983).

[17] Roe, P.L.: Approximate Rieman Solvers, Parameter Vectors and Difference Schemes. J. Comp. Phys., vol. 22, pp. 357, (1981).

[18] Liou, M. S.: On a new Class of Flux Splitting Schemes. Lecture Notes in Physics, vol 414, pp. 115-119, Springer Verlag Berlin, (1992).

[19] van Leer, B.: Towards the Ultimate Conservative Difference Scheme. A second-order sequel to Godunov's method. J. Comp. Phys. vol.32, pp.101-136, (1979).

[20] N. Maman and B. Larrouturou.: Dynamical mesh adaptation for two-dimensional reactive flow simulations. In: A.S. Arcila, J. Häuser, P.R. Eiseman, J.F. Thompson (Ed.): Numerical Grid Generation in Computational Fluid Dynamics and Related Fields, pp. 55-66, North-Holland, Amsterdam, (1991).

[21] S. Taneda: J. Phys. Soc. Japan, Vol 11, pp 1104-1108, (1956).

FLOW SIMULATIONS IN AERODYNAMICALLY HIGHLY LOADED TURBOMACHINES USING UNSTRUCTURED ADAPTIVE GRIDS

C.Roehl and H.Simon
Institute of Turbomachinery,
University of Duisburg,
47048 Duisburg, Germany

SUMMARY

A method for simulations of the compressible flow in aerodynamically highly loaded turbomachines on unstructured adaptive grids is presented. The discretization of the conservation laws in space is executed by the finite volume method. For high efficiency an implicit solution procedure is used. The flux vector of the inviscid flow is calculated by a modified difference splitting method described by Reichert and Simon [9]. The generation of the initial grid and the grid adaptation are executed with the advancing-front method according to Peraire et al. [14] creating a grid consisting of triangles. The grid adaptation is based on an error estimation. In the case of turbulent flows, quadrilateral grid elements are also used for the efficient resolution of boundary layers. Examples are presented for inviscid and viscous subsonic and transonic flows. A test case for supersonic flow is described in [11].

INTRODUCTION

Numerical flow simulations are becoming more and more important for the analysis of the compressible flow in turbomachines and are frequently used as a powerful tool in the development process of turbomachinery components. Following the development of a program package to simulate such flows with the aid of structured grids [7–10], a program system based on unstructured grids is under investigation to fill the gaps in the application of the structured grid code. In the case of transonic flows in turbomachines, the characteristic lengths of different flow phenomena (shocks, flow separations, secondary flows, etc.) differ by several orders of magnitude and have to be considered in the grid generation method. Especially in such applications unstructured grids can be employed economically owing to their great flexibility. For this reason, a finite volume time stepping scheme on unstructured adaptive grids has been developed.
The first part of the paper deals with the discretization of the conservation equations and the turbulence modelling. In the main part, the method of the grid generation and adaptive regriding is described. Finally, the solutions of test cases are presented to demonstrate the efficiency of the code.

NUMERICAL SOLUTION

The Governing Equations and Turbulence Models

The flow of a perfect gas in an arbitrary fixed domain V with surface S may be described by the conservation equations

$$\int_V \frac{\partial \mathbf{u}}{\partial t} dV = -\oint_S \mathbf{F} d\mathbf{s}, \quad \text{with} \quad \mathbf{u} = \begin{bmatrix} \rho \mathbf{v} \\ \rho \\ u_t \end{bmatrix} \qquad (1)$$

being the vector of the conservative variables volume specific momentum $\rho \mathbf{v}$, volume specific mass ρ and volume specific total energy u_t. The flux tensor $\mathbf{F} = \mathbf{F}^I - \mathbf{F}^V$ describes the convective ($\mathbf{F}^I$) and diffusive ($\mathbf{F}^V$) transport of the conservative variables across the surface S per unit of time t.

For the prediction of turbulent flows, several turbulence models have been considered. First the turbulence models have been applied with the above-mentioned code on structured grids. Some results and the experience gained from these investigations are described by Lenke et al. [7, 8]. Based on this work four turbulence models are implemented, the algebraic model of Baldwin and Lomax [3], the Standard $k - \epsilon$ model of Spalding and Launder [15] with wall-functions, a low-Reynolds number form of the $k - \epsilon$ model devised by Lam and Bremhorst [6] and the $k - \omega$ model [19]. The Baldwin-Lomax model is not applicable for complex flows (e.g. flow separation, recirculation). With this model, lines of nodes ordered in increasing distances from the wall and spanning the boundary layer region are required. By using structured grids these lines are generally available but in unstructured grids some extra work is necessary to find the lines. The standard form of the $k - \epsilon$ model is a powerful tool for the prediction of a wide variety of turbulent flows but it requires the implementation of wall-functions to bridge the viscous sublayers near solid walls. The application of these wall-functions is only valid for attached boundary layers. The low-Reynolds number form of the $k - \epsilon$ model is intended to overcome the problems connected with the introduction of wall-functions. The $k - \omega$ model is able to predict boundary layer transition. The $k - \epsilon$ models and the $k - \omega$ model do not need any knowledge of the grid structure in the near-wall region but they require two extra conservation equations for k and ϵ or ω, respectively.

The Discretization Method

The discretization of the conservation equations in space is executed by the Finite Volume Method. The control volumes are generated by joining the centroids of the cells adjacent to the grid points (node centered scheme). The surface S of the control volume V is divided into a finite number of planes represented by the plane normal vectors $\mathbf{s}^i$ (Figure 1). Using proper averages for the unknown vector $\mathbf{u}$ in the domain (o) and for the flux-tensor $\mathbf{F}$ in the surfaces ($\bullet$) we obtain the ordinary differential equation

$$\frac{d\mathbf{u}}{dt} = -\frac{1}{V} \sum_i \mathbf{f}^{iN} |\mathbf{s}^i| \quad \text{with} \quad \mathbf{f}^{iN} = \mathbf{F}^N \mathbf{s}^i / |\mathbf{s}^i| \qquad (2)$$

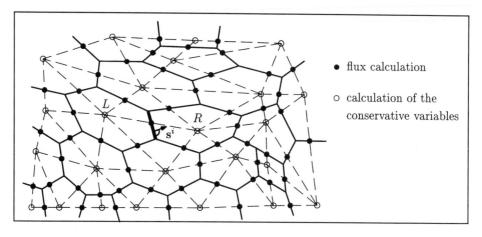

Figure 1: Arrangement of the control volumes.

referred to the numerical flux vector. This is a general finite volume formulation of the Navier-Stokes equation (1). The accuracy of Eqn. (2) depends on the evaluation method for the flux vector.

In our simulations, only steady-state solutions are considered. For high efficiency, an implicit solution procedure is used. A right preconditioned Bi-CGSTAB method introduced by van der Vorst [17], in which the symmetric collective point Gauss-Seidel method is used as a preconditioner, is applied to the discrete conservation equations linearized in time ($\mathbf{Ax} = \mathbf{b}$). In this linear system of equations $\mathbf{A}$, $\mathbf{x}$ and $\mathbf{b}$ represent the coefficient matrix, the rate of change of the conservative variables and the steady operator, respectively. The numerical flux tensor implemented on the left side of this system is evaluated using Roe flux difference splitting [13]. The numerical flux vector $\mathbf{f}^{iN,I}$ implemented on the right side of the system, describing the convective transport of the conservative variables, is computed by the difference splitting method outlined by Reichert and Simon [9]. This method is derived from the methods introduced by Osher and Solomon [12] and Roe [13]. The state vector which is needed at the control volume face for calculating this flux vector depends firstly on the state vectors of the neighbouring control volumes and secondly on the spatial gradients corresponding to the state vectors of the neighbouring control volumes. With these values two interpolation procedures, one interpolation more from the 'left' (L) and one interpolation more from the 'right' (R) supply two different state vectors at the control volume face (w^+, w^-, Figure 2). The interpolation is done using the van Albada limiter [1] or a polinominal fitting. It is carried out in characteristic variables w, which are those components of the vector $\mathbf{u}$, to which known propagation speeds can be related. In this method, one state w^+ or w^- is preferred (upwind).

The flux implemented on the right side of the linear equation system, describing the diffusive transport, is calculated with arithmetically averaged values for the conservative variables.

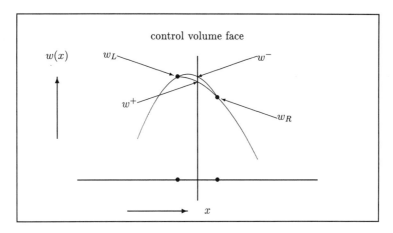

Figure 2: Definition of states used in the flux calculation.

Computation and Limitation of the Spatial Gradients

The spatial gradients of the conservative variables are required for the interpolation procedure in the flux calculation and for the Neumann boundary conditions. The definition of the spatial gradient for a scalar variable u in a volume V with the surface S described by the surface vector **s**

$$\mathbf{grad}\ u = \lim_{V \to 0} \frac{1}{V} \oint_S u_s d\mathbf{s}$$

can be approximated for a small volume V divided into a finite number of planes by

$$\mathbf{grad}\ u \approx \frac{1}{V} \sum_i u_{s_i} \mathbf{s}_i$$

with u_{s_i} being a proper average of the variable u in the surface i. A limitation of the spatial gradients is necessary to avoid the creation of new extrema. The limitation process is based on the method introduced by Barth and Jespersen [4]. For each grid point P the maximum (u_{max}) and the minimum (u_{min}) values of the conservative variables u of the inspected grid point P and its neighbouring points are computed. For the places M_i of the flux calculations surrounding a grid point P values u_{M_i} are calculated with the conservative variables u_P and the spatial gradient $\mathbf{grad}\ u_P$ appertaining to point P to define limiting values φ_i at the same places (Figure 3):

$$\varphi_i = \begin{cases} \min\left(1, \frac{u_{max}-u_P}{u_{M_i}-u_P}\right) & \text{if } u_{M_i} > u_P \\ \min\left(1, \frac{u_{min}-u_P}{u_{M_i}-u_P}\right) & \text{if } u_{M_i} < u_P \\ 1 & \text{if } u_{M_i} = u_P \end{cases}$$

The limiting value φ_{limit} for the spatial gradient of a conservative variable of a grid point P is calculated by

$$\varphi_{limit} = \min(\varphi_i) \qquad i = 1, ..., \text{maxpt}$$

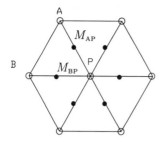 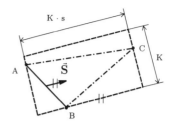

Figure 3: Limitation of the spatial gradients of the conservative variables.

Figure 4: Definition of a new grid point C.

with maxpt being the number of neighbouring grid points of point P. Finally the limited spatial gradient $\mathbf{grad}u_P^{new}$ of a conservative variable u is set to

$$\mathbf{grad}u_P^{new} = \varphi_{limit} \cdot \mathbf{grad}u_P^{old} \qquad 0 \leq \varphi_{limit} \leq 1$$

with $\mathbf{grad}u_P^{old}$ the old gradient and φ_{limit} the limiting value.

MESH GENERATION AND ADAPTATION

The generation of the initial grid and the grid adaptation are coupled to CAD software. Accordingly, the geometry is analytically described and the grid points can be defined exactly on the boundary surfaces. Both for the generation of the initial grid and for the grid adaptation the same algorithm is used which is based on the advancing-front method according to Peraire et al. [14]. In both cases a *new* generation of the grid is accomplished consisting of *triangles*. The definition of new grid points is derived from three characteristic grid parameters of the background grid:

- element size K
- stretching vector $\vec{S}$
- stretching ratio $s = |\vec{S}|$.

With these grid parameters a rectangle for a point connection A-B can be designed to define the location of the third point C of a triangle ABC (Figure 4). The orientation of one side of the rectangle is fixed by the stretching vector. The second side of the rectangle is of course normal to the element stretching vector. The lengths of these sides are defined by the element size K and the element stretching ratio s. The grid parameters are differently calculated depending on generating the initial grid or adapting a grid.
By generating the initial grid, the stretching vector is chosen normal to the point connection A-B. The stretching ratio is set to one. The element size is approximately set to the distance from point A to point B. By this way a smooth unstretched grid is created.

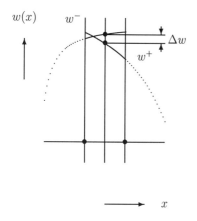
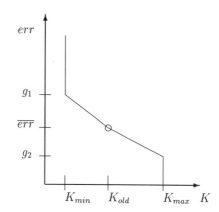

Figure 5: Interpolation procedure at the control volume face.

Figure 6: Transformation of the estimated error err into the element size K.

In the case of grid adaptation the element size is calculated by an estimation of the error in connection with the interpolation procedure for the characteristic variables from the control volume centres to the control volume face for calculating the flux. This interpolation procedure is carried out twice by using the difference splitting method, one interpolation more from the 'left' and one interpolation more from the 'right' (Figure 5). Two interpolated states are obtained at the control volume face (w^+, w^-). In the flow simulation procedure one state is preferred. In the grid adaptation the difference Δw of the two interpolated states is used after a transformation into the conservative variables and a certain flux is calculated. The estimated error corresponds to the total value of the dimensionless solution vector of the conservation law stored for each grid point. This error is transformed into the element size K.

The purpose of a grid adaptation is to obtain a uniformity of the error in the whole grid. Hence the arithmetic average $\overline{err}$ of the estimated errors and the standard deviation

$$\sigma = \sqrt{\frac{1}{maxpt} \sum_{i=1}^{maxpt} (\overline{err} - err_i)^2}$$

for all grid points $maxpt$ are inspected. Additionally four thresholds are defined for the transformation:

$$g_1 = \overline{err} + \sigma, \qquad g_2 \approx 0.5 \cdot \overline{err}, \qquad K_{min} \qquad \text{and } K_{max}.$$

For all grid points with estimated errors err greater than g_1, the element sizes K are minimized to the threshold K_{min} defined by the user. For all grid points with estimated errors err smaller than g_2, the element sizes K are maximized to the threshold K_{max} defined by the user too. For uniformity of the error, the element sizes of the grid points whose estimated errors err are equal to the arithmetic average $\overline{err}$ are not changed. Between the mentioned values a linear distribution is done (Figure 6).

The fundamentals of the calculation of the stretching vector are the spatial gradients

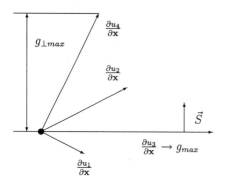
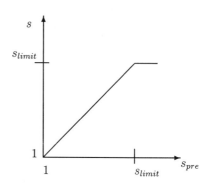

Figure 7: Local gradients $\frac{\partial u_i}{\partial \mathbf{x}}$ of the vector **u**.

Figure 8: Calculation of the stretching ratio s.

$\frac{\partial u_i}{\partial \mathbf{x}}$ of the unknown vector **u** in dimensionless form. For each grid point a gradient is computed for each component i of the vector **u**. In two-dimensional flows, four gradients result with possibly different vectors (Figure 7). The gradient with the greatest total value is determined:

$$g_{max} = \left| \frac{\partial u_i}{\partial \mathbf{x}} \right|_{max}.$$

The direction of the stretching vector $\vec{S}$ is set to be normal to this gradient:

$$\vec{S} \perp \left(\frac{\partial u_i}{\partial \mathbf{x}} \right)_{max}.$$

The modulus $|\vec{S}|$ of the stretching vector $\vec{S}$ is the stretching ratio s. For calculating this value also the greatest component of the local gradients in the direction of $\vec{S}$, $g_{\perp max}$, is considered. The stretching ratio s is preliminarily computed by

$$s_{pre} = \frac{g_{max}}{g_{\perp max}}.$$

As a result of this, large stretching is avoided in regions with different directions of the spatial gradients. If the denominator ($g_{\perp max}$) approaches zero, a stretching limitation is applied by the value s_{limit} to avoid infinite stretching. For infinitesimal gradients the stretching ratio is limited to one. With the values s_{limit}, s_{pre} and $s = 1$, the stretching ratio is calculated as shown in Figure 8.

A smoothing process can be executed for the grid parameters stored at the grid points of the so-called background grid. In this method the grid parameters of the inspected grid point P and its neighbouring points are considered to compute new grid parameters for P. The new element size for P is set to be the minimum of firstly the arithmetic value $\overline{K}$ calculated by the element sizes of P and its neighbouring points and secondly the element size of grid point P before smoothing K_{old} ($K_{new} = \min(\overline{K}, K_{old})$). The smoothing process for the element stretching vector $\vec{S}$ uses informations of the same grid points. For this, it is necessary that the angles between the stretching vector of P and those of the neighbouring points are smaller than $|90°|$. The new element stretching vector for P is set to be equal to the arithmetic average computed by the element stretching vectors of the inspected grid points. The new element stretching ratio s is the total value of the new element stretching vector.

Grid Smoothing

Owing to the application of the advancing front method, a grid smoothing procedure is necessary to decrease skewness. Using a Laplacian filter, the grid points of the triangular grid are moved N times inside the area of their surrounding triangles with the following equation taken from [18]:

$$\mathbf{x}_P^{new} = \mathbf{x}_P^{old} + \frac{\omega}{m} \sum_{i=1}^{m} \left[\mathbf{x}_{P_i}^{old} - \mathbf{x}_P^{old} \right]$$

where $\mathbf{x}_P^{old}$ is the old position of point P, $\mathbf{x}_P^{new}$ is the new position of point P, m is the number of neighbouring points P_i ($i = 1, ..., m$) with coordinates $\mathbf{x}_{P_i}^{old}$ and ω is the relaxation parameter. A good choice of N is 20 and of ω 0.2.

Grid Generation for Turbulent Flows

For the efficient resolution of boundary layers in particular for turbulent flows quadrilateral elements are also used in the grid. This grid generation procedure is also coupled to CAD software. The domain of the region to be computed is divided into well-defined subregions. For each subregion, signed for a triangular or quadrilateral grid, a grid generation takes place. Finally, the single grids are combined to one unstructured grid.

The application of quadrilateral elements in the boundary layer region has additional advantages. Firstly, the total number of places where fluxes are calculated is reduced. Secondly, the introduction of the Neumann boundary conditions is relatively simple. Using triangles, the problem arises which grid points are qualified to give informations to the boundary points.

NUMERICAL RESULTS

Inviscid Flow Through the Plane Turbine Cascade SE 1050

To demonstrate the efficiency of the adaptation method, the inviscid flow through the plane turbine cascade SE 1050 [16] is chosen. Figure 9 shows the initial grid and two adapted grids. A special grid configuration is applied which is similar to a combination of H- and C-grids for structured grids. This grid type gives a good resolution in the vicinity of the trailing edge. The respective results of the flow simulations represented as lines of constant Mach numbers are also shown in Figure 9. The inlet Mach number is about 0.3. The flow is accelerated to a Mach number of 1.3 in the outlet region. At the trailing edge on the pressure side of the profile a compression shock is formed which is reflected at the suction side. The resolution of the shock is very poor for the initial grid. In a first adaptation, grid points are concentrated in the region of the shock from the pressure to the suction side of the profile. On the other hand, the total number of grid points is reduced. The stretching ratio for the triangles is limited to 10. In a second adaptation the resolution of the shock is better owing to the concentration of grid points in the shock region. A comparison of the inviscid flow simulation represented by the density distribution with the interferometric picture taken from [16] shows already a good agreement (Figure 10).

Laminar and Turbulent Flow Over a Flat Plate

For the validation of viscous flow calculations, the boundary layer of a flat plate has been simulated. In case of the laminar flow a triangular grid with about 1500 grid points is used (Figure 11a). The Reynolds number is chosen as 1000 and the inlet Mach number is set to 0.2. The comparison of the calculated velocity profile with the Blasius solution [5] shows a good agreement (Figure 11b).

The first tests for the implemented turbulence models are done for turbulent flow over a flat plate. In this case a combined grid is used (Figure 12a). In the region of the boundary layer quadrilaterals are placed. The total number of grid points is about 2000. The Reynolds number is set to 10^6. The comparison of the solution with the wall-function (Figure 12b) shows that the elementary Baldwin-Lomax turbulence model produces the best results for this simple flow case.

Transonic Flow Through a Radial Inflow Turbine Guide Vane

For the simulation of the viscous flow through a component of a turbomachine a radial inflow turbine guide vane designed by Reichert and Simon [10] is chosen. In Figure 13 a computational grid consisting of about 4000 grid points is shown. The grid consists of quadrilaterals in the region of the boundary layers and of triangles in the remaining part. Simulations are done for the Baldwin-Lomax, Lam-Bremhorst and k-ω models. The Reynolds number is set to $5 \cdot 10^5$. The respective results are shown in Figure 14 as lines of constant Mach numbers. The results differ mainly in the region of the trailing edge and in the boundary layer thickness.

CONCLUSION

A method for flow simulations on unstructured adaptive grids has been described. In addition to some basic examples, applications to turbomachinery components are also presented for inviscid and viscous subsonic and transonic flows. In conjunction with

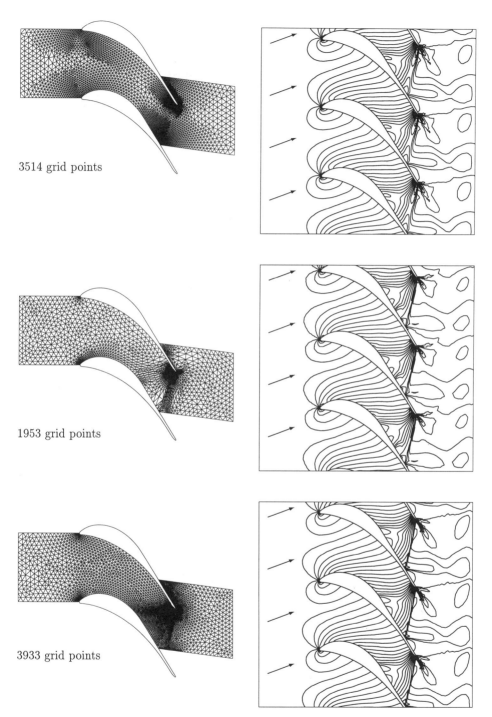

Figure 9: Sequence of grids for the simulation of the flow through SE 1050 turbine cascade and Mach number contours for their computed solutions [11].

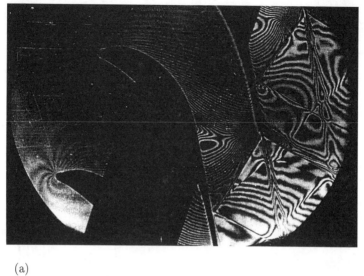

(a)

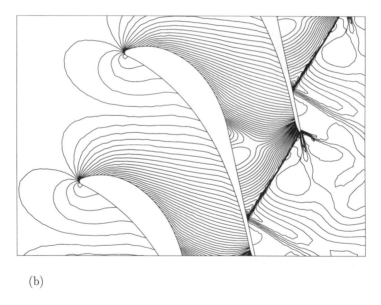

(b)

Figure 10: (a) Interferometric picture [16] and (b) density distribution from the simulation of the inviscid flow through SE 1050 turbine cascade (3933 grid points).

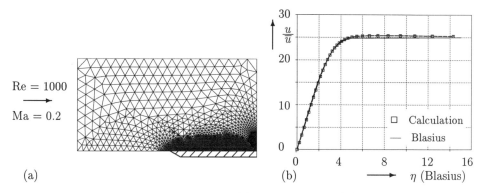

Figure 11: Laminar flow over a flat plate: (a) triangular grid; (b) velocity profile at the end of the plate in coordinates defined by Blasius; with u being the component of the velocity parallel to the plate, $\bar{u}$ the averaged inlet velocity and η a specific wall distance.

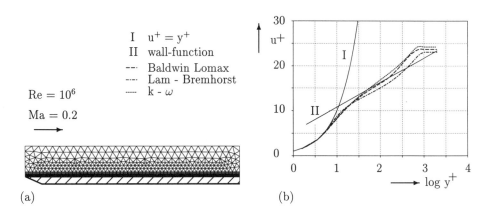

Figure 12: Turbulent flow over a flat plate. Influence of the different turbulence models: (a) combined grid; (b) dimensionless velocity profiles compared with the wall-function.

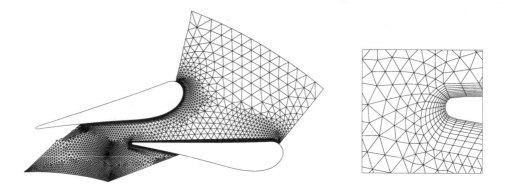

Figure 13: Computational grid for the flow through a radial inflow turbine guide vane. View on the trailing edge.

further test cases, it can be concluded that the adaptation method establishes very good results especially for flows with large gradients, e.g. transonic or supersonic flow with compression shocks. On the other hand, the method has to be improved in boundary layer regions. As a first step, this problem has been solved by the introduction of quadrilaterals in the grid of these regions to ensure an efficient resolution of the boundary layer especially for turbulent flows. Additional work is necessary to optimize the grid adaptation of the so-called combined grids.

It may be concluded that nowadays such CFD codes are qualified to analyse the complex flow in turbomachinery components. The codes are able to consider the main effects of the three-dimensional turbulent transonic flow even in the rotating parts of turbomachinery [2]. Further research is proposed to improve the accuracy of the computational results, especially regarding transition from laminar to turbulent flow, flow separation and overall flow losses. Nevertheless the codes can be applied efficiently during an early stage of the design process and provide the designer with some guidelines to improve the performance of the turbomachine in question.

REFERENCES

[1] VAN ALBADA, G.D., VAN LEER, B., ROBERTS, W.W.: *A Comparative Study of Computational Methods in Cosmic Gas Dynamics.* Astronomy and Astrophysic, **108** (1982).

[2] AMEDICK, V., SIMON, H.: *Numerical Simulation of the Three-Dimensional Turbulent Flow in a Turbine Rotor with Conical Walls.* 6th International Symposium on Transport Phenomena and Dynamics of Rotating Machinery (ISROMAC-6), Honolulu, Hawaii(1996).

[3] BALDWIN, B.S., LOMAX, H.: *Thin Layer Approximations and Algebraic Model for Separated Turbulent Flow.* AIAA Paper 78–257 (1978).

[4] BARTH, T.J., JESPERSEN, D.C.: *The Design and Application of Upwind Schemes on*

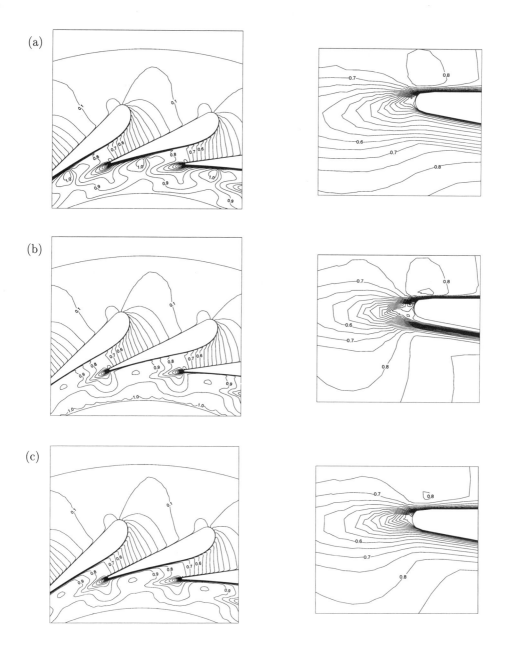

Figure 14: Turbulent flow through a radial inflow turbine guide vane. Mach number distribution with (a) Baldwin-Lomax model, (b) Lam-Bremhorst model and (c) $k - \omega$ model. View on the trailing edge.

Unstructured Meshes. AIAA Paper 89–0366 (1989).

[5] BLASIUS, H.: *Grenzschichten in Fluessigkeiten mit kleiner Reibung.* Z. Math. Phys. **56** (1908) 1–37

[6] LAM, C.K.G., BREMHORST, K.: *A Modified Form of the $k - \epsilon$-Model for Predicting Wall Turbulence.* Journal of Fluids Engineering, **103** (1981).

[7] LENKE, L. J., REICHERT, A. W., SIMON, H.: *Comparison of Viscous Flow Field Calculations Applicable to Turbomachinery Design with Experimental Measurements.* Seventh International Conference on Computational Methods and Experimental Measurements CMEM 95 (1995).

[8] LENKE, L. J., REICHERT, A. W., SIMON, H.: *Viscous Flow Field Calculations for the VKI-1 Turbine Cascade Using Different Turbulence Models.* ASME Paper 95–GT–91 (1995).

[9] REICHERT, A. W., SIMON, H.: *Numerical Investigations to the Optimum Design of Radial Inflow Turbine Guide Vanes.* ASME Paper 94–GT–61 (1994).

[10] REICHERT, A. W., SIMON, H.: *Design and Flow Field Calculations for Transonic and Supersonic Radial Inflow Turbine Guide Vanes.* ASME Paper 95–GT–97 (1995).

[11] ROEHL, C., SIMON, H.: *Simulations of Transonic Flows in Turbomachines Using a Novel Adaptation Method with Unstructured Grids.* 6th ISCFD Lake Tahoe, Nevada (1995).

[12] OSHER, S., SOLOMON, F.: *Upwind Schemes for Hyperbolic Systems of Conservation Laws.* Mathematics of Computations **38** (1982) 339-377.

[13] ROE, P.L.: *Approximate Riemann Solvers, Parameter Vectors and Difference Schemes.* Journal of Computational Physics **43** (1981) 357-372.

[14] PERAIRE, J., VAHADI, M., MORGAN, K., ZIENKIEWICZ, O.C.: *Adaptive Remeshing for Compressible Flow Computations.* Journal of Computational Physics **72** (1987).

[15] SPALDING, D. B., LAUNDER, B. E.: *Turbulence Models and Their Application to the Prediction of Internal Flows.* Heat and Fluid Flow **2** (1972).

[16] ŠŤASTNÝ, M., ŠAFAŘÍK, P.: *Experimental Analysis Data on the Transonic Flow Past a Plane Turbine Cascade.* ASME Paper 90–GT–313 (1990).

[17] VAN DER VORST, H. A.: *Bi-CGSTAB, a Fast and Smoothly Converging Variant of Bi-CG for the Solution of Nonsymmetric Linear Systems.* SIAM Journal Sci. Statist. Comput. **13** (1992) 631–644.

[18] WEATHERILL, N. P.: *Grid Generation.* Numerical Grid Generation, von Karman Institute for Fluid Dynamics, Lecture Series 1990-06, Rhode Saint Genèse, Belgium (1990).

[19] WILCOX, D. C.: *A Half Century Historical Review of the k-ω Model.* AIAA- 91- 0615, Reno, Nevada (1991).

FLOW SIMULATION IN A HIGH-LOADED RADIAL COMPRESSOR

W. Evers, M. Heinrich, I. Teipel and A. R. Wiedermann
University of Hannover, Institute for Mechanics
Appelstr. 11, D-30167 Hannover, Germany

SUMMARY

A two- and three-dimensional Euler and Navier-Stokes code has been developed and successfully used for the computation of the flow field in a high loaded centrifugal compressor. For the purpose of comparison, the algebraic turbulence model of Baldwin and Lomax, a modification of the Baldwin-Lomax model with an extension according to Goldberg and Chakravarthy [1] for the determination of separated flow regions and the two-equation $k - \epsilon$ model according to Kunz and Lakshminarayana [2, 3] are applied to simulate the flow field of the diffuser with the two-dimensional Navier-Stokes code. The three-dimensional solver in addition to the extended and original Baldwin-Lomax model has been applied to obtain the three-dimensional flow field of the diffuser and the impeller.

NOMENCLATURE

$\mathcal{D}, \mathcal{E}$	=	functions of the $k - \epsilon$ model
e	=	specific internal energy
E_{rot}	=	relative total specific internal energy
$\vec{E}$	=	flux vector in direction of the curvilinear coordinates
f_2, f_μ	=	functions of the $k - \epsilon$ model
F_{Kleb}	=	Klebanoff intermittency function
F_{Wake}	=	wake function
G	=	Gaussian distribution
J	=	Jacobian
k	=	turbulent kinetic energy $(= k^\star/(p/\rho)^\star_{tot,\infty})$
l	=	length scale of the Baldwin-Lomax model
n	=	wall distance
p	=	static pressure $(= p^\star/p^\star_{tot,\infty})$
P	=	production rate of k
Pr	=	Prandtl number
q_i	=	Cartesian component of heat transfer
$\vec{Q}$	=	vector of variables of state
r	=	radius $(= r^\star/r^\star_{DE})$
Re	=	Reynolds number $(= r^\star_{DE}\sqrt{(p\rho)^\star_{tot,\infty}}/\mu^\star_{l,\infty})$
R_T	=	local Reynolds number
$\vec{S}$	=	source term vector
t	=	time $(= t^\star\sqrt{(p/\rho)^\star_{tot,\infty}}/r^\star_{DE})$
u	=	relative velocity in x-direction, $u = u_1$ $(= u^\star/\sqrt{(p/\rho)^\star_{tot,\infty}})$
u_s	=	wall friction velocity, velocity scale

U_i	=	contravariant velocities
v, w	=	relative velocities in y and z direction, $v = u_2$, $w = u_3$ (see u)
$\vec{w}$	=	velocity vector
x, y, z	=	relative Cartesian coordinates, $x = x_1$, $y = x_2$, $z = x_3$ ($x_i^\star / r_{DE}^\star$)
δ_{ij}	=	Kronecker delta
ϵ	=	dissipation rate of k ($= \epsilon^\star r_{DE}^\star / \left[(p/\rho)_{tot,\infty}^\star \right]^{1.5}$)
κ	=	isentropic coefficient
μ_l	=	dynamic viscosity ($= \mu_l^\star / \mu_{l,\infty}^\star$)
μ_t	=	turbulent viscosity
ξ_i	=	generalized curvilinear coordinates
ρ	=	density ($= \rho^\star / \rho_{tot,\infty}^\star$)
τ_{ij}	=	Cartesian stress tensor component
τ_w	=	wall shear stress
Ψ	=	function in Eqn. (1)
ω	=	vorticity scale
Ω	=	angular velocity of relative frame of reference ($= \Omega^\star r_{DE}^\star / \sqrt{(p/\rho)_{tot,\infty}^\star}$)

Superscripts and subscripts

$\sim$	=	density weighted value
$-$	=	time averaged value
$+$	=	modified value
$\star$	=	dimensionalized value
a, b, BL	=	layer pointer in the algebraic turbulence models
c	=	inviscid
DE	=	diffuser exit
i	=	layer pointer in the algebraic turbulence models
i, j, k	=	axis pointer
K	=	suction duct upstream the compressor
o	=	layer pointer in the algebraic turbulence models
tot	=	total value
w	=	wall
ν	=	viscous
0	=	impeller exit
∞	=	diffuser inlet

INTRODUCTION

A highly vectorized code solving the Reynolds-averaged Navier-Stokes equations and the Euler equations for compressible fluids has been developed on a Siemens-Fujitsu S400/40 vector-computer with 5 GigaFLOPS peak performance. The solver has been applied for the simulation of steady flow fields in a centrifugal compressor. When simulating the flow field in high-loaded turbomachinery by means of the Navier-Stokes equations, it is necessary to take the turbulence into account. Ever increasing computer capacities give rise to more and more complicated turbulence models, e.g. Reynolds stress models. These extensive models are, of course, more accurate than algebraic, first-order or simple second-order closure models, but in return the amount of computation time required sometimes increases too much, especially when three-dimensional flow problems are to be calculated. Therefore, the simple methods are still a good and fast choice for many practical tasks [4] and they have been used in this work. One problem that restricts the application of most algebraic models is the occurrence of backflow regions. Goldberg and Chakravarthy [1] introduced an extension to the model of Baldwin and Lomax for the prediction of separation bubbles. This modification was successfully applied by Compton et. al. [5] to an external afterbody flow with jet exhaust.

In this paper, the modified algebraic model is used for 2D simulations of the steady internal flow field in an aerodynamic diffuser of a high-loaded radial compressor. The computations are compared with 2D simulations using the two-equation $k - \epsilon$ turbulence model of Kunz and Lakshminarayana and the original Baldwin-Lomax model.

The developed Euler solver and the 3D version of the Navier-Stokes code with the algebraic turbulence models have been used for the simulation of the 3D flow fields in the diffuser and the impeller of the radial compressor. To validate the results of this study, distributions of the static pressure have been compared with experimental data given by Jansen [6]. Further results from this project are published in [7, 8]. An article by Roehl and Simon [9], dealing with a related topic, is also presented in this NNFM volume.

DESCRIPTION OF THE NUMERICAL SCHEME

The developed procedure is an explicit central-finite-difference time-integration scheme. It is first-order accurate in time and second-order accurate in space. The Reynolds-averaged equations (1) are formulated in a steadily rotating relative frame of reference in normalized and conservative manner [10]. The system of flow equations is transformed from the Cartesian coordinate system ($x = x_1, y = x_2, z = x_3$) to a generalized curvilinear system (ξ_1, ξ_2, ξ_3). The angular velocity of the relative frame of reference is directed in the z-axis and has the value Ω.

$$\frac{\partial \vec{Q}}{\partial t} + \frac{\partial \vec{E}_{c,i}}{\partial \xi_i} = \frac{\partial \vec{E}_{\nu,i}}{\partial \xi_i} + \vec{S} \tag{1}$$

with

$$\vec{Q} = \frac{1}{J} \begin{bmatrix} \overline{\rho} \\ \overline{\rho}\tilde{u} \\ \overline{\rho}\tilde{v} \\ \overline{\rho}\tilde{w} \\ \overline{\rho}\tilde{E}_{rot} \end{bmatrix} \quad ; \quad \vec{E}_{c,i} = \frac{1}{J} \begin{bmatrix} \overline{\rho}\tilde{U}_i \\ \overline{\rho}\tilde{u}\tilde{U}_i + \frac{\partial \xi_i}{\partial x}\overline{p} \\ \overline{\rho}\tilde{v}\tilde{U}_i + \frac{\partial \xi_i}{\partial y}\overline{p} \\ \overline{\rho}\tilde{w}\tilde{U}_i + \frac{\partial \xi_i}{\partial z}\overline{p} \\ (\overline{\rho}\tilde{E}_{rot} + \overline{p})\tilde{U}_i \end{bmatrix} \quad ; \quad \tilde{E}_{rot} = \tilde{e} + \frac{1}{2}\left[\vec{\tilde{w}} \cdot \vec{\tilde{w}} - \Omega^2(x^2 + y^2)\right] \quad ;$$

$$\vec{E}_{\nu,i} = \frac{1}{Re}\frac{1}{J} \begin{bmatrix} 0 \\ \frac{\partial \xi_i}{\partial x}\tau_{11} + \frac{\partial \xi_i}{\partial y}\tau_{12} + \frac{\partial \xi_i}{\partial z}\tau_{13} \\ \frac{\partial \xi_i}{\partial x}\tau_{21} + \frac{\partial \xi_i}{\partial y}\tau_{22} + \frac{\partial \xi_i}{\partial z}\tau_{23} \\ \frac{\partial \xi_i}{\partial x}\tau_{31} + \frac{\partial \xi_i}{\partial y}\tau_{32} + \frac{\partial \xi_i}{\partial z}\tau_{33} \\ \frac{\partial \xi_i}{\partial x}\Psi_1 + \frac{\partial \xi_i}{\partial y}\Psi_2 + \frac{\partial \xi_i}{\partial z}\Psi_3 \end{bmatrix} \quad ; \quad \vec{S} = \frac{1}{J} \begin{bmatrix} 0 \\ \overline{\rho}\Omega(\Omega x + 2\tilde{v}) \\ \overline{\rho}\Omega(\Omega y - 2\tilde{u}) \\ 0 \\ 0 \end{bmatrix} \quad ;$$

$$J = \left|\frac{\partial(\xi_1, \xi_2, \xi_3)}{\partial(x, y, z)}\right| \quad ; \quad \tilde{U}_i = \frac{\partial \xi_i}{\partial x}\tilde{u} + \frac{\partial \xi_i}{\partial y}\tilde{v} + \frac{\partial \xi_i}{\partial z}\tilde{w} \quad ; \quad \Psi_i = \tilde{u}\tau_{i1} + \tilde{v}\tau_{i2} + \tilde{w}\tau_{i3} - q_i \quad ;$$

$$\tau_{ij} = (\mu_t + \mu_l)\left[\left(\frac{\partial \tilde{u}_i}{\partial x_j} + \frac{\partial \tilde{u}_j}{\partial x_i}\right) - \frac{2}{3}\delta_{ij}\frac{\partial \tilde{u}_k}{\partial x_k}\right] - \frac{2}{3}\delta_{ij}Re\,\bar{\rho}\,\tilde{k}$$

and

$$q_i = -\frac{\kappa}{\kappa - 1}\left(\frac{\mu_l}{Pr} + \frac{\mu_t}{Pr_t}\right)\frac{\partial}{\partial x_i}\left(\frac{\bar{p}}{\bar{\rho}}\right).$$

The dynamic viscosity coefficient μ_l is predicted by Sutherland's law [10] and the value of the static pressure is evaluated from the equation of state for a perfect gas:

$$\bar{p} = (\kappa - 1)\bar{\rho}\tilde{e}. \qquad (2)$$

In the case of Euler applications, the viscous fluxes $\vec{E}_{\nu,i}$ are omitted.

The discretized difference equations are solved by a hybrid multistep Runge-Kutta scheme [11]. To prevent the numerical solution from oscillations, especially in the vicinity of shocks and from high-order oscillation modes, a second- and fourth-order artificial dissipation [12] is introduced. To speed up the convergence of the scheme, the following acceleration techniques are employed: local time-stepping (2D and 3D code), implicit smoothing of the residuals with constant [13] or weighted coefficients [14] (2D and 3D solver) and a multigrid method [14, 15] (only 3D applications). Furthermore, the original five-step Runge-Kutta scheme has been reduced to a three-step scheme according to Leicher [16].

TURBULENCE MODELS

In solving the Reynolds-averaged Navier-Stokes equations, it is necessary to introduce a turbulence model to determine the eddy viscosity and the turbulent viscosity coefficient μ_t in the viscous flux vectors of Eqn. (1). Three different models have been applied, as follows.

The Baldwin-Lomax Model

The Baldwin-Lomax model [17] is one of the best known algebraic models. According to Cebeci [18], the boundary layer is divided up into an inner and an outer layer.

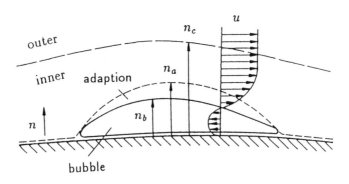

Fig. 1: Structure of a bubble along a wall.

Length and velocity scales of the eddy viscosity are evaluated directly by means of the flow data and the coordinates of the determined flow point. To predict the turbulent viscosity coefficient $\mu_t = \mu_{t,i}$ in the inner layer, Eqn. (3) is applied with $n_b \equiv 0$ (see Fig. 1):

$$\mu_{t,i} = Re \, \bar{\rho} \, \ell^2 \, |\omega| \tag{3}$$

using

$$\ell = k \, n_{BL} \left[1 - exp\left(\frac{-n^+}{A^+}\right)\right] \quad ; \quad |\omega| = \sqrt{\left(\frac{\partial \tilde{v}}{\partial x} - \frac{\partial \tilde{u}}{\partial y}\right)^2} \quad ;$$

$$n^+ = \left(\frac{\sqrt{Re \, \bar{\rho}_w \, \tau_w}}{\mu_w}\right) n_{BL} \quad and \quad n_{BL} = n - n_b \quad .$$

In the outer layer $\mu_t = \mu_{t,o}$ is calculated by means of Eqn. (4); n_{BL} and n^+ are defined in Eqn. (3) and $\vec{w}_{diff}$ is the velocity difference across the boundary layer:

$$\mu_{t,o} = Re \, K \, C_{cp} \, \bar{\rho} \, F_{wake} \, F_{Kleb} \tag{4}$$

with

$$F_{wake} = min \begin{cases} n_{BL}|_{Fmax} \, F|_{Fmax} \\ C_{wk} \, n_{BL}|_{Fmax} \, \vec{w}_{diff}^2 / \, F|_{Fmax} \end{cases} \quad ;$$

$$F = n_{BL}|\omega|\left[1 - exp\left(\frac{-n^+}{A^+}\right)\right] \quad ; \quad F_{Kleb} = \left[1 + 5.5 \left(\frac{n_{BL} \, C_{Kleb}}{n_{BL}|_{Fmax}}\right)^6\right]^{-1} \quad ;$$

$A^+ = 26$; $C_{cp} = 1.6$; $C_{Kleb} = 0.3$; $C_{wk} = 0.25$; $k = 0.4$ and $K = 0.0168$.

Modification by Goldberg and Chakravarthy

Goldberg and Chakravarthy have extended the original model of Baldwin and Lomax by adding two new layers to the inner and outer layers in the case of a separated flow region: a backflow zone (bubble) and an adaption layer. They suggested a Gaussian distribution for the turbulent viscosity in the bubble region [1] and employed the adaption layer to ensure a smooth transition of the eddy viscosity from the backflow regime to the inner layer [5]. The surface pointing to the inner flow in which no velocity parallel to the surface occurs (outer bubble surface) is treated as a reference wall for the Baldwin-Lomax model application according to Compton et. al. [5]. The viscosity coefficients $\mu_t = \mu_{t,i}$ in the inner layer and $\mu_t = \mu_{t,o}$ in the outer layer are evaluated by means of Eqns. (3) and (4), but now n_b represents the distance of the outer bubble surface from the wall (see Fig. 1). In the zone of the recirculating flow Eqn. (5) is applied:

$$\mu_{t,b} = C_1 \, u_s \, n_b \, \sqrt{\bar{\rho}_w \, \bar{\rho} \, Re} \, \left(A\left(\frac{n}{n_b}\right) + B\right) \sqrt{G(n)} \tag{5}$$

making use of

$$u_s = \sqrt{\frac{\omega|_{\omega_{max}} \, \mu|_{\omega_{max}}}{\rho|_{\omega_{max}}}} \quad ; \quad G(n) = \frac{1 - exp\left[-\Phi\left(\frac{n}{n_b}\right)^2\right]}{1 - exp[-\Phi]} \quad ;$$

$A = -0.151$; $B = 0.684$; $C_1 = 0.353$ and $\Phi = 0.5$.

To realize an adaption from $\mu_{t,b}$ to $\mu_{t,i}$ the values $\mu_{t,i}$, which are smaller than $\mu_{t,a}$ according to Eqn. (6) are replaced by $\mu_{t,a}$ as long as the first value of $\mu_{t,i}$ is larger than $\mu_{t,a}$:

$$\mu_{t,a} = C_1\, u_s\, n_b\, \sqrt{\overline{\rho_w}\,\overline{\rho}\, Re}\, (A+B)\, \sqrt{G(n_b)} \quad . \tag{6}$$

Thus, along one wall of a channel the viscous coefficient μ_t follows Eqn. (7):

$$\begin{array}{llllll}
\mu_t = \mu_{t,b} & \text{while} & 0 & \le n < & n_b & (\text{bubble}) \;, \\
\mu_t = \mu_{t,a} & \text{while} & n_b & \le n < & n_a & (\text{adaption layer}) \;, \\
\mu_t = \mu_{t,i} & \text{while} & n_a & \le n < & n_c & (\text{inner layer}) \;, \\
\mu_t = \mu_{t,o} & \text{while} & n_c & \le n & & (\text{outer layer}) \;.
\end{array} \tag{7}$$

3D Approach of the Algebraic Models

For the computation of multi-dimensional flow structures, it is necessary to combine the μ_t values related to the surrounding walls in a suitable manner. In this paper the resulting coefficient to each point in the flow field is determined by weighting the wall-related μ_t values with the inverse-related wall distances and summing up all values of the 'smallest' related layer type. For the four points in Fig. 2 this leads to Eqns. (8) - (11).

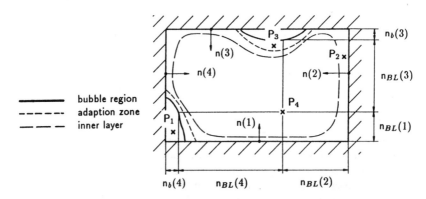

Fig. 2: Section perpendicular to the main flow of a channel with a separation region.

Point P_1:
$$\mu_t = \frac{\dfrac{\mu_{t,b}(1)}{n(1)} + \dfrac{\mu_{t,b}(4)}{n(4)}}{\left(\dfrac{1}{n(1)} + \dfrac{1}{n(4)}\right)} \quad . \tag{8}$$

Point P_2:
$$\mu_t = \frac{\sum\limits_{i=1}^{3} \dfrac{\mu_{t,i}(i)}{n(i)}}{\sum\limits_{i=1}^{3} \dfrac{1}{n(i)}} \quad . \tag{9}$$

Point P_3:
$$\mu_t = \mu_{t,a}(3) \quad . \tag{10}$$

Point P_4:

$$\mu_t = \frac{\sum\limits_{i=1}^{4} \frac{\mu_{t,o}(i)}{n_{BL}(i)}}{\sum\limits_{i=1}^{4} \frac{1}{n_{BL}(i)}} \quad . \tag{11}$$

The vorticity scale used for the prediction of μ_t in the 3D approach is given by the following expression:

$$|\omega| = \sqrt{\left(\frac{\partial \tilde{w}}{\partial y} - \frac{\partial \tilde{v}}{\partial z}\right)^2 + \left(\frac{\partial \tilde{u}}{\partial z} - \frac{\partial \tilde{w}}{\partial x}\right)^2 + \left(\frac{\partial \tilde{v}}{\partial x} - \frac{\partial \tilde{u}}{\partial y}\right)^2} \quad . \tag{12}$$

$k - \epsilon$ Model

The model employed was developed by Kunz and Lakshminarayana [2]. It takes account of compressibility effects in a restricted manner only. Density fluctuations and pressure diffusion terms are neglected in the transport equation of the kinetic turbulent energy k and the equation for the turbulent dissipation rate ϵ. This model has been applied successfully in the computation of turbomachinery flows [3]. To ensure a good vectorization level, the turbulence transport equations (13) are solved in combination with the other flow equations (1). In Eqn. (13) the maximum value of pointer i is $i = 2$, because in this work the model has been used only for the simulation of two-dimensional flow fields:

$$\frac{\partial \vec{Q}}{\partial t} + \frac{\partial \vec{E}_{c,i}}{\partial \xi_i} = \frac{\partial \vec{E}_{\nu,i}}{\partial \xi_i} + \vec{S} \tag{13}$$

by means of

$$\vec{Q} = \frac{1}{J}\begin{bmatrix} \bar{\rho}\tilde{k} \\ \bar{\rho}\tilde{\epsilon} \end{bmatrix} \quad ; \quad \vec{E}_{c,i} = \frac{1}{J}\begin{bmatrix} \bar{\rho}\tilde{k}\tilde{U}_i \\ \bar{\rho}\tilde{\epsilon}\tilde{U}_i \end{bmatrix} \quad ;$$

$$\vec{E}_{\nu,i} = \frac{1}{Re}\frac{1}{J}\begin{bmatrix} (\mu_l + \frac{\mu_t}{Pr_k}) \left\{(\nabla \xi_i \nabla \xi_1)\frac{\partial \tilde{k}}{\partial \xi_1} + (\nabla \xi_i \nabla \xi_2)\frac{\partial \tilde{k}}{\partial \xi_2}\right\} \\ (\mu_l + \frac{\mu_t}{Pr_\epsilon}) \left\{(\nabla \xi_i \nabla \xi_1)\frac{\partial \tilde{\epsilon}}{\partial \xi_1} + (\nabla \xi_i \nabla \xi_2)\frac{\partial \tilde{\epsilon}}{\partial \xi_2}\right\} \end{bmatrix} \quad ;$$

$$\vec{S} = \frac{1}{Re}\frac{1}{J}\begin{bmatrix} P - Re\,\bar{\rho}\,\tilde{\epsilon} + \mathcal{D} \\ (C_1 P - Re\,C_2 f_2 \bar{\rho}\,\tilde{\epsilon})\frac{\tilde{\epsilon}}{\tilde{k}} + \mathcal{E} \end{bmatrix} \quad ; \quad \mathcal{D} = -\frac{2\mu_l \tilde{k}}{n^2} \quad ;$$

$$\mathcal{E} = -\frac{2\mu_l \tilde{\epsilon}}{n^2} exp\left(-0.5 n^+\right) \quad ; \quad n^+ = \frac{n}{\mu_l}\sqrt{Re\,\bar{\rho}\,\tau_w} \quad ;$$

$$P = 2\mu_t \left[\left(\frac{\partial \tilde{u}}{\partial x}\right)^2 + \left(\frac{\partial \tilde{v}}{\partial y}\right)^2\right] - \frac{2}{3}\left[Re\,\bar{\rho}\tilde{k} + \mu_t\left(\frac{\partial \tilde{u}}{\partial x} + \frac{\partial \tilde{v}}{\partial y}\right)\right]\frac{\partial \tilde{u}_k}{\partial x_k} + \mu_t\left(\frac{\partial \tilde{u}}{\partial y} + \frac{\partial \tilde{v}}{\partial x}\right)^2 \quad ;$$

$$f_2 = 1 - \frac{2}{9}exp\left(\frac{-R_T^2}{36}\right) \quad ; \quad R_T = Re\,\frac{\bar{\rho}\tilde{k}^2}{\mu_l \tilde{\epsilon}} \quad ;$$

$$C_1 = 1.35 \quad ; \quad C_2 = 1.80 \quad ; \quad Pr_k = 1.0 \quad \text{and} \quad Pr_\epsilon = 1.3 \quad .$$

The turbulent viscosity finally results from a nonlinear relationship, Eqn. (14), between k and ϵ with the coefficient $C_\mu = 0.09$:

$$\mu_t = Re \frac{C_\mu f_\mu \overline{\rho} \widetilde{k}^2}{\widetilde{\epsilon}} \quad \text{with} \quad f_\mu = 1 - exp\left(-0.0115\, n^+\right) \quad . \tag{14}$$

Since it is difficult to 'cold start' this model [2], the difference equations for k and ϵ are solved first in a so-called bypass mode by using, e.g., the Baldwin-Lomax model to predict a stable flow regime.

GEOMETRY AND RESULTS

The geometry of the numerically investigated compressor is given in Fig. 3, which shows the two main sections of the compressor: the impeller and the vaned diffuser. A more detailed description was given by Jansen [6].

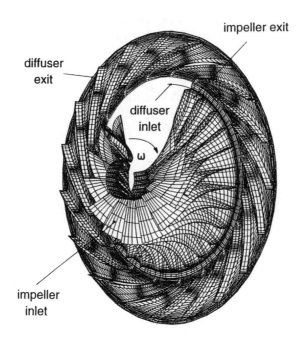

Fig. 3: Geometry of the impeller and the diffuser.

In Fig. 4 velocity vector plots in the diffuser are shown. They were computed with the 2D Navier-Stokes solver and the three different turbulence models (operating conditions: reduced mass flow rate 5.49 kg/s and 14 600 rpm, choking line). The complete flow fields predicted with the different models are very similar to each other, therefore only that calculated by the extended Baldwin-Lomax model has been plotted. Region A shows

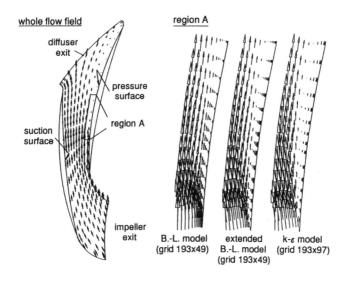

Fig. 4: Computed (2D) velocity vectors in one diffuser channel (mass flux 5.49 kg/s; 14 600 rpm): (a) B.-L. model; (b) extended B.-L. model; (c) $k - \epsilon$ model.

vectors in the vicinity of the flow separation at the pressure surface of the diffuser vanes. For the Baldwin-Lomax model only a very small backflow zone is obtained, whereas the other models yield larger recirculating regions. It is remarkable that the extended Baldwin-Lomax model needs only 49 grid points across the channel to achieve nearly the same resolution as the $k - \epsilon$ model using 97 points. The effect of the larger bubble can be seen in Fig. 5, where the static pressure along the meridional section is shown versus the aspect ratio r/r_0. The shock wave appears nearly at the same position when using the extended Baldwin-Lomax model (extended B.-L. model) or the $k - \epsilon$ model. The smaller pressure increase downstream of the shock related to the $k - \epsilon$ model in comparison

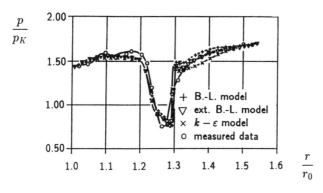

Fig. 5: Static pressure ratio p/p_K in the diffuser meridional section (mass flux 5.49 kg/s and 14 600 rpm): experiment by Jansen and 2D NS simulations.

with the increase obtained by the extended Baldwin-Lomax model may be a result of a sightly larger computed recirculating zone in case of the two-equation model. The pressure corresponding to the orginal Baldwin-Lomax model (B.-L. model) is slightly too large. Apart from these small differences, good agreement of the computed data with the measurement can be seen. Distributions (2D) of the static pressure ratio p/p_K across the whole width of a diffuser channel received by the different turbulence models are presented in Fig. 6. By comparing Figs. 4 and 6, it is obvious that the backflow region begins immediately behind the shock. The shock may be identified by means of the 'fat' black line downstream of the lowest pressure value $p/p_K \approx 0.7$ (see also Fig. 5). Furthermore, a stronger influence of the bubble towards the isobars in the vicinity of the separation zone can be realized when looking at the $k - \epsilon$ model. The distributions according to the Baldwin-Lomax model and the extended Baldwin-Lomax model are nearly identical apart from a small difference in the shock position.

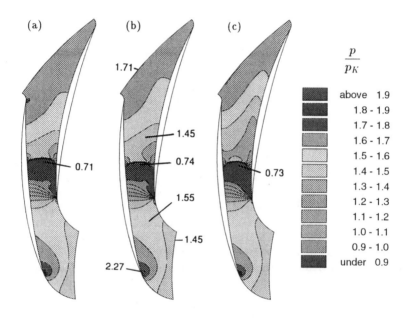

Fig. 6: Computed (2D) static pressure ratio p/p_K over a whole diffuser channel (mass flux 5.49 kg/s; 14 600 rpm): (a) B.-L. model; (b) extended B.-L. model; (c) $k - \epsilon$ model.

In Fig. 7, the static pressure distribution along the meridional section of the diffuser is given for an operating point at the surge line. The presented plots are based on experimental data at the shroud [6] and simulated 2D pressure ratios determined with the extended algebraic model and the two-equation model. Because no separation was detected, the computations with the two algebraic models led to an identical solution. Obviously, the simulated data shown are also very similar. Therefore, they also have in common a too low pressure — compared with the measured data — in the aspect ratio range 1.15 − 1.3. The diffuser blades start with an aspect ratio r/r_0 of 1.15. Considering this value, the too low pressure ratios may be a consequence of a too low predicted

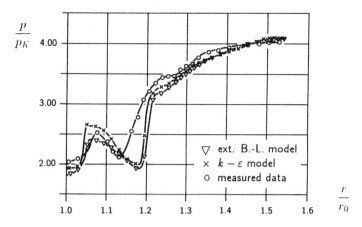

Fig. 7: Static pressure ratio p/p_K in the diffuser meridional section (mass flux 9.04 kg/s and 22 000 rpm): experiment by Jansen and 2D simulations.

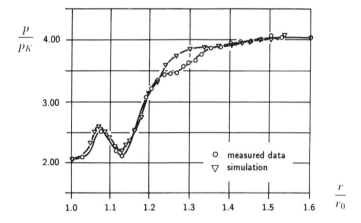

Fig. 8: Static pressure ratio p/p_K in the diffuser meridional section at the shroud (mass flux 9.04 kg/s and 22 000 rpm): experiment by Jansen and 3D NS simulation.

viscosity in front of the leading edge of the blades because, e.g., no hub and shroud influences are taken into account. This statement may be justified by the curves drawn in Fig. 8. The figure illustrates the pressure distribution according to the same operating conditions as predicted with the 3D Navier-Stokes solver. The calculated values differ from the measurement in this figure only in a small section of an aspect ratio range between 1.23 and 1.34.

Another reason why the 2D results do not agree with the experiment is illustrated in Fig. 9. This figure shows 3D streamlines related to the same operating conditions at surge. In the upper part the channel geometry is shown under a view angle of 60° against the shroud surface. In the second plot an angle of 90° is chosen, which means the channel can be seen from the side. The Navier-Stokes simulation has been made with a grid

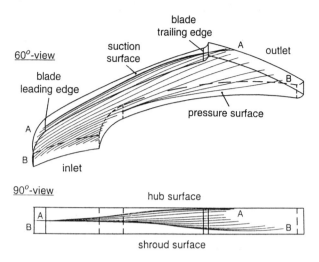

Fig. 9: 3D NS streamline distribution in the diffuser (mass flux 9.04 kg/s and 22 000 rpm).

using $97 \times 37 \times 33$ points. The streamlines show a strong three-dimensional structure of the flow field. They all start at the impeller exit in the centre between the diffuser's hub and shroud. However, between the vaned part and the exit section of the diffuser they can be found across the whole diffuser channel.

In Fig. 10 a comparison of the static pressure p/p_K in the centre between the hub and the shroud along the meridional section of the diffuser is given according to 3D simulations (operating point: mass flux 5.49 kg/s and 14 600 rpm). Whereas the 3D Euler solution [7] can only give a qualitative distribution of the pressure, the 3D Navier-Stokes computation (with the extended B.-L. model) is in good agreement with the measurement [6].

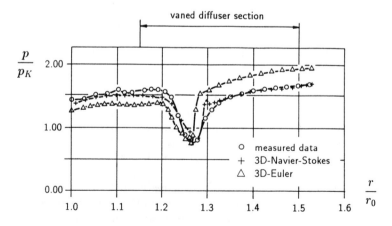

Fig. 10: Static pressure ratio p/p_K along the diffuser meridional section in the channel centre (mass flow rate 5.49 kg/s and 14 600 rpm): experiment by Jansen and 3D simulations.

A difference between the 2D results presented in Fig. 6 and the 3D Navier-Stokes simulation is the shock position. For the 2D case the shock is located at an aspect ratio $r_{2D}/r_0 \approx 1.3$ and for the 3D solver r_{3D}/r_0 is about 1.28.

The next result of the diffuser computation, Fig. 11, presents the mass flow rates for three different speeds of rotation: 14 600, 18 000 and 22 000 rpm. The operating points were predicted with the 3D Navier-Stokes solver. The simulations yielded data which are in good agreement with the experimental results by Jansen [6].

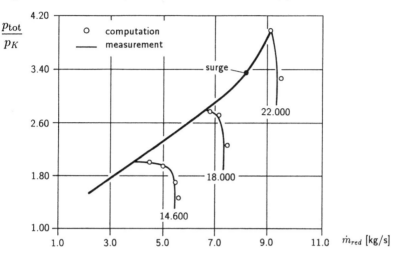

Fig. 11: Characteristics of the total pressure ratio versus mass flow rate of the diffuser at three different speeds of rotation: measurement by Jansen and 3D Navier-Stokes computations.

Finally, a few results are given concerning the impeller prediction. In Fig. 12 the static pressure ratio p/p_K in the centre between the impeller blades at the shroud versus an aspect ratio s/s_0 is given for three different mass fluxes at 14 600 rpm; s_0 is the length of the grid line between the inlet and the exit of the impeller. The data were computed with the 3D Navier-Stokes solver on a grid with 89 x 45 x 45 points per impeller channel, no splitter blades being taken into account. In Fig. 13 the measured mean distributions [6] along the shroud surface are shown. By comparing these two figures, good qualitative agreement of the behaviour of the pressure distributions related to the mass flux is obtained.

Nevertheless, the pressure in the region directly downstream of the impeller inlet is too low compared with the experiment. The reason for this pressure difference can be found in Fig. 14, where the pressure distribution in the midspan surface of the impeller predicted with the 3D Euler code is shown for one simulation with and another computation without splitter blades. The right-hand side of the figure shows a pressure ratio of about 1.0 in the region between the blade leading edges in front of the splitter blades. The pressure on the left-hand side in the same region is approximately 10 % lower (rotor without splitter blades). Furthermore, the pressure of the rotor with splitter blades is 5 % higher at the exit of the impeller than for the other impeller.

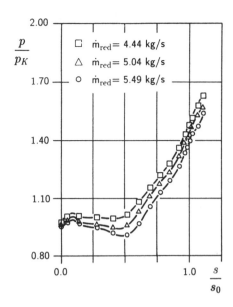
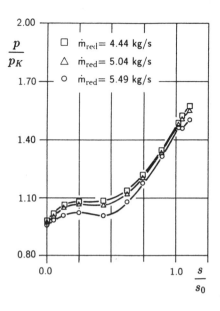

Fig. 12: Static pressure ratio p/p_K at the shroud in the centre of an impeller channel at 14 600 rpm: 3D NS simulations without splitter blades.

Fig. 13: Static mean pressure ratio p/p_K at the shroud surface of the impeller at 14 600 rpm: measurement Jansen.

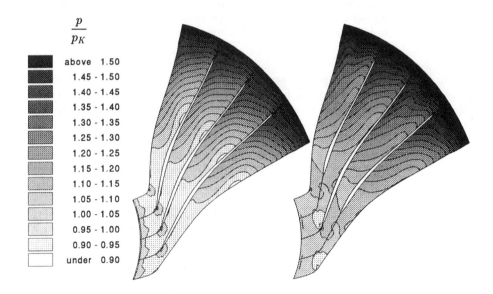

Fig. 14: Static pressure ratio p/p_K in the midspan surface of the impeller (mass flow rate 5.04 kg/s and 14 600 rpm). 3D Euler simulations: left-hand side without and right-hand side with splitter blades.

REFERENCES

[1] Goldberg, U. C.; Chakravarthy, S. R.: Separated Flow Prediction Using a Hybrid k-L / Backflow Model, AIAA Journal, Vol. 28, No. 6, June 1990.

[2] Kunz, R.; Lakshminarayana, B.: Explicit Navier-Stokes Computation of Cascade Flows Using the $k - \epsilon$ Turbulence Model, AIAA Journal, Vol. 30, No. 1, Jan. 1992.

[3] Kunz, R. ; Lakshminarayana, B.: Three-Dimensional Navier-Stokes Computation of Turbomachinery Flows Using an Explicit Numerical Procedure and a Coupled $k-\epsilon$ Turbulence Model, ASME Paper 91-GT-146, 1991.

[4] Michelassi, V.; Liou, M.-S.; Povinelli, L.A.: Implicit Solution of Three-Dimensional Internal Turbulent Flows, NASA TM-103099, 1990.

[5] Compton, W. B.; Abdol-Hamid, K. S.; Abeyounis, W. K.: Comparison of Algebraic Turbulence Models for Afterbody Flows with Jet Exhaust, AIAA Journal, Vol. 30, No. 1, Nov. 1992.

[6] Jansen, M.: Untersuchungen an beschaufelten Diffusoren eines hochbelasteten Radialverdichters, Dissertation, Univ. Hannover, 1982.

[7] Evers, W.: Zur Berechnung der dreidimensionalen reibungsbehafteten Strömung in transsonischen Kanälen und Leitradgittern, Dissertation, Univ. Hannover, 1992.

[8] Teipel, I.; Wiedermann, A.; Evers, W.: Simulation of Flow Fields in High-Loaded Centrifugal Compressors, published in: Notes on Numerical Fluid Mechanics, Vol. 38, pp. 364-378. Editor: Hirschel, E., Vieweg, Braunschweig, 1993.

[9] Roehl, C.; Simon, H.: Flow Simulations in Aerodynamically Highly Loaded Turbomachines Using Unstructured Adaptive Grids, in this publication.

[10] Hirsch, C.: Numerical Computation of Internal and External Flows, Wiley, New York, 1988.

[11] Radespiel, R.: A Cell-Vertex Multigrid Method for the Navier-Stokes Equations, NASA TM-101557, 1989.

[12] Jorgenson, P.; Turkel, E.: Central Difference TVD and TVB Schemes for Time Dependent and Steady State Problems, AIAA Paper 92-0053, 1992.

[13] Jameson, A., Baker, T. J.: Solution of the Euler Equations for Complex Configurations, AIAA Paper 83-1929, 1983.

[14] Arnone, A.; Liou, M. S.; Povinelli, L. A.: Multigrid Calculations of Three-Dimensional Viscous Cascade Flows, ICOMP-91-18, NASA TM 105257, 1991.

[15] Jameson, A.: MAE Report 1651, 1984, published in: Lecture Notes in Mathematics, Vol. 1127, pp. 156-242. Editor: Brezzi, F., Springer, Berlin, 1985.

[16] Leicher, S.: Numerical Solution of Internal and External Inviscid and Viscous 3-D Flow Fields, AGARD-CP-412, Vol. 1, 1986.

[17] Baldwin, B. S., Lomax, H.: Thin Layer Approximation and Algebraic Model for Separated Turbulent Flow, AIAA Paper 78-257, Jan. 1978.

[18] Cebeci, T.: Calculation of Compressible Turbulent Boundary Layers with Heat and Mass Transfer, AIAA Paper 70-741, 1970.

Efficient FEM Flow Simulation on Unstructured Adaptive Meshes

H. GREZA, S. BIKKER AND W. KOSCHEL
Institute for Jet Propulsion and Turbomachinery
Aachen University of Technology, 52056 Aachen, Germany

Summary

An unstructured finite element method for the transient solution of the 2D and 3D compressible Navier–Stokes equations using triangular respectively tetrahedral elements is presented. Different methods including a domain splitting and a multigrid scheme have been developed to accelerate the convergence and improve the performance of the solver. Furthermore efficient solution procedures have been implemented on vector and parallel machines. Methods for the automatic generation of unstructured and hybrid meshes allow for the application to arbitrarily shaped computational domains. In order to define the domain boundaries an interface program enables the coupling of the mesh generator with CAD–systems. The mesh quality is improved by a multi–stage adaptive smoothing process and fast tree–search algorithms are applied to reduce the generation times. Efficient adaptation techniques including enrichment, movement, and remeshing procedures enable the directional refinement and coarsening of the mesh.

1 Introduction

The application of unstructured grids in conjunction with a powerful mesh generation scheme enables both the use of complex boundaries and the incorporation of effective adaptation procedures but inevitably results in a limited efficiency of the flow solver. To eliminate this disadvantage various methods for improving of the convergence behaviour and for the implementation on high–performance computers have been investigated.

2 Numerical Flow Simulation

2.1 Governing Equations

The three–dimensional Navier–Stokes equations governing the flow of a viscous compressible fluid are considered in their conservative form,

$$\frac{\partial U}{\partial t} + \frac{\partial F_j}{\partial x_j} = 0 , \quad U = \begin{pmatrix} \rho \\ \rho u_i \\ \rho e_t \end{pmatrix} , \quad F_j = \begin{pmatrix} \rho u_j \\ \rho u_i u_j + p\delta_{ij} + \sigma_{ij} \\ (\rho e_t + p)u_j + \sigma_{jk}u_k + q_j \end{pmatrix} , \quad (1)$$

where ρ denotes the density, p the pressure, u_i the Cartesian velocity components and e_t the specific total energy of the fluid. U is the solution vector and F_j represents the component of the convective and viscous fluxes in the x_j–direction. The system of equations is completed by the state equations for a perfect gas,

$$p = \rho R T , \quad e = c_v T \quad \Longrightarrow \quad e_t = e + 1/2\, u_i u_i = (\kappa - 1)^{-1} p/\rho + 1/2\, u_i u_i , \quad (2)$$

where κ is the ratio of the specific heats and R the gas constant. For a Newtonian fluid the stress tensor σ_{ij} and the heat flux vector q_i are given by

$$\sigma_{ij} = \frac{2}{3}\mu \frac{\partial u_m}{\partial x_m}\delta_{ij} - \mu \left(\frac{\partial u_i}{\partial x_j} + \frac{\partial u_j}{\partial x_i} \right) \quad \text{and} \quad q_i = -\lambda \frac{\partial T}{\partial x_i} = -\frac{\kappa}{\kappa - 1} R \frac{\mu}{Pr} \quad (3)$$

where μ denotes the coefficient of viscosity and Pr the Prandtl number. The extension to turbulent flows is considered in [10]. High–temperature gas models and the simulation of chemical reacting flows are presented in [3].

2.2 Discretization

2.2.1 Two–Step Scheme The domain Ω is subdivided into finite elements using piecewise linear shape functions N_I. The approximation to the solution vector U is expressed by $\hat{U}(\underline{x},t) = N_I(\underline{x})U_I(t)$ where the summation extends over all nodes I of the mesh. For the time discretization of the fluid equations a two–step scheme is employed. The weighted residual form of the first step, which does not involve boundary conditions or diffusion effects, is given by

$$\int_\Omega P_e \hat{U}^{n+\frac{1}{2}} \, d\Omega = \int_\Omega P_e \hat{U}^n \, d\Omega - \frac{\Delta t}{2} \int_\Omega P_e \frac{\partial \hat{F}_k^n}{\partial x_k} \, d\Omega \qquad (4)$$

where P_e denotes the piecewise constant shape function associated with element e. After inserting the function approximations $\hat{U}^{n+\frac{1}{2}} = P_e U_e^{n+1/2}$, $\hat{U}^n = N_I U_I^n$ and $\hat{F}_k^n = N_I F_{k,I}^n = N_I F_k(U_I^n)$ the integrals appearing in Eq. (4) can be evaluated for each element separately providing a solution $U_e^{n+1/2}$ at the intermediate time level $t^{n+\frac{1}{2}}$. The second step computes a piecewise linear solution at the advanced time level t^{n+1} by use of the Galerkin method. The weak formulation of a suitable variational statement can be written as

$$\int_\Omega N_I \frac{\Delta \hat{U}}{\Delta t} \, d\Omega = \int_\Omega \frac{\partial N_I}{\partial x_k} \hat{F}_k^{n+\frac{1}{2}} \, d\Omega - \int_\Gamma N_I \hat{F}_k^{n+\frac{1}{2}} n_k \, d\Gamma \qquad (5)$$

with $\Delta \hat{U} = \hat{U}^{n+1} - \hat{U}^n$. Herein n_k denotes the outward normal unit vector on the boundary Γ of the domain Ω. By applying the approximations $\Delta \hat{U} = N_J \Delta U_J$ and $\hat{F}_k^{n+1/2} = P_e F_{k,e}^{n+1/2} = P_e F_k(U_e^{n+1/2})$ all integrations in Eq. (5) can be performed exactly. Looping over the individual elements leads to an assembling process in the form

$$M_{IJ} \frac{\Delta U_J}{\Delta t} = R_I \overset{I \equiv i_e}{\underset{J \equiv j_e}{\iff}} \sum_e \int_{\Omega_e} N_i N_j d\Omega_e \frac{\Delta U_j}{\Delta t} = \sum_e \int_{\Omega_e} \frac{\partial N_i}{\partial x_k} d\Omega_e \, F_{k,e}^{n+\frac{1}{2}} - \sum_e \int_{\Gamma_e \in \Gamma} N_i n_k d\Gamma_e \, F_{k,e}^{n+\frac{1}{2}} \qquad (6)$$

where the local nodes i and j of element e send contributions to the corresponding global nodes I and J. The fluxes $F_{k,e}^{n+1/2}$ of the boundary integral in Eq. (6) are corrected according to a linearized characteristics analysis normal to the boundary.

2.2.2 Runge–Kutta Scheme To avoid the coupling of the temporal and spatial discretization caused by the two–step scheme described in Sec. 2.2.1 a five–step Runge–Kutta time–stepping scheme [8] has been implemented. Using a Galerkin method based on a weak variational formulation corresponding to Eq. (5) and inserting linear finite element shape functions N_i both for the weighting functions and for all function approximations [10] leads to an assembling process of element contributions similar to Eq. (6),

$$U_I^{p+1} = U_I^0 - \alpha_p \Delta t_I \left((M_{II}^L)^{-1} R_I^p + \left(D_I^{(2)} - D_I^{(4)} \right)^{\min\{p,1\}} \right), \quad p = 0, ..., 4, \qquad (7)$$

$$R_I^p = \sum_e^{I \equiv i_e} \int_{\Omega_e} \frac{\partial N_i}{\partial x_k} N_j \, d\Omega_e \, F_k(U_j^p) - \sum_e^{I \equiv i_e} \int_{\Gamma_e \in \Gamma} N_i N_j n_k \, d\Gamma_e \, F_k(U_j^p), \quad U_I^0 = U_I^n, \; U_I^{n+1} = U_I^5,$$

where $D_I^{(l)}$ represents the artificial dissipation terms (Sec. 2.3.1). According to Sec. 2.4.3 the system matrix M_{IJ} has been replaced by the diagonal form $M_{II}^L = \sum_J M_{IJ}$. The values of the coefficients α_p in Eq. (7) are taken as $\alpha_p = (1/4, 1/6, 3/8, 1/2, 1)$.

2.3 Damping Methods

2.3.1 Artificial Dissipation The introduction of symmetric shape functions represents a central discretization which requires the addition of artificial dissipation to stabilize the scheme for shock capturing. For the two–step scheme the solution is modified by $U_I^{n+1}|_s = U_I^{n+1} + \Delta t_I D_I^{(2)}$. The term $D_I^{(2)}$ [9] consists of a second order difference operator

$$D_I^{(2)}\Big|^m = C_D^{(2)} (M_{II}^L)^{-1} \sum_e^{I \equiv i_e} \frac{S_e^m}{\Delta t_e} \left(M_{ij} - M_{ij}^L\right) U_j^m , \quad S_I^m = \left|\frac{\sum_e^{I \equiv i_e}(M_{ij} - M_{ij}^L)p_j^m}{\sum_e^{I \equiv i_e}|(M_{ij} - M_{ij}^L)p_j^m|}\right| , \quad (8)$$

where the pressure switch $S_e^m = 1/3 \sum_i S_i^m$, $0 \leq S_e^m \leq 1$, depends on the smoothness of the solution. For the five–step scheme (7) additionally a biharmonic operator $D_I^{(4)}$ is applied,

$$D_I^{(4)}\Big|^m = (M_{II}^L)^{-1} \sum_e^{I \equiv i_e} \frac{f_e^{(4)}}{\Delta t_e} \left(M_{ij} - M_{ij}^L\right) \left((M_{II}^L)^{-1} \sum_e^{I \equiv i_e} \left(M_{ij} - M_{ij}^L\right) U_j^m\right)_{j_e \equiv I} , \quad (9)$$

where the coefficient $f_e^{(4)} = \max\{0, C_D^{(4)} - C_D^{(2)} S_e^m\}$ ensures that the term $D_I^{(4)}$ is not employed in regions where $D_I^{(2)}$ is significant. $C_D^{(2)}$ and $C_D^{(4)}$ are user specified constants.

2.3.2 FCT–Algorithm When computing hypersonic flows the artificial viscosity model of Sec. 2.3.1 tends to produce oscillations in the vicinity of strong discontinuities within the solution. A shock capturing technique according to the method of Flux–Corrected Transport [6] was integrated into the finite element approach [10] providing high resolution of discontinuous flow effects for compressible high–speed computations. The FCT algorithm can be compared to a second-order hybrid scheme which changes self–adjusted to a first–order scheme only in regions with high gradients. This is achieved by averaging the fluxes obtained for a high– and a low–order scheme in such a way that no oscillations within the numerical solution arise. The second–order two–step algorithm (Sec. 2.2.1) is taken for the high–order scheme. The same scheme using the lumped mass matrix and an adapted amount of artificial viscosity provides the first–order scheme.

2.4 Solution of the Discretized Equations

2.4.1 Transient Solution When comnputing true transient solutions equation system (6) is solved iteratively without the necessity of assembing the global matrix M_{IJ},

$$\Delta U_I^\nu = (M_{II}^L)^{-1} \left(\left(\sum_e^{I \equiv i_e} (M_{ij}^L - M_{ij}) \Delta U_j^{\nu-1} \right) + \Delta t\, R_I \right) , \quad \nu = 1, 2, 3 , \quad \Delta U_j^0 = 0 , \quad (10)$$

where three iterations are sufficient and $M_{II}^L = \sum_J M_{IJ}$ denotes the lumped mass matrix.

2.4.2 Stability The explicit character of the solution scheme requires the application of the *CFL*–criterion. A stability analysis [10] leads to a restriction of the permissible time–step Δt, $\Delta t \leq CFL\, h_e((u_i u_i)^{1/2} + a)^{-1}$, $CFL|_{MIJ}^{2step} = 1/\sqrt{3}$, where a denotes the local speed of sound and h_e represents the minimum element height.

2.4.3 Convergence Acceleration Convergence acceleration for steady state solutions is achieved by employing the diagonal matrix $M_{II}^L = \sum_J M_{IJ}$ instead of the consistent system matrix M_{IJ} in Eqs. (6) and (7) respectively. This leads to a *CFL*–number of $CFL|_{ML_{II}}^{2step} = 1$ for the two–step and $CFL|_{ML_{II}}^{5step} = 4$ for the five–step scheme. Furthermore convergence is accelerated by inserting local time–steps Δt_e and Δt_I in Eqs. (5), (6) and (7). The elemental and nodal values of the maximum permissible time–step are determined by applying the *CFL*–criterion of Sec. 2.4.2 locally. Additionally a multigrid method (Sec. 3), domain splitting (Sec. 2.4.4), enthalpy damping and implicit residual averaging (see [10]) are used. An implicit solution algorithm is presented in [2].

2.4.4 Domain Splitting When employing efficient mesh adaption methods the element size usually varies by orders of magnitude. The use of the global time–step Δt (Sec. 2.4.2) leads to inaccuracies because very small CFL-numbers are imposed on the large elements (see [10]). For that reason an algorithm subdivides the domain according to the smallest allowable time–step Δt_{min} and advances the solution in each subregion i with the appropriate time–step $2^{i-1}\Delta t_{min}$. Simultaneously a reduction of the computational costs is achieved by the application of bigger time–steps for the larger elements. For a correct treatment of the inner boundaries between the subdomains two rows of elements are overlapped (Fig. 1). A comparison of the convergence obtained by using the single- and the multi–domain scheme for the steady–state solution of a supersonic inviscid flow past a wedge is given in Fig. 3 which shows the L_2-norm of the density residuals. As an example of a transient solution the subsonic viscous flow behind a cylinder resulting in a Karman vortex–street has been examined. Fig. 2 depicts the finite element mesh, the domain partition with the overlap regions between the subregions and the resulting distribution of the Mach number isolines. Compared to the single–domain run a reduction of the CPU-time by a factor of 1.4 has been achieved. A detailed description of the examples and the influence of the domain splitting on the accuracy is presented in [2].

3 Multigrid Solution

A direct multigrid method for viscous flow computations accelerates the convergence to steady state by operating on a sequence of fine and coarse grids. The implementation allows for the application of completely unrelated meshes and an arbitrary number of grid planes. The advantages of the time–stepping on the coarse meshes are on the one hand larger elements permitting a larger time–step and on the other hand less grid points causing less computational work. The five–step scheme of Sec. 2.2.2 has been integrated because it is characterized by a strong damping of high–frequency error modes (see [8]).

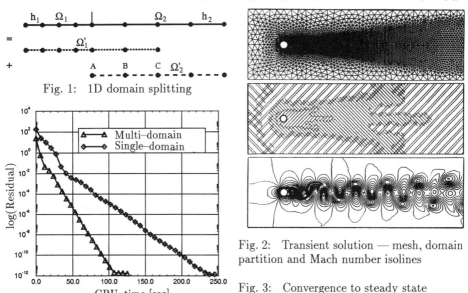

Fig. 1: 1D domain splitting

Fig. 2: Transient solution — mesh, domain partition and Mach number isolines

Fig. 3: Convergence to steady state

The time-stepping on the coarse meshes efficiently damps out the low-frequency error components. The multigrid algorithm uses a V-cycle performing one time-step per level when proceeding from the finest to the coarsest mesh (Fig. 4). For each level the flow variables U_I and the residuals R_I are transferred onto the next coarser mesh. When the coarsest grid has been reached, the corrections $\Delta U_I = U_I^{n+1} - U_I^n$ are transferred back onto the finer grids and are used to modify the fine grid solutions. To ensure an accurate and conservative data transfer between a coarse mesh C and the corresponding fine mesh F the flow variables and the corrections are interpolated whereas the residuals are distributed,

$$U_{i_C}^\star = \sum_{j_F=1,\ldots,3}^{x_{i_C} \in e_F} N_{j_F}(\underline{x}_{i_C}) U_{j_F} \, , \quad \Delta U_{i_F}^\star = \sum_{k_C=1,\ldots,3}^{x_{i_F} \in e_C} N_{k_C}(\underline{x}_{i_F}) \Delta U_{k_C} \, , \quad R_{k_C}^\star = \sum_{i_F}^{x_{i_F} \in e_C} N_{k_C}(\underline{x}_{i_F}) R_{i_F} \, , \quad (11)$$

using linear shape functions N_i for both cases (see Fig. 5). The transfer coefficients arising from Eqs. (11) are calculated efficiently in a preprocessing stage applying the alogorithms described in Sec. 5.6. The time-stepping scheme (7) is modified by adding a fine-to-coarse defect correction $R_I^\star - R(U_I^\star)$ to the residual $R(U_I^p)$. This formulation [8] guarantees that the coarse-grid corrections vanish if the fine-grid residuals become zero for a converged solution. The meshes are generated independently such that each grid has approximately four times the number of elements as the next coarser mesh. A sequence of adapted meshes employed for the multigrid computation of a subsonic inviscid flow past a cylinder is shown in Fig. 6. The convergence history of the multigrid and the fine grid solution

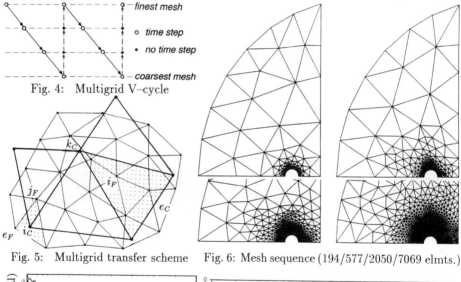

Fig. 4: Multigrid V-cycle

Fig. 5: Multigrid transfer scheme   Fig. 6: Mesh sequence (194/577/2050/7069 elmts.)

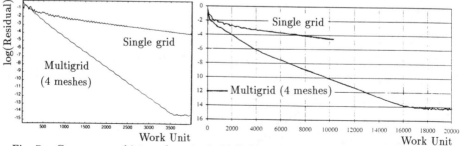

Fig. 7: Convergence history for a inviscid (left) and a viscous (right) computation

in dependence on work units is depicted in Fig. 7. The convergence is measured by the L_2-norm of the density residuals and one work unit means a normalized CPU-time. Fig. 7 also shows the improvement of the convergence for the viscous calculation of a subsonic flow past a flat plate with a Reynolds number of 5000. It has been proved that the solution quality of the finest mesh is not influenced by the multigrid scheme.

4 Simulation on High–Performance Computers

4.1 Implementation on Vector Machines

The algorithm can be vectorized on machines supporting gather- and scatter-loops by use of special hardware routines. The scatter assembling process of Eqs. (6) and (7) can lead to vector dependences resulting from multiple access of elements to common nodes. These recurrences can be suppressed by renumbering the elements and subdividing the entire scatter loop into smaller ones without dependences. Such a modification [10] yields a vectorization rate of 99% and a speed-up factor of 40 on a Siemens S600/20.

4.2 Implementation on Parallel Machines

For the parallelization of the explicit two–step algorithm of Sec. 2.2.1 local communication structures can be applied. Only in case of transient computations one global communication per time–step has to be performed. When distributing the elements to the different processors the domain decomposion method employed has to ensure a good load balancing. Furthermore a minimization of the inner boundaries leads to a reduction of the additional computational costs arising from both double calculations on the boundary points and communication between neighbouring processors. The following approaches for the partitioning of unstructured triangular or tetrahedral meshes are available:

- **Orthogonal recursive section (ORS)** This simple but efficient geometrical method recursively subdivides the domain along the orthogonal coordinate axes.
- **Eigenvalue recursive section (ERS)** The ERS algorithm is based on a spectral decomposition und uses the second eigenpair of the Laplacian of the dual graph. The second eigenvector itself gives some directional information and the differences of the vector components describe distances in the graph.
- **Simulated annealing (SA)** In order to find a good separator this heuristic approach minimizes a cost function using random numbers.
- **Parallel domain splitting** A master process controls the domain splitting procedure described in Sec. 2.4.4. Subsequently submaster processes control the further subdivision by the ORS method until the desired number of subdomains is reached.

The resulting domain decompositions for the ORS, ERS, and SA method are shown in Fig. 8 for the mesh of Fig. 25. Though the SA procedure gave the best results concerning smooth boundaries with a small number of communication points (Fig. 8), the ORS or ERS method is prefered due to the computational costs of the partitioning. A subsequent smoothing of the inner boundaries by exchanging elements between adjacent subdomains reduces the number of communication points by 10%. Under PARIX–Fortran both a synchronous and an asynchronous communication model has been implemented. By separating the elements connected to subdomain boundaries the pass through the remaining

Fig. 8: Domain decomposition by ORS, ERS, and SA (39/33/29% communication points)

elements can be performed in parallel to the local communication leading to to a higher efficiency than the simpler synchronous structure. Fig. 9 contains a comparison of the speed–ups and efficiencies achieved by the synchronous and asynchronous communication for different mesh sizes and numbers of processors applying the ORS or ERS decomposition method. The gain by using the parallel version of the domain splitting scheme is depicted in Fig. 10. In Tab. 1 the speed–ups obtained for different harware configurations are listed. A detailed description of the methods and results is given [1] and [2].

5 Computational Meshes

5.1 Representation of the Domain Boundaries

The geometrical description of 2D domain boundaries consists of an assembly of line segments (Fig. 11). The orientation of the line segments indicates on which side of the line the domain to be gridded is situated. Analogously the boundaries of 3D domains are subdivided into surface segments, which again are decomposed into the line segments connecting neighbouring surfaces (Fig. 11). The orientation of the surface segments is given by the orientation of the corresponding line segments. For the representation of the 2D and 3D line segments the following line types are available (Fig. 12): (L1) a straight line, (L2) prescribed nodal coordinates, (L3) an arc segment, (L4) a parabolic line, (L5) a periodical line, and (L6) a multi–section polynomial line. For the surface definition the following surface segment types are implemented (Fig. 13): (S1) a plane surface (defined by arbitrary line segments), (S2) a cylindrical surface (defined by line segments of type (L1) and (L3)), (S3) a triangular parabolic surface (defined by three parabolic line segments), (S4) a periodical surface, and (S5) a multi–patch polynomial surface. The segments types (L6) and (S5) offer the ability to represent arbitrarily shaped lines and surfaces defined by $\underline{x}(u) = \sum_{m=0}^{n_u} \underline{a}_m u^m$ and $\underline{x}(u,v) = \sum_{n=0}^{n_v} \sum_{m=0}^{n_u} \underline{a}_{mn} u^m v^n$ respectively with arbitrary polynomial degrees n_u and n_v. The values of $\underline{a}_m$ and $\underline{a}_{mn}$ are specified by the user (see Sec. 5.2).

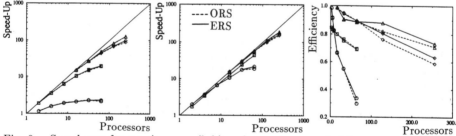

Fig. 9: Speed–ups for synchronous (left) and asynchronous (mid) communication and efficiencies (right) for the asynchronous case (○747/ □ 7919/◇93258/△173160 elements)

Tab. 1: Speed–ups for different hardware configurations

Machine	Processor(s)	CPU-time	Speed-up
Parsytec SC256	1×2.2 MFlops	798 min	1
IBM 3090	1×133 MFlops	23.3 min	34
Parsytec SC256	256×2.2 MFlops	4.26 min	187
Siemens VPP500	1×1.6 GFlops	1.95 min	409
Siemens S600/20	1×5 GFlops	1.38 min	578
Siemens VPP500	4×1.6 GFlops	0.56 min	1425

Fig. 10: Speed–ups for parallel domain splitting

5.2 Coupling with CAD–Systems

In order to support the definition of the domain boundaries by the user, the mesh generator can be coupled with CAD–systems like *IDEAS* or *CATIA* via the interface data structure VDAFs, which is standardized according to DIN 66301. A conversion program transfers the VDAFs data format to the assembly of line and surface segments described in Sec. 5.1. Beside a simple example first results for a complex application are presented in [4]. This case consists of the 3D description of a car interior. The corresponding VDAFs file contains 10 MB of data and leads to the generation of 381 lines and 170 surfaces.

5.3 Automatic Mesh Generation

5.3.1 Control of the Mesh Characteristics The shape of triangular elements is controlled by the local mesh parameters element size δ, stretching factor s and stretching direction α (Fig. 14). For tetrahedral meshes apart from the element size δ the mesh parameters consist of two stretching factors s_i and two stretching directions $\underline{\alpha}_i$. A spatial distribution of these parameters is provided by a background grid, which is made up of linear elements completely covering the computational domain. An initial mesh may be constructed by prescribing a more or less coarse background grid to achieve the desired variation in element size and stretching (Figs. 15 and 16). The stretching of the elements is achieved by a local transformation to an unstretched space. Fig. 17 depicts this transformation for a constant distribution of the mesh parameters. The generation of equilateral elements (Secs. 5.3.3 and 5.3.5) with the locally interpolated value of the element dimension δ is then performed in the unstretched space.

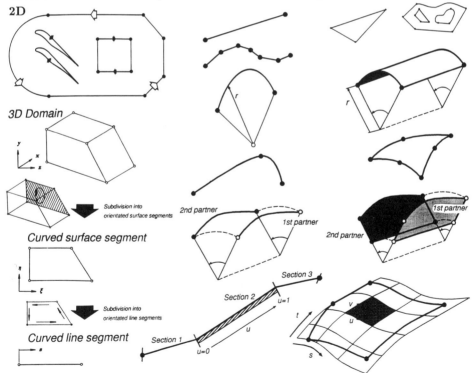

Fig. 11: 2D/3D domain definition   Fig. 12: Line segments   Fig. 13: Surface segments

5.3.2 Discretization of Boundary Lines

Each line segment is considered individually to perform the discretization into sides in a quasi 1D domain. First the line is subdivided into n_u sections of constant length δ_u with the mesh parameters being interpolated at the mid points u_i. The number of boundary sides n_s is calculated by taking the integer value of $r_s = \delta_u \sum_{i=1}^{n_u} \delta_c^{-1}(u_i)$. A local transformation of the tangent to the boundary line to an unstretched space (Fig. 17) is used to determine the desired nodal spacing δ_c. Finally the nodes are placed at the positions u_i where $n_s r_s^{-1} \int_0^{u_i} \delta_c^{-1}(u_i) du$ is equal to i with $i = 1, ..., n_s$.

5.3.3 Generation of Triangular Meshes

The mesh generation scheme utilizes the concept of an advancing generation front [9], which enables the automatic triangulation of arbitrarily shaped computational domains. The initial front is made of the orientated sides arising from the discretization of the boundary lines (Sec. 5.3.2). During the generation process new elements and points are introduced simultaneously permitting significant changes in the local mesh structure. The actually smallest front side is choosen as the base side AB to form a new triangle either with the newly introduced point C or an already existing point (Fig. 18). After the generation of each element the front is updated conserving the orientation. Detailed information on this method may be taken from [4].

5.3.4 Discretization of Boundary Surfaces

Each surface segment in turn is transformed to a quasi 2D domain. The corresponding discretized line segments (Sec. 5.3.2) act as the initial front for the subsequent triangulation. After applying the 2D version of the grid generator the surface mesh is transfered back to physical space. The mapping between the generation plane and the 3D domain consists of two transformations.

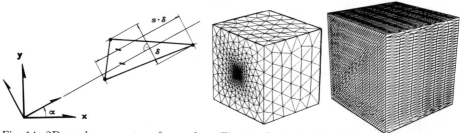

Fig. 14: 2D mesh parameters δ, s and α Fig. 15: Control of the 3D mesh characteristics

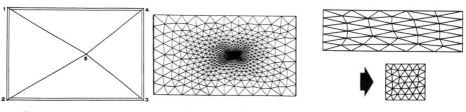

Fig. 16: 2D background grid and resulting mesh   Fig. 17: Local stretching

Fig. 18: 2D element generation and advancing front

First the surface segment is transformed to unit triangle or square using parabolic shape functions. A subsequent linear transformation ensures an accurate reproduction of the angles and lengths. During the process the mesh parameters are interpolated from the 3D background grid and transformed to the 2D domain. For analytically defined surfaces the transformations can be performed exactly.

5.3.5 Generation of Tetrahedral Meshes   After discretizing the line and surface segments (Secs. 5.3.2 and 5.3.4) all boundary faces are entered in the initial front. The 3D element generation procedure in principle follows the method of the 2D scheme (Sec. 5.3.3) with the front now containing the faces of tetrahedrons instead of sides.

5.3.6 Hybrid Mesh Generation   To achieve a predefined mesh topology, which is required mainly when implementing certain turbulence models, methods for the generation of structured subgrids have been included. Proceeding from solid walls quadrangles or wedges are generated layer by layer using a front method. The structured mesh lines are aligned normal and tangential to the corresponding boundary segment. In case of intersections with already existing mesh parts or boundaries the element generation is omitted (Fig. 19). Finally the remaining domain is meshed by the unstructured scheme preserving the geometrical flexibility of the overall procedure. The unstructured flow solver of Sec. 2 demands the subsequent subdivision of the structured elements into triangles or tetrahedrons. Parts of the 2D and 3D structured subgrids are incorporated in the line and surface discretizations of adjacent segments if they do not represent a solid wall (Fig. 20). Examples of 2D and 3D hybrid meshes are shown in Figs. 21 and 29.

5.4 Mesh Adaption

5.4.1 Adaptive Remeshing   The adaptive remeshing methods considered here offers the ability to improve the solution quality in a computationally efficient manner. During the flow analysis the grid adaption is achieved by repeatedly regenerating the complete computational mesh enabling the directional refinement and coarsening independent of the previous mesh. The process is based on information provided by the computed solution on

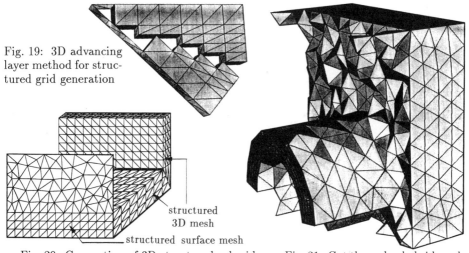

Fig. 19: 3D advancing layer method for structured grid generation

Fig. 20: Connection of 3D structured subgrids

Fig. 21: Cut through a hybrid mesh

the present grid. The new mesh is constructed using the previously described generation scheme allowing a significant variation in element size and stretching of the elements in the vicinity of one–dimensional flow features. The initial computational mesh is now acting as a background grid providing the spatial distribution of the mesh parameters. The method of error estimation is based on a scalar flow variable f (usually ρ or Ma) to give some indication of the error magnitude and direction. The error indicator [9] is derived from the estimation of the 1D interpolation error in the form $E = \delta_x^2 |d^2 \hat{f}/dx^2|$. The condition of a uniformly distributed error indicator within the domain leads to the optimal nodal values of the mesh parameters. The extension to 2D and 3D applications is accomplished by determining the principle directions of the tensor of the second derivatives and applying the 1D error indicator to each principle direction seperately. A blending function is used to blend excessively large mean values e.g. in the vicinity of shocks. Fig. 22 shows the adaptation for the computation of a inviscid shock reflection at a solid wall for the 2D and 3D case. The adaptation process is controlled by user specified global mesh parameters as illustrated in Tab. 2.

5.4.2 Mesh Enrichment and Coarsening The mesh enrichment method employs a scalar error indicator based on the second derivatives of a certain scalar flow variable. Additionally a directional error indicator using the gradients of the flow solution is available. For the 2D and the 3D case the elements are refined by introducing new points at the middle of the element sides. An appropriate amount of elements has to be added to guarantee a valid connectivity. A detailed description and applications of the mesh enrichment techniques is given in [1] and [2]. The mesh coarsening is performed by removing nodes and the corresponding number of sides and elements. The procedure leads to extreme mesh distortions and is therefore used only in conjunction with the smoothing methods (see Sec. 5.5).

5.4.3 Adaptive Mesh Movement In contrast to the previously described methods the mesh is adapted by moving the nodes using a spring system analogy. All sides of the mesh are replaced by springs with the nodal forces being set to the difference between the desired and the existing side length (Fig. 23). The optimal side length is calculated as described in Sec. 5.3.2 applying the directional error indicator of Sec. 5.4.1. To bring the spring system into equilibrium the assembled equation system is solved by iteration.

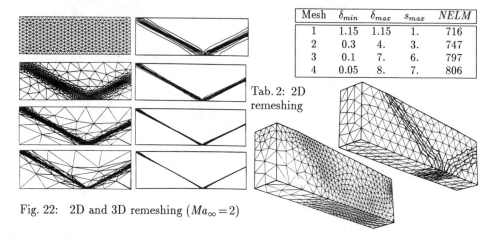

Mesh	δ_{min}	δ_{max}	s_{max}	NELM
1	1.15	1.15	1.	716
2	0.3	4.	3.	747
3	0.1	7.	6.	797
4	0.05	8.	7.	806

Tab. 2: 2D remeshing

Fig. 22: 2D and 3D remeshing ($Ma_\infty = 2$)

5.5 Mesh Smoothing

It turns out that some more or less ill–distorted elements may arise from the generation and adaptation methods. These irregularities deteriorate the solution quality and the convergence behaviour of the flow solver. For this reason a multi–stage smoothing procedure is applied after the mesh generation:

• **Optimizing the Element Connectivities** In the first stage the element connectivities are optimized with the point positions being fixed. All nodes adjacent to less than five elements are removed by the scheme described in Sec. 5.4.2. Subsequently the element sides are swapped within an iteration process aiming at the optimal case with each node being surrounded by six elements (Fig. 24). Up to the present for the 3D case only the connectivity of the surface meshes is optimized.

• **Optimizing the Point Positions** In the subsequent stage the nodes are moved without changing the mesh topology. Beside a simple purely geometrical algorithm, which iteratively moves each node in the centre of the sourrounding points, the adaptive mesh movement procedure presented in Sec. 5.4.3 is used to equalize mesh distortions.

The efficiency of the smoothing methods is demonstrated in Fig. 25, where the second mesh of Fig. 22 is shown before and after optimization. The improved shape of the triangles after the application of the smoothing procedures is evident. Additionally the so–called flow lines of the mesh become more apparent after smoothing (see e.g. Figs. 25, 28 and 29). These flow lines are made of the assembly of connected and nearly equal directed element edges. They should pass through the domain as smooth as possible without interruptions and end at the domain boundary.

5.6 Reduction of Generation Times

The generation and smoothing of unstructured grids in general leads to a quadratic development of the generation time in dependence on the mesh size. An almost linear behaviour (Fig. 26) has been achieved by the use of efficient data structures and fast search algorithms.

• **Efficient Data Structures** The most effective way to accelerate the mesh generation is to store the different data elements point, side, face, and tetrahedron in appropiate data structures to perform the different search operations involved in an optimum manner. The implemented nD–tree structures are applicable to data fields of an arbitrary dimension n. Furthermore connectivity lists like a face/element list are used.

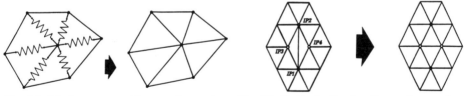

Fig. 23: Mesh movement by a spring system   Fig. 24: Optimization by diagonal swapping

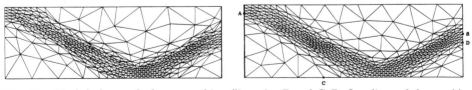

Fig. 25:   Mesh before and after smoothing (lines A—B and C–D: flow lines of the mesh)

- **Fast Search Algorithms**   Linear sorted data are stored in an explicit binary tree structure to enable effective search operations. Quadtree and octree structures represent the extension to data fields with 2D and 3D structure. Multi-dimensional tree structures are applied to solve geometric searching and intersection problems. As search keys both nodal numbers and coordinates are employed. The tree search algorithms are completed by the use of the connectivity lists to give a direct information about adjacent data elements.

6 Examples and Conclusions

The efficiency of the described methods has been demonstrated by several 2D and 3D applications presented in [1], [2], [3], [4], and [10]. The first example presented in this paper consists of a transsonic inviscid flow computation [2] for a NACA 0012 airfoil (Fig. 27). Next the investigation of the inviscid flow past the tail of a hypersonic aircraft [3] is illustrated in Fig. 28. Besides the inflow and outflow boundaries of the outer flow additional boundaries have been prescribed to simulate the inlet and outlet of the propulsion system. The highly resolved shocks originating from the surface as well as the jet boundary of the high enthalpy flow and the expansion waves are clearly visible. Finally Fig. 29 shows the examination [3] of the viscous flow field past an aircraft during take–off. Besides defining the boundary conditions analogously to the previous case a moving ground has been implied. The examples illustrate the large variation in element size within the domain which has been achieved by applying unstructured meshes in connection with the adaptive remeshing and smoothing procedures of Sec. 5.

It has been demonstrated that unstructured grids may be employed advantageously for the accurate simulation of both geometrically as well as physically complex flow fields in a computationally efficient manner.

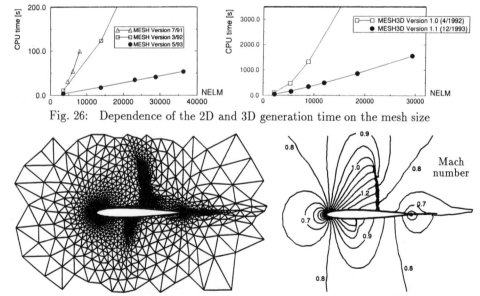

Fig. 26:   Dependence of the 2D and 3D generation time on the mesh size

Fig. 27:   Adapted computational mesh and Mach number isolines for a NACA 0012 airfoil (incidence $\alpha_\infty = 1.25°$, freestream Mach number $Ma_\infty = 0.8$)

Acknowledgements

Work on parallelization was funded by the Deutsche Forschungsgemeinschaft DFG within the program *Flow Simulation on High–Performance Computers*. The development of the 3D mesh generation methods was supported by the German Minister of Science and Technology BMFT within the program *HTGT–TURBOTECH* in connection with the Daimler Benz AG.

References

[1] **S. Bikker, W. Koschel:** Domain Decomposition Methods and Adaptive Flow Simulation on Unstructured Meshes, Notes on Num. Fluid Mech., Vol. 50, 13–24, Vieweg, 1994.

[2] **S. Bikker:** Beitrag zur Strömungssimulation auf Hochleistungsrechnern mit Finite Elemente Methoden, Dissertation, RWTH Aachen, 1996.

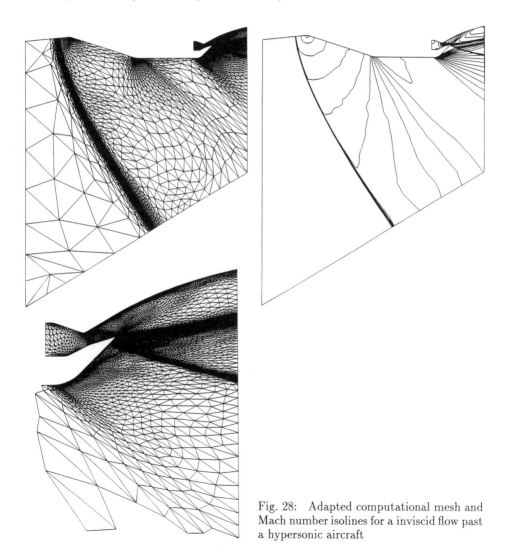

Fig. 28: Adapted computational mesh and Mach number isolines for a inviscid flow past a hypersonic aircraft

[3] **U. Fox:** Finite–Elemente–Simulation von Hochtemperatur–Düsenströmungen, Dissertation, RWTH Aachen, 1996.

[4] **H. Greza:** Adaptive 3D–Generierung für FEM–Verfahren zur Lösung der Euler– und Navier–Stokes–Gleichungen, IST FB 95–03, Inst. f. Strahlantriebe u. Turboarbeitsmaschinen, RWTH Aachen, BMFT 0326801S, TIB Hannover, 1995.

[5] **R. Löhner, K. Morgan, O.C. Zienkiewicz:** The Use of Domain–Splitting with an Explicit Hyperbolic Solver, Comp. Meth. Appl. Mech. Eng., Vol. 45, 313-329, 1984.

[6] **R. Löhner, K. Morgan, J. Peraire, M. Vahdati:** Finite Element Flux–Corrected Transport (FEM--FCT) for the Euler and Navier–Stokes Equations, Int. J. for Num. Meth. in Fluids, Vol. 7, 1093–1109, 1987.

[7] **R. Löhner:** Some Useful Data Structures for the Generation of Unstructured Grids, Comm. Appl. Num. Meth., Vol. 4, 123–135, 1988.

[8] **D.J. Mavriplis:** Multigrid Solution of the Two–Dimensional Euler Equations on Unstructured Triangular Meshes, AIAA Journal, Vol. 26, No. 7, 824-831, 1988.

[9] **J. Peraire, J. Peiro, L. Formaggia, K. Morgan, O.C. Zienkiewicz:** Finite Element Euler Computations in Three Dimensions, Int. J. Num. Meth. Eng. 26, 2135–2159, 1988.

[10] **W. Rick:** Adaptive Galerkin Finite Elemente Verfahren zur numerischen Strömungssimulation auf unstrukturierten Netzen, Diss., RWTH Aachen, Verlag Shaker, Aachen 1994.

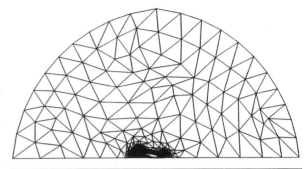
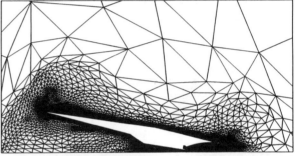
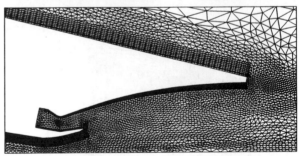
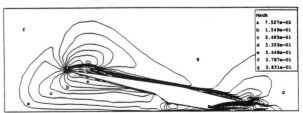

Fig. 29: Hypersonic aircraft during take–off — adapted computational mesh (22477 elements, 11610 nodes) with enlargements and Mach number isolines for a viscous computation (incidence $\alpha_\infty = 12°$, ground speed $v_G = 120 m/s$)

Numerical Methods for Simulating Supersonic Combustion

J. Grönner, E. von Lavante, M. Hilgenstock, M. Kallenberg
University of Essen, Lehrstuhl für Strömungsmaschinen
Schützenbahn 70, D-45127 Essen

Summary

An efficient computer program was developed for the computation of supersonic combustion problems. Several test cases showed the capabilities of implementations of different upwind schemes in calculating chemical reactions. The FAS multigrid procedure accelerated in some cases the convergence to steady state. The numerical results were validated by the corresponding experiments for one of the more demanding cases. A reasonable qualitative as well as quantitative agreement was achieved. A relatively high degree of numerical efficiency was achieved by parallel implementation of the code, using domain decomposition.

Introduction

The research and development of several hypersonic flight vehicles in the USA and Europe has brought many new computational methods for the prediction of hypersonic flow fields. The consideration of new physical phenomena in this regime, like, for example, chemical reactions and vibrational excitation, leads to the solutions of inhomogeneous Navier-Stokes equations with a source term and additional equation of mass conservation for each chemical species. The difficulties in numerically solving the resulting system of governing equations arise from the extreme stiffness of the equation system, due to the very short characteristic time scales associated with the chemistry, and the uncertainties about the material, physical and chemical properties of the participating species.
While the main attention was paid to the analysis of hypersonic flows, the problem of simulating supersonic flows in typical combustion chambers of high speed flight vehicles was treated less extensively. The main goal of the present research was to investigate the possibilities of efficient simulation of complex flows in typical combustion chambers of propulsion units for high supersonic and hypersonic flight regimes. One of the goals of the present work was to address various numerical issues using a few well published test cases. From the many possibilities, recently published in open literature, following problems were selected:

- transverse hydrogen injection in a supersonic airstream

- 2-D scramjet modell (Wada et al [11])

The first test case has also been extensively experimentally investigated by the ISL (German-French Research Institute, Saint Louis), allowing a direct comparison between the experimental data and the numerical results. Three different schemes for the spatial discretization were worked out and tested. Several numerical treatments of the source terms, arising due to the chemical species production, were investigated. The different possibilities of the temporal discretization, leading to stable numerical algorithms, were compared as well, and will be discussed.

Algorithms

In this work, the flow was assumed to be compressible, viscous, and a mixture of thermally perfect species. Due to the relatively low temperature and high pressure in the present configurations, the gas mixture can be treated as in vibrational equilibrium. The governing equations were in this case the compressible Navier-Stokes equations for n_s species:

$$\frac{\partial \hat{Q}}{\partial t} + \frac{\partial \hat{F}}{\partial \xi} + \frac{\partial \hat{G}}{\partial \eta} = \frac{S}{J} \qquad (1)$$

where $\hat{F}$ and $\hat{G}$ are the flux vectors in the corresponding ξ and η directions, Q is the vector of the dependent variables and J is the jacobian of the transformation of coordinates. S is the vector of the chemical source terms. The details of the governing equations are given in [12].

A simple model according to the Fick's law for the binary diffusion coefficient was used, along with the Sutherland equation for the viscous coefficient. The chemical reactions were realized with an 8-reaction model of Evans and Schexnayder [3], for the H_2-air combustion system.

Three different upwind methods were extended to chemical reactions systems. These methods are based on the work of Roe (flux-difference splitting) [9], van Leer (flux-vector-splitting) [6] and Liou (AUSM Advection-Upstream-Splitting-Method) [7].

Roe Scheme

The present research originally started with a numerical scheme based on Roe's Flux Difference Splitting in finite volume form. In previous numerical predictions, this scheme was highly effective in providing accurate viscous results at a wide range of Mach numbers. In the present version, the reconstruction of the cell-centered variables to the cell-interface locations was done using a monotone interpolation as introduced by Grossmann and Cinella in [4]. The interpolation slope was limited by an appropriate limiter, according to the previously published MUSCL type procedure (see, for example, [12]).

Liou Scheme

The AUSM is relatively new. It separates the corresponding flux into three parts, the convective part, the pressure part and the viscous part. As usual, the viscous part is evaluated with central differences, whereas the other parts are upwind differenced. The damping behaviour is proportional to the signal speed, in our case the Mach number in normal direction to the cell interface, and, therefore, the numerical damping decreases in the case of vanishing signal velocity. This behaviour causes pressure oscillations, especially in shear layers. In order to improve the damping characteristics of this scheme in the case of vanishing signal velocities, von Lavante and Yao [8] have proposed a modification of the original Liou scheme. In this modified form the signal velocity is the product of the Mach number and the density at the cell interface. The damping of the continuity equation is now equal to the van Leer scheme. The details of this scheme, with the corresponding modifications, are given by Hilgenstock et. al. [13]. This scheme is automatically positivity preserving in the sense of Larrouturou [5].

van Leer Flux Vector Splitting

The split fluxes are constructed according to the formulation given by Shuen in [10], and will not be repeated here. The fluxes are constructed from the variables Q_L and Q_R, extrapolated from the left or right of the cell interface, depending on the sign of the split fluxes $F^\pm$, using the same MUSCL interpolation as in the Roe Scheme.

Temporal Integration

The governing equations were integrated by a semi-implicit method, with different multi-stage Runge-Kutta type schemes used for the explicit operator of the fluid-dynamics part. Following an idea of Bussing and Murmann [2], only the chemical source terms were treated implicitely,

$$\left\{ I - \Delta t \, \Theta \, \frac{\partial S^n}{\partial Q^n} \right\} \frac{\partial Q^n}{\partial t \, J} = \hat{S}^n - \frac{\partial \hat{F}^n}{\partial \xi} - \frac{\partial \hat{G}^n}{\partial \eta} \qquad (2)$$

with the relaxation parameter Θ. In most of the present computations, a two-stage Runge-Kutta procedure with $\Theta = 1$ seemed to be the best choice.

The numerical effort to invert the Matrix $B = I - \Delta t \, \Theta \, \frac{\partial S^n}{\partial Q^n}$ depends on the formulation of the Jacobian of the chemical source terms. Several different forms of the Jacobian matrix, with increasing complexity and accuracy, were implemented and compared. The most obvious choice is to invert the full $n_s \times n_s$ matrix B. This, however, is a problem from the numerical point of view, since the inversion is CPU time consuming, and the matrix B usually illconditioned. This approach worked, but was rather inefficient. The next possibility to simplify the matrix B consists of dropping all the off diagonal terms, while keeping only the diagonal terms. In our case of eight reactions with seven species, this turned out to be an effective means of accelerating the convergence, with stability limits given by the acoustic wave speeds.

Using a multi-block grid structure resulted in a flexible code with the possibilty of working with different chemical models (nonequilibrium, equilibrium, frozen) in different blocks. Besides, some of the blocks were selectively refined, depending on the evolving results. The present geometrical treatment of the computational domain was simple, yet flexible enough.

Multigrid Acceleration

The still relatively slow rate of convergence was accelerated using the standard FAS multigrid procedure. The residuals and the chemical production terms were restricted using

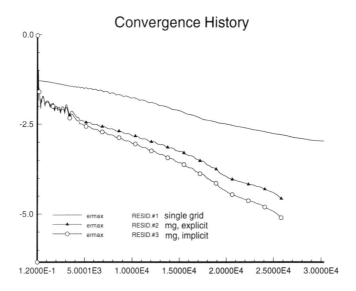

Fig. 1: Multigrid convergence

simple summation; in the restriction of the dependent variables Q, a volume weighted averaging was employed. A bilinear interpolation was used in the prolongation. A detailed description of the FAS scheme can be found in [12]. A comparison of the rates of convergence with and without the MG is presented in Fig.1 for steady flow simulations in a supersonic diffusor, with premixed hydrogen-air mixture of equivalence ratio $\Phi = 0.1$, entering the diffusor at a Mach number $M = 2.5$ and temperature $T = 900K$.

Parallel Implementation

Ideally, since the present simulations of the turbulent flows did not use any wall functions and the simulation extended all the way to the wall, the resolution should be of order of magnitude y^+, making very fine computational grids necessary. Since, in some cases, the regions of high gradients of the flow variables displayed no preferential direction, this very high resolution should be applied not only normal to the solid walls, but in all spacial directions considered in that particular simulation. Additionally, the computational grid should be uniformly distributed. Clearly, even if the above requirements are somewhat relaxed, an extremely high number of grid points (or cells in the finite volume method) have to be utilized. The corresponding computations can be carried out only on the largest computers available. An adequate performance for the present computations is currently offered only on massively parallel computers.

Early in this work, it was decided to implement a data parallel structure, since the multiblock grid system already had data exchange between the blocks built in. In a fashion similar the work of Pokorny, Faden and Engel [1], the parallelization was carried out using domain decomposition. This was accomplished with the help of the MPI library, utilizing the standard point to point communications calls. Only few global operations had to be used. The relative performance per processor unit (PU), shown in Fig. 2, indicates that the scalability of the resulting code is fairly good.

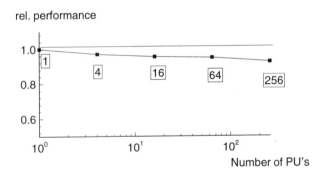

Fig. 2: Relative performance per processor unit

Here, the MFlops per PU, devided by the single PU rating, as implemented on a Cray T3D computer, are displayed for 4, 16, 64 and 256 PUs. In each of these computations, the grid on each processor had the same size of 32x32 internal cells.

The computations with the maximum resolution were carried out on a Cray T3D using 256 PUs. In the two-dimensional case, the total grid size was 1024x1024 internal cells. A similar computation was also done on an Parsytec Xplorer parallel computer with 8 PUs. Most of the results shown below were obtained on this system.

Results

Transverse H_2 Jet in a Supersonic Gas Stream

The geometry and the boundary conditions of this configuration are decribed in detail by von Lavante et. al. [12] . Here, a hydrogen jet is injected into a two-dimensional channel with parallel walls, containing either nitrogen or air at supersonic inflow Mach numbers. At the inflow, the Mach number was $M = 3.1$ for the case of nitrogen main flow, and $M = 2.9$ for the air-hydrogen system. The static pressure in the channel was $p = 0.13MPa$ and the temperature at the same location was $T = 1300K$. The hydrogen jet enters at sonic conditions through a slot of 1.3 mm width; its temperature was $T = 600K$. The Reynolds number was in this case $Re_h = 9\ 10^5$, based on the channel hight. The smallest usefull computational grid had a minimum of 128x128 cells, allowing a reasonable resolution of the important regions. As a typical example, the 16 blocks grid is displayed in Fig. 3. This case is of particular interest, since it has been frequently used in numerical simulations by several authors and was (and is still being) extensively experimentally investigated.

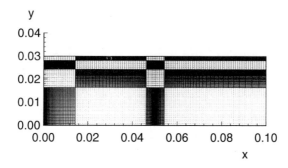

Fig. 3: 16 blocks grid

As the numerical experiments, discussed above, indicated, the van Leer's FVS scheme was too dissipative for accurate predictions of the chemical reactions in the regions where the viscous effects were dominant, so that only the Roe scheme was applied in this test case. The position of the H_2 jet is also visible in Fig. 5, where the pressure contours are shown. Clearly visible are the weak leading edge shock, the strong separation shock, the bow shock, the barrel shock and the recompression shock. The separation shock and the

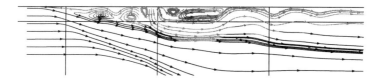

Fig. 4: Streamlines near h_2-injection

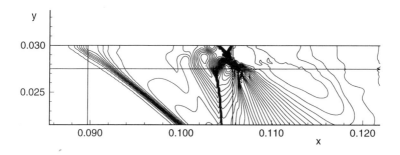

Fig. 5: Pressure contours near injection hole

bow shock merge and reflect at the lower wall. This reflected shock then reflects again off the shear layer between the air and the product gases at the upper wall. A detailed view of the streamlines in the vicinity of the injection port are shown in Fig.4, coloured by the H_2O concentration, pointing out the position of the reaction zone. Several separation vortices and the oscillatory behaviour of the shear layer close to the wall indicate the unsteady character of the flow. In accordance with the experimental measurements, published in [12], the H_2 is carried upstream of the injection opening by recirculating fluid in the boundary layer. Downstream of the region where most of the H_2O production occures, the flow is basically chemically frozen, with H_2O being convected. A small part of the produced water is convected into the boundary layer ahead of the jet by the recirculation, present at this location. The velocity vectors are shown in detail in Fig. 6. The flow at the shear layer-shock interaction was in this case weakly unsteady, with shedding of small vortices. These are consequently convected downstream. The computed velocities fall well within the limits of the experimentally determined values.

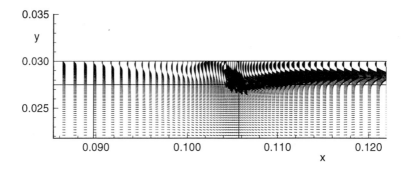

Fig. 6: Velocity vectors near injection hole

Two Dimensional Scramjet

The flow in a two-dimensional model scramjet configuration, shown by Wada et al [11], is presented last. The inflow Mach number was 4, and the Reynolds number based on the maximum hight was $Re_h = 3\ 10^6$. The resulting pressure contours for the three schemes are given in Fig. 7. These results support the conclusions drawn in the previous sections.

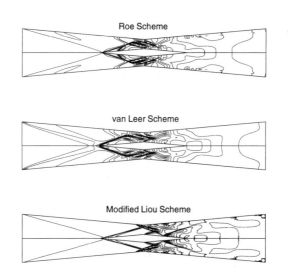

Fig. 7: Scramjet - pressure contours

Conclusions

In the present work, aimed at finding a simple numerical method for simulations of compressible, viscous flows in chemical nonequilibrium, three different spatial discretization schemes were tested on a few supersonic configurations with increasing complexity. The van Leer FVS scheme, known to be very dissipative, was too inaccurate to be of practical use. The prolific Roe scheme gave consistantly good results on adequate computational grids, but displayed a distinct lack of robustness. The relatively untested AUSM scheme, with the proper modifications, gave some promising results, while being simple and comparatively reliable. The CPU time savings due to its simplicity are, however, negligible compared with the effort spent on the modeling of the other real gas effects.

The simplest possibility of removing the exponential stiffness of the governing equations by implicit formulation was the most effective one. The diagonal form of the jacobian matrix $\frac{\partial S}{\partial Q}$ worked well, making CFL numbers, based on the acoustic wave speeds, of $O(1)$ possible.

The FAS multigrid procedure effectively accelerated the rate of convergence in the simple steady case, but was not suited for unsteady computations in the present form.

References

[1] Pokorny, S., Faden, M., and Engel, K., " Development of a Simulation System for Three-Dimensional Unsteady Turbomachinery Flows ", Notes on Numerical Fluid Mechanics, Vol. 38, Vieweg Verlag, 1993.

[2] Bussing, T.R.A. and Murmann, E.M., " Finite Volume Method for the Calculation of Compressible Chemically Reacting Flows ", AIAA Paper 85-0331, Jan. 1985

[3] Evans, J.S., Schexnayder, C.J., " Influence of Chemical Kinetics and Unmixedness on Burning in Supersonic Hydrogen Flames", AIAA-Journal, Febr. 1980, pp. 188-193

[4] Grossmann, B., and Cinella, P., " Flux-Split Algorithms for Flows with Nonequilibrium Chemistry and Vibrational Relaxation ", J. Comp. Phys., vol. 88, pp. 131-168, 1990

[5] Larrouturou, B., and Fezoui, L. " On the Equations of Multi-Component Perfect or Real Gas Inviscid Flow ", Nonlinear Hyperbolic Problems, Lecture Notes in Mathematics, 1402, Springer Verlag, Heidelberg 1989

[6] van Leer, B., " Flux-Vector-Splitting for the Euler Equations ", Institute for Computer Applications in Science and Engineering, Hampton, VA, Rept. 82-30, Sept. 1982

[7] Liou, M.S., " On a New Class of Flux Splitting ", Proceedings , 13th International Conference on Numerical Methods in Fluid Dynamics, Rome 1992, pp. 115-119

[8] von Lavante, E. and Yao, J. " Simulation of Flow in Exhaust Manifold of an Reciprocating Engine ", AIAA-93-2954

[9] Roe, P.L., Pike, J., " Efficient Construction and Utilisation of Approximate Riemann Solutions ", Computing Methods in Applied Sciences and Engineering, VI, pp. 499-516, INRIA, 1984

[10] Shuen, Jiang-Shun, " Upwind Differencing and LU Factorization for Chemical Nonequilibrium Navier-Stokes Equations ", Journal of Comp. Phys., vol. 99, pp. 213-250, 1992

[11] Wada, Y. Ogawa, S. and Ishiguro, T., " A Generalized Roe's Approximate Riemann Solver for Chemically Reacting Flows ", AIAA 89-0202

[12] von Lavante, E., Hilgenstock, M. and Groenner, J., " Simple Numerical Method for Simulating Supersonic Combustion ", AIAA Paper 94-3179

[13] Hilgenstock, M., von Lavante, E. and Groenner, J., " Efficient Computations of Navier-Stokes Equations with Nonequilibrium Chemistry ", ASME Paper 94-GT-251

Aeroelastic Computations of a Fokker–Type Wing in Transonic Flow

D. Nellessen, G. Britten, S. Schlechtriem, J. Ballmann

Lehr- und Forschungsgebiet für Mechanik
RWTH-Aachen, University of Technology
52056 Aachen, Germany

Summary

A computational method for the treatment of solid fluid interaction (SOFIA) has been developed to study the aeroelastic characteristics of wings and rotors by direct numerical simulation. It consists of an aerodynamic part to compute the airloads and a structural part to determine the structural deformation. The unsteady, three–dimensional flow is modeled by the Euler equations and calculated using an implicit solver. The elasticity is modeled by a generalized Timoshenko–Flügge beam, which includes the static and dynamic coupling of bending and torsional modes. A finite–element–method is implemented to predict the deformation. Both solvers are coupled in the time domain. SOFIA has been used to investigate a realistic Fokker–type wing in transonic flow. The steady lift distributions have been calculated for the rigid and the elastic wing. Comparison shows remarkable differences between wings with sweep–back and wings with no sweep.

Symbols, structure:

$\underline{u}_S$	displacement
$\underline{\varphi}$	rotation
$\underline{\gamma}$	shear angle
$EA, E\underline{\underline{I}}_B$	tension and bending stiffness
GA, GI_{D11}	shear and torsional stiffness
$\underline{\underline{K}}$	shear coefficient tensor
$\rho A, \underline{\underline{\Theta}}_S$	mass per unit length, rotatory inertia
$\underline{u}_S^0, \underline{v}^0$	initial conditions (displacement)
$\underline{\varphi}^0, \underline{\omega}^0$	initial conditions (rotation)
$\underline{u}_S^a, \underline{\varphi}^a$	clamped boundary conditions
$\underline{M}^a, \underline{N}^a, \underline{Q}^a$	free boundary conditions
$\delta \mathcal{I}^e$	extended variation
L_B, l_B	Lagrangian function, density
Z	functional
δA^{aero}	virtual aerodynamic work

Symbols, flow:

ρ	density
$\underline{v}$	velocity
p	pressure
e	total specific energy
$\underline{\lambda}$	surface velocity
$\underline{n}$	normal vector
V	volume
∂V	surface of V
l_y	local chord length
h	flight altitude
α	angle of attack
$\overline{\varphi}$	angle of sweep-back
$O(x, y, z)$	cartesian Euler system
$\delta(\ldots)/\delta t$	special time derivative
c_L	lift coefficient

501

$\tilde{m}$	torsional moment per unit length	c_m	momentum coefficient
$\tilde{p}$	transverse load per unit length	c_p	pressure coefficient
		M	Mach number
$O(x'_1, x'_2, x'_3)$	cartesian co-ordinate system	κ	ratio of specific heats
$O(\xi, \eta, \zeta)$	cross–section fixed, Lagrangian system		
$\vert 1$	spatial derivative in ξ–direction		
$t, (\dot{\ldots})$	time, time derivative		

Physical Model

Non-Stationary Flow in a Moving Grid

The underlying physical model for the computation of the flow field does not include molecular transport phenomena, radiation and body forces, since the solution of the complete Navier–Stokes equations needs to much computer time for aeroelastic applications [1, 2]. Furthermore, the fluid is assumed to be a perfect gas. As the wing is deformed by the unsteady aerodynamic loads, the boundary condition at the wing's surface depends on time: The component of the relative flow velocity vector normal to the surface has to be zero consistently. Therefore, the governing equations are solved for control volumina dependent on time. Within the finite-volume approximation the nodes discretizing the wing's surface at the inner boundary are fixed at this surface throughout the motion of the structure, whereas the nodes at the outer boundary which is the boundary of the computational domain remain fixed in space. Thus, the nodes of the finite volumes within the flow field have to be moved in such a way that a body adjusted structure of the grid is ensured during deformation. The balance equations read in conservative form

- mass:
$$\frac{\delta}{\delta t} \int_{V(t)} \rho dV + \int_{\partial V(t)} \rho(\tilde{v} - \tilde{\lambda}) \cdot \tilde{n} dS = 0, \tag{1}$$

- momentum:
$$\frac{\delta}{\delta t} \int_{V(t)} \rho \tilde{v} dV + \int_{\partial V(t)} \tilde{v}[\rho(\tilde{v} - \tilde{\lambda}) \cdot \tilde{n}] dS = - \int_{\partial V(t)} p \tilde{n} dS, \tag{2}$$

- energy:
$$\frac{\delta}{\delta t} \int_{V(t)} e dV + \int_{\partial V(t)} e(\tilde{v} - \tilde{\lambda}) \cdot \tilde{n} dS = - \int_{\partial V(t)} p(\tilde{v} \cdot \tilde{n}) dS. \tag{3}$$

Therein $\tilde{\lambda}$ denotes the velocity of the volume's surface and $\delta(\ldots)/\delta t$ is the time derivative in a volume-fixed frame. Only two of the thermodynamic variables density, pressure and internal energy are independent. With the assumption of thermally and calorically perfect gases one obtains for the total energy

$$e = \frac{p}{\kappa - 1} + \frac{1}{2}\rho(\tilde{v} \cdot \tilde{v}). \tag{4}$$

Underlying Beam Theory

The wing is modeled by a Timoshenko beam with generally non–coinciding centerlines of mass, bending and torsion. Its length l is measured along the Lagrangian co–ordinate ξ of the beam axis. The co–ordinate system $O(\xi,\eta,\zeta)$ corresponds with the co–ordinate system $O(x'_1,x'_2,x'_3)$ in the undeformed configuration and depends on the sweep angle (see figure 3). Six functions of spatial co–ordinate ξ and time t are introduced, which represent the three translational and the three rotational degrees of freedom of each cross–section, and determine the location of the beam in space and time. For many engineering purposes the number of independent functions is reduced by neglecting the shear deformation. Then the bending angle and the corresponding translational degree of freedom are coupled kinematically. Furthermore, in many applications the rotatory inertia of the bending modes are neglected. This results in the well–known Euler and Bernoulli beam theory [3]. Although the results of this model do not differ very much from results with the Timoshenko approximation in case of static problems of slender beams, computations of dynamical problems may show remarkable discrepancies, as the Euler-Bernoulli beam theory leads to a phase velocity which increases with decreasing wavelength out of bound. Moreover the group velocity with which the energy is transported is twice as fast as the phase velocity (anomalous dispersion).

Variational Principle

The formulation of a well–posed initial boundary value problem includes the determination of the initial conditions and the boundary values of the unknown variables. With the use of Lagrangian multipliers the variational equation is extended by the functionals Z_{AB} and Z_{RB} so that the variation finally includes the differential equations, the initial conditions and boundary conditions also [4],[5]. In the same way the winglet is taken into account by the functional Z_W. The variational problem

$$\delta \mathcal{I}^e(\underset{\sim}{u}_S, \underset{\sim}{\varphi}) = \int_{t_a}^{t_e} \delta \int_0^l l_B d\xi + \delta A^{aero} Dt + \delta(Z_{AB} + Z_{RB} + Z_W) = 0 \tag{5}$$

with the secondary conditions

$$\delta \underset{\sim}{u}(\xi, t_e) = \underset{\sim}{0}, \quad \delta \underset{\sim}{\varphi}(\xi, t_e) = \underset{\sim}{0} \tag{6}$$

is called equivalent to the well–posed initial boundary value problem, if it yields the appertaining differential equations, initial conditions and boundary conditions including the equations concerning the winglet (t_a, t_e denote the boundaries of the considered time interval). For the Lagrangian density one obtains

$$l_B = \frac{1}{2}\rho A \dot{\underset{\sim}{u}}_S \dot{\underset{\sim}{u}}_S + \frac{1}{2}\dot{\underset{\sim}{\varphi}} \underset{\approx}{\Theta}_S \dot{\underset{\sim}{\varphi}} + \rho A \underset{\sim}{u}_S \underset{\sim}{g} - \frac{1}{2}G A \underset{\sim}{\gamma} \underset{\approx}{K} \underset{\sim}{\gamma} - \frac{1}{2}\varphi_{1|1} G I_{D11} \varphi_{1|1} - \frac{1}{2}E A u_{B1|1}^2 - \frac{1}{2}\underset{\sim}{\varphi}_{|1} E \underset{\approx}{I}_B \underset{\sim}{\varphi}_{|1} \tag{7}$$

(S and B are the centers of gravity and bending, respectively). Applying Timoshenko's theory the shear angles are represented as

$$\gamma_2 = u_{S2|1} - \varphi_3 - (\zeta_{SD}\varphi_1)_{|1}, \quad \gamma_3 = u_{S3|1} + \varphi_2 + (\eta_{SD}\varphi_1)_{|1} \tag{8}$$

where ζ_{SD} and η_{SD} are the cartesian co–ordinate differences between the center of gravity and the shear center in a cross–section. The virtual work of the airloads δA^{aero} is

$$\delta A^{aero} = \int_0^l \left(\underset{\sim}{p} \delta \underset{\sim}{u}_S + \underset{\sim}{m} \delta \underset{\sim}{\varphi}_S \right) d\xi. \tag{9}$$

The aerodynamic forces per unit length are calculated in the cross–section fixed frame and transformed into the reference co–ordinate system. At the beginning of each time interval the generalized deflections and velocities have to be prescribed in the expression

$$Z_{AB} = \int_0^l \frac{1}{2} \left\{ \rho A \left[\left(\underset{\sim}{u}_S - \underset{\sim}{u}_S^0 \right) \underset{\sim}{\dot{u}}_S + \underset{\sim}{u}_S \underset{\sim}{v}_S^0 \right] + \left[\left(\underset{\approx}{\varphi} - \underset{\approx}{\varphi}^0 \right) \underset{\approx}{\Theta}_S \underset{\sim}{\dot{\varphi}} + \underset{\approx}{\varphi} \underset{\approx}{\Theta}_S \underset{\sim}{\omega}^0 \right] \right\} \Big|_{t=t_a} d\xi. \qquad (10)$$

The boundary conditions are divided up into one part where external forces are applied and another part where deflections are prescribed. At position $\xi = l$ boundary conditions for the generalized forces have to be specified while at the position $\xi = 0$ the generalized deflections in the expression Z_{RB} are given

$$Z_{RB} = \int_{t_a}^{t_e} \left[\left(\underset{\sim}{Q}^a + \underset{\sim}{N}^a \right) \underset{\sim}{u}_S + \underset{\sim}{M}^a \underset{\sim}{\varphi} \right]_l - \left[\left(\underset{\sim}{u}_S - \underset{\sim}{u}_S^a \right) \left(\underset{\sim}{Q} + \underset{\sim}{N} \right) + \left(\underset{\sim}{\varphi} - \underset{\sim}{\varphi}^a \right) \left(\underset{\sim}{M}_T + \underset{\sim}{M}_B \right) \right]_0 Dt. \qquad (11)$$

The winglet at the wing's tip is considered as a rigid body, i.e. elastic deformations of the winglet are neglected. This leads to appropriate terms for the kinetic and potential energy in the additional term Z_W in equation (5):

$$Z_W = \int_{t_a}^{t_e} \frac{1}{2} \left(m_W \underset{\sim}{\dot{u}}_{WS} \underset{\sim}{\dot{u}}_{WS} + \underset{\sim}{\dot{\varphi}}_W \underset{\approx}{\Theta}_{WS} \underset{\sim}{\dot{\varphi}}_W \right) + \left(m_W \underset{\sim}{u}_{WS} \underset{\sim}{g} \right) Dt. \qquad (12)$$

Further additional masses like nacelle and pylons can be treated similarly. As the winglet is assumed to be rigid and fixed to the wing no additional degree of freedom is introduced. Its movement follows the wing due to kinematical constraints.

Solution Strategy

The governing equations of the aerodynamic and structural parts are solved simultaneously in the time domain using the SOFIA solver. The time accurate integration–scheme has been succesfully applied to unsteady aeroelastic problems concerning wings [6, 7] and rotor blades [8, 9]. The central time loop of SOFIA consists of 3 parts:

ODISA (One-DImensional Structural Analysis) calculates the deformation of the structure which depends on the actual airloads.

The points at the inner boundary of the CFD–grid, i.e. at the structure's surface, are moved due to the computed deformation of the structure. In contrast to the points at the inner boundary the grid points at the outer boundary are fixed in the inertia frame. GRIDGEN (GRID GENerator) computes the new distribution of grid points so that a structured grid with the aforementioned characteristics is maintained. The velocities of the grid points result immediately from the difference between current and preceding location divided by the time interval.

Finally the flow field is computed for the current time. The main algorithm is called INFLEX (INstationäre FLußEXtrapolation). Based on the solution of the Euler equations for the conservative variables the airloads are determined.

A subiteration level has been implemented which computes the time step a second time. During this second calculation an average value of the aerodynamic forces of the current and the preceding time step is used. This iteration can be repeated until a certain error bound has been reached. Of course the computer time increases linearly with the (number of) subiteration levels. Numerical experiments have shown that in case of carefully chosen time steps – i.e. all the relevant time scales are taken into account properly – even one subiteration is sufficient.

The solution strategy is illustrated in figure 1. The wing is depicted in the reference positions as well as in the deformed positions at time levels n and $(n + 1)$.

Numerical Method: Flow

For the numerical integration of the strong conservation form of the Euler equations an implicit relaxation scheme is used [10]. The unfactored Euler equations are solved by applying a Newton iteration method. Relaxation is performed with a point Gauß-Seidel algorithm. The combination of a Newton method with a point Gauß-Seidel algorithm leads to a robust numerical scheme. Concerning the resolution of pressure and shock waves a characteristic variable splitting technique is employed. A detailed validation can be found in [11].

Grid Generation

An elliptic grid generator is employed to calculate the grid at each time step. A system of 3 elliptic partial differential equations of 2nd order (Poisson- and Laplace- equation) is solved. These PDE's are discretized by central differences. The resulting system of linear equations is solved iteratively by applying a Gauß-Seidel- algorithm [12]. As each grid deviates only slightly from the grid at the preceding time step, only few iterations are necessary.

- ODISA $\Rightarrow \underline{u}_S^{n+1}, \underline{r}_{SP}^{n+1} = \underline{r}_{SP}^{n+1}(\underline{\varphi}^{n+1})$
- INFLEX $\Rightarrow -(p \Delta A \, \underline{e}_\perp)^{n+1}$

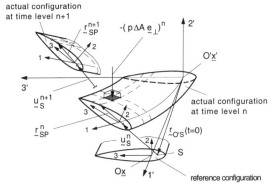

Figure 1: Actual and reference configuration

Numerical Method: Structure

A system of ordinary differential equations (ODEs) which is second order in time to determine the generalized deflections is derived by applying Hamilton's principle and the method of Ritz/Kantorowitsch. Linear damping is included (Rayleigh-damping). Discretization is done by isoparametric, two–noded elements. A reduced integration scheme avoids shear locking [13]. The set of ODEs is integrated by Newmark's method, where the resulting linear system of equations is solved directly with a LU-decomposition. The external forces are assumed to vary linearly during a time-step. Alternatively, the system of ODEs is diagonalized by solving the generalized eigenvalue problem (EVP) and the time integration is done by the evaluation of Duhamel's integral.

Results

Considered flight conditions

A steady state flight at an altitude of $8km$ (density: $0.525 kg/m^3$, velocity of sound: $310 m/s$) with a Mach number of $M_\infty = 0.75$ and an angle of attack of $\alpha_\infty = 3°$ is considered. The wing is not twisted in the undeformed configuration. Figure 2 displays a plane view of the wing and the geometry of the profile (RAE 2822) and figure 3 depicts the co–ordinate systems in which the flow field and the structural deformation are calculated. Furthermore the spanwise positions of the bending center and of the shear center are shown in this figure. As these centers do not coincide in each cross–section the torsional and bending modes are coupled statically. Since we are only interested in asymptotically static deflections, the complete dynamics is not discussed here. In case that the centers of gravity coincide with the bending centers, the eigenfunctions of some of the first vibritional modes exhibit the behaviour illustrated in figure 4. On the left hand side eigenfunctions belonging to bending modes are plotted, while the images on the right hand side belong to torsional modes. To elucidate the deformations the lines of constant deflection in z–direction are plotted additionally.

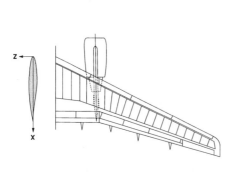

Figure 2: Plan view of the wing and geometry of the profile (RAE 2822)

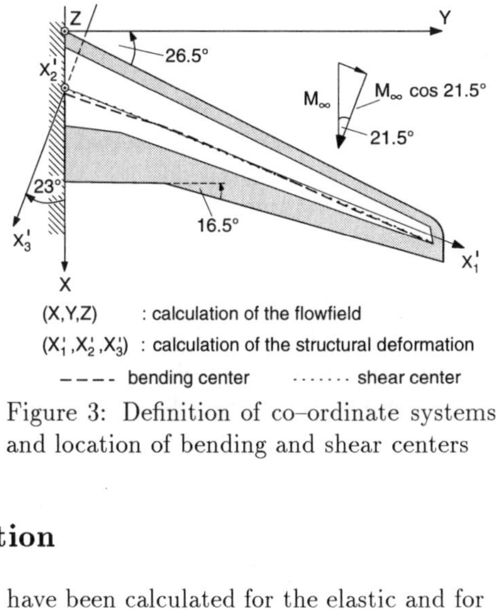

Figure 3: Definition of co–ordinate systems and location of bending and shear centers

Comparison of the lift distribution

Using the SOFIA code the lift distributions have been calculated for the elastic and for the rigid wing. Both calculations were performed for a non–swept (**rig3, ela3**) and for a 21.5° swept–back (**rig4, ela4**) wing. The discretizations of these two geometries are illustrated in figure 5. In figure 6 the results for the lift distributions are shown. In case of the non–swept wing the elasticity has only a minor influence on the lift distribution. This is due to the high torsional stiffness of the investigated wing. But in case of a swept–back wing the bending deformation leads to remarkable changes of the angle of attack. Therefore, for the swept–back wing the lift distribution differs significantly between the elastic (ela4) and the rigid (rig4) case. In the figures 8 and 9 the distributions of the pressure coefficient on the wing's upper surface are plotted for both cases using the same gray scale values for the pressure coefficient levels and contour lines. Due to the elasticity the transonic pocket at the wing's tip disappears. A quantitative comparison is given in figure 7, where the pressure coefficient at the wing's root and at 70% of the span is shown. Again, the structural deformation has only a small effect on the lift distribution in the non–swept case, whereas in the swept–back case the elastic deformation is quite remarkable, e.g. at 70% of the wing's span the transonic shock vanishes completely due to the elastic deformation. That corresponds to a loss of lift or effective span.

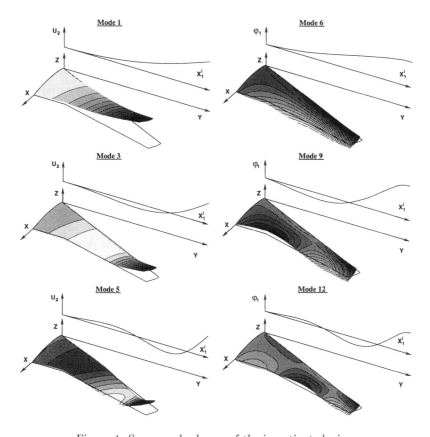

Figure 4: Some mode shapes of the investigated wing

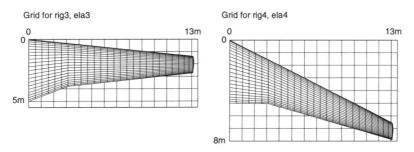

Figure 5: Plan view of the dicretized wing (not all grid points are plotted)

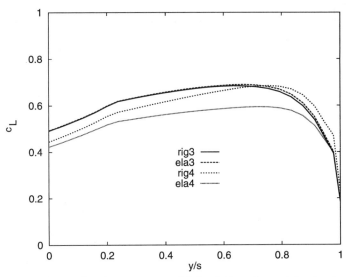

Figure 6: Lift distribution for the non–swept (rig3, ela3) and swept–back wing (rig4, ela4)

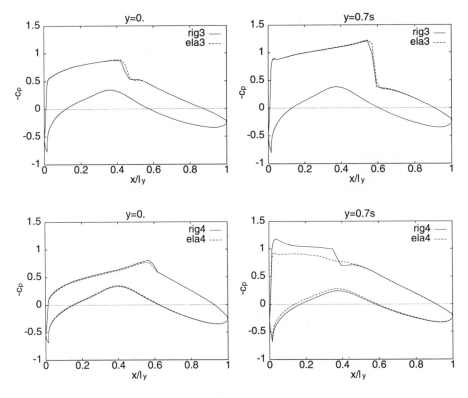

Figure 7: Pressure coefficient at the wing's root and at 70% of the span for the non–swept (rig3, ela3) and for the swept–back wing (rig4, ela4)

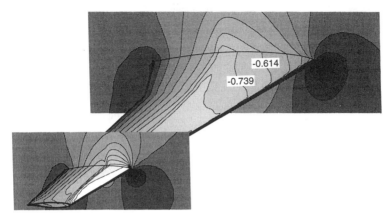

Figure 8: Distribution of the pressure coefficient c_p for the swept–back rigid wing (rig4)

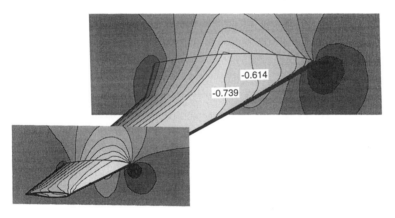

Figure 9: Distribution of the pressure coefficient c_p for the swept–back elastic wing (ela4)

Divergence

In case of small deflections the aerodynamic lift and pitching moment

$$c_L = c_{L0} + c_{L\alpha}\alpha_{\infty,eff}, \quad c_m = c_{m0} + c_{m\alpha}\alpha_{\infty,eff} \qquad (13)$$

can be linearized with respect to the undeformed configuration. Thus, the elastic structural forces as well as the aerodynamic forces depend approximately linearly on the deformation. In contrast to the elastic structural forces the constants of proportionality of the aerodynamic forces depend on the flight dynamic pressure. The flight dynamic pressure which equalizes the constants of proportionality is the critical value bringing about divergence.

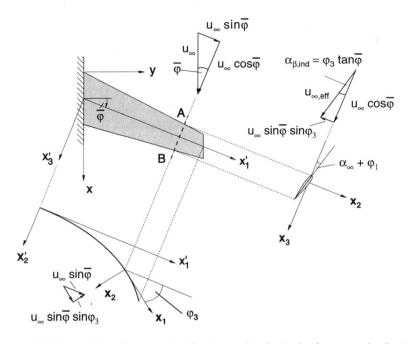

Figure 10: Influence of bending on the effective angle of attack of a swept–back wing

In case of divergence the deformations increase out of bound (within the linear theory). The effective free stream velocity in the plane of the cross–section AB in figure 10 is $u_\infty \cos \overline{\varphi}$ for the undeformed configuration. In the deformed configuration the cross–section AB is rotated about the x_3–axis by the bending angle φ_3. Consequently, the component $u_\infty \sin \overline{\varphi}$ induces a downwash $u_\infty \sin \overline{\varphi} \sin \varphi_3$, and the effective angle of attack becomes

$$\alpha_{\infty,eff} = \alpha_\infty + \varphi_1 - \tan \overline{\varphi} \varphi_3 \tag{14}$$

within the linear theory. This yields the linearized aerodynamic coefficients

$$c_L = c_{L0} + c_{L\alpha}(\alpha_\infty + \varphi_1 - \tan \overline{\varphi} \varphi_3), \quad c_m = c_{m0} + c_{m\alpha}(\alpha_\infty + \varphi_1 - \tan \overline{\varphi} \varphi_3), \tag{15}$$

which are introduced into equation (5). To estimate the critical flight dynamic pressure for static divergence, only the torsional deformation and the bending deformation perpendicular to the undisturbed flow are considered. Inertia effects and shear deformation are neglected. Taking the shear center D as the reference point in each cross–section (instead of the center of gravity) equation (5) yields

$$\delta \Pi = \delta \left\{ -\int_0^l \frac{1}{2} EI_{B33} \varphi_{3|1}^2 + \frac{1}{2} GI_{D11} \varphi_{1|1}^2 d\xi \right\} + \delta A^{aero} = 0. \tag{16}$$

with the virtual work of the aerodynamic forces

$$\begin{aligned}\delta A^{aero} &= \delta \left\{ \int_0^l p_{20} u_{D2} + m_{10} \varphi_1 + m_1' \frac{\varphi_1^2}{2} d\xi \right\} \\ &+ \int_0^l -m_1' \tan \overline{\varphi} \varphi_3 \delta \varphi_1 + p_2'(\varphi_1 - \tan \overline{\varphi} \varphi_3) \delta u_{D2}) d\xi. \end{aligned} \tag{17}$$

For the last equation the abbreviations

$$p_{20} = q_\infty (l_y c_{L0} + l_y c_{L\alpha} \alpha_\infty), \qquad p'_2 = q_\infty l_y c_{L\alpha},$$
$$m_{10} = q_\infty (l_y^2 c_{m0} + l_y^2 c_{m\alpha} \alpha_\infty) + \zeta_{QD} p_{20}, \qquad m'_1 = q_\infty l_y^2 c_{m\alpha} + \zeta_{QD} p'_2 \qquad (18)$$

have been introduced. The boundary conditions for the virtual displacements are

$$\delta u_{D2}(0) = \delta u_{D2}(l) = \delta \varphi_1(0) = \delta \varphi_1(l) = \delta \varphi_3(0) = \delta \varphi_3(l) = 0. \qquad (19)$$

With test functions and free coefficients we make the ansatz

$$\varphi_1 = \sum_{i=1}^{imax1} c_{1i} f_{1i}, \quad f_{1i} = (l-\xi)^{i+1} - l^{i+1}, \qquad \varphi_3 = \sum_{i=1}^{imax3} c_{3i} f_{3i}, \quad f_{3i} = (l-\xi)^{i+2} - l^{i+2}, \qquad (20)$$

for $0 \leq \xi \leq l$, such that it satisfies the geometrical and the dynamical boundary conditions. The deflection

$$u_{D2|1} = \varphi_3 + (\zeta_{DD} \varphi_1)_{|1} = \varphi_3 \qquad (21)$$

results immediately from the integration of the bending angle. By introducing the ansatz into the variational principle (16) one obtains a linear, non-homogeneous system of equations for the unknown coefficients c_{1i}, c_{3i}. In case of a singular coefficient matrix in this system of equations the generalized deflections can increase out of bound. Thus, the condition of a vanishing determinant leads to the critical flight dynamic pressure of divergence.

The derivatives $c_{L\alpha}, c_{m\alpha}$ of the linearized aerodynamic forces can be obtained by varying the angle of incidence α_∞ of the rigid wing, e.g.

$$c_{m\alpha}(y) = \frac{\Delta c_m(y)}{\Delta \alpha_\infty}. \qquad (22)$$

Since the torsional deformation is very important, the flight dynamic pressure is non-dimensionalized by the torsional stiffness, and according to Laidlaw (1958) [14] the characteristic number

$$\Pi_G = \frac{q_\infty l^4}{\overline{GI}_{D11}}. \qquad (23)$$

is introduced. It has a value of about 10 for the investigated wing and the flight conditions prescribed in this section.

Calculation of the critical flight dynamic pressure

First of all the non–swept wing (rig3, ela3) is considered. Figure 11 shows $\Pi_{G,crit}$ dependent on the number $imax1$ of functions in equation (20) for φ_1. A number of four functions is sufficient to determine $\Pi_{G,crit}$.

The critical value $\Pi_{G,crit}$ has been calculated for the swept–back wing (rig4, ela4), too. Both critical values for the non–swept wing (**Ritz3**, indicated with the diamond) and for the swept–back wing (**Ritz4**, indicated with the rectangle) are marked in figure 12. As can be seen the critical value of Π_G is much higher in case of sweep-back compared to no sweep.

The aerodynamic forces, equation (15), depend explicitly on the angle of sweep–back and on the aerodynamic derivatives, which also depend on of the angle of sweep–back. The critical value $\Pi_{G,crit}$ has been calculated for different values of the angle of backward sweep, but keeping the aerodynamic derivatives constant, see figure 12. For the curve Ritz3 the derivatives of the non–swept wing have been taken, whereas the derivatives of the swept–back wing have been used to calculate the curve Ritz4.

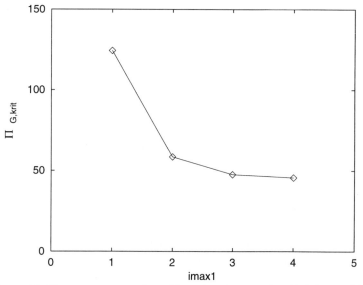

Figure 11: $\Pi_{G,crit}$ versus number of functions in equation (20) for φ_1 (ela3)

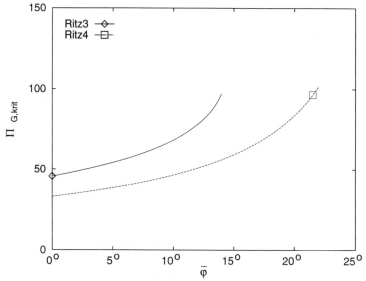

Figure 12: $\Pi_{G,crit}$ versus angle of sweep–back

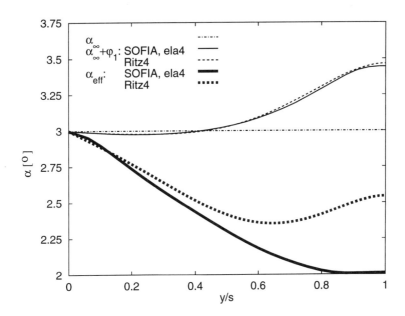

Figure 13: Spanwise distribution of the angle of attack. Results with Ritz' method (Ritz4) utilizing linearized aerodynamics versus results with SOFIA (ela4)

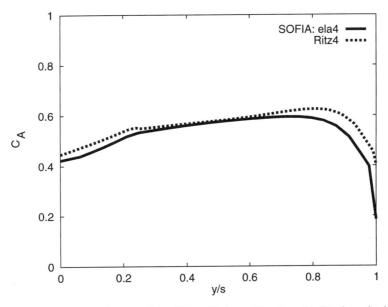

Figure 14: Spanwise distribution of the lift coefficient. Results with Ritz' method (Ritz4) utilizing linearized aerodynamics versus results with SOFIA (ela4)

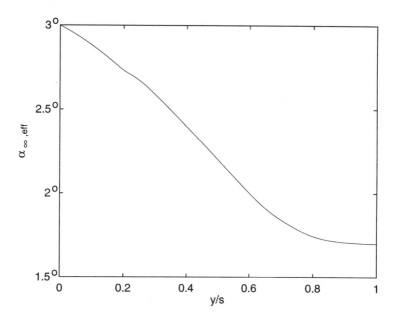

Figure 15: Spanwise distribution of the angle of attack for the non-swept wing at the limit of divergence calculated with SOFIA. $\Pi_{G,crit}$ was estimated with Ritz' method utilizing linearized aerodynamics.

Comparison of the linear Ritz–method and SOFIA

The Ritz–method described above has been used to calculate the deformation for the swept–back wing (Ritz4). Figure 13 shows the angle of attack along the spanwise direction. The torsional deformation calculated with SOFIA and Ritz' method nearly coincide (see $\alpha_\infty + \varphi_1$ for both methods). However, the effective angle of attack which depends on the bending angle shows remarkable differences. The lift coefficient differs at the root and at the tip of the wing (see figure 14). This is due to the differences concerning the effective angle of attack and due to three–dimensional effects which are not included in the linear aerodynamic model.

In addition a computation has been performed using the nonlinear SOFIA code where the flight dynamic pressure is equal to the critical one which was estimated with the linear Ritz method. Considering linear theory one would expect that the deformations increase out of bound. In figure 15 the angle of attack along the spanwise direction for the non–swept wing at the (linear) critical flight dynamic pressure is shown. For this pressure value the torsional angle remains smaller than 1.5° due to the nonlinear effects! (Of course, this might not be the actual maximum deformation in the nonlinear case, since the nonlinearity may cause a shift.)

Conclusions

A numerical method (SOFIA) has been presented to compute simultaneously the transonic flow field about an elastic wing and the wing's deformation. The equations describing the flow field (Euler equations) and the structural equations for a generalized Timoshenko beam are coupled in the time domain and integrated time accurately. A Fokker-type wing has been investigated. Results are shown for static aeroelastic problems. For the swept-back wing a significant change in the lift coefficient is observed, if the wing's elasticity is taken into account.

A Ritz-method including a linear model for the aerodynamic forces has been used to estimate the critical flight dynamic pressure of divergence. As could be expected, the critical pressure increases with increasing angle of sweep-back. A simulation performed with the SOFIA-code using the critical pressure from the linear thoery has revealed that the maximum torsional angle is about 1.5° in the nonlinear case.

Acknowledgements

This work was supported by the Deutsche Forschungsgemeinschaft which is gratefully acknowledged. We would like to thank Dr. A. Brenneis and Dr. A. Eberle from DASA for providing the computer code INFLEX and Mr. Clausen from DASA for providing the structural data. Computations were performed using the facilities of the Rechenzentrum der RWTH Aachen.

References

[1] K. Engel, F. Eulitz, S. Pokorny, M. Faden, 3-D Navier-Stokes Solver for the Simulation of the Unsteady Turbomachinery Flow on a Massively Parallel Hardware Architecture, in this publication.

[2] M. Wierse: Higher Order Upwind Schemes om Unstructered Grids for the Nonstationary Compressible Navier-Stokes Equations in Complex Timedependent Geometries in 3-D, in this publication.

[3] M. Botz, P. Hagedorn, On the Dynamics of Multibody Systems with Elastic Beams, Advanced Multibody System Dynamics pp. 217-236, Kluwer Academic Publisher, Netherlands, 1993.

[4] J. Ballmann, Grundlagen der Finite- Element- Methode I, Prinzipe der Dynamik für elastische Körper, Vorlesungsskript, Lehr- und Forschungsgebiet für Mechanik der RWTH-Aachen, 1975.

[5] K. Washizu, Variational Methods in Elasticity and Plasticity, Second edition 1975, Pergamon Press.

[6] D. Nellessen, S. Schlechtriem and J. Ballmann, Numerical Simulation of Flutter Using the Euler Equations, Notes on Numerical Fluid Mechanics, 1992.

[7] D. Nellessen, Schallnahe Strömungen um elastische Tragflügel, Dissertation, Rheinisch-Westfälische Technische Hochschule Aachen, Mathematisch-Naturwissenschaftliche Fakultät, 1995.

[8] S. Schlechtriem, D. Nellessen and J. Ballmann, Elastic Deformation of Rotor-Blades Due to BVI, 19th European Rotorcraft Forum, Paper No. B1, 1993.

[9] S. Schlechtriem, D. Nellessen, J. Ballmann, A Numerical Investigation of the Influence of Active Control Movements on Vibration and BVI-Noise, Proceedings of the 20th European Rotorcraft Forum, Paper No. 100, 1994.

[10] S.R. Chakravarthy, Relaxation Methods for Unfactored Implicit Upwind Schemes, AIAA Paper 84-0165, 1984.

[11] A. Brenneis and A. Eberle, Evaluation of an Implicit Euler Code Against Two and Three-Dimensional Standard Configurations, in: AGARD CP-507: Transonic Unsteady Aerodynamics and Aeroelasticity, Paper No. 10, March 1992.

[12] W. Schwarz, Elliptic System for Three- Dimensional Configurations Using Poisson Equations, Numerical Grid Generation in Computational Fluid Dynamics, 1st. ed., Pineridge Press, 1986, pp 341-352.

[13] T. J. R. Hughes, The Finite Element Method, Prentice- Hall Inc., Englewood Cliffs New Jersey, 1987.

[14] W.R. Laidlaw, *The Aeroelastic Design of Lifting Surfaces*, Notes prepared for M.I.T. Summer Course on Aeroelasticity, June–July 1958, printed by North American Aviation.

THREE-DIMENSIONAL NUMERICAL SIMULATION OF THE AEROTHERMODYNAMIC REENTRY

E. Laurien and J. Wiesbaum [1]

Institute for Fluid Mechanics, Technical University of Braunschweig
Bienroder Weg 3, D-38106 Braunschweig, Germany

SUMMARY

The aerothermodynamics of a spacecraft under the conditions of reentry into the earth' atmosphere is simulated numerically. The present method uses an unstructured tetrahedral grid with adaptive refinement to resolve local relaxation areas in the flow. Results for the geometry of an elliptical forebody at 30 deg. angle of attack are obtained under radiation adiabatic noncatalytic or fully catatylic wall conditions. Implementation and efficiency aspects of this method on high-performance computers are investigated.

1 INTRODUCTION

Aerothermodynamic flows around reentry vehicles such as the Space-Shuttle or HERMES are characterized by a strong bow shock, very high temperatures of the air (up to $12000\,K$) behind the shock and large surface heat flux [1, 2, 3]. Under these conditions air must be treated as a nonequilibrium real gas, with molecular vibration and chemical reactions between the air components. As the heat shield surface is hot catalytic effects become important, as well as radiation of heat from the body surface.

The three-dimensional numerical simulation of these flows has been a challenge for numerical simulation methods for many years. Problems encountered have been the accurate and efficient treatment of the nonequilibrium real gas effects, the development of efficient second order shock capturing methods for very strong shocks, the approximation of viscous effects, and the implementation of appropriate chemical wall boundary conditions. Geometries of interest are blunt bodies at high angles of attack.

In the present paper a generic forebody of the shape of an ellipsoid with the half axes $2.4\,m$, $2.0\,m$, and $0.6\,m$ at $30°$ angle of attack is considered as a model of the nose region of a HERMES-like configuration. The problem is treated as stationary with radiation-adiabatic boundary condition, i.e. the surface heat flux is in quilibrium with radiation cooling ($\epsilon = 0.85$) and the surface temperature is obtained as a result of the simulation. Inflow conditions were chosen on a point on the HERMES trajectory at an altitude of $69\,km$. The inflow temperature, pressure, density, velocity, viscosity and heat conductivity at this trajectory point can be taken from the following table :

T_∞	p_∞	ρ_∞	U_∞	μ_∞	k_∞
$225.5\,K$	$6.915\,\frac{N}{m^2}$	$1.068 \cdot 10^{-4}\,\frac{kg}{m^3}$	$6108.4\,\frac{m}{s}$	$1.47 \cdot 10^{-5}\,\frac{kg}{m\,s}$	$2.05 \cdot 10^{-2}\,\frac{J}{m\,s\,K}$

The Mach number is 20.3 and with the length of the forebody as the characteristic length the Reynolds number is $1.06 \cdot 10^5$.

---

[1] new address : HILTI AG, Liechtenstein

The purpose of this paper is to present a numerical method [4] (a two-dimensional version has been described in [5]) to simulate this problem on high-performance computers, which were state of the art at the time of the present priority research programme. Our work shows, that physically meaningful results can be obtained when modern numerical techniques such as FCT shock capturing, local refinement of unstructured grids, and vectorization and parallelization methods are combined.

2 BASIC EQUATIONS

The Navier-Stokes equations closely coupled with the chemical rate equations are used. This system can be numerically stiff, resulting in very small allowable time steps and slow convergence of a computation. To study the stiffness for our particular case we have investigated the relation between typical time scales of the flow, the vibration excitation of molecules and the chemical reactions in [6]. From these investigations it is concluded, that for the particular point on the trajectory considered here (B in Fig. 1), the flow must be treated as in vibrational equilibrium and chemical nonequilibrium. The regions of equilibrium and nonequilibrium for a body of $1\,m$ nose diameter are shown in Fig. 1.

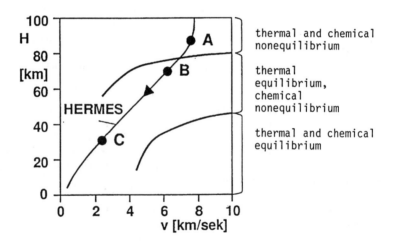

Figure 1: Regions of equilibrium and nonequilibrium real gas effects in the altitude(H) vs. velocity(v) plane and the HERMES trajectory.

In [7] the stiffness problem is discussed using experiences from axisymmetrical simulations (reentry of capsules and axisymmetric models of reentry vehicles). It is shown that on the trajectories of reentry vehicles the time scales of flow and chemistry are of the same order of magnitude and stiffness due to chemical reactions close to equilibrium has to be expected for capsules only. Finally axisymmetric simulations show, that even at the stagnation point the flow is not in chemical equilibrium (as suggested in textbooks, e.g. [2], when diffusion of species is neglected) and stiffness will therefore not be a problem.

The basic equations are the three-dimensional Navier-Stokes equations for a mixture of $m = 6$ species N_2, N, O_2, O, NO and NO^+

$$\frac{\partial \mathbf{U}}{\partial t} + \sum_{i=1}^{3} \frac{\partial \mathbf{F_i}}{\partial x_i} + \sum_{i=1}^{3} \frac{\partial \mathbf{G_i}}{\partial x_i} = \mathbf{W} \qquad (1)$$

with

$$\mathbf{U} = \begin{pmatrix} \rho_{N_2} \\ \rho_N \\ \rho_{O_2} \\ \rho_O \\ \rho_{NO} \\ \rho_{NO^+} \\ \rho \cdot u_1 \\ \rho \cdot u_2 \\ \rho \cdot u_3 \\ \rho \cdot e_{\text{tot}} \end{pmatrix}, \quad \mathbf{F_i} = \begin{pmatrix} \rho_{N_2} \cdot u_i \\ \rho_N \cdot u_i \\ \rho_{O_2} \cdot u_i \\ \rho_O \cdot u_i \\ \rho_{NO} \cdot u_i \\ \rho_{NO^+} \cdot u_i \\ \rho \cdot u_i \cdot u_1 + \delta_{1i} \cdot p \\ \rho \cdot u_i \cdot u_2 + \delta_{2i} \cdot p \\ \rho \cdot u_i \cdot u_3 + \delta_{3i} \cdot p \\ u_i \cdot (\rho \cdot e_{\text{tot}} + p) \end{pmatrix}. \qquad (2)$$

and

$$\mathbf{G_i} = \frac{1}{Re} \cdot \begin{pmatrix} \frac{1}{Sc} \cdot j_{N_2,i} \\ \frac{1}{Sc} \cdot j_{N,i} \\ \frac{1}{Sc} \cdot j_{O_2,i} \\ \frac{1}{Sc} \cdot j_{O,i} \\ \frac{1}{Sc} \cdot j_{NO,i} \\ \frac{1}{Sc} \cdot j_{NO^+,i} \\ \tau_{i1} \\ \tau_{i2} \\ \tau_{i3} \\ u_l \cdot \tau_{li} + \frac{1}{Pr\,Ec} \cdot q_{\text{kond},i} + \frac{1}{Sc\,Ec} \cdot q_{\text{diff},i} \end{pmatrix}, \quad \mathbf{W} = \begin{pmatrix} w_{N_2} \\ w_N \\ w_{O_2} \\ w_O \\ w_{NO} \\ w_{NO^+} \\ 0 \\ 0 \\ 0 \\ 0 \end{pmatrix} \qquad (3)$$

Here $\mathbf{U}$ is the vector of conservative variables, $\mathbf{F_i}$ and $\mathbf{G_i}$ the vectors of convective and diffusive fluxes in coordinate direction $i = 1, 2, 3$. The total energy e_{tot} is the sum of the species internal energies due to translation, molecular rotation and vibration, the kinetic energy and the chemical heat of formation, $j_{m,i}$ the fluxes due to mass diffusion, τ_{ij} the shear stresses, q_{kond} and q_{diff} the heat fluxes due to conduction and mass diffusion and Pr, Ec, and Sc are the Prandtl-, Eckert- and Schmidt-numbers, respectively. $\mathbf{W}$ is a source term containing species mass sources from chemical rate equations.

The radiation adiabatic boundary condition at the wall (Index w) is

$$-k \cdot \frac{\partial T}{\partial x_n} + \sum_{m=1}^{n} h_m \cdot j_{m,x_n} \bigg|_{normal} = \epsilon \sigma T_w^4 \qquad (4)$$

with $\sigma = 5.66 \cdot 10^{-8} \frac{W}{m^2} K^4$ and h_m the heat of formation of species m. Because the chemical boundary condition for realistic ceramic material is yet unknown, we investigate the limiting cases of (i) a noncatalytic wall with zero concentration gradients normal to the wall (i.e. $j_{normal} = 0$) and (ii) a fully catalytic wall with all chemical reactions being in equilibrium at the wall. The complete basic equations and boundary conditions are described in much more detail in [4, 5], see also [1].

3 NUMERICAL SIMULATION METHOD

3.1 GENERATION OF TETRAHEDRAL GRIDS

Our numerical method is based on unstructured locally refined grids consisting of tetrahedra. The computation is started using a coarse initial grid, which is generated as follows : A distribution of points (nodes) is defined on the body surface using the curvilinear coordinates φ and Ψ of Fig. 2, which shows one of four surface sectors. These points are located on lines $\Psi = const.$, their number on each individual line increasing with the value of φ. Then the shooting method is applied to generate nodes in the interior of the flow field along a coordinate ξ, which begins at the wall perpendicular to it.

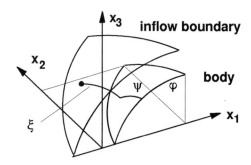

Figure 2: Cartesian and curvilinear coordinates used for grid generation.

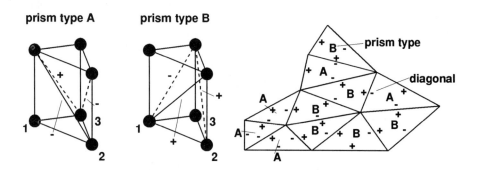

Figure 3: Types of prism subdivision into tetrahedra and application to each column of prisms defined by the surface triangulation.

After all nodes have been generated, triangles are defined on the body surface using Delaunay triangulation [8]. Each triangle is located at the bottom of a 'column' of prisms, each of which is then subdivided into three tetrahedra resulting in a subdivision of the quadrangular prism sides by diagonals in one way (+) or the other (-). To ensure compatibility between neighbouring prisms two types of subdivision (A with -,-,+) and (B with +,+,-), shown in Fig. 3 are applied to the individual column. The computation begins on a relatively coarse initial grid consisting of 9960 tetrahedral elements.

3.2 THREE-DIMENSIONAL TAYLOR-GALERKIN METHOD

We use the Taylor-Galerkin two-step method as described in [9, 10, 11, 12, 13]. The present three-dimensional method was developed from a two-dimensional implementation described in [5, 7, 14, 15].

In the three-dimensional code we were able to use several modules of the two-dimensional version almost unchanged. Many DO-loops in the program have been implemented with variable upper limits, e.g. the number of nodes per element (3 for a triangle, 4 for a tetrahedron) and the space dimension itself (2 or 3). The definition of some quantities to characterize the grid do not depend on the dimension, e.g. the number of nodes and the number of elements. The chemical model is dimension-independent, too.

In order to avoid unphysical oscillations or numerical instabilities near the bow shock, the technique of flux-corrected transport (FCT) [10] is applied. This method has the advantage of being independent of the governing equations, so modifications of the real gas model (e.g. by either treating the vibration excitation in nonequilibrium or assuming chemical equilibrium, see [7]) or perfect gas computations could be made without changing the FCT part of the program.

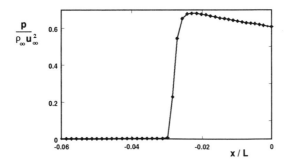

Figure 4: Pressure along a coordinate x across a Mach-20 shock.

The three-dimensional shock-capturing capability of our code is demonstrated : The pressure across a Mach-20-shock rises by a factor of about 500 across the shock. Fig. 4, shows, that the shock is captured within only 3 elements with very little upstream influence.

3.3 ADAPTIVE GRID REFINEMENT

Most of the advantages of the finite-element method compared to methods on structured grids become evident, when the grid is adaptively refined in regions of required higher resolution. We have developed a three-dimensional refinement algorithm as follows : First all node connections (edges of the tetrahedra) to be subdevided are marked. Then new nodes are generated half way between the two bounding nodes of an edge. If the connection is a line of the curved body

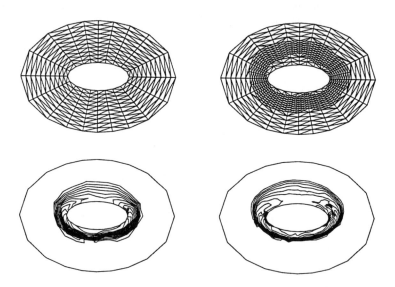

Figure 5: Coarse initial grid at the exit plane and adaptively refined grid (upper right), lower row : corresponding temperature contours.

surface, care must be taken to ensure, that the new node is actually on the body contour. We determine the position of new nodes on the basis of the curvilinear coordinates described above.

As a further step of the refinement algorithm those tetrahedra have to be eliminated, which have at least one refined edge, and new tetrahedra are generated in that place. In our implementation the refinement is based on edges (lines), sides (of triangular shape) and elements (tetrahedra). After the modification of the element information, the sides are modified by subdivision of one triangle into two (if one edge is refined), three(if two edges are refined) or four (if all three edges are refined) subtriangles. Based on the modified side information, the elements are finally generated by subdivion of one element in up to 8 (if all sides are subdivided into 4 subtriangles) sub-elements. The resulting refined grid is checked for consistency by the condition

$$nno - nco - ntri - nel = 1 \qquad (5)$$

with the number of nodes (nno), connections (nco), triangles ($ntri$) and elements (nel) [4].

An example of a refined triangular grid in the outflow plane is shown in Fig. 5 together with temperature contours. In this test the no slip condition at the wall has been replaced by the kinematic flow condition to avoid refinement near the body surface. The gradient of the Mach number was used as a sensor for refinement. This test confirms, that the gradients computed on a coarse grid become sharper as expected, when new nodes are added.

3.4 TEST OF THE NUMERICAL METHOD

Several tests of the two-dimensional version have been performed, see [7]. Our three-dimensional code was tested as follows : The flow around the spherical nose (diameter L) of the reentry body RAM-C II [16] was simulated at an altitude of $71\,km$, a Mach number of 26 and a Reynolds number of 5040. The inflow data are :

T_∞	p_∞	ρ_∞	U_∞	μ_∞
$212\,K$	$4.177\frac{N}{m^2}$	$6.844 \cdot 10^{-4}\frac{kg}{m^3}$	$7600\frac{m}{s}$	$1.53 \cdot 10^{-5}\frac{kg}{m\,s}$

The wall is assumed to be at $T_w = 1500\,K$ and noncatalytical. The temperature and mass fraction of O_2 along the stagnation streamline are compared with an axisymmetric solution by KLOMFASS et al. [17], see Fig. 6.

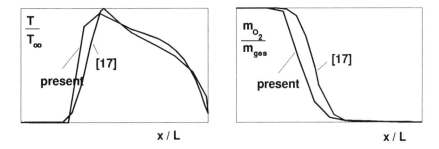

Figure 6: Comparison of a solution of the present method with [17] for the RAM-C-II test case (stagnation streamline, coordinate x).

Along the stagnation streamline the temperature rises sharply across the shock and decreases downstream to the prescribed wall value, the O_2-mass fraction decreases due to the nonequilibrium dissociation of oxygen. The beginning of the temperature increase due to the bow shock is in good agreement with [17], but our method resolves the shock with a steeper gradient. Therefore oxygen dissociation takes place in a region located upstream of the corresponding region in [17]. The overall agreement, however, is good.

3.5 COMPUTATIONAL ENVIRONMENT

The code was developed on the CONVEX C-210 of the Institute of Fluid Mechanics, vector computer with 50 Mflop/s peak performance and 256 MByte of memory. Two-dimensional calculations and 3-D tests were performed in an interactive mode [15] with quick visualization of results on a workstation.

All three-dimensional calculations were performed on the Siemens/Fujitsu S 400/40 of the Regional Computer Center of Lower Saxony (RRZN) in Hannover in batch modus. At the time of the computations only a 64 kbit/s link was available between Braunschweig and Hannover, which led to severe restrictions due to long transfer times of the result data for postprocessing in Braunschweig.

4 SIMULATION RESULTS

4.1 CONVERGENCE HISTORY

The L_2-norm of the total density residual in the flow field is shown in Fig. 7 vs. the number of time steps. The calculation is started on the coarse grid with an isothermal noncatalytic wall boundary condition. After the residual has dropped more than 3 orders of magnitude (after approximately 400 time steps), the grid is adaptively refined. This corresponds to the first increase of the residual in Fig. 7.

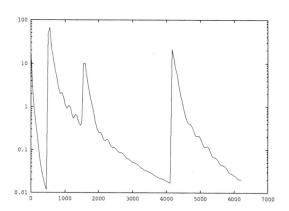

Figure 7: Residual vs. the number of time steps.

Then further timesteps are performed, with a slower decrease of the residual per time step on the finer grid, see Fig. 7. The second increase of the residual corresponds to a change in the boundary condition from isothermal to radiation adiabatic and the third to a change from noncatalytic to fully catalytic wall. If the program is started directly with the desired boundary conditions it converges much slower or not at all.

4.2 WALL HEAT FLUX

For the final solution the wall heat flux is plotted as a function of the coordinate s along the symmetry line as shown in Fig. 8 for the cases of fully catalytic and noncatalytic walls. The wavyness in the curves shown is caused by the computation not being fully converged. Values for realistic heat shield material must lie between these limiting cases. Our results show that a design on the basis of a 'worst case' (i.e. fully catalytic wall) assumption leads to an inaccuracy of about 17 per cent. The corresponding temperatures in the stagnation points are 1987 K for fully catalytic and 1892 K for the noncatalytic wall. At the stagnation point the mass fractions in the noncatalytic case are $m_{N2}/m_{tot} = 0.598$ and $m_N/m_{tot} = 0.16$, oxygen is totally dissociated.

Our numerical resolution of the flow is regarded to be sufficient, with 15 nodes across the boundary layer. However due to restrictions of 200 MByte maximum memory we were not able to refine the grid further and check grid dependency.

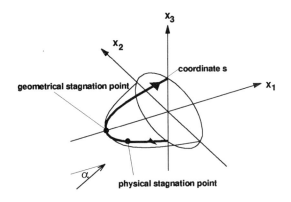

Figure 8: Coordinate s along the symmetry line of the ellipsoid.

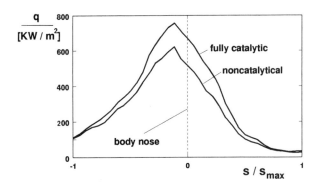

Figure 9: Heat flux along the symmetry line.

5 ASPECTS OF IMPLEMENTATION ON HIGH-PERFORMANCE COMPUTERS

5.1 VECTORIZATION

The vectorization technique used in our code is based on a well-known technique [18] and is described in detail in [19]. The elements (tetrahedra) are sorted into groups so that each pair of elements of the same group has no node in common. This helps in resolving recurrences which occur when computing nodal values of flow variables by accumulating data from adjacent nodes. The process of assigning triangles to groups can be implemented with time complexity $O(nel)$ (nel denoting the number of elements) in the following way:

We assume that the triangles are numbered from 0 to $M - 1$. First of all we take a reasonable number $N > 1$ of groups (let the groups be numbered from 0 to $N - 1$) which is supposed to be sufficient for placing each triangle in exactly one group so that each pair of triangles having a node or edge in common will belong to different groups. For an unstructured grid of triangles a lower limit for that number is given by the maximum number of triangles a node can belong to. For our problem 60 was sufficient.

We then try to put each triangle $m(= 0, .., M-1)$ in one of the G_N groups starting each trial at group $G_{mod(m,N)}$. When there are triangles in group $G_{mod(m,N)}$, which prevent us from inserting triangle m, we go to the next group $G_{mod(m,N)+1}$, then $G_{(m,N)+2}$, and so on. After reaching G_{N-1} we continue with G_0, then G_1 etc. If there is no group left where the triangle m can be put in we increase N by 1 and start again from the very beginning.

The reason why we rotate the starting group through all of the groups is the even filling we aim at for all groups at the end of the insertion. The process of checking whether a tetrahedron can be inserted in a given group where tetrahedra were put in can be rather time consuming if one goes through all tetrahedra of that group step by step. In order to speed-up this procedure (and bring the time complexity down to $O(nel)$) we introduce a $(nno \times N)$-matrix (represented as Boolean array in order to save storage), where nno denotes the number of all grid nodes, and - in case of a tetrahedra is being put in group G_j - mark all 3 entries of column j corresponding to the 3 nodes of the tetrahedron. It is now possible to perform the checking process for the given tetrahedron for group j by examining at most 3 entries of column j. This gives time complexity $O(nel)$.

5.2 PARALLELIZATION ON A SHARED MEMORY VECTOR-MULTIPROCESSOR

Especially for computationally intensive applications parallelization can reduce the elapsed time required to run a program although the CPU-time may increase. Nonetheless this may result in faster turnaround times for a parallel job in a multi-user environment.

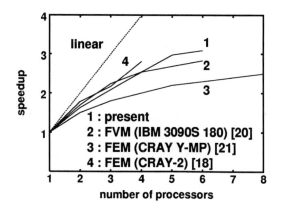

Figure 10: Speedup vs. number of processors.

As a first step towards parallelization we investigated the speedup of our code on the IBM 3090/600 VF of the Technical University of Braunschweig, a shared memory multiprocessor system with 6 vector processors. IBM released the VS FORTRAN 2.5 compiler which offers the opportunity for running FORTRAN code in parallel and hence supports language constructs for explicitly expressing parallelism.

Results and benchmark test of our two-dimensional code have been reported in [15, 19]. Speedup measurements in a dedicated runtime environment vs. the number of processors are summarized in Fig. 10, and compared with results of other authors. The comparison shows, that the speedup obtained with the present method can well compete with the work of other authors. However, this technique does not appear to be efficient for more than three processors, because the speedup curve decreases in slope and almost no further speedup is obtained with more than 6 processors. This can be explained as follows:

The speedup S defined as the ratio of elapsed time with a single processor t_1 and with p processors t_p is related to the degree of parallelization f (fraction of operation, which run in parallel) by Amdahl's law:

$$S = \frac{t_1}{t_p} = \frac{p}{1 + f(p-1)} \quad . \tag{6}$$

For our code comparison of Fig. 10 with graphs of eq. (6) yields $f \approx 0.1$, i.e. approximately 10% of our code runs in a serial manner (including some system overhead). It can easily be seen from Amdahl's law, that the speedup is only satisfactory for large numbers of processors (e.g. more than 6), when f is very small (e.g. $f = 0.01$).

Therefore more work would have to be put into further parallelization. From our experience, however, this cannot be done 'by hand' with reasonable amount of effort. For a discussion of efficient parallelization methods for our problem see [22]. In that paper the application of the domain decomposition method for distributed memory parallel computers and workstation clusters is given special attention.

6 CONCLUSIONS

The aerothermodynamics of a spacecraft under the conditions of reentry into the earth's atmosphere is simulated numerically. The present work shows, that physically meaningful results can be obtained with state of the art numerical methods and high-performance computers. Further investigations with our code would be desirable, in particular:

- How accurate is our result for the heat flux compared with free-flight experiments? How does a correct chemical wall model look like and how close is the result obtained with such a model to the limiting cases of fully catalytic or noncatalytic walls?
- More information and tests about grid dependency of our solution would be desirable.
- Since the most promising computer architecture today seems to be the distributed memory parallel computer, the technique of domain decomposition in connection with the message passing concept should be used for parallelization.

ACKNOWLEDGEMENT

This work was sponsored by the German Research Foundation (DFG) within the priority research program 'Flow Simulation on High-Performance Computers'. The authors are indebted to Prof. H. Oertel jr. for initiating this work and for his ongoing support. The first author would like to mention the many fruitful discussions within the priority research program, which helped him in the further understanding of various issues of computational fluid dynamics [23].

REFERENCES

[1] H. Oertel Jr., M. Böhle, J. Delfs, D. Hafermann und H. Holthoff: Aerothermodynamik, Springer-Verlag, Berlin, 1994.

[2] J.D. Anderson jr.: Hypersonic and High Temperature Gas Dynamics, McGraw-Hill, New York, 1989.

[3] C. Park: Nonequilibrium Hypersonic Aerothermodynamics, John Wiley and Sons, New York 1990.

[4] J. Wiesbaum: Dreidimensionale Numerische Simulation des aerothermodynamischen Wiedereintritts im kontinuumsmechanischen Bereich, Dissertation, Technische Universität Braunschweig, ZLR-Bericht 95-03, 1995.

[5] H. Holthoff: Numerische Simulation des Wärmeübergangs in Hyperschallströmungen, Dissertation, Technische Universität Braunschweig, ZLR-Bericht, 1995.

[6] E. Laurien, M. Böhle, H. Holthoff und J. Wiesbaum: Modelluntersuchungen zur numerischen Berechnung von Wiedereintrittsströmungen, DGLR-Jahrbuch I, 613-620, 1991.

[7] E. Laurien, M. Böhle, H. Holthoff und J. Wiesbaum: Numerische Simulation von Wiedereintrittsströmungen mit dem Taylor-Galerkin Finite-Elemente Verfahren, DGLR-Kongreß 28.9 - 1.10 1993, Göttingen.

[8] D. F. Watson: Computing the n-dimensional Delaunay tessellation with application to Voronoi polytopes, The Computer Journal Vol. 24 NO. 2 167-172 1981.

[9] R. Löhner, K. Morgan, and O.C. Zienkiewicz: An Adaptive Finite-Element Procedure for Compressible High Speed Flows, Comp. Meth. Appl. Mech. and Eng. 51, 441-465 (1985).

[10] R. Löhner: Finite-Element Flux-Corrected-Transport (FEM-FCT) for Euler and Navier-Stokes Equations, International Journal for Numerical Methods in Fluids 7, 1093-1109 (1987).

[11] K. Morgan, J. Peraire: Finite-Element Methods for Compressible Flows, VKI-LS 1987-04, 1987.

[12] M. Lötzerich: Beitrag zur Strömungsberechnung in Turbomaschinen mit Hilfe der Methode der Finiten-Elemente, Dissertation, RWTH Aachen, 1987.

[13] A. Vornberger: Strömungsberechnung auf unstrukturierten Netzen mit der Methode der Finiten-Elemente, Dissertation, RWTH Aachen, 1989.

[14] E. Laurien, M. Böhle, H. Holthoff, J. Wiesbaum, and A. Lieseberg: Finite-Element Algorithm for Chemically Reacting Hypersonic Flows, AIAA 92-0754 (1992).

[15] J. Wiesbaum, H. Holthoff, E. Laurien und W. Rönsch: Experiences with the Taylor-Galerkin Finite-Element Method for Hypersonic Aerothermodynamics on Supercomputers, in : E. H. Hirschel (ed.) : Flow Simulation with High-Performance Computers I, Notes on Numerical Fluid Mechanics 38, 379-392, Vieweg, Braunschweig 1993.

[16] G.V. Candler R.W. MacCormack: The Computation of Hypersonic Ionized Flows in Chemical and Thermal Nonequilibrium, AIAA 88-0511, 1988.

[17] A. Klomfaß, S. Müller, J. Ballmann: Modellierung Hypersonischer Strömungen, Sonderforschungsbereich 253, Inst. f. Geometrie u. Praktische Mathematik, RWTH Aachen, 1993.

[18] J. Erhel: Finite-Element Methods on Parallel and Vector Computers, Application in Fluid Dynamics, Proceedings of the International Conference on Supercomputing, Athens 1987, Springer, Heidelberg, New York, 1988, 768-781.

[19] W. Rönsch, J. Wiesbaum, H. Holthoff, and E. Laurien: Parallelization of the Taylor-Galerkin Finite-Element Method for Shared Memory Vector Multiprocessors, Proc. 1993 Summer Meeting of ASME Fluid Engineering Division 'CFD Algorithms and Applications for Parallel Processors', June 21-23, 1993.

[20] Agarwal and J.C. Lewis: Computational Fluid Dynamics on Parallel Processors, Computing Systems in Engineering 3, 251-259 (1992).

[21] A.F. Junior and M.F. Ahmad: PERFES- Parallel Finite-Element Solvers for Flow-Induced Fracture, Computing Systems in Engineering 3, 379-392 (1992).

[22] E. Laurien: Parallelisierung adaptiver Simulationsverfahren in der Strömungsmechanik, Proc. Workshop über Wissenschaftliches Rechnen, 2.-4. Juni 1993, Braunschweig.

[23] H. Oertel jr. und E. Laurien: Numerische Strömungsmechanik, Springer-Verlag, Berlin, 1995.

Investigations of Hypersonic Flows past Blunt Bodies at Angle of Attack

S. Brück, G. Brenner,[*] D. Rues, D. Schwamborn
DLR Institute of Fluid Mechanics, Bunsenstr. 10, D–37073 Göttingen

Abstract

The hypersonic, chemically reacting flow past a generic HERMES geometry is simulated numerically. Interest is focused on the influence of the modeling of chemical kinetics and the dynamics of energy exchange in flows with complex viscous and inviscid interactions such as shock/boundary- layer and shock/shock interactions. Results for the axisymmetric flow past a flared hyperboloid are obtained with the NSHYP code to study the principal effects and to validate the present model by comparing with Direct Monte Carlo Simulations and experimental results obtained in the high enthalpy wind tunnel HEG. Furthermore, the generic shape of a double ellipsoid is chosen to demonstrate the performance of the CEVCATS code for three-dimensional simulations.

Introduction

For the design of more reliable geometries of reusable reentry vehicles the effects of chemical and thermal nonequilibrium are important. The overestimated flap efficiency at the design of the Space-Shuttle is a good example for the importance of these effects. With the beginning of the project we had been able to calculate three-dimensional flows on a CRAY YMP vector computer with the assumption of a perfect gas and the model of chemical and thermal equilibrium. Due to the increased computational effort the implemented models for chemical and thermal nonequilibrium could only be applied to two-dimensional and axisymmetric flow. The use of codes on the new vector computer (NEC SX3) of the DLR has required an adaptation of the programs, which was done for the perfect gas version of the NSHYP code. We decided to implement the validated models for nonequilibrium flows in the existing aerodynamic code CEVCATS (see Kroll et al. [7]), where we have focussed from the beginning on highest performance and flexibility, which now allows to simulate three-dimensional flows with detailed chemistry past complex configurations.

---

[*] Present address: Department of Fluid Mechanics, University Erlangen, Germany.

Simulation Model

Governing Equations

Here the equations are discussed which describe the viscous flow including chemical reactions in nonequilibrium. For the system of conservation equations two formulations have been used and implemented into the flow solvers NSHYP and CEVCATS. The conservation of the total mass yields,

$$\frac{\partial \rho}{\partial t} + \frac{\partial \rho u_j}{\partial x_j} = 0, \tag{1}$$

as well as the conservation of the species mass

$$\frac{\partial \rho_s}{\partial t} + \frac{\partial \rho_s u_j}{\partial x_j} + \frac{\partial \rho_s v_j}{\partial x_j} = \dot{\omega}_s^c, \tag{2}$$

where $\dot{\omega}_s^c$ is the production term of the species s and n_s the total number of species. Only $n_s - 1$ additional equations have to be considered if the total mass conservation in equation (1) is included which holds for the CEVCATS code while the NSHYP code treats only the mass-species equations. The momentum equation is written as

$$\frac{\partial \rho u_i}{\partial t} + \frac{\partial}{\partial x_j}(\rho u_i u_j + p\delta_{ij}) + \frac{\partial}{\partial x_j}\tau_{ij} = 0, \tag{3}$$

and the energy equation as

$$\frac{\partial e}{\partial t} + \frac{\partial}{\partial x_j}((e+p)u_j) + \frac{\partial}{\partial x_j}(\dot{q}_j^t + \dot{q}_j^\nu) + \frac{\partial}{\partial x_i}u_j\tau_{ij} + \frac{\partial}{\partial x_j}\sum_s \rho_s v_{sj} h_s = 0. \tag{4}$$

The equation for the vibrational energy is

$$\frac{\partial e_{s'}^\nu}{\partial t} + \frac{\partial}{\partial x_j}e_{s'}^\nu u_j + \frac{\partial}{\partial x_j}\varepsilon_{s'}^\nu \rho_{s'} v_{s'j} + \frac{\partial}{\partial x_j}\dot{q}_{s'j}^\nu = \dot{\omega}_{s'}^c \varepsilon_{s'}^\nu + \dot{\omega}_{s'}^\nu. \tag{5}$$

In the above equations Einstein's summation convention is used and ρ, ρu_i, e, e^ν denote total density, components of the momentum, total energy and vibrational internal energy, respectively.

Thermodynamic and Caloric State Equations

We consider a mixture of monatomic and diatomic, electrically neutral gases. It is assumed, that the density of this gas is sufficiently low. Thus, each component can be considered as a perfect gas (weakly interacting gas). A common translational-rotational temperature T has been defined to characterize the equilibrium contribution to the internal energy ε_s^t. The nonequilibrium part of the internal energy ε_s^ν is characterized by T_s^ν. In the present model, this is due to the contributions of the vibrational modes. For the internal energy of the species s it follows

$$\varepsilon_s = \varepsilon_s^t(T) + \varepsilon_s^\nu(T_s^\nu). \tag{6}$$

The energy of the mixture ε is the mass-averaged energy of all species, hence

$$\varepsilon = \sum_s \alpha_s \varepsilon_s^t + \sum_{s'} \alpha_{s'} \varepsilon_{s'}^\nu \text{ with } \alpha_s = \frac{\rho_s}{\rho} \text{ and } \rho = \sum_s \rho_s. \tag{7}$$

The equilibrium contributions to the internal energy can be obtained from the caloric equation of state. Since, in the present model, the translational and rotational modes are fully excited, the corresponding equation simplifies to

$$\varepsilon_s^t = c_{v,s}(T - T_0) + \Delta h_s^0 \text{ with} \tag{8}$$

$$c_{v,s} = \frac{f}{2} R_s. \tag{9}$$

Here f is the number of degrees of freedom ($f = 3$ for monatomic species, $f = 5$ for diatomic species) and R_s is the gas constant for species s.

The pressure of the mixture is calculated according to Dalton's law as the sum of the partial pressures,

$$p = \sum_s p_s = \rho T \sum_s \alpha_s R_s. \tag{10}$$

The enthalpy of the species is

$$h_s = \varepsilon_s + R_s T. \tag{11}$$

The calculation of the vibrational energy of the species is based on the assumption of the harmonic oscillator. For the vibrational energy per unit mass follows

$$\varepsilon_{s'}^\nu = R_{s'} \frac{\Theta_{s'}^\nu}{exp\left(\frac{\Theta_{s'}^\nu}{T_{s'}^\nu}\right) - 1}, \tag{12}$$

where Θ_s is the characteristic temperature for vibration.

The transition of energy between the vibrational and the translational modes is approximated by the Landau-Teller formula

$$\dot{\omega}_{s'}^\nu = \rho_{s'} \frac{\varepsilon_{s'}^{\nu*} - \varepsilon_{s'}^\nu}{\bar{\tau}_{s'} + \tau_{cs'}}. \tag{13}$$

Here, $\varepsilon_{s'}^{\nu*}$ is the vibrational energy evaluated at equilibrium, i.e. the translational-rotational temperature. The averaged relaxation times $\bar{\tau}_{s'}$ are obtained from correlations of Millikan & White [8]. Since the relaxation times, obtained with this approach are unreasonably short for high temperatures, a correction as suggested by Park [9] is employed, which describes the correction term $\tau_{cs'}$ in the Landau-Teller formulation in equation (13).

Transport Properties

The calculation of the transport phenomena is based on the hypotheses of Stokes, Fourier and Fick. The fluxes of the momentum, energy and mass are calculated from the gradients of the respective physical properties, mass-averaged velocity, temperature and concentrations. The viscosity of the single species is obtained from the curve fits of Blottner et al. [2], the mixture viscosity from Wilke's mixture rule. The thermal conductivity is obtained based on the Eucken correction. A constant Lewis number of $Le = 1.2$ is used to calculate the diffusion coefficient.

Modeling of the Chemical Reactions

The increase or decrease of the species concentrations due to chemical reactions is given by the source terms

$$\dot{\omega}_s^c = M_s \sum_r (\nu_{sr}'' - \nu_{sr}') k_r^f \left(\prod_{i=1}^{ns} \left(\frac{\rho_i}{M_i}\right)^{\nu_{ir}'} - \frac{1}{K_s^{eq}} \prod_{i=1}^{ns} \left(\frac{\rho_i}{M_i}\right)^{\nu_{ir}''} \right). \tag{14}$$

Here, k_r^f is the forward reaction velocity of the r-th reaction, K_s^{eq} is the equilibrium constant of the s-th elemental reaction, ν_{sr}'' and ν_{sr}' are the stoichiometric coefficients and M_s are the molar masses. The temperature dependence of the reaction velocities is described by the modified Arrhenius formula

$$k_r^f = a_r T^{b_r} exp\left(-\frac{E_{ar}}{k_B T}\right), \tag{15}$$

where k_B is the Boltzmann constant. The parameters a_r and b_r as well as the activation energy E_{ar} tabulated for most species of technical interest are taken from Park [9]. In the present investigations 5 species (N_2, O_2, NO, N, O) and the 17 dissociation and exchange reactions have been considered.

Boundary Conditions

Here, only the boundary conditions for the species concentrations and the temperature are discussed, for further information see Brenner [3]. For the species concentration two different boundary conditions are used representing the two states in between which the real catalysis mechanism would be expected. For the fully catalytic wall we assume the equilibrium value for the concentrations

$$\alpha_s = \alpha_{s,eq} \tag{16}$$

and for the noncatalytic wall

$$\frac{\partial \alpha_s}{\partial x_j} n_j = 0. \tag{17}$$

In addition to an isothermal wall the radiation adiabatic wall is introduced

$$-\sigma \epsilon T^4 + \sum_s h_s \rho v_j n_j + \dot{q}_j^t n_j + \dot{q}_j^\nu n_j = 0, \tag{18}$$

where σ is the Stefan-Boltzmann constant and ϵ the emission rate, which is 0.85 for the present investigation.

Numerical Method

The described models for the detailed nonequilibrium chemistry and vibrational excitation were implemented into an implicit finite-difference method based on an upwind TVD discretisation (NSHYP). The equation system is solved by a Gauss-Seidel line relaxation method (see Brenner [3]). For three dimensional simulations the existing model was implemented into the CEVCATS code. This program solves the time-dependent Euler/Navier-Stokes equations with a finite-volume method based on an upwind flux-vector splitting scheme. Due to the stiffness of the source terms a point-implicit treatment is required. Up to now only the simulation of chemical nonequilibrium and thermal equilibrium has been implemented. The present implementation allows the simulation of any set of nonequilibrium chemical reactions with a moderate number of species.

Results

Flow past the Hyperboloid-Flare Configuration

The Hyperboloid-Flare configuration is a generic axisymmetric geometry with the contour of the windward centerline of the HERMES 1.0 at an angle of attack of $30°$ and a flap angle of $15°$. The flow past this configuration has been investigated in several cold and hot hypersonic wind tunnels in Europe with respect to the separation length near the hinge line and the resulting flap efficiency. In the present investigation this geometry was chosen to validate the numerical method under defined conditions with experimental results, to demonstrate the effects of different modeling of chemistry and vibrational excitation and to compare with another numerical method based on the description of discrete particles and their collisions (DSMC).

Validation with wind-tunnel experiments in HEG

The "HochEnthapie Kanal Göttingen" of the DLR is a free-piston driven shock tunnel which produces flows at enthalpies form 10 to 25 MJ/kg for a testing time of about 1-2 ms. In this facility it isn't possible to reproduce simultaneously all free-flight relevant parameters. But some parameters e.g. the binary scaling parameter $\rho \cdot l_{ref}$, can be reproduced and used for the validation of computational codes. Only if a precise reproduction of the flows in the experimental facility like in HEG is possible, a prediction of flows around real geometries at free-flight conditions is credible. This strategy has also been applied for the present investigations with the NSHYP and the CEVCATS code.

The typical flow field past the hyperboloid-flare configuration can be seen in figure 1 where the numerically calculated schlieren picture is compared with the picture recorded at HEG and documented by Kastell et al. [6]. The most important flow features are pointed out and a good qualitative agreement can be observed. Differences in the flow field can be found near the hinge line where the numerical simulation underestimates the experimentally observed separation length. This leads to a slightly different shock configuration where the separation shock collides with the flare shock in front of the intersection with the bow shock while in the experiment all these shocks come together at one point. At the trailing edge of the flare the flow expands. Differences in the separation region can also be observed in the comparison of the wall values of pressure (figure 2) and heat flux (figure 3). Four computations have been performed with different catalysis models and different numbers of grid points to prove grid convergence. Besides small differences in the hinge region the pressure distributions are nearly the same for all computations and agree with the experimental results. The heat-flux distributions are shown in figure 3. The solutions representing a noncatalytic wall are very close to experimental results. Thus, it can be assumed that the wall is noncatalytic on the forebody.

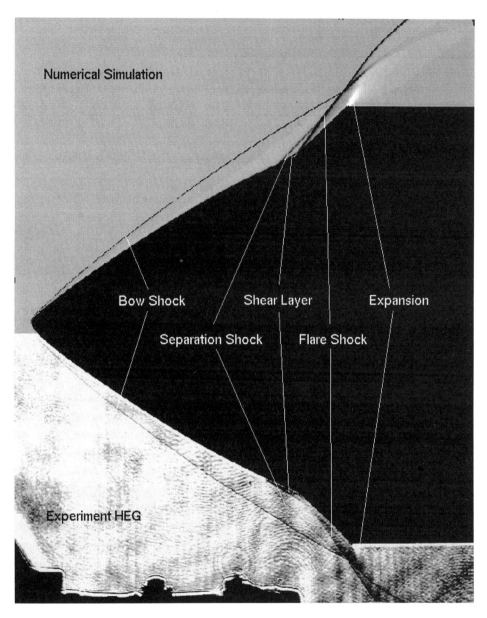

Figure 1: Calculated and optical schlieren picture for the flow past the hyperboloid-flare configuration in the HEG facility.

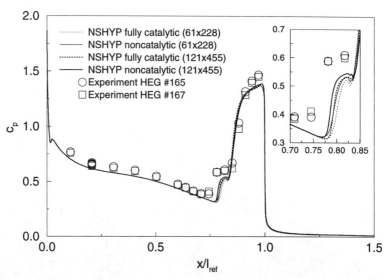

Figure 2: Wall pressure along the body surface for different grid resolutions and wall catalysis modeling compared with experimental HEG results.

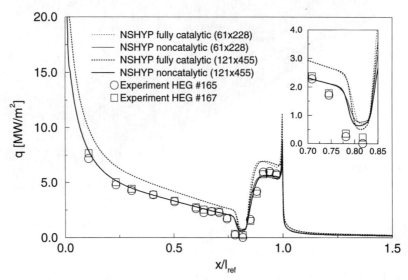

Figure 3: Heat flux along the body surface for different grid resolutions and wall catalysis modeling compared with experimental HEG results.

Comparison with a DSMC method for conditions in 87 km altitude

Figure 4 shows the comparison of the pressure and heat-flux coefficients for simulations obtained with the Navier Stokes code NSHYP (model I) and a Direct Simulation Monte Carlo method (DSMC) for reentry conditions in 87 km altitude. A higher trajectory point had to be chosen because a DSMC computation for 77 km altitude was not feasible due to the decreasing Knudsen number. More informations can be found in Bergemann [1]. Besides the different approaches the same chemical and thermodynamic models have been applied in the Navier-Stokes and DSMC code. The presented results show an excellent agreement.

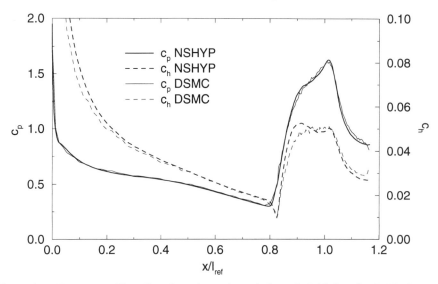

Figure 4: Pressure and heat flux along the surface of a hyperboloid-flare for the Navier-Stokes code NSHYP and a DSMC method.

Comparison of different physico-chemical models for free-flight conditions

The hyperboloid-flare configuration has also been used for sensitivity studies of the NSHYP code concerning the modeling of chemistry and the excitation of the vibrational modes. Here free-flight conditions at the reentry point at 77 km altitude are carried out (see Brenner et al. [4]). The pressure (figure 5) and the heat-flux distribution (figure 6) for the assumption of a mixture of chemically frozen gases (perfect gas), chemical and thermal nonequilibrium (model I), chemical nonequilibrium and thermal equilibrium (model II) and chemical and thermal equilibrium (model III) are compared.

The latter model leads to a significant overestimation of the peak pressure after the shock interaction compared to the nonequilibrium solutions (see figure 5). Furthermore, the position of the peak pressure is shifted upstream. The same holds for the heat-flux distribution shown in figure 6 where also significantly overestimated loads are observed. For this kind of flows chemical nonequilibrium effects seem to be very important for the prediction of surface loads and the flap efficiency. The influence of thermal nonequilibrium is to be less important for these flow regimes.

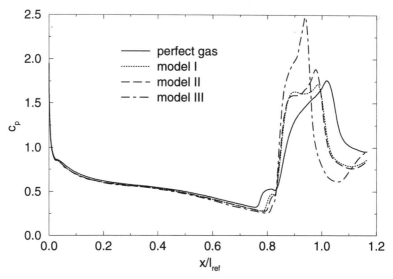

Figure 5: Surface-pressure distribution for the perfect-gas simulation, chemical and thermal nonequilibrium (model I), chemical nonequilibrium and thermal equilibrium (model II) and chemical and thermal equilibrium (model III).

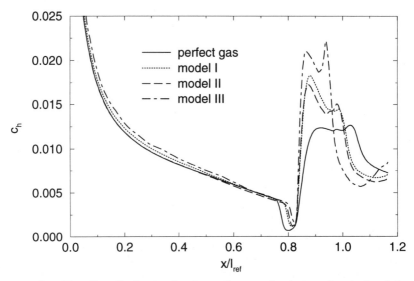

Figure 6: Heat-flux distribution for the perfect-gas simulation, chemical and thermal nonequilibrium (model I), chemical nonequilibrium and thermal equilibrium (model II) and chemical and thermal equilibrium (model III).

Flow past a Double Ellipsoid

The double-ellipsoid configuration is a generic geometry for the investigation of the flow past the forebody of the HERMES reentry vehicle. It was used as test-case configuration at the workshops on "Hypersonic Flows for Reentry Problems" held in Antibes, see Désidéri et al. [5].

The chosen free-stream conditions correspond to the workshop test case VI.10 ($M_\infty = 25$, $75\,km$ altitude) however the scale of the geometry has been changed in the present investigation to more realistic size. A realistic scaling is very important for the assumption of a radiation adiabatic wall, see Riedelbauch [10], as well as for chemical and thermal relaxation processes. Instead of a reference length (half width) of 0.15 m a length of 1 m has therefore been chosen.

The CEVCATS code has been used for the calculation of the three-dimensional flow field with the modeling of chemical nonequilibrium and thermal equilibrium. A grid with $121 \times 49 \times 61$ number of grid points in streamwise, wall-normal and circumferential direction, respectively, was used applying a multi-grid strategy for convergence acceleration.

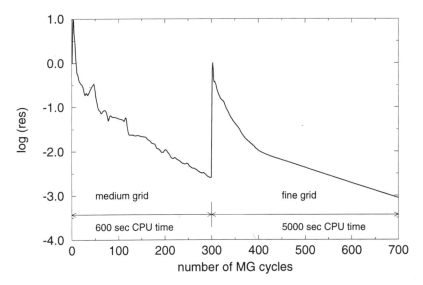

Figure 7: Typical convergence history of a computation of the nonequilibrium flow past a double ellipsoid.

The computations were made on the NEC SX3 vector computer of DLR with a performance of 1.6 GFLOPS. Each simulation required ~ 6000 sec CPU time, 800 Mbyte memory and 150 Mbyte external memory (XMU). Four simulations were performed with different catalysis modeling (fully catalytic, noncatalytic) and different formulations for the wall temperature (isothermal wall with $T_{wall} = 1500\,K$, radiation adiabatic wall).

The species concentration on the stagnation streamline for the different catalysis models are shown in figure 8. The molecular nitrogen and oxygen in the free stream is dissociated behind the bow shock due to the increased temperature. The concentration of oxygen equilibrates after a distance corresponding to 20 % of the shock stand-off distance while the concentration of nitrogen equilibrates later. For the fully catalytic wall the atomic species recombine completely.

The influence of the different boundary conditions can be seen in figure 9 where the temperature distributions along the stagnation streamline are shown. They all represent temperature profiles typical for a chemical nonequilibrium flow . Differences can be observed for the different catalysis models while the influence of the temperature boundary conditions is only felt close to the wall.

The temperature distributions along the symmetry line on the windward and leeward side for the assumption of a radiation adiabatic wall are shown in figure 10. The maximum temperature in the stagnation point region predicted with the assumption of a fully catalytic and radiation adiabatic wall gives the upper possible limit of the temperature loads. For this assumption a maximum temperature of 2000 K is observed while for the noncatalytic wall only 1600 K is predicted. On the leeward side the radiation adiabatic wall temperature is quite low. This boundary condition is considered more realistic than a high constant wall temperature. At the canope an increase of the wall temperature is observed.

In comparison with pressure distributions along the symmetry line it can be seen that the locations of the maximum temperature and pressure loads are slightly different. This effect has also been observed by other authors, e.g. Riedelbauch [10].

The flow fields in the symmetry plane and surface distributions are shown in figure 12. Figures 12a and 12b show the species concentration for N and O assuming a noncatalytic wall and below (figures 12c and 12d) the respective distributions for the fully-catalytic wall are compared. The dissociation in the bow shock region shows nearly the same behavior while recombination is found for the fully-catalytic wall solution. A weak shock can be seen generated due to the canopy geometry as a concentration of isolines.

In figure 12e the surface distribution of the radiation-adiabatic-wall temperature on the leeward side of the configuration is compared for different catalysis models (top: noncatalytic, bottom: fully catalytic) with the corresponding skin-friction lines in figure 12f. The flow structure (separation and reattachment) is very similar for both simulations while the noncatalytic solution shows a larger separation in front of the canopy. The same effect is also observed for the hyperboloid-flare configuration.

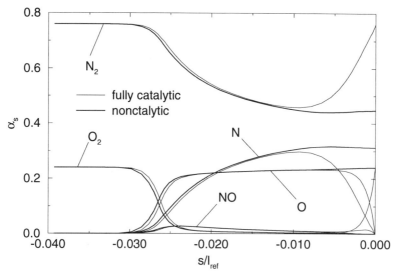

Figure 8: Composition along the stagnation streamline of the double ellipsoid for different wall-catalysis modeling.

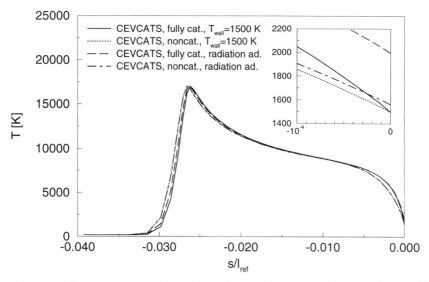

Figure 9: Temperature distribution along the stagnation streamline of the double ellipsoid for different boundary conditions.

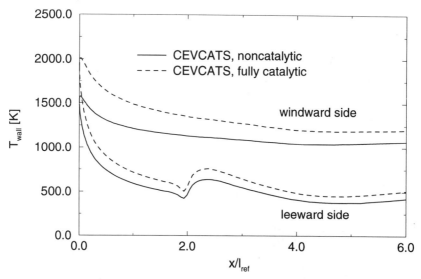

Figure 10: Distributions of the radiation-adiabatic wall temperature in the symmetry plane of the double ellipsoid for different catalysis models.

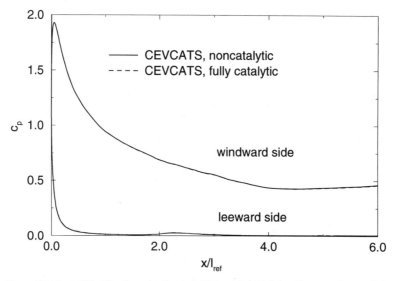

Figure 11: Pressure Distributions in the symmetry plane of the flow on the surface of the double ellipsoid.

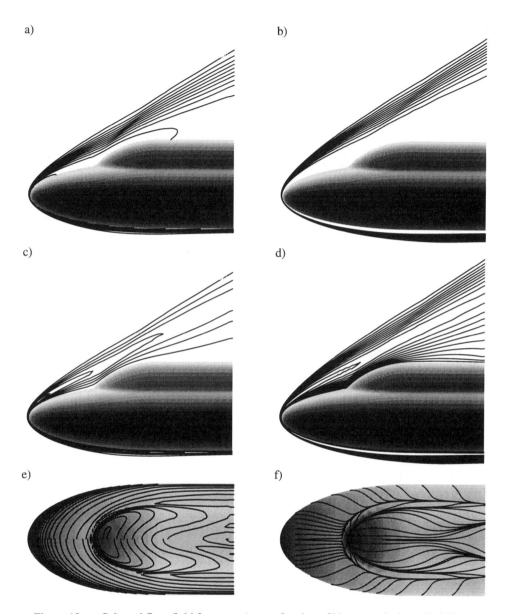

Figure 12: Selected flow-field features: a) mass fraction of N, noncatalytic wall, $\Delta N = 2\%$; b) mass fraction of O, noncatalytic wall, $\Delta O = 2\%$; c) a) mass fraction of N, fully catalytic wall, $\Delta N = 2\%$; d) mass fraction of O, fully catalytic wall, $\Delta O = 2\%$; e) radiation adiabatic wall temperature, noncatalytic (up) and fully catalytic wall (down); f) skin friction lines, noncatalytic (up) and fully catalytic wall (down).

Concluding Remarks

The flow past the hyperboloid-flare configuration has been investigated with respect to the validation of the numerical method for axisymmetric flows (NSHYP) including chemical and thermal nonequilibrium, where a good agreement with experimental results is achieved. Different models have also been investigated for free-flight conditions. Here, a detailed formulation of the chemistry is necessary while the modeling of the vibrational relaxation is less important. The results for free-flight conditions are also reproduced by a DSMC method.

The geometry of a double ellipsoid has been used to demonstrate the performance of the validated CEVCATS code including chemical nonequilibrium. Different boundary conditions have been employed to point out the influence on heat flux and/or temperature distributions. With this code the investigation of more complex geometries is possible with a reasonable amount of computational resources.

References

[1] F. Bergemann. Gaskinetische Simulation von kontinuumsnahen Hyperschallströmungen unter Berücksichtigung von Wandkatalyse. DLR-FB 94-30, Göttingen, 1994.

[2] F.G. Blottner, M. Johnson, and M. Ellis. Chemically Reacting Viscous Flow Program for Multi-Component Gas Mixtures. Technical Report:SC-RR-70-754, SANDIA, 1971.

[3] G. Brenner. Numerische Simulation von Wechselwirkungen zwischen Stößen und Grenzschichten in chemisch reagierenden Hyperschallströmungen. DLR-FB 94-04, Göttingen, 1994.

[4] G. Brenner, W. Kordulla, and S. Brück. Further Simulations of Flows Past Hyperboloid-Flare Configurations. DLR IB 221-93 A 13, Göttingen, 1993.

[5] J.-A. Désidéri, R. Glowinski, and J. Périaux, (eds.) *Hypersonic Flows for Reentry Problems*, Volume I–III. Springer, Heidelberg, 1990/91.

[6] D. Kastell and G. Eitelberg. Flow Visualization in High Temperature, High Velocity Flows with a combined Holographic Interferometer and Laser-Schlieren System. In *Proc. of 7th Int. Symp. on Flow Visualization (ISFV)*, Seattle WA, September 1995.

[7] N. Kroll and R. Radespiel. An Improved Flux Vector Split Discretization Scheme for Viscous Flows. DLR-FB 93-53, Braunschweig, 1994.

[8] R.C. Millikan and D.R. White. Systematics of Vibrational Relaxation. *Journal of Chemical Physics*, 139:3209–3212, 1963.

[9] C. Park. *Nonequilibrium Hypersonic Aerothermodynamics*. John Wiley & Sons, New York, 1989.

[10] S. Riedelbauch. Aerothermodynamische Eigenschaften von Hyperschallströmungen über strahlungsadiabate Oberflächen. DLR-FB 91-42, Göttingen, 1992.

IV. RESULTS OF BENCHMARK COMPUTATIONS

Benchmark Computations of Laminar Flow Around a Cylinder

M. Schäfer[a] and S. Turek[b]

(With support by F. Durst[a], E. Krause[c] and R. Rannacher[b])

[a]Lehrstuhl für Strömungsmechanik, Universität Erlangen-Nürnberg
Cauerstr. 4, D-91058 Erlangen, Germany
[b]Institut für Angewandte Mathematik, Universität Heidelberg
INF 294, D-69120 Heidelberg, Germany
[c]Aerodynamisches Institut, RWTH Aachen
Wüllnerstr. zw. 5 u. 7, D-52062 Aachen, Germany

SUMMARY

An overview of benchmark computations for 2D and 3D laminar flows around a cylinder is given, which have been defined for a comparison of different solution approaches for the incompressible Navier-Stokes equations developed within the Priority Research Programme. The exact definitions of the benchmarks are recapitulated and the numerical schemes and computers employed by the various participating groups are summarized. A detailed evaluation of the results provided is given, also including a comparison with a reference experiment. The principal purpose of the benchmarks is discussed and some general conclusions which can be drawn from the results are formulated.

1. INTRODUCTION

Under the DFG Priority Research Program "Flow Simulation on High Performance Computers" solution methods for various flow problems have been developed with considerable success. In many cases, the computing times are still very long and, because of a lack of storage capacity and insufficient resolution, the agreement between the computed results and experimental data is - even for laminar flows - only qualitative in nature. If numerical solutions are to play a similar role to wind tunnels, they have to provide the same accuracy as measurements, in particular in the prediction of the overall forces.

Several new techniques such as "unstructured grids", "multigrid", "operator splitting", "domain decomposition" and "mesh adaptation" have been used in order to improve the performance of numerical methods. To facilitate the comparison of these solution approaches, a set of benchmark problems has been defined and all participants of the Priority Research Program working on incompressible flows have been invited to submit their solutions. This paper presents the results of these computations contributed by altogether 17 research groups, 10 from within of the Priority Research Program and 7 from outside. The major purpose of the benchmark is to establish, whether constructive conclusions can be drawn from a comparison of these results so that the solutions can be improved. It is not the aim to come to the conclusion that a particular solution A is better than another solution B; the intention is rather to determine whether and why certain approaches are superior to others. The benchmark is particularly meant to stimulate future work.

In the first step, only incompressible laminar test cases in two and three dimensions have been selected which are not too complicated, but still contain most difficulties representative of industrial flows in this regime. In particular, characteristic quantities such as drag and lift coefficients have to be computed in order to measure the ability to produce quantitatively accurate results. This benchmark aims to develop objective criteria for the evaluation of the different algorithmic approaches. For this purpose, the participants have been asked to submit a fairly complete account of their computational results together with detailed information about the discretization and solution methods used. As a result it should be possible, at least for this particular class of flows, to distinguish between "efficient" and "less efficient" solution approaches. After this benchmark has been proved to be successful it will be extended to include also certain turbulent and compressible flows.

It is particularly hoped that this benchmark will provide the basis for reaching decisive answers to the following questions which are currently the subject of controversial discussion:

1. Is it possible to calculate incompressible (laminar) flows accurately and efficiently by methods based on explicitly advancing momentum?

2. Can one construct an efficient solver for incompressible flow without employing multigrid components, at least for the pressure Poisson equation?

3. Do conventional finite difference methods have advantages over new finite element or finite volume techniques?

4. Can steady-state solutions be efficiently computed by pseudo-time-stepping techniques?

5. Is a low-order treatment of the convective term competitive, possibly for smaller Reynolds numbers?

6. What is the "best" strategy for time stepping: fully coupled iteration or operator splitting (pressure correction scheme)?

7. Does it pay to use higher order discretizations in space or time?

8. What is the potential of using unstructured grids?

9. What is the potential of a posteriori grid adaptation and time step selection in flow computations?

10. What is the "best" approach to handle the nonlinearity: quasi-Newton iteration or nonlinear multigrid?

These questions appear to be of vital importance in the construction of efficient and reliable solvers, particularly in three space dimensions. Everybody who is extensively consuming computer resources for numerical flow simulation should be interested.

The authors have tried their best in presenting and evaluating the contributed results in as much detail as possible and hope that all participants of the benchmark will find themselves correctly quoted.

2. DEFINITION OF TEST CASES

This section gives a brief summary of the definitions of the test cases for the benchmark computations, including precise definitions of the quantities which had to be computed and also some additional instructions which were given to the participants.

2.1 Fluid Properties

The fluid properties are identical for all test cases. An incompressible Newtonian fluid is considered for which the conservation equations of mass and momentum are

$$\frac{\partial U_i}{\partial x_i} = 0$$

$$\rho \frac{\partial U_i}{\partial t} + \rho \frac{\partial}{\partial x_j}(U_j U_i) = \rho \nu \frac{\partial}{\partial x_j}\left(\frac{\partial U_i}{\partial x_j} + \frac{\partial U_j}{\partial x_i}\right) - \frac{\partial P}{\partial x_i}$$

The notations are time t, cartesian coordinates $(x_1, x_2, x_3) = (x, y, z)$, pressure P and velocity components $(U_1, U_2, U_3) = (U, V, W)$. The kinematic viscosity is defined as $\nu = 10^{-3}\,\mathrm{m^2/s}$, and the fluid density is $\rho = 1.0\,\mathrm{kg/m^3}$.

2.2 2D Cases

For the 2D test cases the flow around a cylinder with circular cross–section is considered. The geometry and the boundary conditions are indicated in Fig. 1. For all test cases the

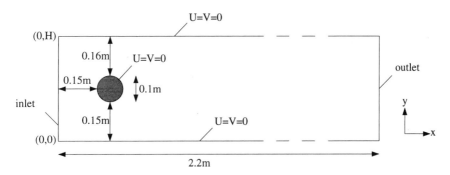

Figure 1: *Geometry of 2D test cases with boundary conditions*

outflow condition can be chosen by the user.

Some definitions are introduced to specify the values which have to be computed. $H = 0.41\,\mathrm{m}$ is the channel height and $D = 0.1\,\mathrm{m}$ is the cylinder diameter. The Reynolds number is defined by $Re = \overline{U}D/\nu$ with the mean velocity $\overline{U}(t) = 2U(0, H/2, t)/3$. The drag and lift forces are

$$F_D = \int_S (\rho \nu \frac{\partial v_t}{\partial n} n_y - P n_x) \, dS \quad , \quad F_L = -\int_S (\rho \nu \frac{\partial v_t}{\partial n} n_x + P n_y) \, dS$$

with the following notations: circle S, normal vector n on S with x-component n_x and y-component n_y, tangential velocity v_t on S and tangent vector $t = (n_y, -n_x)$. The drag and lift coefficients are

$$c_D = \frac{2 F_w}{\rho \bar{U}^2 D} \quad , \quad c_L = \frac{2 F_a}{\rho \bar{U}^2 D}$$

The Strouhal number is defined as $St = Df/\bar{U}$, where f is the frequency of separation. The length of recirculation is $L_a = x_r - x_e$, where $x_e = 0.25$ is the x-coordinate of the end of the cylinder and x_r is the x-coordinate of the end of the recirculation area. As a further reference value the pressure difference $\Delta P = \Delta P(t) = P(x_a, y_a, t) - P(x_e, y_e, t)$ is defined, with the front and end point of the cylinder $(x_a, y_a) = (0.15, 0.2)$ and $(x_e, y_e) = (0.25, 0.2)$, respectively.

a) Test case 2D-1 (steady):

The inflow condition is

$$U(0, y) = 4 U_m y (H - y)/H^2, \ V = 0$$

with $U_m = 0.3 \, \text{m/s}$, yielding the Reynolds number $Re = 20$. The following quantities should be computed: drag coefficient c_D, lift coefficient c_L, length of recirculation zone L_a and pressure difference ΔP.

b) Test case 2D-2 (unsteady):

The inflow condition is

$$U(0, y, t) = 4 U_m y (H - y)/H^2, \ V = 0$$

with $U_m = 1.5 \, \text{m/s}$, yielding the Reynolds number $Re = 100$. The following quantities should be computed: drag coefficient c_D, lift coefficient c_L and pressure difference ΔP as functions of time for one period $[t_0, t_0 + 1/f]$ (with $f = f(c_L)$), maximum drag coefficient c_{Dmax}, maximum lift coefficient c_{Lmax}, Strouhal number St and pressure difference $\Delta P(t)$ at $t = t_0 + 1/2f$. The initial data $(t = t_0)$ should correspond to the flow state with c_{Lmax}.

c) Test case 2D-3 (unsteady):

The inflow condition is

$$U(0, y, t) = 4 U_m y (H - y) \sin(\pi t/8)/H^2, \ V = 0$$

with $U_m = 1.5 \, \text{m/s}$, and the time interval is $0 \leq t \leq 8 \, \text{s}$. This gives a time varying Reynolds number between $0 \leq Re(t) \leq 100$. The initial data $(t = 0)$ are $U = V = P = 0$. The following quantities should be computed: drag coefficient c_D, lift coefficient c_L and pressure difference ΔP as functions of time for $0 \leq t \leq 8 \, \text{s}$, maximum drag coefficient c_{Dmax}, maximum lift coefficient c_{Lmax}, pressure difference $\Delta P(t)$ at $t = 8 \, \text{s}$.

2.3 3D Cases

For the 3D test cases the flows around a cylinder with square and circular cross-sections are considered. The problem configurations and boundary conditions are illustrated in Figs. 2 and 3. The outflow condition can be selected by the user. Some definitions are

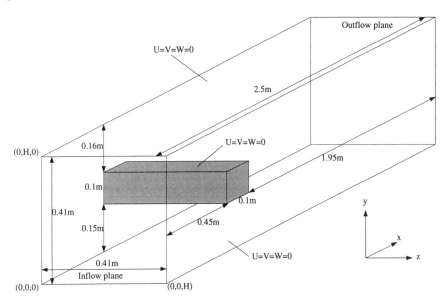

Figure 2: *Configuration and boundary conditions for flow around a cylinder with square cross-section.*

introduced to specify the values which have to be computed. The height and width of the channel is $H = 0.41$ m, and the side length and diameter of the cylinder are $D = 0.1$ m. The characteristic velocity is $\overline{U}(t) = 4U(0, H/2, H/2, t)/9$, and the Reynolds number is defined by $Re = \overline{U}D/\nu$. The drag and lift forces are

$$F_D = \int_S (\rho\nu \frac{\partial v_t}{\partial n} n_y - p n_x) \, dS \quad , \quad F_L = -\int_S (\rho\nu \frac{\partial v_t}{\partial n} n_x + P n_y) \, dS$$

with the following notations: surface of cylinder S, normal vector n on S with x-component n_x and y-component n_y, tangential velocity v_t on S and tangent vector $t = (n_y, -n_x, 0)$. The drag and lift coefficients are

$$c_D = \frac{2F_w}{\rho \overline{U}^2 DH} \quad , \quad c_L = \frac{2F_a}{\rho \overline{U}^2 DH}$$

The Strouhal number is $St = Df/\overline{U}$ with the frequency of separation f, and a pressure difference is defined by $\Delta P = \Delta P(t) = P(x_a, y_a, z_a, t) - P(x_e, y_e, z_e, t)$ with coordinates $(x_a, y_a, z_a) = (0.45, 0.20, 0.205)$ and $(x_e, y_e, z_e) = (0.55, 0.20, 0.205)$.

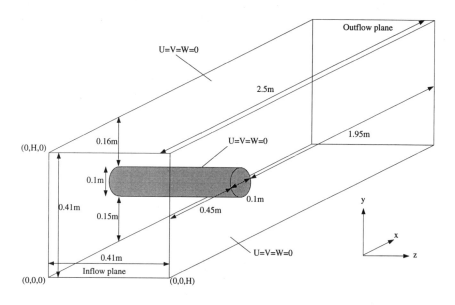

Figure 3: *Configuration and boundary conditions for flow around a cylinder with circular cross-section.*

a) Test cases 3D-1Q and 3D-1Z (steady):

The inflow condition is
$$U(0,y,z) = 16U_m yz(H-y)(H-z)/H^4, \; V = W = 0$$
with $U_m = 0.45\,\text{m/s}$, yielding the Reynolds number $Re = 20$. The following quantities should be computed: drag coefficient c_D, lift coefficient c_L and pressure difference ΔP.

b) Test cases 3D-2Q and 3D-2Z (unsteady):

The inflow condition is
$$U(0,y,z,t) = 16U_m yz(H-y)(H-z)/H^4, \; V = W = 0$$
with $U_m = 2.25\,\text{m/s}$, yielding the Reynolds number $Re = 100$. The following quantities should be computed: drag coefficient c_D, lift coefficient c_L and pressure difference ΔP as functions of time for <u>three</u> periods $[t_0, t_0 + 3/f]$ (with $f = f(c_L)$), maximum drag coefficient c_{Dmax}, maximum lift coefficient c_{Lmax} and Strouhal number St. The initial data $(t = t_0)$ are arbitrary, however, for fully developed flow.

c) Test cases 3D-3Q and 3D-3Z (unsteady):

The inflow condition is
$$U(0,y,z,t) = 16U_m yz(H-y)(H-z)\sin(\pi t/8)/H^4, \; V = W = 0$$

with $U_m = 2.25$ m/s. The time interval is $0 \leq t \leq 8$ s. This yields a time-varying Reynolds number between $0 \leq Re(t) \leq 100$. The initial data $(t = 0)$ are $U = V = P = 0$. The following quantities should be computed: drag coefficient c_D, lift coefficient c_L and pressure difference ΔP as functions of time for $0 \leq t \leq 8$ s, maximum drag coefficient c_{Dmax}, maximum lift coefficient c_{Lmax} and pressure difference $\Delta P(t)$ for $t = 8$ s.

2.4 Instructions for Computations

The following additional instructions concerning the computations were given to the participants:

- In the case of the steady calculations 2D-1, 3D-1Q and 3D-1Z, the results have to be presented for three successively coarsened meshes (notation: h_1, h_2 and h_3 with finest level h_1).

- Any iterative process used for the steady computations should start from zero values.

- In the case of the unsteady calculations 2D-2, 2D-3, 3D-2Q, 3D-2Z, 3D-3Q and 3D-3Z, the results have to be presented for three successively coarsened meshes (notation as in the steady case) with a finest time discretization (notation: Δt_1) and also for two successively coarsened time discretizations (notation: Δt_2 and Δt_3) together with the finest mesh h_1.

- The finest spatial mesh h_1, the finest time discretization Δt_1 and the coarsening strategies can be chosen by the user.

- The convergence criteria for the iterative method in the steady case and for each time step in the unsteady cases (in connection with implicit methods) can be chosen by the user.

- The outflow condition can be chosen by the user.

- If possible, the calculations should be performed on a workstation. For all computers used, the theoretical peak performance and the MFlop rate for the *LINPACK1000-benchmark* (in 64-bit arithmetic) should be provided. The LINPACK1000 value should be obtained with the same compiler options as used for the flow solver.

- In addition to the benchmark results a description of the solution methods should be given.

3. PARTICIPATING GROUPS AND NUMERICAL APPROACHES

In Table 1 the different groups that provided results for the present benchmark computation are listed, and the individual test cases for which results were provided are also indicated. In Table 2 the numerical methods and implementations of the participating groups are summarized. Only the major features which are the most important for the evaluation of the results are given. The following abbreviations are used in the table: Finite difference method (FD), Finite volume method (FV), Finite element method (FE), Navier-Stokes equations (NS) and Multigrid method (MG). PEAK means the peak performance in MFlops and LINP the Linpack1000 MFlop-rate.

Table 1: *Participating groups and test cases for which results were provided. The p indicates that only parts of the required results for the corresponding test case were given, and x indicates a full set of results*

Participants/test cases	2D			3D					
	1	2	3	1Q	1Z	2Q	2Z	3Q	3Z
1) RWTH Aachen, Aerodynamisches Institut E. Krause, M. Weimer, M. Meinke	x	x	x	x	x			p	p	p
2) ASC GmbH (TASCflow) F. Menter, G. Scheuerer					x				
3) TU Berlin Inst. für Strömungsmechanik F. Thiele, L. Xue	p	p	p	p	p	p	p	p	p
4) TU Chemnitz, Fakultät für Mathematik A. Meyer, S. Meinel, U. Groh, M. Pester	x	x	x						
5) Daimler-Benz AG (STAR-CD) F. Klimetzek				p					
6) Univ. Duisburg, Inst. für Verbrennung und Gasdyn. D. Hänel, O. Filippova	x	x	p	p	p	p	p	p	p
7) Univ. Erlangen, Lehrstuhl für Strömungsmechanik F. Durst, M. Schäfer, K. Wechsler	x	x	x	x	x		x		x
8) Univ. Freiburg, Inst. für Angewandte Mathematik E. Bänsch, M. Schrul	x	x	x	x	x			p	p
9) Univ. Hamburg, Inst. für Schiffbau M. Perić, S. Muzaferija, V. Seidl	x	x	x		x	p			
10) Univ. Heidelberg, Inst. für Angewandte Mathematik R. Rannacher, S. Turek	x	x	x	x	x	x	x	x	x
11) TU Karlsruhe, Inst. für Hydromechanik W. Rodi, M. Pourquie				x				p	
12) Univ. Karlsruhe, Inst. für Therm. Strömungsmasch. C.-H. Rexroth, S. Wittig	x								
13) Kyoto Inst. of Tech., Dept. of Mech. and Syst. Eng. N. Satofuka, H. Tokunaga, H. Hosomi	x	x							
14) Univ. Magdeburg, Inst. für Analysis und Numerik L. Tobiska, V. John, U. Risch, F. Schieweck	x		x						
15) TU München, Inst. für Informatik C. Zenger, M. Griebel, R. Kreißl, M. Rykaschewski	x	x	x	x		p		p	
16) UBW München, Inst. f. Strömungsmech. u. Aerodyn. H. Wengle, M. Manhart				x		p			
17) Univ. Stuttgart, Inst. für Computeranwendungen G. Wittum, H. Rentz-Reichert	x								

Table 2: *Numerical methods and implementation of participating groups*

	Space discretization	Time discretization	Solver	Implementation
1	FD, blockstructured non-staggered QUICK upwinding	fully implicit 2nd ord. equidistant	artificial compressibility expl. 5-step Runge-Kutta FAS-MG (steady) line-Jacobi (unsteady)	serial Fujitsu VPP500 1600 PEAK
2	FV, blockstructured 2nd ord. upwindig	implicit Euler equidistant	ILU with algebraic MG for linear problems	serial IBM RS6000/370 37 LINP
3	FV, blockstructured non-staggered QUICK upwinding	fully implicit 2nd ord. equidistant	stream function form fixed-point iteration ILU for lin. subproblems	serial SGI-Indigo2 75 PEAK parallel Cray T3D/16 16x88 LINP
4	FE, blockstructured 4Q1-Q1 BTD stabilisation	Projection 2 (Gresho) Crank-Nicolson (diff.) explicit Euler (conv.) adaptive	pseudo time step (steady) PCG for lin. subproblems hierarch. preconditioning	parallel GC/PP32 32x13.9 LINP
5a	FV, unstructured 1st ord. upwind	STARCD software	pressure correction	serial HP 735 13 LINP
5b	FV, unstructured CDS	STARCD software	pressure correction	serial HP 735 13 LINP
6	Lattice BGK equidistant orth. grid	explicit equidistant	gaskinetic solution of BGK-Boltzmann equation evol. of distribution funct.	serial HP735 13 LINP
7a	FV, blockstructured non-staggered CDS with def. corr.	Crank-Nicolson equidistant	nonlinear MG SIMPLE smoothing ILU for lin. subproblems	serial HP735 13 LINP
7b	FV, blockstructured non-staggered CDS with def. corr.	fully implicit 2nd ord. equidistant	nonlinear MG SIMPLE smoothing ILU for lin. subproblems	parallel GC/PP128,32,8 128x13.9 LINP
8a	FE, unstructured P2-P1 (Taylor-Hood) CDS	2nd order fract. step operator splitting equidistant	nonlinear GMRES PCG for lin. subproblems	serial SGI R4000 8.3 LINP
8b	FE, unstructured P2-P1 (Taylor-Hood) CDS adaptive refinement	2nd order fract. step operator splitting equidistant	nonlinear GMRES PCG for lin. subproblems	serial SGI R4400 13.2 LINP IBM RS6000/590 58 LINP
9a	FV, blockstructured CDS	fully implicit 2nd ord. equidistant	SIMPLE ILU for lin. subproblems	serial IBM RS6000/250 34 LINP
9b	FV, unstructured CDS	fully implicit 2nd ord. equidistant	SIMPLE ILU-CGSTAB for lin. subproblems	serial IBM RS6000/590 90 LINP

Table 2: *(continued)*

	Space discretization	Time discretization	Solver	Implementation
10	FE, blockstructured Q1(rot)-Q0 adaptive upwind	2nd order fract. step projection method adaptive	fixed-point iteration MG for lin. NS with Vanka smoother (steady) MG for scalar lin. subproblems (unsteady)	serial IBM RS6000/590 90 LINP
11	FV, structured CDS with momentum interpolation	explicit 3rd ord. Runge-Kutta equidistant	SIMPLE ILU for lin. subprobl.	serial SNI S600/20 5000 PEAK
12	FV, unstructured non-staggered adapt. 2nd ord. DISC deferred correction	–	SIMPLEC ILU-BICGSTAB for lin. subproblems	serial SUN SS10 5.5 LINP
13a	FD, structured	explicit Euler equidistant	stream function form pseudo time step SOR for lin. subprobl.	serial IBM RS6000/590 90 LINP
13b	FD, structured	explicit 4th order Runge-Kutta-Gill equidistant	stream function form SOR for lin. subprobl.	serial IBM RS6000/590 90 LINP
14a	FE, blockstructured P1-P0 (Crouzeix-Raviart) 1st order upwind	–	nonlinear MG Vanka smoother	serial HP737/125 6.6 LINP
14b	FE, blockstructured Q1(rot)-Q0 1st order upwind		fixed-point iteration MG for lin. NS with Vanka smoother	parallel GC/PP96 96x13.9 LINP
14c	FE, unstructured P1-P0 (Cr.-Rav.) Samarskij upwind adaptive refinement	BDF(2), equidistant	fixed-point iteration GMRES for pressure Schur-complement lin. MG for velocity	parallel GC/PP24 24x13.9 LINP
15a	FD, structured staggered, orthogonal CDS/UDS flux-blend.	explicit Euler adaptive	SOR for pressure	serial HP720 7.4 LINP
15b	FD, structured staggered, orthogonal CDS/UDS flux blend.	explicit Euler adaptive	SOR for pressure	parallel HP720 cluster 8x7.4 LINP
16	FV, blockstructured CDS	explicit 2nd ord. leap-frog time-lagged diff.	pressure correction Gauss-Seidel for lin. subproblems	serial SGI Indigo 9.6 LINP Convex C3820 19.2 LINP
17	FV, unstructured adaptive upwind	–	fixed-point iteration MG for lin. NS $BILU_\beta$ smoother	serial SGI R4400 8.3 LINP

4. RESULTS

The results of the benchmark computations are summarized in Tables 3-11. The number in the first column refers to the methods given in Table 2. The last column contains the performance of the computer used (as given by the contributors), either the Linpack1000 Mflop rate (LINP) or the peak performance (PEAK), which of course should be taken into account when comparing the different computing times. The column "unknowns" refers to the total number, i.e. the sum of unknowns for all velocity components and pressure. The CPU timings are all given in seconds. In the last row of each table estimated intervals for the "exact" results are indicated (as suggested by the authors on the basis of the obtained solutions).

We remark that for the 2D time-periodic test case 2D-2 also measurements were carried out, where the Strouhal number and time-averaged velocity profiles at different locations along the channel are determined experimentally. However, a direct comparison with the numerical results in Table 4 is problematic, because owing to the short distance between the inlet and the cylinder for the computations, the flow conditions in front of the cylinder are slightly different. To give some comparison with method 7a (see Table 2), a computation with a longer distance between the inlet and cylinder was carried out. The experimentally obtained Strouhal number of $St = 0.287 \pm 0.003$ agrees very well with the numerically computed value of $St = 0.289$. A comparison of time-averaged velocity profiles can be seen in Fig. 4, which also are in fairly good agreement.

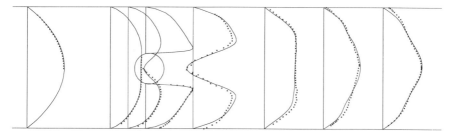

Figure 4: *Comparison of experimental and numerical time-averaged velocity profiles for test case 2D-2 with extended inlet part.*

Table 3: *Results for steady test case 2D-1*

	Unknowns	c_D	c_L	L_a	ΔP	Mem.	CPU time	MFlop rate
1	200607	5.5567	0.0106	0.0845	0.1172	15	788	1600 PEAK
	51159	5.5567	0.0106	0.0843	0.1172	4	273	
	13299	5.5661	0.0105	0.0835	0.1169	1	144	
3a	10800	5.6000	0.0120	0.0720	0.1180	2.5	121	75 PEAK
4	297472	5.5678	0.0105	0.0847	0.1179	137	31000	445 LINP
	75008	5.5606	0.0107	0.0849	0.1184	73	8000	
	19008	5.5528	0.0118	0.0857	0.1199	57	2000	
6	1314720	5.8190	0.0110	0.0870	0.1230	40	80374	13 LINP
	332640	5.7740	0.0030	0.0830	0.1230	10	10461	
	85140	5.7890	-0.0060	0.0870	0.1230	2.6	1262	
7a	294912	5.5846	0.0106	0.0846	0.1176	75	192	13 LINP
	73728	5.5852	0.0105	0.0845	0.1176	19	47	
	18432	5.5755	0.0102	0.0842	0.1175	5	13	
8a	20487	5.5760	0.0110	0.0848	0.1170	9.0	2574	8.3 LINP
	6297	5.5710	0.0130	0.0846	0.1160	2.9	362	
	2298	5.4450	0.0200	0.0810	0.1110	1.3	109	
9a	240000	5.5803	0.0106	0.0847	0.1175	53	9200	34 LINP
	60000	5.5786	0.0106	0.0847	0.1173	10	1400	
	15000	5.5612	0.0109	0.0848	0.1166	2.5	200	
10	2665728	5.5755	0.0106	0.0780	0.1173	350	677	90 LINP
	667264	5.5718	0.0105	0.0770	0.1169	89	169	
	167232	5.5657	0.0102	0.0730	0.1161	22	52	
	42016	5.5608	0.0091	0.0660	0.1139	5	18	
12	32592	5.5069	0.0132	0.0830	0.1155	18	1796	5.5 LINP
	26970	5.5125	0.0056	0.0827	0.1154	15	1099	
	22212	5.6026	-0.0031	0.0815	0.1167	13	3437	
13a	25410	5.6145	0.0159	0.8315	3.0002	4	14203	90 LINP
	12738	5.6114	0.0169	0.8224	2.9943	2	3018	
	6562	5.7377	0.0514	0.8107	3.2277	1		
14a	3077504	5.6323	0.0137	0.0782	0.1159	214	15300	6.6 LINP
	768704	5.6382	0.0102	0.0775	0.1156	53	5490	
	191840	5.5919	-0.0009	0.0750	0.1143	13	2800	
14b	30775296	5.5902	0.0108	0.0853	0.1174	5340	1534	1334 LINP
	7695104	5.6010	0.0110	0.0844	0.1174	1341	400	
	1922432	5.6227	0.0113	0.0833	0.1172	338	119	
14c	797010	5.5708	0.0167	0.0837	0.1168	460	8000	334 LINP
	363457	5.5598	0.0142	0.0835	0.1166	230	3290	
	176396	5.5106	0.0046	0.0835	0.1150	110	2560	
15a	432960	5.5602	0.0329	0.0730	0.1054	4.4	179986	7.4 LINP
	108240	5.6300	0.0751	0.0720	0.1037	1.1	13593	
	27060	5.7769	0.2085	0.0680	0.0998	0.3	688	
17	111342	5.5610	0.0107		0.1170	87	2568	8.3 LINP
	60804	5.5520	0.0102		0.1168	47	1092	
	19416	5.5160	0.0099		0.1158	15	373	
	lower bound	5.5700	0.0104	0.0842	0.1172			
	upper bound	5.5900	0.0110	0.0852	0.1176			

Table 4: *Results for time–periodic test case 2D-2*

	Unknowns		c_{Dmax}	c_{Lmax}	St	ΔP	Mem.	CPU time	MFlop rate
	Space	Time							
1	267476	67	3.2224	0.9672	0.2995	2.4814	—	—	1600 PEAK
	267476	34	3.2030	0.9223	0.2941	2.4664	—	—	
	267476	18	3.1605	0.8026	0.2901	2.4466	—	—	
	68212	67	3.2171	0.9591	0.2995	2.5009	—	—	
	17732	68	3.2168	0.9295	0.2979	2.5573	—	—	
3	12800	34	3.2200	0.9720	0.2960	2.4700	2.5	789	75 PEAK
4	297472	670	3.2460	0.9840	0.2985	2.4900	137	6600	445 LINP
	297472	338	3.2710	0.9800	0.2959	2.4870	137	3400	
	297472	172	3.3200	0.9720	0.2907	2.4810	137	1700	
	75008	670	3.2410	0.9910	0.2985	2.5020	73	2350	
	19008	674	3.2320	1.0260	0.2967	2.5320	57	1350	
6	332640	12000	4.1210	1.6120	0.3330	3.1420	10	10086	13 LINP
	85140	6000	4.7330	2.0600	0.3380	3.4300	2.6	1259	
7a	294912	36	3.2358	1.0069	0.3003	2.4892	75	6167	13 LINP
	294912	19	3.2356	1.0000	0.2973	2.4871	75	6391	
	294912	10	3.2152	0.9028	0.2881	2.4715	75	4994	
	73728	36	3.2443	1.0261	0.2994	2.4929	19	1946	
	18432	36	3.2706	1.0695	0.2968	2.5035	5	445	
8a	29084	66	3.2240	1.0060	0.3020	2.4860	11	4992	8.3 LINP
	29084	33	3.2470	1.0740	0.3030	2.5010	11	3777	
	29084	16	3.2900	1.2500	0.3130	2.5700	11	3217	
	8764	66	3.1740	0.9640	0.3000	2.4630	3.6	1000	
	2978	70	2.8920	0.5540	0.2890	2.2870	1.5	339	
9a	240000	5000	3.2267	0.9862	0.3017	2.4833	53	32500	34 LINP
	60000	10000	3.2232	0.9830	0.3012	2.4773	10	8550	
	60000	5000	3.2232	0.9832	0.3012	2.4773	10	4500	
	60000	2500	3.2232	0.9836	0.3012	2.4773	10	3400	
	15000	5000	3.2058	0.9651	0.2994	2.4587	2.5	3240	
10	667264	612	3.2314	0.9999	0.2973	2.4707	128	8545	90 LINP
	667264	204	3.2351	1.0123	0.2957	2.4734	128	2850	
	667264	68	3.2771	1.1205	0.2997	2.4961	128	1065	
	167232	188	3.2498	1.0081	0.2927	2.4410	32	655	
	42016	164	3.2970	0.8492	0.2713	2.3423	8	147	
13b	25410	6755	3.1822	1.0692	0.2960	2.6066	5.1	44710	90 LINP
	25410	3877	3.1895	1.0883	0.2968	2.6057	4.8	27175	
	25410	1678	3.2043	1.1268	0.2979	2.5307	4.7		
	12738	6799	3.1945	1.1233	0.2941	2.6140	2.9	13045	
	6562	7223	3.1317	1.2961	0.2768	3.0253	1.8		
15a	432960	7790	3.0804	0.7256	0.2778	2.1330	4.4	108844	7.4 LINP
	108240	4003	3.1677	0.6880	0.2646	2.0954	1.1	34876	
	108240	3859	3.1096	0.8249	0.2841	2.1105	1.1	58003	
	27060	1985	3.2544	0.5658	0.2336	1.9727	0.3	3796	
	27060	1670	3.1759	0.7656	0.2740	1.9961	0.3	4188	
	lower bound		3.2200	0.9900	0.2950	2.4600			
	upper bound		3.2400	1.0100	0.3050	2.5000			

Table 5: *Results for unsteady test case 2D-3*

	Unknowns		c_{Dmax}	c_{Lmax}	ΔP	Mem.	CPU time	MFlop rate
	Space	Time						
1	267476	400	2.9387	0.3504	-0.1048	—	—	1600 PEAK
	68212	800	2.9459	0.4492	-0.1057	—	—	
	17732	800	2.9532	0.3908	-0.1007	—	—	
3	12800	800	2.9600	0.4300	-0.0976	2.5	7567	75 PEAK
4	297472	16000	2.9715	0.4806	-0.1101	137	160000	445 LINP
	297472	8000	2.9984	0.4794	-0.1035	137	88000	
	297472	4000	3.0508	0.4750	-0.1018	137	48000	
	75008	16000	2.9660	0.4903	-0.1098	73	47000	
	19008	16000	2.9551	0.5228	-0.1061	57	31000	
6	332640	36960	3.8420	1.1100	0.0200	10	19772	13 LINP
	85140	9460	4.5310	1.7610	0.0090	2.6	2511	
7a	294912	800	2.9520	0.4793	-0.1086	75	43119	13 LINP
	294912	400	2.9520	0.4787	-0.1016	75	29165	
	294912	200	2.9512	0.4021	-0.1047	75	22141	
	73728	800	2.9511	0.4711	-0.0995	19	10003	
	18432	800	2.9461	0.4638	-0.1024	5	2847	
8a	21508	1600	2.9200	0.4910	-0.1110	8	44028	8.3 LINP
	21508	800	2.9210	0.5390	-0.1140	8	31481	
	21508	400	2.9230	0.7250	-0.1160	8	25512	
	5822	1600	2.8160	0.3560	-0.1060	2.5	9294	
	1705	1600	2.7220	0.0055	-0.1220	1.1	2109	
9a	240000	5000	2.9505	0.4539	-0.1095	53	220000	34 LINP
	60000	10000	2.9483	0.4651	-0.1062	10	92000	
	60000	5000	2.9483	0.4630	-0.1062	10	64000	
	60000	2500	2.9482	0.4575	-0.1039	10	38000	
	15000	5000	2.9397	0.4349	-0.1095	2.5	22000	
10	667264	4540	2.9538	0.4782	-0.1053	128	62734	90 LINP
	667264	1612	2.9566	0.5533	-0.1029	128	22431	
	667264	704	3.0650	0.8443	-0.1090	128	11832	
	167232	4068	2.9776	0.4768	-0.1097	32	14005	
	42016	2908	3.0949	0.3223	-0.0951	8	2532	
14c	638880	800	3.0599	0.6326	-0.1100	550	740000	334 LINP
	858848	800	3.1441	0.5266	-0.1142	850	660000	668 LINP
15a	432960	9060	2.8916	0.2649	-0.0987	4.4	237397	7.4 LINP
	108240	4070	2.8927	0.3171	-0.0956	1.1	29140	
	108240	4020	3.0134	0.2921	-0.0945	1.1	25697	
	27060	2857	3.1817	0.2702	-0.1138	0.3	3371	
	27060	2013	3.0098	0.3973	-0.0941	0.3	2541	
	lower bound		2.9300	0.4700	-0.1150			
	upper bound		2.9700	0.4900	-0.1050			

Table 6: *Results for steady test case 3D-1Q*

	Unknowns	c_D	c_L	ΔP	Mem.	CPU time	MFlop rate
1	2530836	7.6415	0.0673	0.1740	251	1975	5000 PEAK
	657492	7.6029	0.0665	0.1738	72	702	1600 PEAK
3	634872	7.6100	0.0642	0.1730	72	1935	1408 LINP
5a	1472000	7.9200	0.0645	0.1751	121	127984	13 LINP
	184000	8.0400	0.0642	0.1722	17	3805	
	23000	7.6600	0.0720	0.1609	3	73	
5b	1472000	7.4400	0.0615	0.1721			
	184000	7.2800	0.0582	0.1673			
	23000	6.7400	0.0615	0.1509			
6	6303750	8.0930	0.0700		43	168657	13 LINP
7a	454656	7.5395	0.0797	0.1715	115	9525	13 LINP
	56832	7.1280	0.0861	0.1616	13	1280	
	7104	6.4590	0.0988	0.1385	3	88	
8a	362613	7.6480	0.0670	0.1751	126	46970	13.2 LINP
	73262	7.6530	0.0590	0.1766	28	6590	
8b	97822	7.6340	0.0660	0.1742	38	8648	13.2 LINP
10	6094976	7.6148	0.0600	0.1729	690	8244	90 LINP
	768544	7.5622	0.0503	0.1683	88	1267	
	97736	7.3069	0.0348	0.1590	10	380	
11	1425600	7.7583	0.0511	0.1744	100	2538	5000 PEAK
	460800	7.7673	0.0406	0.1721	38	536	
	128000	7.2372	0.0602	0.1611	17	86	
15b	6724000	6.0770	0.0859	0.0825	64	10600	52 LINP
	1681000	5.5060	0.1420	0.0796	16	1400	
16	2007040	7.3700	0.0619	0.1720	25	45000	19.2 LINP
	405503	7.2500	0.0549	0.1680	6	11000	9.6 LINP
	lower bound	7.5000	0.0600	0.1720			
	upper bound	7.7000	0.0800	0.1800			

Table 7: *Results for steady test case 3D-1Z*

	Unknowns	c_D	c_L	ΔP	Mem.	CPU time	MFlop rate
1	2426292	6.1295	0.0093	0.1693	233	2097	5000 PEAK
	630564	6.1230	0.0095	0.1680	71	1238	1600 PEAK
2	555000	6.1440	0.0074	0.1604	122	8731	26 LINP
	276800	5.8600	0.0042	0.1616	67	6094	
3	608496	6.1600	0.0095	0.1690	74	4150	1408 LINP
6	6303750	6.2330	-0.0040		43	221706	13 LINP
7b	12582912	6.1932	0.0093	0.1709	3571	2630	1779 LINP
	1572864	6.1868	0.0092	0.1703	518	1120	445 LINP
	196608	6.1366	0.0098	0.1673	71	460	111 LINP
8a	362613	6.1430	0.0084	0.1694	126	51280	13.2 LINP
	73262	6.0990	0.0067	0.1695	28	7178	
9	2355712	6.1800	-0.0010	0.1691		62000	90 LINP
	753664	6.1720	0.0090	0.1680		6000	
	94208	6.1310	0.0100	0.1605		950	
10	6116608	6.1043	0.0079	0.1672	700	8440	90 LINP
	771392	5.9731	0.0059	0.1605	89	1466	
	98128	5.8431	0.0061	0.1482	11	290	
	lower bound	6.0500	0.0080	0.1650			
	upper bound	6.2500	0.0100	0.1750			

Table 8: *Results for time–periodic test case 3D-2Q*

	Unknowns							
	Space	Time	c_{Dmax}	c_{Lmax}	St	Mem.	CPU time	MFlop rate
3	634872	188	4.3170	0.0495	0.3130	74	3368	1408 LINP
	634872	95	4.3170	0.0495	0.3210	74	2754	
6	6303750	18000	4.5870	-0.0050	–	43	168657	13 LINP
10	6094976	142	4.3923	0.0146	0.2777	840	29428	90 LINP
	6094976	124	4.3932	0.0191	0.2806	840	29945	
	6094976	84	4.4071	0.0896	0.2400	840	30372	
	768544	–	4.4819	0.0036	–	105	–	
	97736	–	4.5529	-0.0080	–	13	–	
16	2007040	1726	4.6738	0.0389	0.3488	25	20040	19.2 LINP
	405503	833	4.8808	0.0392	0.3610	6	10020	9.6 LINP
	lower bound		?	?	?			
	upper bound		?	?	?			

Table 9: *Results for time–periodic test case 3D-2Z*

	Unknowns Space	Time	c_{Dmax}	c_{Lmax}	St	Mem.	CPU time	MFlop rate
1	630564	177	3.3018	-0.0014	0.3390	78	26115	1600 PEAK
3	608496	–	3.2250	-0.0142	–	74		1408 LINP
	608496	–	3.2250	-0.0142	–			
6	6303750	18000	3.7920	-0.0210	–	43	142646	13 LINP
7b	12582912	93	3.3052	-0.0105	0.3409	3571	24459	1779 LINP
	1572864	378	3.3057	-0.0118	0.3172	518	9487	445 LINP
	1572864	261	3.3054	-0.0118	0.2250	518	2740	445 LINP
	1572864	126	3.3050	-0.0018	0.2400	518	1956	445 LINP
	196608	–	3.3121	-0.0150	–	71	–	111 LINP
9b	2355712		3.2968					90 LINP
	753664		3.3254					
	94208		3.3284					
10	6116608	128	3.2950	-0.0081	0.2912	840	31145	90 LINP
	6116608	120	3.2970	-0.0025	0.2830	840	31730	
	6116608	80	3.3200	0.0480	0.2684	840	21586	
	771392	68	3.3801	0.0086	0.2343	105	2163	
	98128	–	3.4593	-0.0102	–	13	–	
	lower bound		3.2900	-0.0110	0.2900			
	upper bound		3.3100	-0.0080	0.3500			

Table 10: *Results for unsteady test case 3D-3Q*

	Unknowns Space	Time	c_{Dmax}	c_{Lmax}	ΔP	Mem.	CPU time	MFlop rate
1	657492	800	4.3804	0.0308	-0.1392	78	121960	1600 PEAK
3	634872	1600	4.3030	0.0476	-0.1361	74	51253	1408 LINP
	634872	800	4.3020	0.0473	-0.1354	74	37241	
6	6303750	18000	4.8680	0.0310		43	168657	13 LINP
8a	362613	1000	4.5530	0.0137	-0.1436	126	398000	58 LINP
8b	228451	1000	4.5080	0.0432	-0.1427	105	915000	13.2 LINP
10	6094976	772	4.4086	0.0133	-0.1264	840	164749	90 LINP
	6094976	392	4.5698	0.0262	-0.1213	840	89679	
	6094976	82	5.5709	0.1230	0.0183	840	35600	
	768544	696	4.5223	0.0061	-0.1113	105	22747	
	97736	588	4.5820	0.0033	-0.0718	13	3031	
11	3712800	7720	4.3400	0.0500	-0.0810	105	5711	5000 PEAK
	1523200	7720	4.4000	0.0480	-0.1160	48	2741	
	352000	7720	4.3600	0.0680	-0.1090	18	706	
	lower bound		4.3000	0.0100	-0.1400			
	upper bound		4.5000	0.0500	-0.1200			

Table 11: *Results for unsteady test case 3D-3Z*

	Unknowns		c_{Dmax}	c_{Lmax}	ΔP	Mem.	CPU time	MFlop rate
	Space	Time						
1	630564	800	3.2826	0.0027	-0.1117	79	156460	1600 PEAK
3	608496	1600	3.2590	0.0026	-0.1072	74	76142	1408 LINP
	608496	800	3.2590	0.0026	-0.1157	74	50764	
6	6303750	18000	4.1600	0.0200		43	142646	13 LINP
7b	1572864	1600	3.3011	0.0026	-0.1102	518	149923	445 LINP
	1572864	800	3.3008	0.0026	-0.1105	518	93055	445 LINP
	1572864	400	3.3006	0.0026	-0.1107	518	62026	445 LINP
	196608	1600	3.3053	0.0028	-0.1066	71	63057	111 LINP
8a	362613	1000	3.2340	0.0028	-0.1114	126	347000	58 LINP
8b	199802	1000	3.2120	0.0122	-0.1112	105	846000	13.2 LINP
	98637	1000	3.2350	0.0123	-0.1114	39	243000	
10	6116608	668	3.2802	0.0034	-0.0959	840	164837	90 LINP
	6116608	272	3.3748	0.0360	-0.0603	840	77538	
	6116608	60	2.7312	0.0069	-0.0682	840	29742	
	771392	724	3.3323	0.0033	-0.0766	105	24745	
	98128	660	3.4200	0.0040	-0.0407	13	5687	
	lower bound		3.2000	0.0020	-0.0900			
	upper bound		3.3000	0.0040	-0.1100			

5. DISCUSSION OF RESULTS

On the basis of the results obtained by these benchmark computations some conclusions can be drawn. These have to be considered with care, as the provided results depend on parameters which are not available for the authors of this report, e.g., design of the grids, setting of stopping criteria, quality of implementation, etc.

For five of the ten questions above the answers seem to be clear:

1. In order to compute incompressible flows of the present type (laminar) accurately and efficiently, one should use implicit methods. The step size restriction enforced by explicit time stepping can render this approach highly inefficient, as the physical time scale may be much larger than the maximum possible time step in the explicit algorithm. This is obvious from the results for the stationary cases in 2D and 3D, and also for the nonstationary cases in 2D. For the nonstationary cases in 3D only too few results on apparently too coarse meshes have been provided, in order to draw clear conclusions. This question requires further investigation.

2. Flow solvers based on conventional iterative methods on the linear subproblems have on fine enough grids no chance against those employing suitable multigrid techniques. The use of multigrid can allow computations on workstations (provided the problem fits into the RAM) for which otherwise supercomputers would have to be used. In the submitted solutions supercomputers (Fujitsu, SNI, CRAY) have mainly been used for their high CPU power but not for their large storage capacities. For example, in test case 3D-3Z

(Table 11) the solutions 1 and 3 require with about 600,000 unknowns on supercomputers significantly more CPU time than the solution 10 with the same number of unknowns on a workstation.

3. The most efficient and accurate solutions are based either on finite element or finite volume discretizations on contour adapted grids.

4. The computation of steady solutions by pseudo time-stepping techniques is inefficient compared with using directly a quasi-Newton iteration as stationary solver.

5. For computing sensitive quantities such as drag and lift coefficients, higher order treatment of the convective term is indispensable. The use of only first order upwinding (or crude approximation of curved boundaries) does not lead to satisfactory accuracy even on very fine meshes (several million unknowns in 2D).

For the remaining five questions the answers are not so clear. More test calculations will be necessary to reach more decisive conclusions. The following preliminary interpretations of the results obtained so far may become the subject of further discussion:

6. In computing nonstationary solutions, the use of operator splitting (pressure correction) schemes tends to be superior to the more expensive fully coupled approach, but this may depend on the problem as well as the quantity to be calculated (compare, e.g., for the test case 2D-3 (Table 5), the solution 14c with 7a and 10). Further, as fully coupled methods also use iterative correction within each time step (possibly adaptively controlled), the distinction between fully coupled and operator splitting approach is not so clear.

7. The use of higher than second-order discretizations in space appears promising with respect to accuracy, but there remains the question of how to solve efficiently the resulting algebraic problems (see the results of 8 for all test cases). The results provided for this benchmark are too sparse to allow a definite answer.

8. The most efficient solutions in this benchmark have been obtained on blockwise structured grids which are particularly suited for multigrid algorithms. There is no indication that fully unstructured grids might be superior for this type of problem, particularly with respect to solution efficiency (compare the CPU times reported for the solutions 7 and 9 in 2D). The winners may be hierarchically structured grids which allow local adaptive mesh refinement together with optimal multigrid solution.

9. From the contributed solutions to this benchmark there is no indication that a-posteriori grid adaptation in space is superior to good hand-made grids (see the results of 14c). This, however, may drastically change in the future, particularly in 3D. Intensive development in this direction is currently in progress.

For nonstationary calculations, adaptive time step selection is advisable in order to achieve reliability and efficiency (see the results of 10).

10. The treatment of the nonlinearity by nonlinear multigrid has no clear advantage over the quasi-Newton iteration with multigrid for the linear subproblems (compare the results of 7 with those of 10). Again, it is the extensive use of well-tuned multigrid (wherever in the algorithm) which is decisive for the overall efficiency of the method.

6. CONCLUDING REMARKS

The authors would like to add some final remarks to the report presented. Although, this benchmark has been fairly successful as it has made possible some solidly based comparison between various solution approaches, it still needs further development. Particularly the following points are to be considered:

1. In the case 3D-3Z it should be the maximum absolute value of the lift which has to be computed as c_L may become negative.

2. In the nonstationary test cases further characteristic quantities (e.g., time averages, pressure values, etc.) should be computed, as in some cases, by chance, "maximum values" may be obtained with good accuracy even without capturing the general pattern of the flow at all.

3. For the nonstationary 3D problems a higher Reynolds number should be considered, since in the present case (Re = 100) the problem may be particularly hard as the flow tends to become almost stationary.

Even in the laminar case, the chosen nonstationary 3D problems showed to be harder than expected. In particular, it was apparently not possible to achieve reliable reference solutions for the test cases 3D-2Q and 3D-2Z. Hence the benchmark has to be considered as still open and everybody is invited to try again.

ACKNOWLEDGMENTS

The authors thank all members of the various groups who have contributed results for the benchmark computations, K. Wechsler for his help in evaluating the results and J. Jovanovic and M. Fischer for carrying out the experiments.

LDA Measurements in the Wake of a Circular Cylinder

F.Durst, M.Fischer, J.Jovanović, H.Kikura and C.Lange
Lehrstuhl für Strömungsmechanik, Universität Erlangen-Nürnberg
Cauerstrasse 4, D-91058 Erlangen

Summary

This paper reports laser-Doppler measurements of the laminar flow field around a circular cylinder placed in a two-dimensional channel flow. The experimental conditions were adjusted to match numerical investigations which have been performed under the frame of the DFG-Schwerpunktprogramm "Strömungssimulation mit Hochleistungsrechnern" as the test case 2.2b. A specially designed two-dimensional channel with transparent test section together with a one-component LDA system permitted detailed velocity measurements in the wake behind the cylinder. Measurements of the shedding frequency as a function of the Reynolds number are presented. These results show very good agreement with the corresponding numerical data. The experimental results also include the mean velocity distributions in the wake region. A comparison of measured and predicted data across the wake region further demonstrate the high reliability of the computational results.

1 Introduction

With advances in the computer technology it became feasible to study flows of engineering interest by applying numerical methods. However, high performance numerical techniques together with fast parallel computers are needed to yield reliable flow predictions. Application of these techniques, as developed in the DFG-Schwerpunktprogramm "Strömungssimulation mit Hochleistungsrechnern", seems to be very effective in providing the complete data that can deepen our insight when applied to many important fields of fluid mechanics. One also observes that new direction in flow research closely follows the computer technology which is in an advanced stage of development. Really large steps forward can be expected with the new parallel computers presently becoming available on the market.

In the frame of the DFG-Schwerpunktprogram "Strömungssimulation mit Hochleistungsrechnern" different test cases were specified in order to demonstrate the performance of the developed computer codes and to cross-check the reliability of the results obtained by using different numerical schemes. For testing the performance of numerical codes in a two dimensional flow, two test cases were specified corresponding to steady ($R_e = 20$) and unsteady ($R_e = 100$) laminar flows past a circular cylinder in a fully developed laminar channel flow. The cylinder is aligned asymmetrically to the axis of the channel in order to promote growth of the disturbances which cause the unsteady nature of the wake at $R_e = 100$.

Some of the main objectives of the present study were as follows:
- To perform LDA measurements behind the circular cylinder placed in a fully developed laminar channel flow as specified by the test case 2.2b.

- To provide from LDA measurements the data for the vortex shedding frequency as a function of Reynolds number.

- To validate the numerical result for the vortex shedding frequency at $R_e = 100$.

- To provide the mean velocity profiles in the wake region behind the circular cylinder at $R_e = 100$.

- To analyse the experimental and numerical data in order to demonstrate reliability of numerical simulations for accurate flow predictions.

The present paper provides some of these results. Further data can be obtained from the authors.

2 Test section and measuring equipment

The flow field behind the circular cylinder, placed in a fully developed laminar channel flow, was investigated using one-component laser-Doppler anemometry as the measuring technique. Experiments were performed in the water channel flow facility

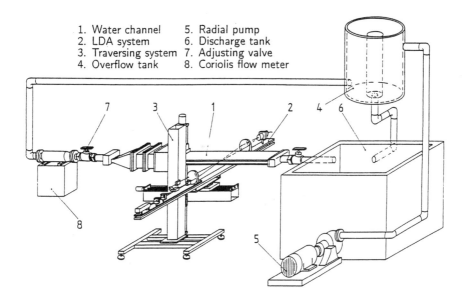

Figure 1: Channel flow and LDA optical system.

shown schematically in Figure 1. This figure shows that the water flow was driven from the discharge to the overflow tank by a radial pump. The constant pressure difference of 6 m water column between the overflow tank and the two-dimensional flow channel ensured well defined flow measurement conditions. The flow rate was maintained by controlling the hydrostatic pressure difference between the inlet and outlet of the channel.

Experimental investigations were performed in a two-dimensional test section of dimensions $l \times b \times H = 1 \times 0.18 \times 0.0102$ m. The entrance to the channel was preceded by a rectangular contraction of upstream dimensions (0.15 × 0.18 m). Upstream of the contraction, an 80 mm long honeycomb with 6 mm diameter cells was inserted to improve flow uniformity. Downstream of the honeycomb, grids of 1 mm mesh size were installed to reduce the free stream turbulence intensity. Two Plexiglas plates spanning the whole length of the channel were used as side-walls to allow optical access of the flow. The height of the channel was kept constant by using precise metal pieces between the base plates of the channel.

The circular cylinder of 2.5mm in diameter was fixed 49 channel heights downstream from the entrance providing fully developed laminar flow conditions in the front of the cylinder. The distances of the cylinder axis from the bottom and top walls were 5.2mm and 5.0mm respectively. The mechanical arrangement of the cylinder with respect to the test section is shown in Figure 2. Measurements to be reported herein were performed in the region 42 to 72 channel heights downstream from the inlet.

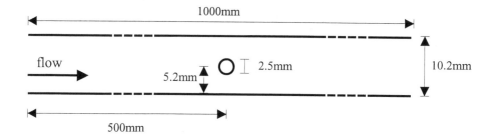

Figure 2: The mechanical arrangement of the cylinder placed in a channel flow.

A TSI 9100-6 system with an in-house made beam expander and a double Bragg cell unit together with an 8 mW He-Ne laser was used in these measurements. A frequency shift of 0.1 MHz was employed. The focal length of the sending lens was 160 mm and the measuring control volume based on the e^{-2} Gaussian intensity cut-off point, was 80 μm in diameter and 520 μm in length. The light scattered in forward direction was collected by the receiving optics and directed on to an appropriately sized pinhole in front of an avalanche photodiode. The output signal of the photodiode was bandpass filtered between 30 kHz and 1 MHz.

The transmitting optics were tilted at an angle of 0.3° towards the channel wall. The receiving optics were also tilted at 1.1° towards the wall to reduce intercepting reflected light from the channel wall. All optical elements were mounted on a three-dimensional traversing mechanism which allowed positioning of the measuring control volume with a resolution of about 5 μm. The distance from the wall was determined by a Mitsutoyo digital gauge of 1 μm resolution.

A TSI Model 1990 counter operated in the total burst mode with at least 32 cycles per burst was used for all measurements. The counter was interfaced to a PC using a Dostec signal processing card. The mean velocity was computed using arrival time averaging. To ensure sufficient data rates seeding particles of about 4.0μm in diameter were used. Data rates of 5 – 100 Hz were typical, depending on the local flow velocity. At every measuring point at least 5000 samples were acquired.

3 Experimental results

3.1 Measurements of Strouhal number

When the LDA system was traversed downstream of the cylinder large and regular sinusoidal oscillations were recorded in the wake of the flow having a cylinder diameter Reynolds number of 100. These oscillations corresponded to the vortices which were shed at a regular rate from the cylinder. It is interesting to note that the clearest sinusoidal signals were found approximately 5 to 10 diameters downstream of the cylinder off-side the center of the wake.

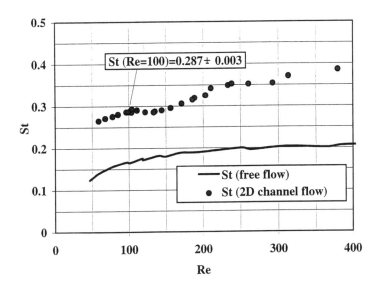

Figure 3: Experimental data for the frequency of the vortex shedding in a channel flow.

To carry out the described LDA measurements, the flow was heavily seeded to ensure the high data rates required for the determination of the vortex shedding frequency. By careful adjustments of the gain factor of the counter processor it was possible to detect about 30 times more particle arrivals then the vortex shedding frequency. In this way the sinusoidal variations of the downstream velocity could be clearly seen on the time varying velocity records.

The vortex shedding frequency expressed in non-dimensional form, as the Strouhal number versus Reynolds number, is shown in Figure 3. The normalization of measurements was performed with respect to a bulk velocity which corresponds to two-thirds of the centerline velocity (U_s) in front of the cylinder

$$U_s = \frac{2}{3} U(X_{norm} = -\infty, Y_{norm} = 0.5), \tag{1}$$

and the cylinder diameter (d)

$$S_t = \frac{fd}{U_s}, \quad R_e = \frac{U_s d}{\nu}, \tag{2}$$

where f and ν are the shedding frequency and kinematic viscosity of the flow medium, respectively.

It can be seen from Figure 3 that we have covered the Reynolds number range $R_e = 50 - 300$. Below $R_e \simeq 50$ no oscillations could be observed in the measured time traces behind the cylinder. Above $R_e \simeq 300$ the signals were chaotic in phase

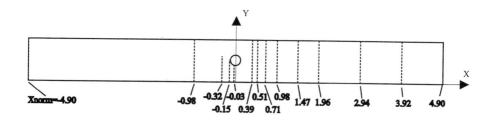

Figure 4: Cross-stream locations for profile measurements.

and amplitude indicating that transition to turbulence had occurred in the separated shear layers behind the cylinder.

In the Reynolds number range from 50 to about 100 the Strouhal number monotonically increased with Reynolds number. A plateau in the Strouhal-Reynolds number relationship existed for Reynolds numbers between 100 and 150 indicating a change in the two-dimensional flow structure. Above $R_e \simeq 150$ transition from the two-dimensional flow state to three-dimensionality of the shed vortex structures seemed to occur and is marked by a further rise of the Strouhal number curve shown in Figure 3. The present experimental results show much higher values of Strouhal number compared to the data obtained by Rosko (1954) under free flow conditions.

The measured value of Strouhal number for $R_e = 100$, $S_t = 0.287 \pm 0.003$ is in good agreement with the corresponding numerical value

$$S_t = 0.286, \tag{3}$$

obtained from the flow computations at LSTM-Erlangen as reported by Schäfer and Turek (1996) using the numerical program FASTEST-2D.

3.2 Mean flow velocity measurements

In this section experimental results of mean velocity measurements at a Reynolds number 100 are presented. These measurements were performed at a number of cross-stream planes in the front and behind the cylinder with the actual location being indicated in Figure 4. For these velocity profile measurements the optical arrangement was modified in order to avoid blockage of one of the beams in the region very close to the cylinder. For this purpose a beam spacer was inserted between Bragg cells and the transmitting lens which made it possible to move one of the beams to pass through the optical axis of the LDA system parallel to the axis of the cylinder. This is indicated in Figure 5.

In the presentations that follow the coordinates are normalized as

$$X_{norm} = \frac{X}{H}, \quad Y_{norm} = \frac{Y}{H} \tag{4}$$

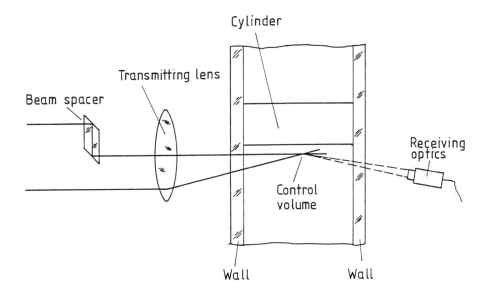

Figure 5: Modified optical system for LDA-measurements close to the cylinder.

where H corresponds to the channel height. The measured mean velocity profiles are normalized with respect to two-thirds of the centerline velocity in front of the cylinder:

$$U_{norm} = \frac{U}{U_s}. \qquad (5)$$

Figure 6 shows the mean velocity profiles plotted for different locations indicated in figure 4. In front of the cylinder ($X_{norm} = -4.90$) the measured data follow the parabolic velocity profile valid for the fully developed laminar channel flow. As the cylinder is approached the flow starts to decelerate at the channel centerline owing to the blockage effect of the cylinder and the flow accelerates near the wall in order to satisfy the integral continuity requirement of the nominally two-dimensional flow. In close proximity of the cylinder, measurements were possible only across one half of the channel due to the optical disturbances caused by the supporting fittings which were holding the mounted obstacle in the channel.

In the near wake region ($X_{norm} = 0.39$) there exists a noticeable asymmetry in the mean velocity data caused by the asymmetry of the cylinder mounting in the channel. These data also indicate a flow reversal in the wake. The development of the velocity profiles further downstream indicates the tendency towards dismissing velocity defect in the wake and slow recovery towards the flow condition upstream of the cylinder.

Figure 7 shows the predictions of Schäfer and Turek (1996) against the measured profiles. The degree of agreement achieved demonstrates the high reliability of the computational results using the numerical program FASTEST-2D.

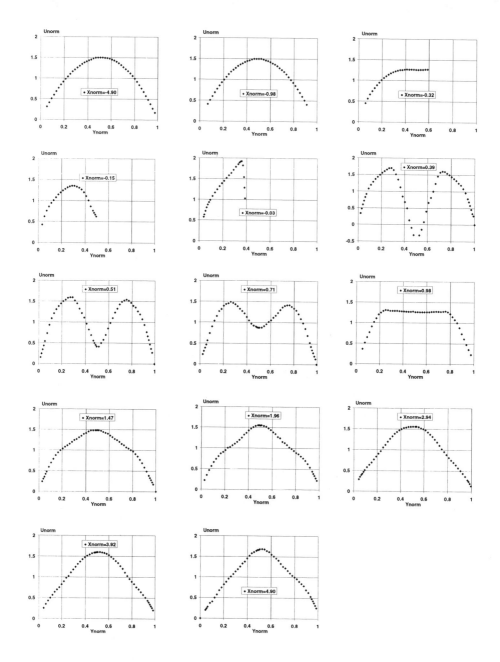

Figure 6: Measurements of the mean velocity profiles in the wake of a circular cylinder at $R_e = 100$.

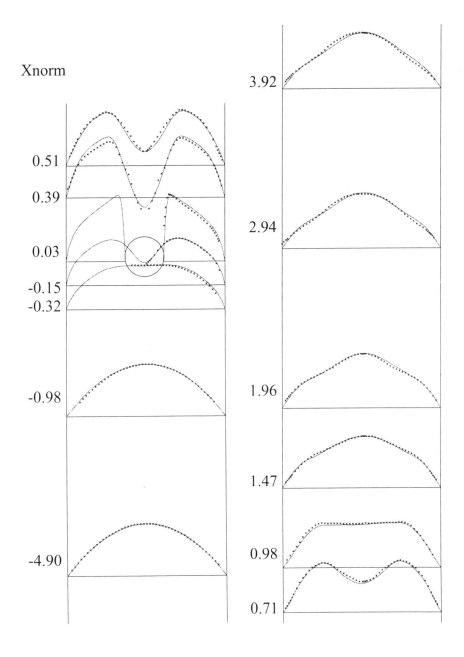

Figure 7: Comparison of measured and predicted mean velocity profiles.

Conclusions and final remarks

The present investigations were carried out within a special project granted just before the completion of the DFG-Schwerpunktprogramm "Strömungssimulationen mit Hochleistungsrechnern" in order to provide data for test cases for the computer programs developed within the above mentioned DFG-Schwerpunktprogramm. In the remaining time, measurements for one Reynolds number were completed and made available for comparisons with numerical predictions. Good agreement between numerical and experimental results were obtained.

In the present paper the experimental set-up is described and modifications of the optical systems for the LDA-measurement are presented. These modifications were necessary in order to permit measurements very close to the cylinder. The shedding frequency could be measured as a function of Reynolds number and data were compared for $R_e = 100$ with the computational results. Very good agreement was obtained.

The authors presently continue their investigations of the cylinder flow inside of a two-dimensional channel to provide further information on the flow field. Further data can be obtained from the authors.

Acknowledgement

The authors gratefully acknowledge the support (Du 101/42-1) given to us by the Deutsche Forschungsgemeinschaft.

References

Rosko, A. 1954 *On the drag and shedding frequency of two-dimensional bluff bodies.* NACA *Rep.* 3169.

Schäfer, M. and Turek, S. 1996 *Benchmark computations of laminar flow around a cylinder.*(this publication)

Addresses of the Editors of the Series "Notes on Numerical Fluid Mechanics"

Prof. Dr. Ernst Heinrich Hirschel (General Editor)
Herzog-Heinrich-Weg 6
D-85604 Zorneding
Federal Republic of Germany

Prof. Dr. Kozo Fujii
High-Speed Aerodynamics Div.
The ISAS
Yoshinodai 3-1-1, Sagamihara
Kanagawa 229
Japan

Prof. Dr. Bram van Leer
Department of Aerospace Engineering
The University of Michigan
3025 FXB Building
1320 Beal Avenue
Ann Arbor, Michigan 48109-2118
USA

Prof. Dr. Michael A. Leschziner
UMIST-Department of Mechanical Engineering
P.O. Box 88
Manchester M60 1QD
Great Britain

Prof. Dr. Maurizio Pandolfi
Dipartimento di Ingegneria Aeronautica e Spaziale
Politecnico di Torino
Corso Duca Degli Abruzzi, 24
I-10129 Torino
Italy

Prof. Dr. Arthur Rizzi
Royal Institute of Technology
Aeronautical Engineering
Dept. of Vehicle Engineering
S-10044 Stockholm
Sweden

Dr. Bernard Roux
Institut de Mécanique des Fluides
Laboratoire Associé au C.R.N.S. LA 03
1, Rue Honnorat
F-13003 Marseille
France

Brief Instruction for Authors

Manuscripts should have well over 100 pages. As they will be reproduced photomechanically they should be produced with utmost care according to the guidelines, which will be supplied on request. In print, the size will be reduced linearly to approximately 75 per cent. Figures and diagrams should be lettered accordingly so as to produce letters not smaller than 2 mm in print. The same is valid for handwritten formulae. Manuscripts (in English) or proposals should be sent to the general editor, Prof. Dr. E. H. Hirschel, Herzog-Heinrich-Weg 6, D-85604 Zorneding.

Notes on Numerical Fluid Mechanics (NNFM)   Volume 52

Series Editors: Ernst Heinrich Hirschel, München (General Editor)
                   Kozo Fujii, Tokyo
                   Bram van Leer, Ann Arbor
                   Keith William Morton, Oxford
                   Maurizio Pandolfi, Torino
                   Arthur Rizzi, Stockholm
                   Bernard Roux, Marseille

Volume 30   Numerical Treatment of the Navier-Stokes Equations (W. Hackbusch / R. Rannacher, Eds.)

Volume 29   Proceedings of the Eighth GAMM-Conference on Numerical Methods in Fluid Mechanics (P. Wesseling, Ed.)

Volume 28   Vortical Solution of the Conical Euler Equations (K. G. Powell)

Volume 27   Numerical Simulation of Oscillatory Convection in Low-Pr Fluids (B. Roux, Ed.)

Volume 26   Numerical Solution of Compressible Euler Flows (A. Dervieux / B. van Leer / J. Periaux / A. Rizzi, Eds.)